$$A = \sqrt{x^2 + y^2}$$

$$z^* = x - jy = A \angle -\theta$$

Exponential and logarithmic functions:

$$e^A e^B = e^{A + B}$$

$$\frac{e^A}{e^B} = e^{A - B}$$

$$\ln AB = \ln A + \ln B$$

$$\ln A^n = n \ln A$$

$$\ln A = 2.303 \log_{10} A$$

If $x \ll 1$, $e^x \cong 1 + x$, $\ln (1 + x) \cong x$

Trigonometry

$$\cos \theta = \sin (\theta + 90°)$$

$$\sin \theta = \cos (\theta - 90°)$$

$$\sin (A \pm B) = \sin A \cos B \pm \cos A \sin B$$

$$\cos (A \pm B) = \cos A \cos B \mp \sin A \sin B$$

If $x \ll 1$, $\sin x \cong x$, $\cos x \cong 1 - x^2/2$

Calculus

$$\frac{d}{dx} e^{ax} = a e^{ax}$$

$$\frac{d}{dx} \ln x = 1/x$$

$$\int e^{ax}\, dx = \frac{1}{a} e^{ax}$$

$$\int \ln x\, dx = x \ln x - x$$

$$\frac{d}{dx} \sin ax = a \cos ax$$

$$\frac{d}{dx} \cos ax = - a \sin ax$$

$$\int \sin ax\, dx = -\frac{1}{a} \cos ax$$

$$\int \cos ax\, dx = \frac{1}{a} \sin ax$$

Taylor's series:

$$f(x + h) = f(x) + h \frac{df}{dx}\bigg|_x + \frac{h^2}{2!} \frac{d^2f}{dx^2}\bigg|_x + \frac{h^3}{3!} \frac{d^3f}{dx^3}\bigg|_x + \cdots$$

ELECTRICAL ENGINEERING:

AN INTRODUCTION

SECOND EDITION

THE OXFORD SERIES IN ELECTRICAL AND COMPUTER ENGINEERING

M.E. Van Valkenburg, Senior Consulting Editor
Adel S. Sedra, Series Editor, Electrical Engineering
Michael R. Lightner, Series Editor, Computer Engineering

ELECTRICAL ENGINEERING:
AN INTRODUCTION

SECOND EDITION

STEVEN E. SCHWARZ
UNIVERSITY OF CALIFORNIA — BERKELEY

WILLIAM G. OLDHAM
UNIVERSITY OF CALIFORNIA — BERKELEY

SAUNDERS COLLEGE PUBLISHING

New York   Oxford

OXFORD UNIVERSITY PRESS

Oxford University Press

Oxford   New York
Athens   Auckland   Bangkok   Bombay
Calcutta   Cape Town   Dar es Salaam   Delhi
Florence   Hong Kong   Istanbul   Karachi
Kuala Lumpur   Madras   Madrid   Melbourne
Taipei   Tokyo   Toronto

and associated companies in
Berlin   Ibadan

Published by Oxford University Press, Inc.,
198 Madison Avenue, New York, New York 10016

Oxford is a registered trademark of Oxford University Press

Cover Credit: © Mel Lindstrom / Tony Stone Worldwide
Library of Congress Catalog Card Number: 92-050559
ISBN: 0-19-510585-0

2 4 6 8 9 7 5 3 1
Printed in the United States of America
on acid-free paper

Contents

III ANALOG SIGNALS AND TECHNIQUES

9   Principles of Analog Systems   333

10   Practical Analog Technology   357

IV DIGITAL SIGNALS AND THEIR USES

11   Digital Building Blocks   391

V   ACTIVE DEVICES AND CIRCUITS

VI MAGNETIC DEVICES, ELECTRIC POWER, AND MACHINES

PREFACE

T his is an introduction of a fairly substantial kind. Not all introductions are, of course. "Pleased ta meetcha," says one person to another, and the next day, very likely, no one even remembers the event. Some introductory courses are like that, too. But in this book we are trying for something else. We mean to introduce electrical engineering in such a way that the basic principles are really learned, so that they can be built upon in future studies. Electrical techniques are essential in all branches of engineering, and of course many techniques of electrical engineering also appear, under different names, in other disciplines. Thus, there is every reason for students to learn more than just the names of things, and acquire substantial knowledge.

In this spirit, we make no attempt to cover everything. Instead, there are in-depth discussions of what we feel are the most important topics, those that are basic, and those that are likely to be of use. Space is then available for derivations, explanations, and motivation. The last, we feel, is especially important. Students need to see *why* they are learning this subject, and how it fits with what they already know, with their other studies, and with their future plans. Whenever possible, students should be given the flavor of engineering design, that distinctive process in our profession in which alternatives are compared and the best one found. Thus, for example, we devote a section to a discussion of when it is better to use individual transistors, and when it is better to design around pre-constructed integrated circuits. The choice of series or parallel connection of Christmas-tree lights is dealt with as early as Chapter 1.

Introductory electrical engineering courses are often taken by students majoring in other, non-electrical engineering fields. It is important to look ahead, to see what kind of knowledge these non-EE majors will eventually need. They will be *users* of electrical devices, systems, and machines. They will need to know what electrical techniques can do, what building blocks are available, and insofar as possible, how to hook things up. What they will probably *not* need is to construct the building blocks themselves. Thus, our table of contents provides, immediately after the introductory chapters on circuit analysis, two chapters on analog systems and two on digital systems. These are the heart of the course, because they explain the techniques that non-electrical engineers are most likely to use. Next come three chapters on transistors and circuit design, but those chapters are intended primarily for electrical engineering majors. (It's a rare mechanical engineer who builds his or her own op-amps, nowadays!) The book then concludes with two chapters on electric power and machines, for those who require that material. The needs of non-electrical engineers are also reflected in our choice of examples. Wherever possible, we have chosen examples that non-EEs will find relevant: for example "Starting your Volkswagen" in Chapter 3, or "Automobile Suspension System" (an application of phasor analysis) in Chapter 7.

For over a decade this book (in its earlier versions) has also been used here at Berkeley in the first course for EE majors. A sophomore course introducing the principles of both passive and active circuits has definite curricular advantages. The upper-division program is constantly getting more crowded, making it beneficial—perhaps even necessary—to begin active circuits in the lower division. Furthermore, it is desirable to get students into realistic engineering early; they shouldn't have to be in college for two years before finding out what their chosen profession is like. And lastly, a course including active concepts is more interesting and more fun, for students and teachers alike. Because this book is in no way "watered down" for non-majors, it is suited for majors as well.

Experience shows that the best way to learn this subject is to work a lot of problems. This book *provides* a lot of problems, carefully chosen and graded in three levels of difficulty (the "unstarred," "one-starred," and "two-starred" problems.) The unstarred problems are introductory and descriptive; most students should do the more serious one-starred problems too. To help readers get started, worked examples are generously scattered throughout the text. The two-starred problems are "industrial strength." Some require mathematical sophistication, while others puzzle the student with conceptual riddles or apparent contradictions. (They will challenge the best students you can find!) Answers to even-numbered problems are provided at the end of the book, so students can check their answers. Instructors can request a copy of the Instructors' Manual, containing solutions for all the problems.

All users of the book will probably want to include Chapters 1 through 8, on basic circuit analysis. (Sections marked with an asterisk can be omitted without loss of continuity.) In courses for non-majors, we recommend proceeding with Chapters 9 through 12, on analog and digital systems. This is enough material to fill a semester at a moderate pace. If more time is available, (or if a faster pace is acceptable,) non-EE majors can add Chapters 16 and 17, on electric power and machines. EE majors would add Chapters 13, 14, and 15 on design of active circuits.

New in This Edition

In order to illustrate the concepts being taught, and show how they relate to the rest of the student's world, sections called "Applications" have been added at the end of each chapter. These sections are written in a less formal voice than is the rest of the book, and are mostly non-mathematical; they are meant to be pleasant interludes amid the rigors of the course. Each Application shows how the material of its chapter can be used in some interesting way. For instance, the Application in the chapter on amplifiers shows how op-amps can be used in a microphone mixer for music recording; the one at the end of Chapter 12 ("Introduction to Digital Systems") explains the modern use of microcomputers for control of "smart" machines. In addition to the Applications, there is new material on SPICE, a new chapter on inductors and capacitors, and expanded and revised chapters on analog systems. In some ways, the first chapters of the course are the most important, because it is there that the student first comes to grips with this new and very abstract subject. Ten years' worth of students' questions, criticisms, and complaints have therefore been used in revising Chapters 1 through 4, in order to make these critical chapters as clear as possible. For additional clarity, two-color printing has been adopted.

Acknowledgments

Over the years we have been assisted by most of our Electrical Engineering colleagues at Berkeley, and many in the Computer Science Division; we thank you all. In preparing the second edition we were helped by Profs. D. Auslander, W. Webster, R. Bea, and D. A. Hodges, and by R. C. Crist and Sy Mouber of the Bay Area Rapid Transit District. Emily Barrosse, Senior Acquisitions Editor at Saunders, made many important suggestions; Laura Shur, Assistant Editor, kept the project moving; and Kimberly A. LoDico, our Project Editor, created the book. Expert word-processing was contributed by Chris Colbert, and hospitality was generously provided by King's College, London, while some of the work was being done.

We would also like to thank the following reviewers for their valuable comments and suggestions during the development of the manuscript:

Roger C. Conant, University of Illinois at Chicago
Min Yen Wu, University of Colorado at Boulder
John L. Mason, Western Michigan University
Charles C. Nguyen, Catholic University
Edgar O'Hair, Texas Tech University
Karl H. Norian, Lehigh University
Lloyd B. Craine, Washington State University
Edward Gardner, University of Colorado at Boulder
Tso-Ping Ma, Yale University
D. A. Calahan, University of Michigan

Steven E. Schwarz
William G. Oldham

INTRODUCTION AND OVERVIEW

E lectrical engineering is a large and active field. Considered as a branch of human knowledge, it is modern, useful, and abstract. It is developing rapidly, and it has great influence on everyday life. Skills in this field are very valuable, and electrical engineering is an excellent career. An acquaintance with the subject is useful to every scientist and engineer, since electronic devices are now vital to every technical field. The steady appearance of powerful new techniques makes the field exciting, and engineers with up-to-date skills are always in demand. Engineers are more than ever sought out and consulted for their knowledge, designing and building innovative, useful products, creating wealth, and changing the world.

It is a large field, but it rests upon a few major principles. The principles have subprinciples, which in turn are based on many facts, which are based on subfacts, which have details—and so on. Before we venture into these thickets, though, it is good to survey the subject as a whole. Thus in this chapter we shall survey the forest; then in Chapters 1 to 17 we can talk about the trees.

0.1 Scope of Electrical Engineering

Electrical engineering includes the two main areas of *information systems* and *power systems*. Under the heading of information systems we include all those applications in which the goal is to store, transmit, or process information (such as words, numbers, or even music). Examples of each are tape recording (information storage), radio broadcasting (communication), and computing (information processing). Information can also be stored, communicated, and processed by nonelectrical means (books, messengers, abacus), but electrons and electromagnetic fields are superbly suited for these jobs because they can move so fast. In information technology, speed is of great importance, and systems are rated by how much information they can handle each second. For example, in television broadcasting, a comparatively modern art, about 400 times as much information is transmitted per second as in the older art of AM radio broadcasting; such increases are typical of advances in the field. Modern information systems are also accurate, versatile, and, thanks to integrated-circuit technology, remarkably inexpensive. These factors have led to a veritable explosion of new applications (pocket calculators, video games, computerized carburetors, laser-disk phonographs), a process that will surely continue.

A closely related field is that of *computer science.* Computers as we know them originated when it was found that certain electronic circuits could act something like brain cells and when properly connected could "think," in a very primitive way. The study of how the circuits should be connected became the field of computer science. Originally the circuits were built with vacuum tubes, but they were bulky, unreliable, used a lot of power, and produced a lot of heat, so that only a few circuits at a time could be used. Thus the question of how to interconnect them was not too complicated. Now, however, the circuits are built in the form of integrated circuits, with perhaps a half million circuits in a single thumbnail-sized semiconductor chip. Thus the question of how to interconnect them has become much more complicated, and computer science has become a sophisticated field in its own right. In digital technology, an electrical quantity, usually voltage, is used to represent a mathematical quantity, such as a number. Our point of view is that so long as voltages and currents are in the discussion, we are talking about electrical engineering. When voltages and currents have disappeared, replaced entirely by the mathematical quantities they represent, the subject has changed to computer science. Although this book does not attempt to survey computer science in detail, we approach the subject when we discuss digital hardware in Chapters 11, 12, and 15.

Whenever there is an electrical current, electrical energy is usually being transmitted from one place to another. One reason for the usefulness of electricity in information systems is that extremely low levels of electrical power are sufficient to convey sizable amounts of information. (For example, the power coming into your TV set from its antenna may be only about 10^{-14} watts! Yet all the information needed for picture and sound are being conveyed with this tiny power.) On the other hand, it is also possible to transmit *large* amounts of power by electrical means—not to transmit information, but rather to transfer energy from one place to another. For example, potential energy may exist in the form of water storage behind a dam. The water falls through turbines, producing mechanical power. No one would dream of transmitting this power to a distant city by means of long rotating mechanical shafts. Instead, it is converted by generators into electrical form, transmitted by high-voltage transmission lines, and eventually brought by distribution systems into homes and industries. The network of wires and machines being used here is called an *electric power system.* In general, the goal of power-systems design is to transmit electrical energy from place to place with high efficiency (that is, without losing much of the power by converting it to heat). In addition, power systems must provide reliability, adaptability to changing conditions, safety, and reasonable cost.

A subject closely related to electrical power is that of *electrical machines*. This term usually refers to apparatus for converting power from mechanical to electrical form (generators) or vice versa (motors). Electrical machines vary widely in size, ranging from enormous generators capable of lighting whole towns down to tiny stepping motors used for turning the hands of wristwatches. The fields of power and machines are among the oldest in electrical engineering, but they show no signs of losing their practical importance.

0.2 Devices, Circuits, Systems, and Integrated Circuits

In the last section, the subject of electrical engineering was divided in terms of its purposes, or applications. It is also possible to divide the subject in other ways, according to the kinds of work done by different specialists. Most engineers do tend to specialize, and the various specialities are then considered to be subfields of the entire subject. Unfortunately, these divisions tend to be rather vague and to overlap each other. On the other hand, it is almost impossible to discuss electrical engineering without these distinctions; thus we shall now introduce some of the most important terms and define them as best we can.

A most important development took place when Lee DeForest introduced the triode vacuum tube in 1906. The basic virtue of this tube was that it made it possible to control large currents or voltages by means of small ones. Thus it gave birth to, among other things, the piece of apparatus we know today as the amplifier. (In a phonograph amplifier, for example, the large currents fed to the loudspeakers are controlled by tiny currents coming from the phonograph pickup.) Components that have this ability (to control large currents or voltages under instructions from small ones) are known as *active devices* or, in present usage, simply devices. Today, vacuum tubes are obsolete, except for a few special purposes, and by "electronic devices" we primarily mean transistors of various kinds.* These are made from semiconductor materials and thus are known as *semiconductor devices*. As a field of specialization, the "devices" area includes such subareas as materials, microfabrication, and device analysis and design, as shown in Fig. 0.1(a). Physics, especially semiconductor physics, tends to play a large part in this work. In this book such devices are introduced in Chapter 13.

Devices are not useful until they are interconnected with other devices and with *circuit elements*, such as resistors, capacitors, and inductors. Such a group of elements is known as a *circuit*. In electrical engineering, the area of study known as "circuits" deals with the analysis of circuits and with ways of designing them to perform desired tasks. Circuits that contain active devices are called *active circuits*; circuits that do not

---

*Diodes, which do not amplify, are also considered to be devices.

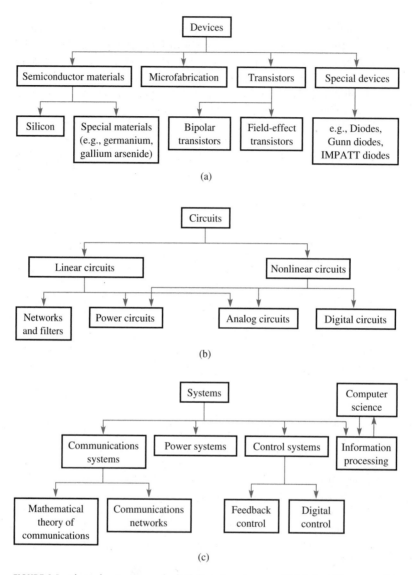

FIGURE 0.1   Electrical engineering can be divided into the major areas of (a) devices, (b) circuits, and (c) systems. Here we illustrate subdivisions of each of the three major areas.

are called *passive circuits*. The subject can also be divided according to whether or not all the devices and elements in a circuit are described by linear equations, in which case the circuit is called a *linear circuit*. If not, it is called a *nonlinear circuit*. In general, nonlinear circuits require elaborate mathematical techniques for design and analysis, often with the help of a computer. All active circuits are nonlinear, but sometimes they can be treated as if they were linear in order to simplify their analysis. The subdivisions of the "circuits" field are shown in Fig. 0.1(b). The electric circuit is probably the most basic and widespread unit in electrical engineering, and it is necessary to gain familiarity with circuits at an early stage. Hence, in this book, Chapters 1 to 8 deal with passive, linear circuits in some detail. Then, after devices have been introduced, we return to deal with active circuits in Chapters 14 and 15.

The word *systems* is especially hard to define. In general, an electrical system is a collection of circuits connected together to perform a complex task. Of course, a set of circuits connected together could also be considered as just a large circuit, so what is the distinction? In our view, a connection that can reasonably be analyzed and completely understood by a single engineer can be considered to be a circuit. Thus a phonograph amplifier, or perhaps even a TV set, could be considered to be a circuit. On the other hand, probably no one person can understand a satellite communications system in all its detail. Instead, systems are understood by breaking them up into building blocks and analyzing the interconnections between the blocks. For example, the designer of a satellite communications system would probably think of transmitters and receivers as building blocks and be concerned with how to use them to transmit maximum information reliably, with minimum cost. The systems designer probably would not be much concerned with the internal details of the transmitters and receivers; these are the responsibility of the circuit designers. Some of the subdivisions of the field of knowledge known as "systems" are shown in Fig. 0.1(c). In this book we shall introduce systems concepts in Chapters 9 to 12.

INTEGRATED CIRCUITS

The development of integrated circuits in the 1960s was as revolutionary as the invention of the triode in 1906. Integrated circuits (or ICs, as they are known) are circuits fabricated in tiny chips of semiconducting material. They are becoming more and more elaborate; in fact, advanced ICs containing hundreds of thousands of transistors might more correctly be called integrated systems. ICs are mass-produced by means of an ingenious technology (*microfabrication*) that makes their production inexpensive, in the same way that printing newspapers or stamping out phonograph records is inexpensive. Moreover, elaborate ICs do not cost much more to make than simple ones (just as it costs about the same to print a page of a book whether there are ten words on it or thousands).

ICs are natural building blocks for systems. Internally they can be very complex, yet to the systems designer, who sees them from outside and thinks of them as units, they can be powerful and yet easy to use. Electrical engineers are increasingly involved with ICs—the system designer with their applications, devices and circuits experts with the design of their increasingly complex insides.

The main result of IC technology has been an increase in the level of sophistication of all systems. In modern work, systems designers can take as building blocks ICs that themselves are as complex as anything their predecessors could build. Through interconnection of these integrated systems, very large systems, such as computers, have become possible. Figure 0.2 illustrates a large system and shows the parts played by devices, circuits, and ICs. In this example a computer, which is a large system (uppermost box), is composed of ICs and other components. Each IC is itself a small system. The ICs are versatile building blocks that can be used in various kinds of systems; they may have been purchased by the computer builder from another manufacturer, where they were designed by that company's engineering staff. The ICs contain numerous circuits, probably designed by one group of engineers, while another group concentrated on the organization of the circuits into the IC's system. Since the circuits contain devices, experts in this field were also required, especially in regard to microfabrication procedures. This is not to suggest that the various designers are completely

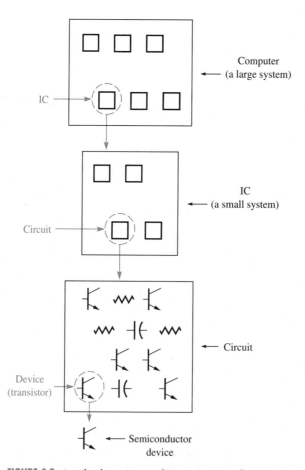

FIGURE 0.2 A modern large system, such as a computer, is often composed of ICs, which are themselves small systems. Internally the ICs are composed of interconnected circuits. These circuits in turn are interconnections of active and passive devices.

specialized; good engineers always try to understand not only their own small part but also the "big picture." Clearly this leads to the best results.

0.3 Analog and Digital Technology

Another distinction that can be drawn inside the field of information systems is that between analog and digital technology. These terms refer to two different ways in which information can be represented in electrical form. A voltage that varies over time and represents information is called a *signal*; the two important types are *analog signals* and *digital signals*. The two kinds of representation make use of different kinds of circuits and thus have motivated the development of two rather different, but parallel, branches of technology. Analog technology is the older, more traditional type; digital technology is comparatively recent, having received its major impetus from the development of integrated circuits. At this writing, analog and digital technologies are of about equal importance. Therefore, in this book, we treat the two with approximately equal emphasis. We shall now discuss the difference between analog and digital signals.

An analog signal is usually a voltage (or current) that is proportional to some physical quantity of interest. For example, imagine an electric thermometer that gives out a voltage that is proportional to the temperature. As time passes, the temperature may change, as shown in Fig. 0.3(a). The thermometer's output voltage then changes proportionally, as shown in Fig. 0.3(b). This time-varying voltage is an analog signal representing the information concerning temperature. Anyone with a voltmeter can measure the voltage and (assuming that the constant of proportionality is known) determine the temperature at any time.

The same information can also be represented in digital form. In this case, the information is converted to numbers, and voltages representing the numbers are transmitted. Measurements of the temperature might, for example, be made at times t_1, t_2, and t_3, as shown in Fig. 0.3(c). (This procedure is called *sampling* the temperature.) Suppose the temperature is 3° at t_1 and 2° at t_2 and t_3. A digital signal representing this information might consist of three pulses of voltage at t_1 and two pulses at t_2 and at t_3, as shown in Fig. 0.3(d). Anyone can observe these voltage pulses and determine the temperatures at the corresponding times.

It may seem that the digital signal is more complicated than the analog signal, but digital signals have great advantages. They are more resistant to errors than analog signals, and they are ideal for representing information that consists of numbers to begin with, as in a computer. Their real advantage, however, arises from the versatility and low cost of the digital circuits used to process them.

It is interesting that digital signals can be used even when the information varies rapidly in time. For example, telephone conversations can be transmitted in digital form, as shown in Fig. 0.4. The information to be transmitted is the time-varying acoustic pressure at the microphone. The microphone generates an analog signal proportional to acoustic pressure, and the analog signal is converted to a sequence of digital pulses by a device known as an *analog-to-digital converter*. Of course the rapidly

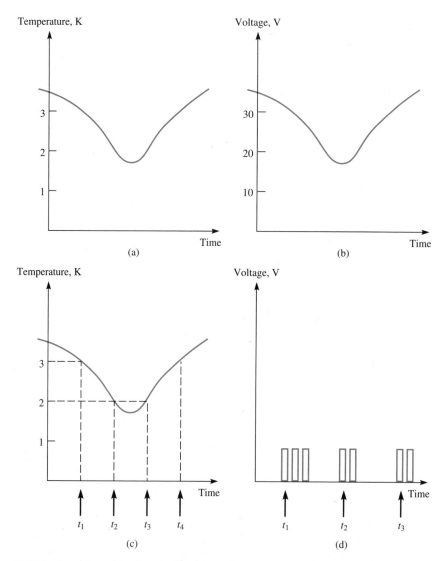

FIGURE 0.3 Analog and digital signals. Shown are (a) the information we wish to represent and (b) a suitable analog signal. (The constant of proportionality is 10 volts per degree.) (c) Alternatively, we can sample the information at times t_1, t_2, and t_3; then a digital signal (d) can be constructed. Counting the number of pulses that arrive at each sampling time tells us the temperature at that time.

varying analog signal must be sampled frequently; 8,000 times per second is typical. At the receiving end a *digital-to-analog converter* is used. This device uses the transmitted digits as instructions, according to which it generates an analog signal closely resembling the original. The reconstituted analog signal then goes to the telephone receiver's earphone, where it is converted back to time-varying acoustic pressure. Note

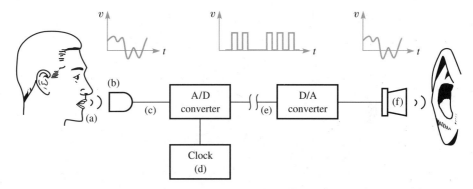

FIGURE 0.4   A digital telephone system. Acoustic waves (a) are converted by the microphone (b) into an analog signal in a wire (c). The analog signal is a time-varying voltage proportional to the acoustic pressure at the microphone. The clock (d) sends periodic impulses to the analog-to-digital (A/D) converter, instructing it to "sample" — that is, to produce a set of pulses representing the numerical value of the analog signal at the instant of sampling (see Figure 0.3). These pulses, which are a digital signal, are sent through the transmission medium (e). At the receiving end they are converted back to an analog signal by a digital-to-analog (D/A) converter and then to acoustic waves by an earphone (f). The transmission medium can be a wire, but usually the pulses are transmitted by radio or optical means.

that although the recipient of the call imagines hearing the caller's voice, what is actually heard is a kind of robot voice, synthesized according to instructions in the form of numbers. It seems remarkable that such a complicated scheme could be useful, but in fact the speed and low cost of modern integrated circuits makes digital transmission entirely practical, and today it is a common technique.

• EXERCISE 0.1

The temperature information in Fig. 0.3(a) is to be transmitted digitally. What number is to be transmitted at the time indicated as t_4 in Fig. 0.3(c)? What digital signal should be sent at time t_4? **Answer:** The number to be transmitted is 3. The signal should consist of three pulses, all the same height.

Circuits used to process information fall into two different classes, depending on whether the information is represented in analog or in digital form. For analog signals one uses continuous-state circuits, also known simply as analog circuits. Digital signals are handled by discrete-state circuits, also called digital circuits. In this book Chapters 9, 10, and 14 deal primarily with analog systems and circuits; Chapters 11, 12, and 15 contain corresponding treatments of digital systems and circuits.

0.4 The Building-Block Approach

We have seen that electrical systems are often large and complex. It is extremely helpful to consider such large systems as being built out of smaller units, which we shall call *building blocks*. The building blocks may themselves be very complicated internally. However, with the building-block point of view we do not concern ourselves with the interiors of the blocks, only with how they perform as seen from the outside.

Each block will have wires or some other kind of connection to the rest of the system. The system designer begins by knowing the *terminal properties* of the blocks—for example, how input signals applied to certain wires result in output signals at other wires. Often information about the terminal properties of the blocks is supplied by their manufacturer. Fortunately, even for complex blocks the terminal properties can be quite simple. For instance, in Chapter 4 we shall see that just three numbers are sufficient to characterize an ideal amplifier block, even though internally the amplifier may contain several sophisticated circuits and a dozen transistors. At any rate, when the blocks' terminal properties are known, it is possible for the system designer to interconnect them to produce the desired results. Conceivably the resulting large system can then itself be regarded as a building block, allowing even larger and more sophisticated systems to be built.

Useful building blocks can be small and simple or large and complex. The main characteristic of a good building block is that its terminal properties be well understood and fairly simple. Integrated circuits are especially suitable, since they are pre-engineered functional blocks with clearly defined terminal properties spelled out on data sheets provided by their manufacturers. But we shall also use the building-block approach in other ways.

In circuit analysis we shall learn to recognize certain small subcircuits on sight and to remember their terminal properties. (For example, the voltage divider, a very simple subcircuit containing two resistors, is described by a single equation that is easily remembered.) Then, when we look at more complex circuits, we can mentally break them down into blocks we recognize. In this way even large, complicated circuits can be quickly understood.

Figure 0.5 uses a radio communications system to illustrate the building-block point of view. To the large-scale-system designer, the system appears as in Fig. 0.5(a). The major blocks—keyboard, transmitter, antennas, receiver, printer—are in all likelihood made by different manufacturers. The system designer will think of them primarily in terms of their specified terminal properties and will select them to achieve the desired information capacity, reliability, and cost. However, the transmitter block was probably considered by the engineers who built it as itself being a collection of blocks, as shown in Fig. 0.5(b). Some of these blocks, such as the low-frequency amplifier, are likely to be ICs; others, like the power amplifier, are not. The low-frequency amplifier IC was probably designed and built by yet another manufacturer, and its designers probably thought of it as being composed of familiar subcircuits, as shown in Fig. 0.5(c). An example of such a subcircuit—one that would be recognized instantly by most circuit designers—is shown in Fig. 0.5(d).

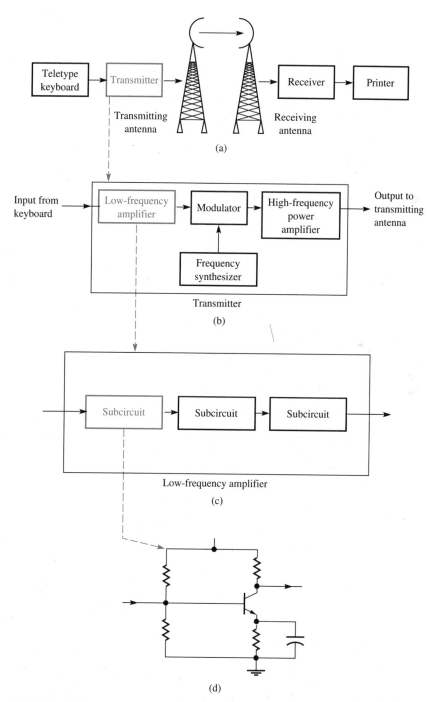

FIGURE 0.5 A radio communications system can be considered as being composed of building blocks (a). The blocks themselves can be broken down into simpler blocks (b), (c), and (d).

0.5 Plan of This Book

Now that we have outlined our subject, let us survey the way this book is arranged. Passive circuits, which are fundamental to almost all branches of the field, are treated in Chapters 1–3 and 5–8. In Chapters 4, 9, and 10 we introduce the important analog building blocks; there we explain how the blocks are described in terms of their terminal properties and how they are used in analog systems. Similarly, in Chapters 11 and

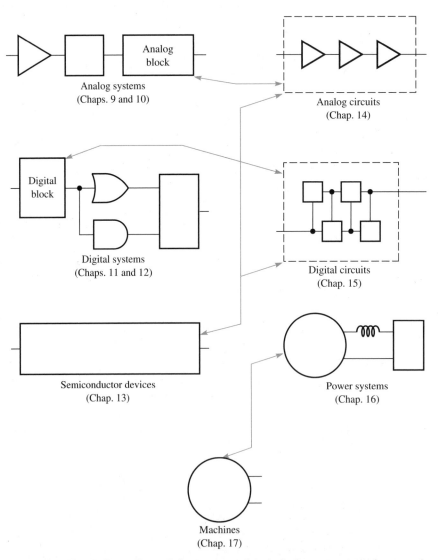

FIGURE 0.6   Schematic diagram illustrating the arrangement of this book. Chapters 1 to 8 (not shown) provide background for all the later chapters. The arrows indicate relationships between subjects (not prerequisites).

12 we introduce digital blocks and their use in digital systems. For the most part we deal in Chapters 9 to 12 with blocks that are conveniently available in IC form; thus we need not necessarily be concerned with what is inside the blocks. On the other hand, with an understanding of the blocks' insides one will be able to use them more expertly; and such knowledge is vital, of course, to those engineers who actually design and build the IC blocks. Thus in Chapter 13 we introduce some of the more important semiconductor devices, and with this background we go on in Chapter 14 to analog circuits and in Chapter 15 to digital circuits. Finally, in Chapters 16 and 17 we leave information systems and discuss power systems and electrical machines. This organization is shown schematically in Fig. 0.6. Chapters 1 to 8, not included in the figure, provide background for all the later parts of the book.

POINTS TO REMEMBER

- Electrical engineering divides by purposes into *information systems* and *power systems*. In the former, electrical means are used to transmit, store, and/or process information. In the latter, electricity is used to convey comparatively large amounts of energy from one place to another and to convert power from one form to another.
- *Electric circuits* are the most fundamental structures of electrical engineering. Circuits are collections of circuit *elements* connected by wires.
- *Active devices* are circuit elements that have the ability to control large currents or voltages under instructions supplied by small ones. Most important today are the *semiconductor devices.*
- *Systems* are collections of circuits that, taken as a whole, are too complex to be studied in full detail by individual designers.
- *Integrated circuits* are small, inexpensive, prefabricated circuits or systems. The term most often applies to circuits built on a single piece of semiconducting material.
- A time-varying voltage (or current) representing information is called a *signal.* Two important kinds are *analog signals* and *digital signals.* Analog signals are usually voltages (or currents) proportional to some physical quantity of interest. Digital signals represent *numbers* that convey the information in question.
- *Building blocks* are subunits that can be adequately described by their simple terminal properties. They can be connected to form larger circuits or systems.

PROBLEMS

0.1 Explain the following terms:
   a. circuit
   b. active device
   c. system
   d. integrated circuit

0.2 Explain the following terms:
  a. passive circuit
  b. analog circuit
  c. digital circuit

0.3 Sketch a graph of an analog signal representing the temperature in Fig. 0.3(a). The constant of proportionality is -2 volts per degree.

0.4 The temperature information of Fig. 0.3(a) is to be represented by a digital signal, with a sampling rate twice as large as in Fig. 0.3(c). (That is, we shall sample twice as often.) Numbers with two decimal places will be transmitted as follows: a group of pulses whose number is equal to the first digit, followed by a short time interval, followed by a group of pulses whose number is equal to the second digit—followed by a longer time interval before the next sampling. The first sampling occurs at t_1. Sketch the digital signal for five samplings.

0.5 Suppose the input analog signal in Fig. 0.4 is a sinusoidal voltage of the form $v(t) = A \sin 10^4 t$, where A is a constant, t is the time in seconds, and the argument of the sine function is given in radians. Assume that the D/A converter gives an output voltage function that consists of the values of the input voltage at the sampling times, connected by straight lines, as shown in Fig. 0.7. Sketch the way Fig. 0.7 would appear for the following sampling rates:
  a. 40,000 per second
  b. 1,000 per second

Estimate, very approximately, the minimum sampling rate needed to assure that the output resembles the input.

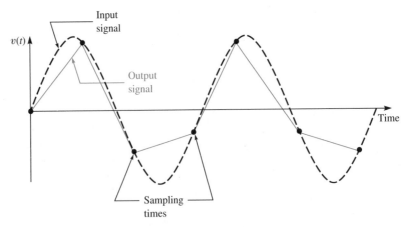

FIGURE 0.7

Introduction to Circuits

CHAPTER 1

PRINCIPLES OF ELECTRIC CIRCUITS

L et us begin our study of electrical engineering with its most characteristic unit, the electric circuit. As we shall use the term, a *circuit* is a collection of objects called *circuit elements,* connected together with wires through which currents can flow. Examples of circuit elements are resistors, capacitors, inductors, diodes, and transistors. Essential as these elements are, they are not useful all by themselves. It is only when they are connected together in a circuit that they can be used.

The subject of circuits includes both *circuit analysis* and *circuit design.* The purpose of circuit analysis is to calculate the voltages and currents that appear in a circuit and then, on the basis of these, to predict the useful performance of the circuit. In circuit design, on the other hand, we begin with a blank sheet of paper, and we create a new circuit to perform some function. (After creating the new circuit we will probably perform a circuit analysis on our design in order to check our work.) Of the two subjects, circuit analysis requires less experience, so it is here that we shall begin. Chapters 1 through 8 of this book deal mainly with analysis. Later chapters introduce some aspects of circuit design.

In the present chapter we shall introduce the fundamental properties of circuits, including their basic governing equations, known as *Kirchhoff's laws.* At the end of the chapter—and of each chapter throughout the book—we shall summarize what we have learned through an application. Our application for this chapter will deal with series and parallel connections and their properties.

1.1 Electrical Quantities

An understanding of circuits must be built on a basic knowledge of electrical physics. One must be familiar with electrical quantities, such as *charge, current,* and *voltage,* which are used in describing the operation of circuits. It is expected that to most readers the material of this section will not be entirely new. Nonetheless, it will serve as review and will also familiarize the reader with our definitions, notation, and point of view.

The branch of physics known as *electricity* seeks to describe electrical phenomena in terms of mathematical equations. The quantities in these equations must describe

TABLE 1.1   The SI Units

Quantity	Symbol	Unit	Abbreviation of Unit
Length	l	meter	m
Mass	m	kilogram	kg
Time	t	second	sec
Energy	E	joule	J
Force	F	newton	N
Power	P	watt	W
Charge	Q	coulomb	C
Current	I or i	ampere	A
Potential (or voltage)	V or v	volt	V
Resistance	R	ohm	Ω
Capacitance	C	farad	F
Inductance	L	henry	H

Note: Other units may be derived from the units in the table by means of decimal-multiplier prefixes, as follows:

Prefix	Abbreviation	Multiplies Unit by
pico-	p	10^{-12}
nano-	n	10^{-9}
micro-	μ	10^{-6}
milli-	m	10^{-3}
kilo-	k	10^{3}
mega-	M	10^{6}
giga-	G	10^{9}

For example, 1 nanosecond (nsec) is 10^{-9} second; 1 kilovolt (kV) is 1,000 V. These units are often more convenient than the basic SI units; for instance, it is handier to write "1.6 μA" than "0.0000016 A." However, it should be remembered that almost all equations in this book are true statements only when all quantities are expressed in the basic SI units. Numerical errors can occur if the prefixed units are carelessly used instead.

specifically electrical conditions and hence are called electrical quantities. As with other physical quantities, electrical quantities must be expressed in a consistent system of *units*. The system of units to be used throughout this book is known as the International System (or Système Internationale), and the units are referred to as SI units. A summary of these units is given in Table 1.1.

The three most fundamental electrical quantities are charge, current, and voltage. We shall now review the meanings of these terms in detail.

CHARGE

Each kind of atomic particle carries a certain amount of *charge*. In the international system of units, charge is measured in *coulombs* (C). Experimentally it is found that the neutron has a charge of zero. The proton has a charge of $+1.6 \times 10^{-19}$ coulomb; this amount of charge is given the symbol q. The electron has a charge which is exactly equal to that of the proton but opposite in sign; that is, the charge of an electron is -1.6×10^{-19} C, or $-q$. It is also found experimentally that the charge q is the smallest unit of charge that exists in nature. Larger charges occur when many charged particles are collected together. However, the coulomb is a large unit. Charges as large as 1 C may occur in electric power systems, but they are almost never found in electronics.

EXAMPLE 1.1

Throughout most of its volume, an impure silicon crystal contains equal densities of electrons (each with charge $-q$) and positive ions (each with charge $+q$). However, there is a small region, with dimensions 5×10^{-4} cm by 1 mm by 1 mm, where only the ions are present. The density of ions everywhere in the crystal is $10^{22}/m^3$. Find the total charge of the crystal.

SOLUTION

In the region where both ions and free electrons are present, their opposite charges cancel, and the charge density is 0. Hence the total charge of the crystal arises from the region containing ions only. The charge density here is given by the number of ions per unit volume times the charge of each ion; the latter quantity is +q, or 1.6×10^{-19} C. Thus

$$\rho = (10^{22}) \cdot (1.6 \times 10^{-19}) \text{ C/m}^3 = 1.6 \times 10^3 \text{ C/m}^3$$

The total charge Q is equal to the charge density (charge per unit volume) ρ times the volume of the charged region. In computing the latter we must convert the given dimensions to SI units:

$$V = (5 \times 10^{-6} \text{ m}) \cdot (10^{-3} \text{ m}) \cdot (10^{-3} \text{ m}) = 5 \times 10^{-12} \text{ m}^3$$

Thus

$$Q = \rho V = (1.6 \times 10^3 \text{ C/m}^3) \cdot (5 \times 10^{-12} \text{ m}^3) = 8 \times 10^{-9} \text{ C}$$

CURRENT

A flow of electrical charge from one region to another is called a *current*. Currents not only have magnitude, but also have direction. To indicate its direction, a current is described by a number with a sign. At the top of Fig. 1.1 the current in the wire has been given the name I_1. We indicate this by writing "I_1" next to the wire; in addition, we draw a small arrow along the wire. The arrow is important. It means that the current *in the direction of the arrow* has the value I_1. The direction of the arrow is called the *reference direction* of I_1. The choice of which direction is the reference direction is arbitrary. This choice is simply part of the process of naming the current.

By definition, a positive current exists when *positive* charges move in the reference direction, as illustrated by Case (a) of Fig. 1.1. In Case (a) the numerical value of I_1 will be positive. However, if positive charges move *opposite* to the reference direction, as in Fig. 1.1(c), the current has a negative value. The current could also arise from motion of negative charges, as in Cases (b) and (d) of Fig. 1.1. When negative charges are moving, the sign of the current is *reversed* from what it would be if positive charges were moving.

Suppose a quantity of charge dQ moves past some point in the reference direction in time dt. Then the value of the current is dQ/dt. If a charge dQ moves opposite to the reference direction in time dt, the current is $-dQ/dt$. The SI unit of current is the *ampere*, 1 ampere (A) representing a current of 1 coulomb per second.

I_1

This arrow indicates the reference direction

| I_1 IS A POSITIVE NUMBER IN THESE CASES: | I_1 IS A NEGATIVE NUMBER IN THESE CASES: |

Case (a): positive charges move in the reference direction

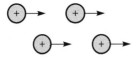

Case (c): positive charges move opposite to the reference direction

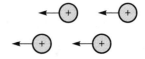

Case (b): negative charges move opposite to the reference direction

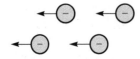

Case (d): negative charges move in the reference direction

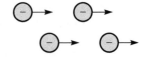

FIGURE 1.1

EXAMPLE 1.2

(a) In Fig. 1.2(a) the current I_x has the reference direction indicated. The current arises from the motion of 10^{13} electrons per second, as shown. What is the value of I_x?

(b) Figure 1.2(b) shows exactly the same charges moving in the same way. However, we have chosen to define a new current I_y whose reference direction is opposite from that chosen for I_x. What is the value of I_y?

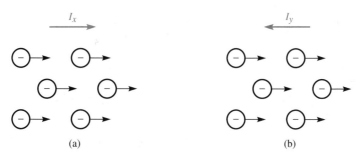

(a)   (b)

FIGURE 1.2

SOLUTION

(a) In this case we have negative charges moving in the reference direction. Hence the numerical value of the current is $+dQ/dt$, where dQ, the charge transferred in time dt, is a negative number. Hence

$$I_x = (-1.6 \times 10^{-19})(10^{13}) = -1.6 \ \mu A$$

(b) In this case the charges are moving *opposite* to the reference direction. Hence the current in the reference direction is $-dQ/dt$, where dQ is a negative number. Thus

$$I_y = -(-1.6 \times 10^{-19})(10^{13}) = +1.6 \ \mu A$$

• EXERCISE 1.1

Suppose electrons are moving in a wire, and the reference direction is from A to B. The numerical value of the current in the reference direction is -3 mA. How many electrons are moving per second? Are they moving from A to B or from B to A? **Answer:** 1.875×10^{16} electrons per second are moving from A to B.

Note: In everyday speech, one often hears expressions like "the current is in the direction from A to B." What is meant is that the sign of the current is the same as if positive charges were moving from A to B.

ELECTRICAL POTENTIAL AND VOLTAGE

When electrical forces act on a particle, it will possess potential energy; the value of this potential energy will depend on where the particle is located. For example, if a positively charged particle is located close to a cluster of other positive charges, it will possess a large amount of potential energy. This energy can be turned into kinetic energy by releasing the particle so it is pushed away by electrostatic repulsion. After a while the particle will have moved to a position where its potential energy is lower, and the potential energy difference will have been converted to kinetic energy ($\frac{1}{2}mv^2$).

In order to describe the potential energy that a particle will have at a point *x,* we define the electrical potential *V:*

$$V(x) = \frac{E(x)}{Q} \tag{1.1}$$

Here $E(x)$ is the potential energy that a particle with charge Q has when it is located at the position *x.* $V(x)$ is called the *electrical potential at point x.*

An important point, one to note carefully, is that the zero point of potential energy can be arbitrarily chosen. The reason for this is that only *differences* in energy have practical meaning. For example, suppose you have to raise a weight from 500 feet above sea level to 1,500 feet. What matters is that you have to raise the weight 1,000 feet. It does *not* matter that the starting point is at 500 feet. In fact, we could change the reference altitude from sea level to, say, the elevation of Denver, Colorado (which we shall pretend is exactly 5,000 feet above sea level). Now our job is to raise the weight from a place $-4,500$ feet above Denver to $-3,500$ feet above Denver. The important quantity, the distance we have to raise the weight, is still the same: $-3,500 - (-4,500) = 1,000$ feet. The situation with electrical potential is similar. We can designate any point to be the point where electrical potential is zero. This point is referred to as the *reference point* or *ground point.* Potentials at other points are then described by comparing them with the potential at the reference point.

Just as with potential energy, it is the *differences* of electrical potential that matter. We refer to a difference in electrical potential as a *potential difference* or *voltage.* These two terms are equivalent. For example, let the electrical potential at point A be called V_A and that at B, V_B, both with respect to the same arbitrarily chosen ground point. Then we define the potential difference (or voltage) V_{AB} by the equation $V_{AB} = V_A - V_B$. We would say in words, "The potential at A is higher than that at B by the voltage V_{AB}." The same physical situation could be described by $V_{BA} = V_B - V_A$, which would be read, "The potential at B is higher than that at A by the voltage V_{BA}." Clearly, $V_{BA} = -V_{AB}$. Voltages can be either positive or negative numbers and are measured in *volts.*

EXAMPLE 1.3

The potential at point C with respect to the reference point is 13 V. The potential at point D is 15 V with respect to the same reference point. What is the potential at point C with respect to point D?

SOLUTION

The potential at point C with respect to point D is $V_{CD} \equiv V_C - V_D$. In this case $V_{CD} = 13 - 15 = -2V$.

GROUND

As mentioned earlier, it is possible to define the potential at any point in a circuit to be equal to zero. Such a point is then known as the *reference node,* or, equivalently, is said to be *grounded.* The chosen reference node is indicated on the circuit diagram by the "ground" symbol $\perp$. In Fig. 1.3, for example, we see that point A is designated as the

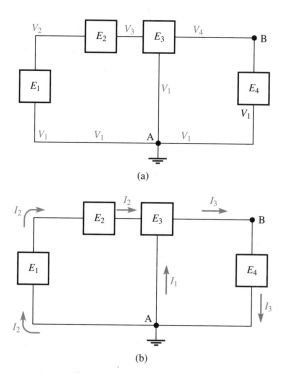

(a)

(b)

FIGURE 1.3

reference node. Defining one point as being at zero potential simplifies the stating of the other potentials. For instance, with point A grounded, the statement "point B is higher in potential than point *A* by 5.4 V" is simplified to "the potential at point B is 5.4 V."

It is usual to express all potentials in a circuit with respect to ground. In everyday language the potential difference between any point and ground is often called the voltage at that point. *In this book the words "voltage at point* A*" always mean "the potential at point* A *with respect to ground." The symbol for this voltage is* V_A.

VOLTAGE SYMBOLS

The symbols used in describing voltages are illustrated by Fig. 1.4. This figure shows a portion of a circuit containing the two points A and B. (The dashed line at the top indicates that the circuit continues in a manner that does not concern us.) The voltage at A (with respect to ground) is V_A, as shown at the right. The + and − signs on each voltage indication are very important. Their meaning is that *the indicated voltage is equal to the potential at the* + *sign minus the potential at the* − *sign.* On the right side of the figure the symbols indicate that $V_A − 0 = V_A$. (Remember that the potential at the ground point is zero.) On the left side of the figure the symbols indicate that $0 − V_A = −V_A$, which is clearly the same thing. Do you understand why V_{AB} and V_{BA} are indicated as they are?

Often more than one point on a circuit diagram is marked with a ground symbol. This is just a kind of shorthand. What is really meant is that all such points are to be regarded as connected together, and the voltage at their common connection point is defined to be 0. This trick makes the circuit diagram simpler because the wires that

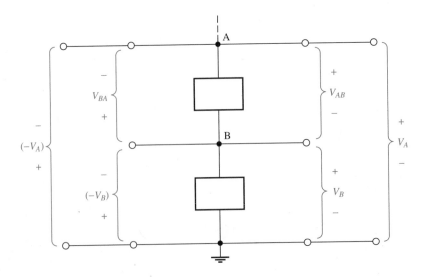

FIGURE 1.4

actually connect all the grounded points do not have to be shown. For example, look ahead to Fig. 1.15. The current I_3 in this figure is actually flowing from point A to point B through circuit element N_1. This can easily be seen when we remember that the two grounded points are connected by a wire not shown explicitly in the diagram.

• EXERCISE 1.2

V_A (the potential at A with respect to ground) in Fig. 1.5 is 4 V, and $V_{BA} = -7$ V. What is V_B? What is V_C? **Answers:** $V_B = -3$ V; $V_C = 0$.

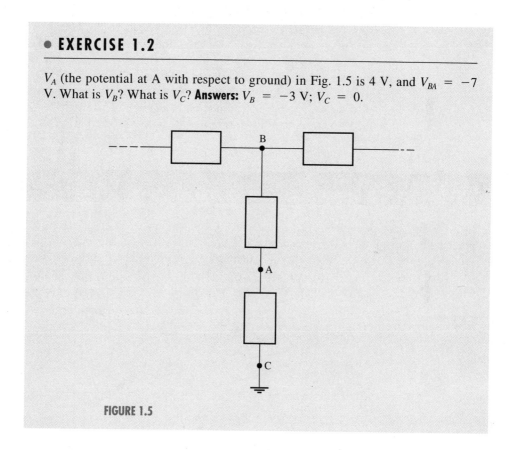

FIGURE 1.5

POWER

The potential V at a certain point is the energy of a unit positive charge located at that point. Thus if we move positive charges from a lower potential to a higher potential, we have to add additional energy to the system. In other words, we "do work" on the system. On the other hand, by allowing positive charges to fall from a higher potential to a lower, we can take energy out of the system. This is like obtaining energy from falling water, as it moves from a higher position to a lower one through a dam.

The energy obtained when a charge Q moves from a place where the potential is V_2 to a place where there is a lower potential V_1 is given by $Q(V_2 - V_1)$. If we do this repeatedly, the energy per second is equal to $(V_2 - V_1)$ times the charge that moves each second. The energy given out per second is called the *power P*. The charge per second is the current *I*. Thus

$$P = (V_2 - V_1)I_{2\to 1} \tag{1.2}$$

where $I_{2\to 1}$ is the current with reference direction from 2 toward 1.

As with voltage and current, the power has associated with it an algebraic sign. This sign can be quite important. Its meaning has to do with the *direction* of power transfer. For instance, suppose we say that the power entering a certain box is *P*. If *P* is a positive number, then energy is going *into* the box, as would be the case if the box contained an energy-consuming device such as a light bulb. However, the value of *P* might turn out to be a negative number. In that case the power entering the box is negative, meaning that energy is coming *out* of the box. This might be the case if the box contained a battery. We shall return to this subject in Section 3.3.

EXAMPLE 1.4

In a certain television picture tube (Fig. 1.6), electrons are released at electrode A, where the potential is V_A, and fall to the screen, where the potential is V_B. The current flowing from A to B is $-I_1$. When the electrons hit electrode B their energy is converted to heat. Find the power being converted from electrical form to heat if $V_A = -15$ kV, $V_B = 0$, and $I_1 = 1$ mA.

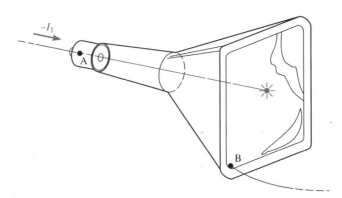

FIGURE 1.6

SOLUTION

In this case we have $I_{A\to B} = -I_1$. (The value of this current is negative because negatively charged electrons are moving from A to B.) Thus the power being dissipated (turned into heat) is

$$P = (-15{,}000 - 0)(-10^{-3}) = 15 \text{ watts}$$

EXAMPLE 1.5

From measurements we find that the dc motor in Fig. 1.7(a) produces a mechanical output of 41.3 horsepower (hp). The voltmeter tells us that $V_{AB} = 442$ V, and the ammeter indicates a current of 83.1 A in the direction of the arrow.

(a) Find the efficiency of the motor, assuming that there is no voltage drop in the ammeter (that is, $V_{AC} = 0$).

(b) A real ammeter does in fact usually have some internal voltage drop. (This will be discussed further in Section 2.6.) Recalculate the efficiency of the motor under the assumption that $V_{AC} = 20$ V. (*Note:* 1 hp $= 746$ W.)

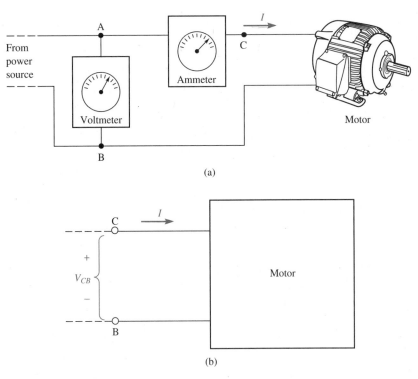

(a)

(b)

FIGURE 1.7(a) and (b)

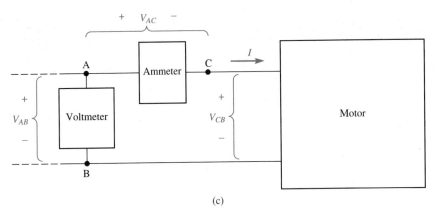

(c)

FIGURE 1.7(c)

SOLUTION

(a) According to Eq. (1.2) the power entering the motor is $P_{IN} = V_{CB}I$, as illustrated in Fig. 1.7(b). Since we are to assume $V_{AC} = 0$, we know that $V_{CB} = V_{AB}$. Thus

$$P_{IN} = (442)(83.1) = 36.7 \text{ kW} = 49.2 \text{ hp}$$

The efficiency E is given by

$$E = \frac{P_{OUT}}{P_{IN}} = \frac{41.3}{49.2} = 0.84 \quad \text{(or 84\%)}$$

(b) In this case the circuit is as shown in Fig. 1.7(c). The electrical power entering the motor is still $V_{CB}I$, but the quantity indicated by the voltmeter is V_{AB}. We are given $V_{AC} = 20$ V, and we can use the relationship

$$V_{CB} = V_{AB} - V_{AC}$$

(can you prove this?). Hence

$$E = \frac{P_{OUT}}{P_{IN}} = \frac{P_{OUT}}{(V_{AB} - V_{AC})I} = \frac{41.3}{(422)(83.1)/(746)} = 0.88 \quad \text{(or 88\%)}$$

1.2 The Electric Circuit

An electric circuit is a closed path or combination of paths through which current can flow. In most cases the greater part of the circuit is composed of good electrical conductors, which we shall call *wires*. (The conductors need not actually be in the form of conventional wires; the term is used here only for convenience.) Good electrical conductors tend to be at nearly the same potential everywhere. Wires are very often "idealized"; that is, it is often assumed that wires are *perfect* conductors and therefore are at *exactly* the same potential everywhere. The reader should remember that in this book, unless otherwise indicated, *all points along a wire are assumed to be at exactly the same potential.*

It is also generally assumed (although this is also an idealization) that wires cannot store any charge within themselves. Therefore all current that enters one end of a wire must exit at the other end.

In addition to wires, a circuit contains *circuit elements,* such as resistors, capacitors, and transistors. A circuit element has two or more terminals where wires may be connected. As with wires, we assume that all current entering one terminal of a two-terminal element must exit at the other terminal. Thus the current through a two-terminal element is a well defined quantity. For each kind of circuit element there is some relationship between the voltages at the terminals and the currents that flow through the terminals. Ohm's law (to be discussed in Section 2.2) is an example of such a relationship.

Electric circuits are shown by circuit diagrams. Some of the basic wiring symbols used in this book's circuit diagrams are listed in Table 1.2. Additional symbols for circuit elements will be introduced in the next chapter (see Table 2.1). Regrettably, symbols used for the same thing by different writers vary, particularly those in different areas of electrical engineering practice. The symbols used in this book are those most often used in electronics at present. Some alternative symbols not used in this book but which may be seen elsewhere are also listed in Table 1.2. Note that the symbol most often used in electronics for "wires not connected" means just the opposite, "wires connected," in the alternative system. Thus in reading circuit diagrams one must be aware of what conventions are being used.

An example of an electric circuit is shown in Fig. 1.3. This circuit contains four circuit elements; three of these have two terminals, and one is a three-terminal element. Since each wire may be at a different potential, there are four different potentials in this circuit. The names of these potentials are indicated on the diagram; they are designated V_1, V_2, V_3, and V_4. Note, however, that all points marked V_1 are at the same potential because the corresponding wires are connected together.

Figure 1.3(b) shows the same circuit as Fig. 1.3(a); in this figure, however, the currents rather than the voltages are labeled. Each circuit element is assumed to have a total charge of zero at all times. Therefore the current entering one terminal of a two-terminal element is the same as the current exiting at the other terminal. From Fig. 1.3(b) we see that there are three different currents in this circuit. The names of these three currents are indicated in the diagram, and arrows show the direction assigned as the reference direction for each current. Note that all the currents marked I_2 must be equal.

When two points are not connected together by any circuit element at all, an *open circuit* is said to exist between them. Putting it another way, the two points are said to be open-circuited. If a wire is broken, an open circuit is created; no current can flow through the open circuit, although there may be a potential difference across it. Conversely, when two points are connected together by an ideal wire, a *short circuit* is said to exist between them; in other words, the two points are short-circuited. By the nature of an ideal wire, no potential difference can exist between points that are short-circuited, although current may flow from one to the other.

TABLE 1.2 Symbols Used in Circuit Diagrams

Symbols Used in This Book	Meaning	Alternate Symbol
	Wires connected.	
	Wires not connected.	
I	The current in the wire has the value I. Its reference direction is indicated by the arrow.	
V_A	The potential at the indicated node is V_A with respect to ground.	
$+$ v $-$	The terminal marked "+" is higher in potential than the terminal marked "−" by the voltage v.	$+$ v $-$
	Ground; the potential at the indicated node is defined to be zero.	

In this book we shall follow general usage by referring to any quantity that is constant in time as a *dc quantity* (for example, "dc voltage," "dc current"). A quantity that varies in time will be called a *time-varying quantity*. Moreover, dc voltages and currents will be represented by uppercase symbols (V_2, I_B); time-varying quantities will be represented by lowercase symbols (v_E, i_2).

BRANCHES AND NODES

Some additional terms are useful in circuit analysis. A point at which two or more circuit elements are connected is called a *node*. A two-terminal circuit element connected between two nodes is called a *branch*. Figure 1.8 shows a circuit with three nodes and four branches. The reader should note that the voltage at point D in this circuit is the

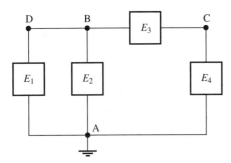

FIGURE 1.8   A circuit containing three nodes and four branches. Node A is the reference node, as indicated by the ground symbol. Points D and B are located at the same node.

same as the voltage at point B, since the two points are connected by a wire. Hence D and B are not two separate nodes; for our purposes points D and B are one and the same node. The thing we call a node includes not only the particular point designated by a dot or letter but also all other points connected to the designated point by wires. This point is often misunderstood. If you are in doubt, mark all points in the circuit you think are nodes. If any two of these are connected by an uninterrupted wire, they are the same node.

1.3 Kirchhoff's Laws

The two statements known as Kirchhoff's laws are fundamental to all problems of circuit analysis. In general, these laws, together with a knowledge of the properties of all circuit elements, are sufficient for calculation of the voltage and current everywhere in a circuit.

The first of Kirchhoff's laws, known as *Kirchhoff's current law,* arises from the physical assumption that no charge can accumulate in a wire: Whatever current flows in one end must flow out the other. More generally, if a certain amount of current flows into a node through one wire, an equal amount must leave the node through other wires. Kirchhoff's current law is a formal statement of this idea: It states that *the sum of all currents entering a node is zero.* Clearly, unless all currents are zero, at least one of the currents entering a node must be negative.

EXAMPLE 1.6

It is known that the current I_1 entering node A in Fig. 1.9 through element E_1 equals $+15$ mA. The current I_2 is known to equal $+32$ mA. What is I_3?

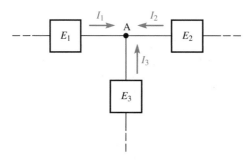

FIGURE 1.9

SOLUTION

According to Kirchhoff's current law, the sum of the currents entering the node is zero. We therefore write

$$I_1 + I_2 + I_3 = 0$$

$$I_3 = -I_1 - I_2 = -15\,\text{mA} - 32\,\text{mA} = -47\,\text{mA}$$

Note that Kirchhoff's current law may also be stated, "The sum of all currents leaving a node is zero." To show this, let us assume that I_1, I_2, and I_3 are the only three currents entering a node. Then by Kirchhoff's current law $I_1 + I_2 + I_3 = 0$. Now if the current entering a node through a certain wire is I_1, the current going outward from the node through that wire is $-I_1$. Therefore the sum of all currents flowing outward is $-I_1 - I_2 - I_3 = -(I_1 + I_2 + I_3) = -(0) = 0$.

• EXERCISE 1.3

In Fig. 1.3(b) let $I_2 = 4.7$ mA and $I_3 = 3.3$ mA. What is I_1? **Answer:** -1.4 mA.

The second of Kirchhoff's laws, known as *Kirchhoff's voltage law,* can be deduced by the following argument: The voltage at any point of a circuit at any instant has a certain value. Suppose we begin at one point of the circuit, pass through a circuit element, and note the decrease, or "drop," in voltage. We then pass through another element, note this second drop in voltage, and add it to the first drop. Suppose now that after passing through several elements we arrive back at the original point. The sum of all the voltage drops that were experienced must now add up to zero, since we have returned to the original potential. The closed itinerary just described is known as a "loop," and the foregoing argument can be formalized into what is known as Kirchhoff's voltage law: *The sum of all the voltage drops around a complete loop is zero.* Clearly some of these voltage drops must have negative values in order that their sum can be zero.*

EXAMPLE 1.7

Let us call the voltage at node B in Fig. 1.10 V_B, at node C, V_C, and at node D, V_D. In terms of these three voltages, write an equation expressing Kirchhoff's voltage law for the loop A → B → C → D → A.

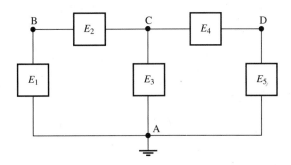

FIGURE 1.10

SOLUTION

We shall add voltage drops as we go around the loop. From the presence of the ground symbol we know that node A is at zero potential. The drop as we go from A to B is $V_A - V_B = 0 - V_B = -V_B$. The drop as we go from B to C is $V_B - V_C$. Adding the voltage drops around the entire loop and setting the sum equal to zero gives

$$-V_B + (V_B - V_C) + (V_C - V_D) + (V_D - 0) = 0$$

The correctness of this equation is of course obvious by inspection. Kirchhoff's voltage law is really just a restatement of the assumption that there is a single, unique voltage associated with each point.

---

*It would be just as reasonable to add potential increases, or *rises,* around a complete loop. Thus an alternative statement of Kirchhoff's voltage law is that the sum of voltage rises around a complete loop is zero.

In Fig. 1.3(a), let $V_1 - V_2 = 6.1$ V, $V_3 - V_1 = 4.4$ V. Find $V_3 - V_2$. **Answer:** 10.5 V.

1.4 Application: Series and Parallel

Two-terminal circuit elements can be connected together in many ways. One very common way is the *series connection* shown in Fig. 1.11(a). Any number of elements can be connected together in series; in this figure we have three. The distinguishing feature of the series connection is that all the current leaving one element enters the next; thus $I_1 = I_2 = I_3$. The potentials V_X and V_Y have to be calculated, using one's knowledge of the circuit elements. If the three elements in the figure are identical, it is reasonable to expect that the voltage drops across them will be the same. In that case we expect that $(V_X - V_A) = (V_Y - V_X) = (V_B - V_Y) = (V_B - V_A)/3$. However, if the elements are not identical, the voltages across them may be different.

Circuit elements connected side-by-side, as in Fig. 1.11(b), are said to be connected *in parallel.* In a parallel connection, it is the voltages across the circuit elements that are the same. In general, the currents through the elements will be different. In the

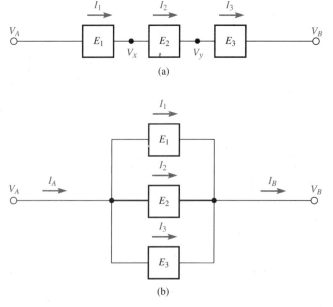

(a)

(b)

FIGURE 1.11

special case of identical elements, the currents would also be the same. From Kirchhoff's current law we know that whether or not the elements are identical, we must have $I_A = I_1 + I_2 + I_3 = I_B$.

A familiar example arises in lighting systems, such as outdoor lighting systems or strings of Christmas tree lights. The N lamps can be connected either in parallel, as in Fig. 1.12(a), or in series, as in Fig. 1.12(b). In the parallel connection, the voltage across each lamp, V_P, is the same as the power-line voltage, V_0. However, in the series connection, the voltage across each lamp, V_S, is less, only V_0/N. If the power of each lamp is to be the same in either case, then the current through each bulb will have to be N times larger in the series connection than in the parallel.

Suppose you are the design engineer. Which connection do you choose? You will want to consider the advantages and disadvantages of each method, such as:

a. When a light bulb fails, it turns into an open circuit. In the parallel connection this does not affect the other bulbs, which remain lit. However, in the series circuit, if one bulb fails, all the bulbs go out. Thus the parallel circuit could be said to be *more reliable.*

b. Suppose the circuit has a large number of bulbs, and one bulb fails. In the parallel circuit it is obvious which one needs to be replaced. However, in the series circuit you cannot tell which bulb is bad, because they all go out. To correct the fault, you will have to go through the entire circuit, testing the bulbs one by one. Thus the parallel circuit is *easier to maintain.*

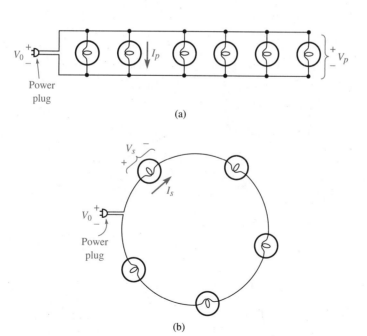

(a)

(b)

FIGURE 1.12

c. Inspecting the figure, we observe that in the parallel circuit there are two copper wires between each lamp, while in the series circuit there is only one. This gives the series connection the advantage of *lower initial cost.* If you are planning to manufacture thousands of light strings, the cost saving here might be important.

d. In the parallel connection, the voltage across each bulb is the full power-line voltage, while in the series connection the voltage across each bulb is N times less. Thus, in normal operation, electrical shock and fire hazards might be less serious with the series connection, and it might be said to have a *safety* advantage. However, the safety advantage here is a bit of an illusion, because in the event of a bulb burnout, the entire line voltage does suddenly appear across the burned-out bulb. The reason this happens will become apparent in the next chapter.

Design choices like this, based on tradeoffs between various advantages and disadvantages, are a challenging part of an engineer's daily work. Knowledge and experience are needed to arrive at the best decision.

POINTS TO REMEMBER

- The current at a point on a wire is equal to the amount of charge passing that point each second. In SI units, charge is measured in coulombs and current in amperes. Currents can be positive or negative numbers. The sign of a current is very important.

- A voltage is a difference in electrical potential between two points. It is measured in volts. Voltages can be positive or negative. The sign is important.

- It is possible to designate any single point in a circuit as "ground," meaning that by definition the potential at that point is zero.

- Power is transferred from one part of a circuit to another whenever a current flows between two points at different potential. Although $|P| = IV$, power also has a *sign*, which indicates the direction of power flow.

- Kirchhoff's current law states that the sum of all currents entering a node is zero. Kirchhoff's voltage law states that the sum of all voltage drops around any closed path is zero.

PROBLEMS

Section 1.1

1.1 A current flows through a wire. If 10^{17} electrons pass through the wire from point B to point A each second, what is the current in the direction from A to B?

1.2 A beam containing two types of particles is moving from A to B. Type I particles have charge $+2q$, and type II particles have charge $-2q$ (where $-q$ is the charge on an electron). Type I particles flow at the rate of 4×10^{15}/sec, type II particles at 6×10^{15}/sec. Find the current flowing in the direction *from* B *to* A.

1.3 In Fig. 1.13(a), the current I_1 is known to be 2 mA. What is I_2?

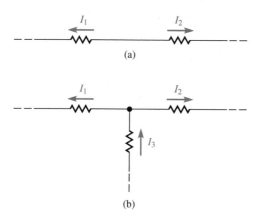

(a)

(b)

FIGURE 1.13

1.4 In Fig. 1.13(b), the current I_1 is known to be 2 mA.
  a. Is it possible to say what I_2 is?
  b. Is it possible to say what $I_2 + I_3$ is?

1.5 The voltage at point A is 6 V with respect to ground, and that at point B is 9 V with respect to ground.
  a. What is the voltage at A with respect to B?
  b. What is V_{BA}?

1.6 Suppose $V_{AB} = -2$ V, $V_{AC} = 5$ V, and $V_{DB} = -4$ V. Point C is at 3.5 V with respect to ground. What is the voltage at D with respect to ground?

1.7 The potential at X is higher than that at Y by -6 V. Moreover, $V_{ZY} = 5$ V and V_X (the voltage at X with respect to ground) is -3 V. What is V_Z?

1.8 Prove that if A, B, and C are any three points in a circuit, $V_{AC} = V_{AB} + V_{BC}$.

1.9 Refer to Fig. 1.4. Let $V_A = 280$ V and let $V_{BA} = -520$ V. What is V_B?

1.10 Draw a circuit (of any shape) containing three nodes A, B, and C. Indicate on your diagram, by means of symbols similar to those of Fig. 1.4, that V_A is less than V_B by an amount V_x, and that the potential at C is less than that at A by an amount V_y. If $V_x = 440$ V and $V_y = 120$ V, what is V_{CB}?

1.11 The power output (as light) of a certain automobile headlight is 10 W, and the lamp is known to be 28% efficient as a converter of electric power to light. The voltage applied to the lamp is 12.0 V.
  a. Construct a diagram showing the lamp, indicating points 1 and 2 and the current $I_{1\to2}$ in Eq. (1.2).
  b. What is the value of $I_{1\to2}$ corresponding to the diagram you have made?

1.12 An electron that is initially stationary falls through a potential increase of 160 V. Find its final velocity. Express this velocity as a fraction of the velocity of light. The mass of an electron is 9.1×10^{-31} kg.

Sections 1.2 and 1.3

1.13 A 25 watt lamp is powered by a 6 volt battery. How many electrons pass through the lamp every second?

1.14 Suppose three branches come together at a certain node. The reference directions for the three currents are all directed into the node. Prove that if at least one of the currents is nonzero, at least one must have a negative value.

1.15 In Fig. 1.14 the box contains any circuit element. Show that $I_1 + I_2 + I_3 + I_4 = 0$.

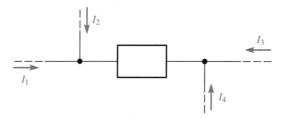

FIGURE 1.14

1.16 In Fig. 1.15 $I_1 = 5$ mA and $I_2 = -3$ mA. Find I_3.

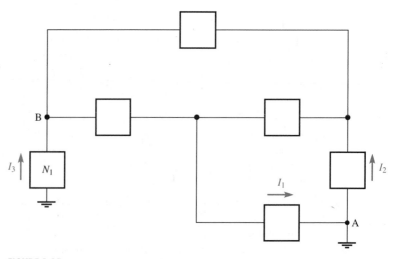

FIGURE 1.15

1.17 In Fig. 1.3(b) let i_1 and i_2 depend on the time t according to $i_1 = 10\cos{(376t)}$ A, $i_2 = 10\sin{(376t)}$ A, where t is given in seconds and angles in radians. Find the times at which $i_3 = 0$.

1.18 In Fig. 1.16:
  a. Show that $V_{12} + V_{23} + V_{34} = V_{16} + V_{65} + V_{54}$.
  b. Show that $V_{12} + V_{34} + V_{56} = -(V_{23} + V_{45} + V_{61})$.
  c. Show that $V_{13} + V_{24} + V_{35} + V_{46} + V_{51} + V_{62} = 0$.

1.19 In Fig. 1.16 a current I (not equal to zero) flows around the loop as shown. Use Eq. (1.2) and Kirchhoff's voltage law to show that the sum of the powers being converted in all six circuit elements is zero. What is the physical meaning of this result?

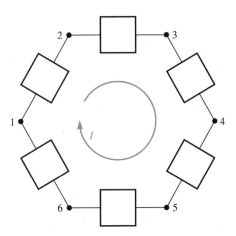

FIGURE 1.16

Section 1.4

1.20 Determine whether the circuit elements in Figs. 1.17(a) and (b) are connected in series or in parallel.

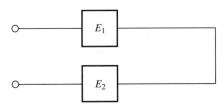

FIGURE 1.17(a)

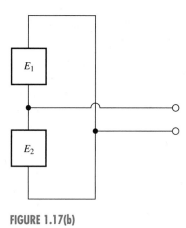

FIGURE 1.17(b)

1.21 Suppose the light strings in Fig. 1.12(a, b) each contain 10 lamps, and each lamp consumes 25 watts. Find the current through each lamp for the parallel and series connections. The supply voltage at the power plug in both cases is 100 V.

2

Introduction to dc Circuit Analysis

uppose we need to know how a given circuit works. We are given a diagram of the circuit, along with complete information about the behavior of all the circuit elements. In general, our problem is then to calculate all the voltages and currents that will appear throughout the circuit. This kind of calculation is known as *circuit analysis.* Once we have found the voltages and currents, we will know whether the circuit does the job we want it to do.

The technique of circuit analysis is based on the use of Kirchhoff's laws. In principle, the steps are quite straightforward, and thus computers, which can be accused of lacking imagination, are nonetheless quite good at circuit analysis (as we shall see in Section 2.7). However, there are many different cases to consider: non–time-varying, sinusoidally time-varying, pulsed signals, linear and nonlinear circuit elements, and so forth. Thus there are a number of approaches and techniques to learn. The effort required to master these techniques will be well spent. Circuit analysis is placed first in this book because it is the electrical engineer's most basic tool, and these skills will be useful in all further electrical engineering work.

The present chapter will deal with analysis of dc circuits. The abbreviation "dc," standing for "direct current," is a time-honored description of circuits in which currents and voltages *do not vary in time.* The most important circuit elements in this chapter are the ideal resistance and ideal voltage and current sources. The reader has probably seen at least the first two of these elements before but will find that we are more careful about their mathematical properties than previously. In particular, we must be precise about the algebraic signs of voltages and currents, which are of great importance in practice.

Having introduced ideal circuit elements, we shall go on to describe two mathematical techniques for circuit analysis, known as the node method and the loop method. We shall then discuss two special subcircuits, the voltage divider and the current divider, which turn up very often and are handy to know. Of course, instead of calculating the currents and voltages in a circuit mathematically, one might simply measure them, and of course in practice measurements often must be performed. Thus we shall briefly describe the way circuit measurements are made. Finally, our "application" for this chapter will be a discussion of circuit diagrams and their uses. This will give an idea of how circuit diagrams and circuit analysis figure in engineering work.

2.1 Ideal Independent Sources

Real electronic circuits contain physical circuit elements such as light bulbs, batteries, or transistors. Such elements are in general quite complicated; to analyze circuits containing them in full detail would be a slow and difficult task. For this reason it is usual to substitute *models* for real circuit elements. Models are composed of simpler, idealized circuit elements with more convenient mathematical properties. Modeling is one of the most important tools of circuit analysis, and examples of its use will be seen extensively in this book. In this chapter we shall introduce some of the most important ideal circuit elements.

The reader should not be confused about the difference between actual physical circuit elements, such as transistors, and ideal circuit elements. Ideal circuit elements do not exist in the real world at all. They are imaginary objects, defined by simple mathematical properties. We use them as stand-ins for real circuit elements because they make it easier to perform circuit analysis. In most cases, sufficiently accurate answers are obtained. However, we should not forget that we are dealing with models for real physical circuit elements. The accuracy of the results depends on the quality of the model and the circumstances in which it is used.

One example of an ideal circuit element that has been seen already is the *ideal wire.* In all our discussions thus far, we have regarded wires as being ideal, in the sense that the potential everywhere on an ideal wire is exactly the same. This is a property of perfectly conducting metals, which have no resistance. Of course the real metals used to make wires do have some resistance, so real wires are not actually ideal. However, in many cases the resistance of real wires has little effect, so little error is introduced by modeling them as though they were ideal. On the other hand, some judgment has to be used. If a wire is 100 miles long, its resistance will probably turn out to be important!

As our next example, let us consider the *ideal independent voltage source,* shown in Fig. 2.1. This source has only one simple property: The voltage that appears across it has the value specified. The plus and minus signs are part of the symbol for this circuit element, and they are important, as they indicate the sign of the voltage. Their meaning is that *the potential at the side marked plus is higher than the potential at the side marked minus by the indicated value.* In Fig. 2.1(a), terminal A is V volts higher in

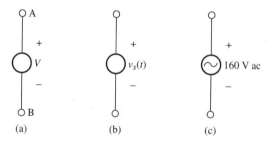

FIGURE 2.1 Circuit symbols for ideal voltage sources. (a) The voltage between the terminals is controlled so that the (+) terminal is higher in potential than the (−) terminal by the value V. (b) The voltage between the terminals has the value $v_s(t)$, which is time-varying. (c) The voltage between the terminals varies sinusoidally in time. The amplitude of the sinusoid is 160 V.

potential than terminal B. Note that V can be either a positive or a negative number. The important thing to observe about the ideal voltage source is its simplicity. It has only one simple property: The potential difference across its terminals has the specified value. Whether current happens to be passing through the element makes no difference; neither does the temperature, or the phase of the moon. This is why we say the circuit element is "ideal." The word "independent" in "independent voltage source" means that the value of the source is specified independently, rather than depending on what is going on elsewhere in the circuit. Dependent sources also exist and will be encountered in Chapter 4.

EXAMPLE 2.1

Find the potential difference V_{AB} ($\equiv V_A - V_B$) in Fig. 2.2. The boxes marked "CE" are circuit elements; their properties do not matter in this example.

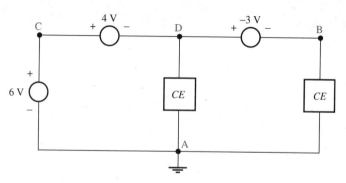

FIGURE 2.2

SOLUTION

The 6 volt voltage source requires that the potential at point C be 6 volts higher than that at point A. The 4 volt source requires that the potential at D be -4 volts higher than at C. In other words, the potential at D must be 4 volts *lower* than at point C. Thus $V_D - V_A = 6 - 4 = 2$ volts; this is the same as saying $V_{AD} = -2$ volts.

Continuing, we see that the potential at D is -3 volts higher than at B; that is, $V_D - V_B \equiv V_{DB} = -3$ volts. Therefore $V_{AB} = V_{AD} + V_{DB} = (-2) + (-3) = -5$ volts.

Notice that although some currents may be flowing through the "CE" circuit elements, this does not affect the result. The known properties of the ideal voltage sources are sufficient to determine the answer.

27 mA   16 cos ωt mA

(a)   (b)

FIGURE 2.3 Ideal current sources. In (a), the current is 27 milliamperes in the direction of the arrow. In (b), the current through the source is time-varying.

A similar circuit element is the *ideal independent current source,* the symbol for which is shown in Fig. 2.3. Again, this circuit element possesses only one simple property: *The current through it has the indicated value.* The reference direction for the known current is indicated by the arrow. This is the only property the circuit element has. One must not be distracted by the presence of other circuit elements that may be present. No matter what the rest of the circuit is like, the ideal current source maintains the indicated current.

EXAMPLE 2.2

Find the current I in Fig. 2.4.

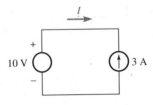

FIGURE 2.4

SOLUTION

The current I is identical with the current through the current source, but its reference direction is opposite. Therefore $I = -3$ A.

The reader should not be confused by the presence of the voltage source. It has no properties that affect the answer to this question.* We must have $I = -3$ A, or the properties of the current source would be violated.

---

* At some point the reader may have heard that "current flows from plus to minus," which seems to be the opposite of what is happening here. This is because the statement that "current flows from plus to minus" is not correct in general; it applies only to currents flowing through resistances. Do not be misled: The ideal current source always states the actual current through that part of the circuit. This is so simple that it can be overlooked!

● EXERCISE 2.1

Find the current I in Fig. 2.5. **Answer:** 8 mA.

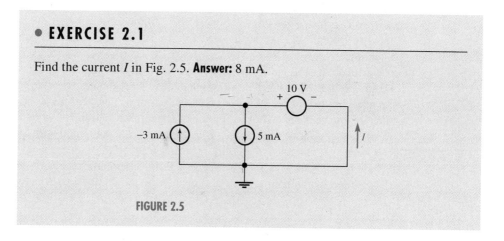

FIGURE 2.5

Sometimes the question is asked, "What happens if an ideal voltage source is connected across a short circuit?" This situation should never arise, because such a circuit would be a logical inconsistency. Of course one can take a pencil and *draw* such a circuit, but it is nonsensical; one might just as well take a pencil and write "two equals three." The point is that to draw such a circuit would be to make two statements that contradict each other. The voltage source means that the potential difference between its terminals has some given value, but if we connect those terminals with an ideal wire we are also saying that the potential difference between the terminals is zero. Similarly, an ideal current source in series with an open circuit also makes no sense.

2.2 Ideal Resistance

An *ideal resistor* is a circuit element characterized by the property that the current through it is linearly proportional to the potential difference across its terminals. Because of this property the ratio V/I has a fixed, constant value for any particular resistor. The value of this ratio is called the *resistance*. In SI units resistance is measured in *ohms* (with the symbol Ω).

The relationship between current and voltage (known as the *I-V relationship*) for an ideal resistance is known as *Ohm's law*. In stating Ohm's law one must pay careful attention to the algebraic signs of the quantities involved. Figure 2.6(a) shows the symbol for a resistor. We have named one terminal A and the other B. We also have defined a current $I_{A \to B}$, the reference direction for which is from A to B. The statement of Ohm's law is then

$$I_{A \to B} = \frac{V_A - V_B}{R} \qquad (2.1)$$

Of course Ohm's law holds regardless of which end of the resistor we call A and which we call B.

It is important that sign errors be avoided. We suggest the following method. First draw an arrow indicating your chosen reference direction for current through the resistor. *The value of this current is the potential at the tail of the arrow minus the potential at the head of the arrow, divided by* R. (This of course means that the *difference* between the two potentials is to be divided by *R*.) Try memorizing this rule; it will help in avoiding errors.

Most real electrical conductors resemble ideal resistors as long as only moderate-sized voltages and currents are applied. However, the resistive properties of different materials vary widely. We have taken the point of view that metal wires are ideal conductors and thus have zero resistance. This approximation is a good one provided that the actual resistance of the wires is negligibly small compared with other resistances in the circuit. A resistor is made by inserting a piece of comparatively resistive material between two wires, as shown in Fig. 2.6(b). The value of the resulting resistance is determined by the composition of the material and its physical dimensions.

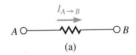

(a)

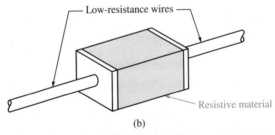

(b)

FIGURE 2.6   Resistance. (a) Circuit symbol; (b) one type of construction.

EXAMPLE 2.3

In Fig. 2.6(a), terminal A is at potential -3 V with respect to ground, and terminal B is at -2 V. The resistance is 10 Ω. What is the current $I_{A \to B}$?

SOLUTION

Applying Ohm's law, we find

$$I_{A \to B} = \frac{(-3) - (-2)}{10} = -0.1 \text{ A}$$

EXAMPLE 2.4

Suppose the current $I_{A \to B}$ in Fig. 2.6(a) is known to be positive. Does the voltage drop observed in moving from A to B have a positive or a negative value?

SOLUTION

Since $I_{A \to B}$ is greater than zero, it must be true that $V_A > V_B$. Hence the voltage drop from A to B is a positive number.   (B → A → neg)

EXAMPLE 2.5

In everyday usage, such expressions as "the direction of the current" or "the actual direction of the current" tell us what direction must be chosen as the reference direction if we wish the numerical value of the current to be *positive*. Suppose in Fig. 2.6(a) that the "actual direction" of current is from B to A, the absolute value of the current is 10 A, and the resistance is 3 Ω. What is V_{AB}?

SOLUTION

Since the "actual current direction" is from B to A, the current $I_{B \to A}$ must be positive. Hence $I_{B \to A} = +10$ A. From Ohm's law,

$$V_{BA} = R I_{B \to A}$$

However, we are asked for V_{AB}. Thus

$$V_{AB} = -V_{BA} = -(3)(10) = -30 \text{ V}$$

EXAMPLE 2.6

References for the voltage V_1 and current I_1 have been chosen as indicated in Fig. 2.7. (The dashed lines indicate that the wires continue to another part of the circuit with which we are not concerned.) Measurements reveal that the value of I_1 is -200 A. If $R_1 = 3 \ \Omega$, what is V_1?

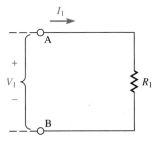

FIGURE 2.7

SOLUTION

According to Eq. (2.1),

$$I_1 = \frac{V_A - V_B}{R_1}$$

The positions of the plus and minus signs above and below the symbol V_1 indicate that V_1 is defined by

$$V_1 = V_A - V_B$$

Thus $V_1 = I_1 R_1 = -600$ V.

● **EXERCISE 2.2**

In the circuit of Fig. 2.8, $R_1 = 3{,}000 \ \Omega$, $R_2 = 4{,}700 \ \Omega$, and $R_3 = 3{,}300 \ \Omega$. From measurements we know that $V_A = -2$ V, $V_B = 3$ V, $V_C = -3$ V, and $V_D = 12.735$ V (with respect to ground). Find I_1, I_2, and I_3. **Answers:** $I_1 = 1.67$ mA; $I_2 = -1.28$ mA; $I_3 = -2.95$ mA.

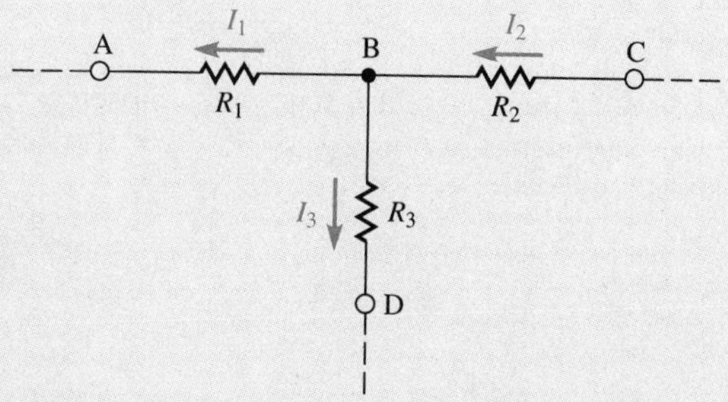

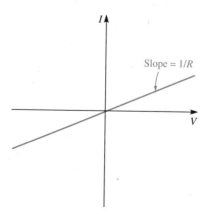

FIGURE 2.8

I-V GRAPH FOR THE IDEAL RESISTANCE

When the current through a circuit element is a function of the voltage across it, we can make a graph of current voltage. This graph is known as the circuit element's *I-V graph*. For an ideal resistance, the *I-V* graph is determined by Ohm's law and is thus a straight line through the origin, as shown in Fig. 2.9. The slope of this line is $1/R$, where R is the resistance.

SERIES AND PARALLEL CONNECTIONS OF RESISTANCES

The idea of series and parallel connections has already been introduced in Section 1.4. A series connection of two resistances is shown in Fig. 2.10(a); Fig. 2.10(b) shows two resistances connected in parallel. We notice that each of these circuits has two terminals for connection to the outside world. Thus each can be regarded as a composite cir-

FIGURE 2.9 The *I-V* graph for an ideal resistance is a straight line through the origin. The slope of the line is $1/R$.

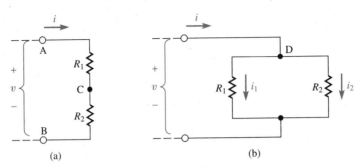

FIGURE 2.10   Composite circuit elements composed of two resistances (a) in series and (b) in parallel.

cuit element, with its own new *I-V* relationship. For example, in Fig. 2.10(a) the terminals of the composite circuit element are A and B, the voltage across these terminals is called *v*, and the current into terminal A is called *i*. Let us now find the *I-V* relationship of the series subcircuit.

We observe that the current flowing through both resistors in Fig. 2.10(a) is the same and is equal to *i*. Using Ohm's law, Eq. (2.1), we find that the voltage at point A, v_A, is higher than that at point C, v_C, by the amount iR_1; in other words, $v_A - v_C = iR_1$. Similarly, $v_C - v_B = iR_2$. The total voltage *v* across the element equals $v_A - v_B$, which is the same as $(v_A - v_C) + (v_C - v_B)$. Therefore $v = iR_1 + iR_2 = i(R_1 + R_2)$. We see by comparing with Eq. (2.1) that the *I-V* relationship of the series combination is the same as that of a single resistor with value $(R_1 + R_2)$. Because of the identical nature of the *I-V* relationships, we say that the series combination is *equivalent* to a single resistor of value

$$R = R_1 + R_2 \quad \text{(resistors in series)} \tag{2.2}$$

Further use will be made of the concept of equivalence in later chapters.

Now consider the parallel combination of two resistors is shown in Fig. 2.10(b). We observe that the voltages appearing across resistors R_1 and R_2 are the same and are equal to *v*. According to Eq. (2.1), $i_1 = v/R_1$ and $i_2 = v/R_2$. Applying Kirchhoff's current law to node D (note that the current entering node D through R_1 is $-i_1$), we have $i - i_1 - i_2 = 0$. Therefore

$$i = i_1 + i_2 = v\left(\frac{1}{R_1} + \frac{1}{R_2}\right) \tag{2.3}$$

$$= v\left(\frac{R_2 + R_1}{R_1 R_2}\right)$$

This may be rewritten

$$i = v \bigg/ \left(\frac{R_1 R_2}{R_1 + R_2}\right) \tag{2.4}$$

Again comparing with Eq. (2.1), we see that the parallel combination is equivalent to a single resistor with value

$$R = \frac{R_1 R_2}{R_1 + R_2} \quad \text{(resistors in parallel)} \tag{2.5}$$

As a shorthand, the resistance given by Eq. (2.5) is sometimes abbreviated as $R_1 \| R_2$; that is, by definition $R_1 \| R_2 \equiv R_1 R_2 / (R_1 + R_2)$. More complex combinations of resistors can be shown to be equivalent to single resistors by repeated use of Eqs. (2.2) and (2.5).

EXAMPLE 2.7

The combination of resistors shown in Fig. 2.11 has two terminals designated A and B. It is desired to replace it with a single resistor connected between terminals A and B. What should the value of this resistor be so that the resistance between the terminals is unchanged?

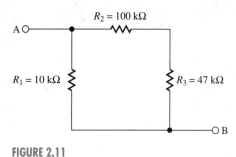

FIGURE 2.11

SOLUTION

On inspection it is seen that R_2 and R_3 are in series and that their series combination is in parallel with R_1. Thus the resistance between A and B is given by

$$R = (R_2 + R_3) \| R_1 = \frac{R_1(R_2 + R_3)}{R_1 + R_2 + R_3}$$

$$= \frac{(10{,}000)(147{,}000)}{157{,}000} = 9{,}360 \ \Omega$$

The three resistors may be replaced by a single resistor of this value without altering the resistance between A and B.

POWER DISSIPATION IN IDEAL RESISTOR

An important property of the resistor is its ability to convert energy from electrical form into heat. In Chapter 1 it was pointed out that when a current I flows from a place at a higher voltage to one where the voltage is lower by an amount V, then the power converted is given by $P = VI$. This is the case in a resistor; the voltage drop V is, according to Ohm's law, equal to IR. Thus when a current I flows through a resistance R, the power converted into heat is

$$P = I^2R \qquad (2.6)$$

Alternatively, since $I = V/R$, we may also state that when a potential difference V exists across a resistance R, the power converted into heat is

$$P = V^2/R \qquad (2.7)$$

In practice, resistors must be designed so that they can dissipate the necessary amount of heat. Generally, the manufacturer states the maximum power dissipation of a resistor in watts. If more power than this is converted to heat by the resistor, the resistor will overheat and be damaged.

EXAMPLE 2.8

The power dissipation of a 47,000 Ω resistor is stated by the manufacturer to be 1/4 W. What is the maximum dc voltage that may be applied? What is the largest dc current that can be made to flow through the resistor without damaging it?

SOLUTION

The maximum voltage is obtained from Eq. (2.7), where $P = 1/4$ W. We have

$$V^2_{max} = RP$$
$$V_{max} = \sqrt{(4.7 \times 10^4)(0.25)} = 108.4 \text{ V}$$

Similarly, to find the maximum current we use Eq. (2.6):

$$I^2_{max} = P/R$$
$$I_{max} = \sqrt{(0.25)/(4.7 \times 10^4)} = 2.3 \times 10^{-3} \text{ A} = 2.3 \text{ mA}$$

IDEAL VOLTMETER AND IDEAL AMMETER

A voltmeter or ammeter is used to indicate, respectively, the voltage or current at some point in a circuit. An *ideal* voltmeter is an imaginary device having the property that it measures the voltage while drawing no current; that is, *the current through an ideal*

TABLE 2.1   Circuit Elements of dc Circuits

Name	Symbol	Property
ideal independent voltage source		the potential difference between the terminals is V
ideal independent current source		the current through the element (with reference direction given by the arrow) is I
ideal resistance		$I = \dfrac{V_A - V_B}{R}$
ideal voltmeter		$I_1 = I_2 = 0$
ideal ammeter		$V_A - V_B = 0$

voltmeter is zero. Similarly, an ideal ammeter possesses the idealized property that *the voltage across the terminals of an ideal ammeter is zero.* Circuit symbols for these and the other circuit elements of this chapter are given in Table 2.1.

2.3 Node Analysis

In circuit analysis the usual goal is to calculate voltages or currents at various places in a circuit. This can be done, provided that the *I-V* relationships of all the elements in the circuit are known. In this section and the next, two general techniques for circuit analysis will be described: the *node method* and the *loop method*. Either method can be used, although at times one may require less effort than the other.

The node method is a technique used to calculate the voltages at certain positions known as *nodes* of the circuit. A node is any point in the circuit at which terminals of two or more different circuit elements are connected together. Figure 2.12(a) shows a circuit containing four nodes, marked A, B, C, and D. In general, the potentials at the different nodes will be different, and these potentials, V_A, V_B, V_C, and V_D, are what we wish to calculate. It should be noted that the node is not just the particular point marked by a dot or letter, but also includes all other points connected to the node by ideal wires. Thus, for example, in Fig. 2.12(b) there are only three nodes; the points marked B and D are the same node. If you are in doubt, mark all points in the circuit you think are nodes. If any two of these are connected by an uninterrupted wire, they are the same node.

THE NODE METHOD

The state of a circuit is known completely if the voltage at all nodes and the current through each of the branches is known. One of the node voltages arbitrarily can be set

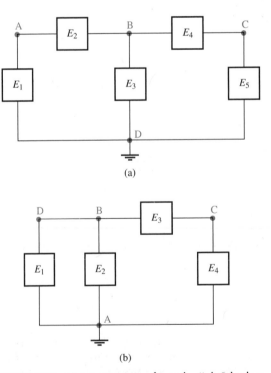

(a)

(b)

FIGURE 2.12 (a) A circuit containing four nodes. Node D has been selected as the reference node, where by definition the potential is zero. (b) A circuit containing *three* nodes.

equal to zero. Thus the number of unknowns to be solved for is equal to the number of branches plus the number of nodes, minus one. In the node method of analysis, one solves first for all unknown node voltages. This is usually done by writing an equation expressing Kirchhoff's current law for each node where the voltage is unknown. These equations state that the sums of the currents flowing into each node are zero. The currents, in turn, are expressed in terms of the unknown node voltages by means of the *I-V* relationships for the branches. The resulting set of simultaneous equations is then solved for the unknown node voltages. Finally, branch currents can then be obtained using the calculated node voltages and the *I-V* relationships of the branches.

As a demonstration of how the node method is used, let us calculate the branch currents and node voltages for the circuit of Fig. 2.13. Before we can proceed, we must locate the node voltages to be found. The four nodes have been designated A, B, C, and D. One node can be chosen arbitrarily as the reference node, which by definition has potential zero. In this case node D has been selected as the reference node and marked with the ground symbol. The potentials of other nodes are then defined by comparison with the potential at node D. For example, a node 2 V lower in potential than node D is said to have a potential of -2 V.

We may now proceed with solution by the node method. There are four node voltages in the problem, but they are not all unknown. Because of our choice of the ground point we know that $V_D = 0$, and because of the voltage source we know also that $V_A = V_0$. This leaves V_B and V_C as unknown node voltages. They may be found by simultaneous solution of the two equations expressing Kirchhoff's current law for nodes B and C.

Let us use the form of Kirchhoff's law that states that the sum of all currents entering node B is zero. The current entering node B through R_1 has, by Ohm's law, the value $(V_A - V_B)/R_1$. Since $V_A = V_0$, we may write this current as $(V_0 - V_B)/R_1$. Similarly, the current entering through R_2 is $(V_C - V_B)/R_2$. The current entering node B through R_4 is $(V_D - V_B)/R_4$, but since $V_D = 0$, this current is simply $(-V_B)/R_4$. The statement of Kirchhoff's current law for node B is then

$$\frac{(V_0 - V_B)}{R_1} + \frac{(V_C - V_B)}{R_2} + \frac{(-V_B)}{R_4} = 0 \tag{2.8}$$

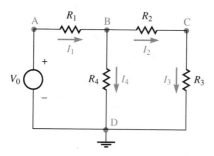

FIGURE 2.13

In similar fashion we may write the node equation for node C:

$$\frac{(V_B - V_C)}{R_2} + \frac{(-V_C)}{R_3} = 0 \tag{2.9}$$

Solving Eqs. (2.8) and (2.9) simultaneously for V_B and V_C, we have

$$V_C = V_0 \frac{R_3 R_4}{R_1 R_2 + R_1 R_3 + R_1 R_4 + R_2 R_4 + R_3 R_4} \tag{2.10}$$

$$V_B = V_0 \frac{R_4(R_2 + R_3)}{R_1 R_2 + R_1 R_3 + R_1 R_4 + R_2 R_4 + R_3 R_4}$$

The node voltages are now all known. To complete the analysis, we can now find the currents that flow through the circuit's *branches*, which are the connections between the nodes. The branch currents can be found by simple applications of Ohm's law; thus $I_1 = (V_0 - V_B)/R_1$, $I_2 = (V_B - V_C)/R_2$, $I_3 = V_C/R_3$, and $I_4 = V_B/R_4$.

Recapitulating, in the node method, equations representing Kirchhoff's current law are written for the nodes whose voltages are unknown and solved simultaneously for the node voltages. Branch currents are then found from the node voltages by applications of Ohm's law.

● EXERCISE 2.3

Find the voltage at node A in Fig. 2.14 by means of the node method. The circuit values are $R_1 = 470\ \Omega$, $R_2 = 330\ \Omega$, $R_3 = 510\ \Omega$, $I_0 = 1$ mA, $V_0 = 0.5$ V.
Answer: 0.504 V.

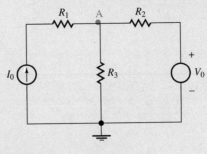

FIGURE 2.14

A case requiring special treatment occurs when two nodes, where the voltages are unknown, are connected together by a voltage source. Consider the circuit shown in

Fig. 2.15(a). Here the voltage source V_2 forms the branch between nodes A and B, where the voltages are unknown. In this case, one cannot write an expression for the current flowing from B to A in terms of V_B and V_A because the current through the voltage source is not a function of the voltages. Thus the sum of currents entering node A or B cannot be written and set equal to zero in the usual way.

To circumvent this difficulty, we mentally draw a dashed line around the two nodes giving difficulty, as shown in Fig. 2.15(b). Using reasoning similar to that of Kirchhoff's current law, we can see that the sum of all currents entering the region bounded by the dashed line must be zero. This must be true: Charge cannot accumulate at the nodes, and all currents flowing into one end of a wire or circuit element must flow out the other end. But it is easy to write the currents entering the dashed region and set this sum equal to zero:

$$\frac{V_1 - V_A}{R_1} + \frac{0 - V_A}{R_2} + \frac{0 - V_B}{R_3} = 0 \qquad (2.11)$$

This provides one equation in the two unknowns, V_A and V_B. A second equation is obtained from the known property of the voltage source V_2:

$$V_A + V_2 = V_B \qquad (2.12)$$

These two unknowns V_A and V_B may now be found by simultaneous solution of Eqs. (2.11) and (2.12).

Often before proceeding with nodal analysis one can simplify a circuit by reducing the number of nodes. This can be done by regarding series configurations of elements as composite elements and using the *I-V* relationships for the composite elements. This eliminates the node between the two series elements that are combined.

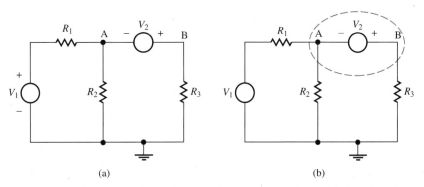

(a)   (b)

FIGURE 2.15   (a) A circuit in which a modified technique must be used for nodal analysis. This technique involves setting equal to zero the sum of all currents entering the region enclosed by the dashed line (b).

EXAMPLE 2.9

Find the voltage V_0 in the circuit in Fig. 2.16(a).

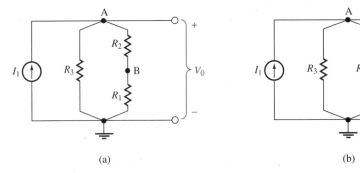

(a)   (b)

FIGURE 2.16

SOLUTION

The voltage V_0 is identical with the node voltage V_A. The voltage at node B is also unknown. Node B can be eliminated from the circuit, however, by replacing the series combination of R_1 and R_2 by the single resistor $R_4 = R_1 + R_2$, which has been shown to have the same *I-V* characteristic. The circuit then becomes that shown in Fig. 2.16(b).

There is now only one unknown node voltage remaining, V_A ($\equiv V_0$). Writing Kirchhoff's current law at node A, we have

$$I_1 - \frac{V_A}{R_3} - \frac{V_A}{R_4} = 0$$

The solution is

$$V_A = I_1 \frac{R_3 R_4}{R_3 + R_4} = I_1 \frac{R_3(R_1 + R_2)}{R_3 + R_1 + R_2}$$

2.4 Loop Analysis

In the loop method of analysis, one defines special currents known as *mesh currents*. These are related to the branch currents in a simple way, so that in the loop method, unlike the node method, the branch currents are obtained first. Finally, the node voltages can be found from the branch currents using the *I-V* relationships of the branches.

Let us again consider the circuit of Fig. 2.13, which was earlier analyzed by the node method. In Fig. 2.17 this circuit is redrawn; two mesh currents, I_1 and I_2, have

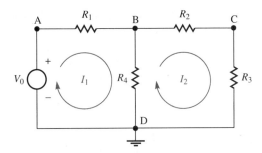

FIGURE 2.17 For loop analysis of this circuit, two mesh currents must be defined.

been selected as shown. The mesh currents are circulating currents assumed to flow around the loops of the circuit. The branch currents can each be expressed in terms of the mesh currents. For instance, in Fig. 2.17 the current flowing from A to B through R_1 is equal to I_1. The current flowing from C to B through R_2 is equal to $-I_2$. When two mesh currents both flow through a branch, the branch current is the algebraic sum of the mesh currents. Thus the current flowing from B to D through R_4 is equal to $I_1 - I_2$.

There is considerable freedom in the choice of mesh currents. However, it is necessary that every branch of the circuit have at least one mesh current flowing through it. The directions of the mesh currents (that is, clockwise or counterclockwise) may be freely chosen, and different directions may be used for different mesh currents in the same circuit. In simple circuits, such as that of Fig. 2.17, one chooses as many mesh currents as there are loops in the circuit—two in this case. In more complex or three-dimensional circuits, the following formula may be used to find the number of mesh currents needed for solution:

(Number of mesh currents)

= (Number of branches) − (Number of nodes) + 1

For instance, in Fig. 2.17 there are four nodes and five branches. Thus the requisite number of mesh currents is 5 − 4 + 1 = 2.

The values of the mesh currents are now the unknowns to be solved for. A number of equations equal to the number of unknown mesh currents is needed. These equations are obtained by writing equations expressing Kirchhoff's voltage law: The sum of the voltage drops around any closed path is zero. Any nonidentical closed paths through the circuit may be used; often for convenience one chooses the paths to be the routes of the mesh currents, but this is not necessary.

As a demonstration of the procedures, let us consider the circuit of Fig. 2.17 with the mesh currents I_1 and I_2 as there defined. We shall first write an equation stating that the sum of the voltage drops experienced, as one follows the closed path A–B–D–A, is equal to zero. The voltage drop in going from A to B through R_1 is $I_1 R_1$. The drop experienced in going from B to D through R_4 is $(I_1 - I_2)R_4$. In going from D to A through the voltage source, a drop of $-V_0$ is experienced. Setting the sum of the voltage drops around the closed path equal to zero, we have

$$I_1 R_1 + (I_1 - I_2)R_4 - V_0 = 0$$

Proceeding similarly around the loop B—C—D—B, we obtain

$$I_2 R_2 + I_2 R_3 + (I_2 - I_1)R_4 = 0 \tag{2.13}$$

Solving simultaneously, we find

$$I_1 = V_0 \frac{R_2 + R_3 + R_4}{R_1 R_2 + R_1 R_3 + R_1 R_4 + R_2 R_4 + R_3 R_4} \tag{2.14}$$

and

$$I_2 = V_0 \frac{R_4}{R_1 R_2 + R_1 R_3 + R_1 R_4 + R_2 R_4 + R_3 R_4} \tag{2.15}$$

The individual branch currents can now simply be obtained from I_1 and I_2, as stated above. Finally, node voltages may be found by using the calculated values of I_1 and I_2 and Ohm's law. For instance,

$$V_C = I_2 R_3 = V_0 \frac{R_3 R_4}{R_1 R_2 + R_1 R_3 + R_1 R_4 + R_2 R_4 + R_3 R_4} \tag{2.16}$$

The result is in agreement with Eq. (2.10), obtained via the node method.

EXAMPLE 2.10

Find the current I and voltage V in Fig. 2.18(a) by means of the loop method.

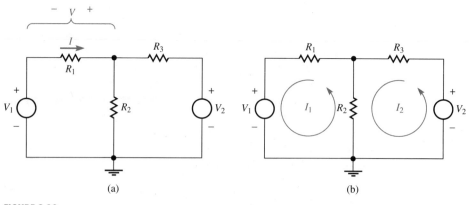

(a)   (b)

FIGURE 2.18

SOLUTION

The first step is to designate mesh currents. It doesn't matter whether these are assumed to circulate clockwise or counterclockwise; we arbitrarily choose them counterclockwise, as shown in Fig. 2.18(b). Adding voltage drops around the mesh containing I_1, we have

$$I_1 R_1 + V_1 + (I_1 - I_2)R_2 = 0$$

Adding voltage drops around the mesh containing I_2, we have

$$(I_2 - I_1)R_2 - V_2 + I_2 R_3 = 0$$

Solving the two equations simultaneously, we obtain

$$I_1 = \frac{R_2 V_2 - (R_2 + R_3)V_1}{R_1 R_2 + R_1 R_3 + R_2 R_3}$$

To complete the solution we note that I, the current asked for, is given by $I = -I_1$. The voltage V is given by $V = +I_1 R_1$ (note the positive sign).

It is necessary to write as many loop equations as there are unknown mesh currents in the problem. Sometimes the value of a mesh current can be seen by inspection. For example, in Fig. 2.19 I_1 is the current that goes through the current source; therefore $I_1 = 20$ mA. There is only one unknown mesh current, I_2, remaining; therefore writing a single statement of Kirchhoff's voltage law around one closed path will be sufficient. Sometimes, too, it is possible to reduce the number of unknown mesh currents by replacing a parallel combination of elements by an equivalent single element. The circuit of Fig. 2.17, for example, can be analyzed with only a single mesh current if R_2, R_3, and R_4 are replaced by a single resistor, which has value $R_5 = R_4 \| (R_2 + R_3)$, connected between B and D.

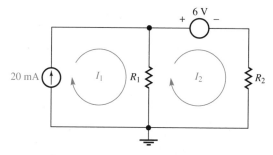

FIGURE 2.19 The value of the mesh current I_1 must be 20 mA, since the current passing through the current source is I_1. There is only one unknown mesh current in this circuit.

• EXERCISE 2.4

Find the current I_2 in Fig. 2.20 by means of the loop method. Let $R_1 = 470\ \Omega$, $R_2 = 330\ \Omega$, $R_3 = 510\ \Omega$, $I_0 = 1\ \text{mA}$, and $V_0 = 0.5\ \text{V}$. **Answer:** $I_2 = -1.20\ \text{mA}$.

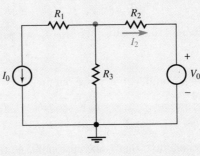

FIGURE 2.20

A case requiring special treatment occurs when two unknown mesh currents both pass through a current source. Consider the circuit of Fig. 2.21(a). Here it is impossible to write Kirchhoff's voltage law for the closed path A–B–D–A because one cannot write the voltage drop across a current source as a function of the current through it. One way to handle the problem is to write Kirchhoff's voltage law for the closed path A–B–C–D–A, thus obtaining an equation in the unknowns I_1 and I_2. A second equation is then obtained from the basic property of the current source: $I_0 = I_2 - I_1$.

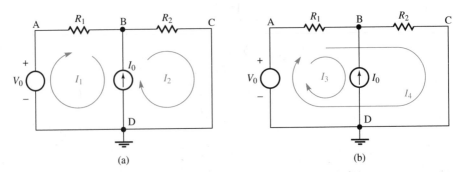

(a)   (b)

FIGURE 2.21   A circuit requiring special treatment when analyzed by the loop method. (a) If two unknown mesh currents pass through the current source, one cannot write the loop equation for the closed path A–B–D–A. (b) The difficulty can be circumvented by a different choice of circulating currents.

A more elegant way of handling the difficulty is to alter the choice of mesh currents to that shown in Fig. 2.21(b). Now only one of the circulating currents, I_3, passes through the current source; therefore $I_3 = -I_0$. There is then only one unknown circulating current, I_4, remaining in the problem.* It may be found by writing a single loop equation for the closed path A–B–C–D–A.

EXAMPLE 2.11

Find the current from B to C through R_2 in the circuit of Fig. 2.21.

SOLUTION

Let us choose circulating currents as in Fig. 2.21(b). Then the current asked for equals I_4, and we see by inspection that $I_3 = -I_0$. Writing an equation expressing Kirchhoff's voltage law for the closed path A–B–C–D–A, we have

$$R_1(I_4 + I_3) + R_2 I_4 - V_0 = 0$$

Replacing I_3 by $-I_0$ and rearranging, we have

$$I_4(R_1 + R_2) = I_0 R_1 + V_0$$

$$I_4 = \frac{I_0 R_1 + V_0}{R_1 + R_2}$$

COMPARISON OF NODE AND LOOP METHODS

Whether analysis is more easily performed by the node method or the loop method depends on the particular circuit under consideration. When one specifically needs to know a node voltage somewhere in the circuit, the node method may take fewer steps; when one needs to know a current, the loop method is favored, all other things being equal. On the other hand, if there are more loops in the circuit than there are nodes, the node method will probably require the solution of fewer simultaneous equations; the converse also is true. For the beginner, one way to avoid errors is to choose a single method (we prefer the node method) and stay with it. The steps of the node and loop methods are recapitulated in Table 2.2.

---

* A mesh is properly defined as a loop that does not contain any other loops inside it; therefore I_4 is not, strictly speaking, a mesh current. Irrespective of terminology, the procedure is clearly correct.

TABLE 2.2   The Loop and Node Methods of Circuit Analysis

I. Node Method

1. Select reference node and locate nodes where voltage is unknown.
2. (a) Express currents into each node as functions of known and unknown node voltages, and (b) write equations stating that the sum of the currents into each node is equal to zero.
3. Solve equations obtained in Step 2 simultaneously for unknown node voltages.
4. Obtain desired branch currents from node voltages found in Step 3 and the *I-V* relationships of the branches.

II. Loop Method

1. Select the proper number of mesh currents such that at least one mesh current passes through each branch.
2. (a) Express voltage drops across each element as functions of known and unknown mesh currents, and (b) write equations stating that sums of voltage drops around closed paths are zero.
3. Solve equations obtained in Step 2 simultaneously for unknown mesh currents.
4. Obtain branch currents in terms of the mesh currents found in Step 3 and obtain desired node voltages from the branch currents and the *I-V* relationships of the branches.

2.5 Voltage and Current Dividers

THE VOLTAGE DIVIDER

A subcircuit known as the *voltage divider* occurs so often that it is worthy of special attention. Figure 2.22 shows a typical voltage divider circuit. The part of the circuit that is properly the voltage divider consists of resistors R_1 and R_2 and is enclosed by the dotted line. The input terminals of the voltage divider are connected, in this case, to an

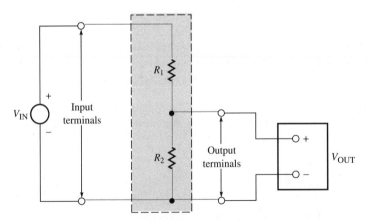

FIGURE 2.22 The voltage divider circuit. The voltage indicated by the ideal voltmeter is a simple fraction of the input voltage.

ideal voltage source, and the output terminals are connected to an ideal voltmeter. By assumption, an ideal voltmeter measures a voltage, but draws *no* current from the circuit to which it is connected. For a given voltage at the input terminals, what voltage is measured by the ideal voltmeter at the output?

Let us answer this question using the node method of solution. Designating the output voltage as V_{OUT}, we use Kirchhoff's current law for the node at the upper output terminal:

$$\frac{V_{IN} - V_{OUT}}{R_1} - \frac{V_{OUT}}{R_2} = 0 \tag{2.17}$$

Solving, we have

$$V_{OUT} = V_{IN} \frac{R_2}{R_1 + R_2} \quad \text{(voltage divider)} \tag{2.18}$$

This is the general voltage-divider formula; we recommend that it be memorized. Very often formal circuit analysis can be avoided when a circuit can be recognized as a simple voltage divider.

The voltage appearing across R_1 in Fig. 2.22 is equal to V_{IN} minus the voltage across R_2. Thus the voltage across R_1 is $V_{IN}[1 - R_2/(R_1 + R_2)] = V_{IN}[R_1/(R_1 + R_2)]$. The symmetry between this expression and Eq. (2.18) is readily apparent: The expression for the voltage across R_2 has R_2 in its numerator; that for the voltage across R_1 has R_1 in its numerator.

In using the voltage-divider formula, one must be careful to include all resistances that are present in the circuit. For instance, in Fig. 2.23 we have a circuit very similar to that of Fig. 2.22, but the voltmeter is not ideal. On the contrary, it is more like an actual physical voltmeter, which consists of a large internal resistance in parallel with

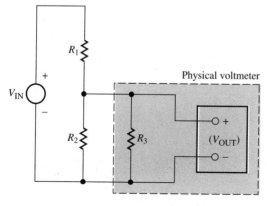

FIGURE 2.23 A voltage divider circuit with a load. The resistance R_3 of the nonideal voltmeter must be taken into account in calculating the output voltage.

an ideal voltmeter. The circuit of Fig. 2.23 is nonetheless a voltage divider, but R_3 must be regarded as being in parallel with R_2 so that the voltage-divider formula may be used. The output voltage for Fig. 2.23 is therefore

$$V_{OUT} = V_{IN} \frac{(R_2 \| R_3)}{(R_2 \| R_3) + R_1} \tag{2.19}$$

Expanding $R_2 \| R_3$, we have

$$V_{OUT} = V_{IN} \frac{R_2 R_3}{R_2 R_3 + R_1 R_2 + R_1 R_3} \tag{2.20}$$

EXAMPLE 2.12

Find the voltage indicated by the ideal voltmeter in the circuit shown in Fig. 2.24.

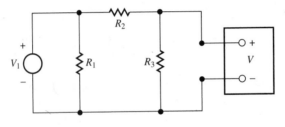

FIGURE 2.24

SOLUTION

We observe that the voltage across the terminals of R_1 is V_1, regardless of the value of R_1 (provided only that $R_1 \neq 0$). Using the voltage-divider formula, we have

$$V = V_1 \frac{R_3}{R_2 + R_3}$$

EXAMPLE 2.13

A dc lighting system contains ten bulbs connected in series; 120 V power is supplied to the system, which is designed to use ten 50-W 12-V bulbs. However, a mistake is made and someone replaces one of the 50-W bulbs with a 100-W 12-V bulb. How much power is actually supplied to the 100-W bulb?

SOLUTION

The 50 W bulbs each have resistance

$$R_{50} = \frac{(12)^2}{50} = 2.88 \; \Omega$$

and the 100 W bulb has resistance

$$R_{100} = \frac{(12)^2}{100} = 1.44 \; \Omega$$

The circuit thus contains a total resistance of $R_T = 9(2.88) + 1.44 = 27.36 \; \Omega$, and the voltage across the 100 W bulb is, from the voltage-divider formula,

$$V_{100} = \frac{R_{100}}{R_T} \cdot 120 = \frac{1.44}{27.36} \cdot 120 = 6.32 \; \text{V}$$

Thus the power delivered to the 100 W bulb is

$$P_{100} = \frac{(V_{100})^2}{R_{100}} = \frac{(6.32)^2}{1.44} = 27.7 \; \text{W}$$

THE CURRENT DIVIDER

A similar subcircuit that can sometimes be recognized in larger circuits is the *current divider*, shown in Fig. 2.25. In this figure the parallel combination of R_1 and R_2 gives a net resistance of $R_1 R_2/(R_1 + R_2)$ across the current source. Therefore by Ohm's law $V_A - V_B = I_{IN} R_1 R_2/(R_1 + R_2)$. The values of I_1 and I_2 can now be found from $I_1 = (V_A - V_B)/R_1$ and $I_2 = (V_A - V_B)/R_2$:

$$I_1 = I_{IN} \frac{R_2}{R_1 + R_2} \tag{2.21}$$

and

$$I_2 = I_{IN} \frac{R_1}{R_1 + R_2} \tag{2.22}$$

Note that although there is a similarity between these equations and Eq. (2.18) for the voltage divider, there is also a subtle difference. In Eq. (2.18) the fraction appearing in the expression for the voltage across R_2 has R_2 in its numerator. In Eq. (2.22) the expression for the current through R_2 has R_1, not R_2, in the numerator. This is as one would expect intuitively; as R_1 is increased, more current is diverted to the path through R_2.

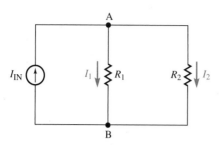

FIGURE 2.25   A current divider circuit. The values of I_1 and I_2 are given by Eqs. (2.21) and (2.22).

2.6 Instrumentation and Measurements

In practical situations it is often necessary to make measurements on electrical circuits. One may need to know the voltage at some point in a circuit or the current flowing through some wire. The subject of electrical measurements is a large one and can be material for a whole book in itself. Here we shall only introduce the subject and present a few of its basic principles.

To make an electrical measurement, one usually employs an instrument designed for measuring the quantity of interest: For instance, one uses a voltmeter to measure voltages and an ammeter to measure currents. The *idealized* forms of these two instruments have already been mentioned in Section 2.2. A *practical* measuring instrument has an important property that makes it different from ideal instruments. Unlike an ideal instrument, *a practical measuring instrument disturbs a circuit to which it is connected.* For example, if we wish to find out the potential difference between points A and B, we may take the terminal marked (+) of a practical dc voltmeter and connect it to point A, while the terminal marked (−) is connected to point B. The meter will then indicate the value of $V_A - V_B$. *But note:* What the meter indicates is the value of $V_A - V_B$ *while the meter is connected.* Because the practical meter disturbs the circuit, the value of $V_A - V_B$ when the meter is removed will probably be somewhat different. Since the meter usually is introduced only temporarily in order to make the measurement, the voltage without the meter is usually the thing one wants to know. To get this information one must either (1) use a meter that is known to disturb the circuit minimally; or else (2) calculate backward from the measured value to find what the voltage must be when the meter is disconnected. The fact that the act of measurement itself alters the system being measured is a basic problem of electrical measurements and in fact of measurements of every kind.

Another basic problem of measurements has to do with *noise.* This term refers to the presence of random signals that obscure the measurement. Small quantities are more easily obscured by noise than large ones. Thus noise usually establishes a limit on, for example, the smallest voltage or smallest current that can be measured. Other kinds of measurement problems exist as well; some quantities under some circumstances simply cannot, for various practical reasons, be measured with existing technology. Progress in measurements, however, is always being made.

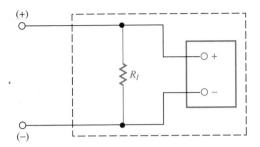

FIGURE 2.26 A practical voltmeter can usually be modeled as a parallel combination of an ideal voltmeter and a resistance.

PRACTICAL VOLTMETERS AND AMMETERS

A practical voltmeter can usually be modeled as a parallel combination of an ideal voltmeter and a resistance, as shown in Fig. 2.26. This resistance (R_I in the figure) is called the *internal resistance* of the practical voltmeter. (The internal resistance of an ideal voltmeter is infinite; that is what makes it ideal.) The value of the internal resistance depends on the way the practical meter is constructed. One common type is the electromechanical meter; in this type R_I is relatively low, in the range of 10^3 to 10^6 ohms. (In general, the larger the voltage range of an electromechanical meter, the larger is R_I.) Another common type, more sophisticated, is the electronic voltmeter. Here transistor circuits are used in order to increase R_I; values of 10^7 ohms or more are common. The way measurements are affected by internal resistance will now be illustrated by an example.

EXAMPLE 2.14

We wish to measure the voltage between point A and ground in the circuit shown in Fig. 2.27(a). Two measuring instruments are available: an electromechanical meter with $R_I = 1,000 \ \Omega$ and an electronic voltmeter with $R_I = 10^7 \ \Omega$. What is the voltage at A when no instrument is connected? Find the voltages that will be indicated by each of the two instruments.

SOLUTION

When no instrument is connected, the voltage at A is easily seen by means of the voltage-divider formula, Eq. (2.18), to be

$$V_A = 1.5 \ \frac{10^4}{10^4 + 10^4} = 0.75 \ \text{V}$$

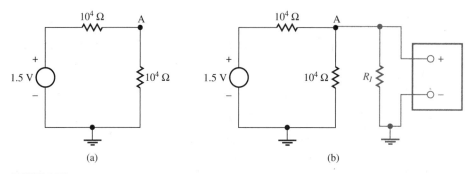

FIGURE 2.27

When a meter is connected, the circuit appears as shown in Fig. 2.27(b). Recall that all grounded points are connected together. We see that R_I is connected in parallel with one of the 10 kΩ resistors. Thus the circuit is still of the voltage-divider form, but now the voltage at A is given by

$$V_A = 1.5 \ \frac{R_I \| 10^4}{(R_I \| 10^4) \ + \ 10^4} = 1.5 \ \frac{10^4 R_I}{2 \ \times \ 10^4 R_I \ + \ 10^8}$$

Note that the ideal voltmeter plays no role in this calculation. It does not affect the circuit because no current passes through it; it merely indicates the place where voltage is being indicated. In this example the dial of the meter will tell us the potential difference between point A and ground, since this is where the terminals of the ideal voltmeter are connected.

When the 1,000 Ω electromechanical meter is used, the above formula tells us that V_A = 1.5 · 0.08 = 0.125 V. This is very different from 0.75 V, the voltage that exists at A when no meter is connected. Clearly this meter has a strong effect on the circuit, and its indication does not tell us the voltage that exists at A when no measurement is being made.

When the 10^7-Ω electronic meter is used, V_A = 1.5 · 0.49975 $\cong$ 0.75 V. This measurement is almost exactly the same as the voltage at A with no meter connected. The 10^7 Ω meter does not disturb the circuit much because it is connected in parallel with a 10^4 Ω resistor; $10^7 \| 10^4 \cong 10^4$, so the presence of the meter does not change the circuit much. In general, a practical voltmeter will not affect a circuit much if its internal resistance is much larger than the resistance across which voltage is being measured. This is why electronic voltmeters, which have large R_I, are often used.

Practical ammeters are used less often than voltmeters because the circuit has to be broken in order for the ammeter to be inserted. (The two measurements are to some degree interchangeable. For example, in order to find the current through a certain resistor, it suffices to measure the voltage across the resistor and apply Ohm's law. The voltage measurement is easier because the circuit does not have to be disassembled.) Like the voltmeter, a practical ammeter can also affect the circuit. The practical ammeter can be modeled as an ideal ammeter in *series* with an internal resistance R_I.

OSCILLOSCOPES

The meters we have spoken of until now are used for measuring constant (dc) voltages and currents. Very often, however, we need to measure voltages that vary rapidly in time. Clearly, this cannot be done by watching a needle on a dial; in modern work observable voltage changes occur in times as short as 10^{-11} sec. To measure time-varying voltages, instruments known as oscilloscopes are used. Fundamentally, an oscilloscope is just another kind of electronic voltmeter (with internal resistance typically on the order of 10^6 to 10^7 ohms). However, the means of information display is different. Instead of a mechanical meter or digital readout, the oscilloscope uses a cathode-ray tube similar to a television screen. On this screen is projected a graph that shows voltage on the vertical axis and time on the horizontal axis. This allows one not only to read off the voltage at any instant but also to observe the general behavior of the voltage as a function of time.

For example, suppose the terminals of an oscilloscope are connected across an unknown ac voltage source, and the display on the oscilloscope screen is as shown in Fig. 2.28(a). This display is referred to as a *sinusoidal waveform*, and by a glance at the

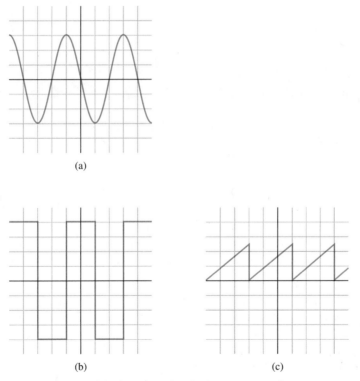

(a)

(b)   (c)

FIGURE 2.28 Typical signal waveforms that may be seen on an oscilloscope screen. (a) Sinusoidal wave; (b) rectangular wave; (c) sawtooth wave.

oscilloscope screen we see that voltage is a sinusoidal function of time. The horizontal and vertical scales of the display are set by the oscilloscope's controls. Suppose these controls are set so that each horizontal division is 5 msec (total width of the screen, 50 msec) and each vertical division is 50 V. Then we read from the oscilloscope that the voltage varies from a maximum of +150 V to a minimum of −150 V and repeats every 20 msec. Two other common waveforms are shown in Figs. 2.28(b) and (c), referred to as a *rectangular wave* and a *sawtooth wave.* Many other waveforms also occur. Displays such as those shown in Fig. 2.28 are easily seen when the voltage is a periodic (repetitive) function of time. This is because the oscilloscope continuously generates new graphs of voltage versus time, one after another. If the voltage is repetitive, the moving light spot repeatedly graphs the same shape, and the stationary waveform is seen.

Of course there also are cases when the voltage is not periodic. For instance, suppose the voltage comes from a microphone into which someone is speaking. In that case the oscilloscope display would look jumbled, because the oscilloscope would present graphs of voltage versus time that were always different. A common way of handling this difficulty is to cause the oscilloscope to make only one single graph, representing the voltage over a single short time period. (This is known as *single-sweep operation.*) Since the display then lasts only a very short time, it is photographed for later inspection.

EXAMPLE 2.15

An oscilloscope displays the repetitive waveform shown in Fig. 2.28(b). The scales are set to 1 msec per horizontal division, 10 mV per vertical division. Find the maximum and minimum voltages and the frequency (number of repetitions per second).

SOLUTION

The maximum vertical height is four divisions, corresponding to a maximum voltage of $4 \cdot 10$ mV = 40 mV. Similarly, the minimum value is −40 mV.

The spacing between identical points on the waveform is four horizontal divisions; thus the time for a repetition is $4 \cdot 1$ msec = 4 msec. In 1 sec the number of repetitions will be 1,000 msec/4 msec; thus the frequency is 250/sec.

NOISE

When the voltage (or current) in any electronic circuit is carefully inspected, one always finds the randomly varying signals known as *noise.* Noise signals can come from accidental coupling to other circuits. (The other circuit may not be manmade;

AM radio noise often comes from lightning bolts occurring far away.) Noise is also generated as an undesired byproduct of the operation of the circuit itself.

As an example, suppose you wish to measure a small dc voltage by means of an electronic voltmeter with a mechanical moving pointer display. (The display has zero at the center; thus both positive and negative voltages can be measured.) You observe that in addition to a constant displacement of the meter needle, caused by the dc voltage being measured, there is also a continual random motion of the needle around the average value. After observing for a while, you conclude that the average meter reading is 20 nV ($20 \cdot 10^{-9}$ V) and that typical noise variations are $\pm$ 5nV around this value. In this case the measurement of the dc voltage, although disturbed by the noise, is still possible. Now suppose, however, that you wish to measure a smaller dc voltage — on the order of 1 nV. In this case the random motions of the needle will be about five times larger than the average displacement, and it will be difficult to determine just what the average value is. Clearly even smaller voltages will be even more difficult to measure. Thus it is the presence of noise that usually determines the smallest value of a quantity that can be conveniently measured.

It is interesting to note that the measurement of a 1 nV dc voltage in the presence of 5 nV of noise is not entirely hopeless. One might make numerous measurements of the voltage over a long period of time. These would vary a good deal, but one could still compute their average. The laws of statistics show that the more measurements that are taken, the closer their average will be to the actual voltage being measured. Thus noise does not necessarily make measurements impossible; it just makes them more time-consuming. Of course, if it turns out that one must collect measurements for a whole year in order to get accurate results, one might incline to say that that particular measurement was nearly impossible!

*2.7 Introduction to *SPICE*

Circuits containing linear circuit elements (such as voltage sources, current sources, and resistances) can be analyzed in straightforward mathematical ways, using either the loop or node methods. However, the actual calculations may be cumbersome. For instance, to solve a circuit containing N nodes, using the node method, it is generally necessary to solve N simultaneous equations; for more than two nodes, the job is probably too large to do by hand. Furthermore, many important circuit elements, such as diodes and transistors, are *nonlinear*. Circuits containing a single nonlinear element can be analyzed approximately, by hand; but when two or more nonlinear elements are present, analysis can be quite difficult. Fortunately, we do not have to rely solely on hand calculation. Computer tools for circuit analysis are readily available.

The best-known family of software tools for circuit analysis is the one known as *SPICE*. It is available in various versions for use in various environments (for example, *PC-SPICE* for personal computers) or special applications (such as *MICROWAVE-SPICE* for high-frequency circuit analysis). Here we shall present the basic method, common to all the various versions. The user should then refer to the documentation of his or her version for particular details.

---

*Sections marked with an asterisk may be omitted without loss of continuity.

In order to analyze a circuit using *SPICE*, we must first provide a complete description of the circuit. This is done through a series of *element* statements, with one such statement for every element in the circuit. These are then followed by *control* statements, which tell the program what it is we wish to calculate. In *SPICE Version 2G* each analysis must begin with a *title* statement and end with an *end* statement, but such details may vary from one version to another.

To begin with, we may wonder how a circuit, which is a two-dimensional picture, can be described through a computer keyboard. The way this is done is to assign a number to each node of the circuit. Each *element* statement then tells us what kind of element we have and to which nodes it is connected. For example, in the circuit of Fig. 2.29 we have located four nodes and numbered them 00 through 03. Note that for purposes of *SPICE*, any place where two or more circuit elements come together is treated as a node and numbered. Nodes such as 01, where the potential is obviously known from the start, are nonetheless included. So is the ground node, which by convention *must* be numbered 00. We also assign a name to each circuit element. The first letter of the name indicates what sort of element it is: "R," for example, indicates a resistor; "V" indicates an ideal voltage source. The *element* statement for R1 then reads as follows:

```
R1   01   02   3.3K
```

If we wish to know the voltage between node 3 and ground, we use the control statement

```
.PRINT DC V(03,00)
```

The entire set of instructions is then

```
TEST CIRCUIT

R1   01   02   3.3K
R2   02   03   6.8K
R3   02   00   2.7K
R4   03   00   4.7K
V1   01   00   DC 10

.PRINT DC V(03,00)

.END
```

The expression "DC" in two of the statements indicates that it is a dc, or constant, voltage source and that dc analysis is to be performed. Control instructions are distinguished by the period with which they begin. Again, these are details that may vary with the software package being used.

The capabilities of *SPICE* go far beyond simple dc circuits. In fact, every type of circuit to be discussed in this book can be analyzed using *SPICE*; it is a convenient, versatile tool. Indeed, the reader may wonder, why bother to learn node or loop analysis at all; why not simply rely on *SPICE*? There are several good reasons: For instance, a computer may not always be available, and anyway, analyzing a simple problem with

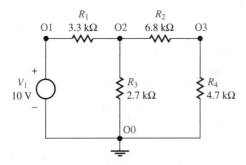

FIGURE 2.29   Numbering of nodes for *SPICE* analysis.

SPICE will usually be slow compared with doing it in one's head. The most important thing is that in design work one needs intuition, which in turn requires familiarity with circuit analysis.

2.8 Application: Circuit Diagrams

Now that we have seen some of the methods used in circuit analysis, let us look ahead to see what a real circuit is like. As an example, consider Fig. 2.30, which contains the schematic diagram for a commercial AM-FM radio receiver. Although this circuit seems complicated, it operates on the same basic principles we have been discussing in this chapter. That is, it is a collection of circuit elements connected together with wires. Altogether the circuit is intended to do a particular job, and to do this job, it has to generate certain voltages and currents. These voltages and currents, when applied to the loudspeaker (right-hand side of the diagram) cause it to produce the sounds that the broadcast station sends.

An inspection of the circuit reveals numerous resistors. There are no ideal voltage or current sources, however. Why? Because this is a practical circuit diagram, containing only real circuit elements; we recall that ideal sources are fictional elements that do not exist in fact. There are also a number of other circuit elements that we have not yet discussed: coils and transformers (which look like loops of wire), capacitors, diodes, and also five transistors (shown by the symbols surrounded by large heavy circles). There is also an integrated circuit (IC), represented by the large vertical rectangle with 16 terminals. The IC is itself a complex electronic circuit, but its insides are not shown in the diagram. The IC is treated as though it is simply one large circuit element. ICs used as components, as in this example, are now very common, and represent a major step forward in electronics. An IC is usually a very powerful building block, but we save the effort of designing it and the cost of assembling it. ICs are very cheap to mass-produce; thus substituting them for circuits that must be assembled leads to great savings in cost. Furthermore, using ICs as components greatly expands the power of the resulting circuits because the building blocks are already very powerful in themselves.

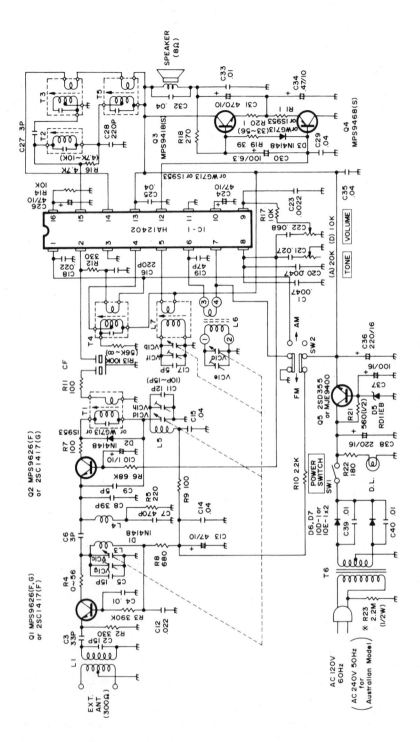

FIGURE 2.30   Circuit diagram of a commercial AM-FM radio receiver. *(Reproduced with permission of Tandy Corporation.)*

Because the circuit is large and complex, one might think that analyzing its performance would be difficult. However, an experienced engineer would be able to understand its operation almost immediately by inspection. That is because it is composed of recognizable parts, or *stages*, that would already be familiar. At the bottom, to the right of the power plug, is a power-supply circuit, intended to convert ac house current to regulated dc. Two transistors used for FM amplification and detection are at the upper left, near the antenna terminals, while the two transistors at the lower right compose an audio output stage that provides the currents needed to run the loudspeaker. Only the IC is inscrutable. Since its contents are not shown, we can only guess what it does; but this is fairly easy to figure out, since we can see what functions are missing from the rest of the circuit. Evidently most of the AM radio receiver is contained inside the IC!

POINTS TO REMEMBER

- Circuits are interconnections of circuit elements. Each circuit element is characterized by particular properties relating the relevant currents and voltages. Ideal circuit elements, which do not actually exist, are assumed to be described by very simple current-voltage characteristics.

- An ideal voltage has the property that the voltage across its terminals has the stated value. This is its *only* property.

- An ideal current source has the property that the current flowing through it always has the stated value.

- An ideal resistor is described by the property known as Ohm's law. In using Ohm's law, the algebraic signs of voltage and current are very important. In order to determine these signs, one must first specify a reference direction for current and a reference polarity for voltage.

- The voltages and currents that exist in a circuit are calculated by the procedure known as circuit analysis. Two basic methods for circuit analysis are the node method and the loop method.

- Circuit analysis is made easier if common subcircuit blocks can be recognized at sight. Perhaps the most common of all such blocks is the voltage divider. A similar block known as the current divider is also seen.

- Voltages and currents in a real circuit can also be determined by measurement. Ideal measuring instruments, which like other ideal elements do not exist, are imagined not to disturb the circuit under test. Real measuring instruments in general do change the voltages and currents in a circuit to which they are connected. These changes must be accounted for as part of the measurement process.

- Voltmeters and ammeters are instruments used to measure voltages and currents. Another important measuring device is the oscilloscope, which displays on its screen a graph of voltage versus time.

PROBLEMS

Section 2.1

2.1 Find the voltage (with respect to ground) at points A and B in Fig. 2.31.

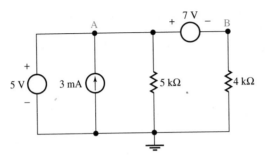

FIGURE 2.31

2.2 Find the currents I_1, I_2, and I_3 in Fig. 2.32.

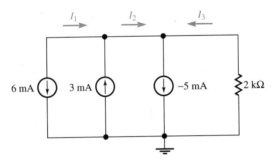

FIGURE 2.32

2.3 A graph of the voltage across a circuit element versus the current through it is known as the element's *I-V graph* (or *I-V characteristic*). Draw *I-V* graphs for
 a. an ideal voltage source with value 2 V.
 b. an ideal current source with value 3 mA.

*2.4 An ideal voltage source and resistor are connected as shown in Fig. 2.33. Sketch a graph of the voltage *V* versus the current *I*.

---

*Problems are unstarred, one-starred, and two-starred, indicating their level of difficulty. The unstarred problems are introductory and descriptive; most students should do these *and* the more difficult one-starred problems. The two-starred problems are "industrial strength," challenging the students' mathematical sophistication as well as their ability to solve conceptual riddles and apparent contradictions.

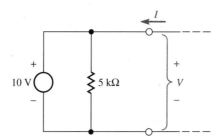

FIGURE 2.33

2.5 State whether or not each of the following represents a logical inconsistency:
- **a.** an ideal voltage source in parallel with an open circuit
- **b.** an ideal current source in parallel with a short circuit
- **c.** an ideal current source in series with an open circuit
- **d.** an ideal voltage source in parallel with an ideal current source
- **e.** an ideal voltage source in series with an ideal current source

Section 2.2

2.6 Find the currents marked "I" in Fig. 2.34(a), (b), and (c).

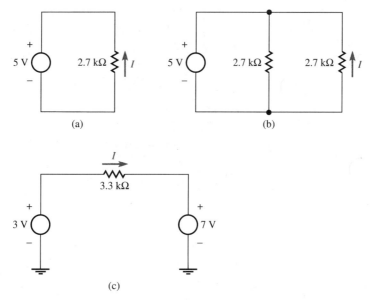

FIGURE 2.34

2.7 A current of 3 mA flows through an 6.8 kΩ resistor in the direction from A to B. Is A or B at higher potential? What is the magnitude of the potential difference?

2.8 Let I_1 be defined as the current through a 470 Ω resistance in the direction from A to B. The potential at A is 6 V and that at B is −4 V. What is I_1? Note that the sign as well as the magnitude of I_1 must be determined.

2.9 A current of 0.8 mA flows through a resistance R from A to B. V_A (the potential at A) is 8 V, and V_B is 3 V. What is R?

2.10 Refer to Fig. 2.7. Let the reference directions of V_1 and I_1 be as shown.
   a. In terms of the symbol I_1, what is the voltage drop as one moves through the resistor from B to A?
   b. Subsequent measurements reveal that $I_1 = -25$ A. If $R_1 = 3.2$ Ω, what is V_{AB}?

2.11 Find the resistance between terminals A and B in Fig. 2.35. Assume $R_1 = 5.6$ kΩ, $R_2 = 4.7$ kΩ, $R_3 = 3.3$ kΩ, and $R_4 = 8.2$ kΩ.

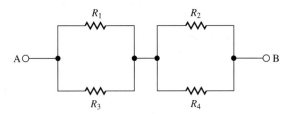

FIGURE 2.35

2.12 Calculate the resistance of a combination of three resistances with values R_1, R_2, and R_3 connected in parallel.

2.13 A manufacturer supplies resistors with the following power-dissipation ratings: 1/4, 1/2, 1, 2, 5, and 10 W. The resistors with higher power dissipation ratings are larger and more expensive. In a particular application you intend to connect a 75-Ω resistor between points A and B, where the potentials, respectively, are −6.3 and 2.1 V. Which resistor should you choose?

*****2.14** In Fig. 2.36 what is the current I_1 through the 100 Ω resistance? What is the current I_2 through the voltage source?

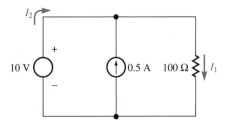

FIGURE 2.36

*2.15
   a. Two resistors, R_1 and R_2, are connected in parallel inside a larger circuit. The total power dissipated in both resistors is P. What is the power dissipated in R_1?
   b. Repeat part (a), assuming that the two resistors are connected in *series*.

Sections 2.3 and 2.4

2.16 In the circuit of Fig. 2.37 calculate the voltage V_1 by the node method. Let $I_0 = 1.4 \text{ mA}$, $R_1 = 2.7 \text{ k}\Omega$, $R_2 = 2.7 \text{ k}\Omega$, and $R_3 = 3.3 \text{ k}\Omega$.

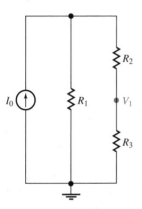

FIGURE 2.37

2.17 In the circuit of Fig. 2.37 calculate the voltage V_1 by the loop method. Use the numerical values given in Problem 2.16.

2.18 Find the current I_1, as indicated in Fig. 2.38, by the loop method. Let $I_0 = 55 \text{ A}$, $R_1 = 2 \Omega$, $R_2 = 1.5 \Omega$, $R_3 = 4.0 \Omega$, and $V_0 = 110 \text{ V}$.

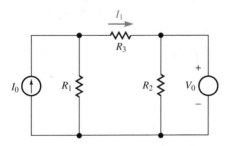

FIGURE 2.38

2.19 Find the current I_1 of Fig. 2.38 using the node method. Use the numerical values given in Problem 2.18.

2.20 Examine the circuit of Fig. 2.39.

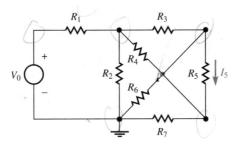

FIGURE 2.39

  a. How many nodes are present?
  b. How many branches?
  c. How many mesh currents are required to analyze this circuit?

2.21 Write a set of loop equations sufficient for a solution for I_5 in Fig. 2.39. It is not asked that you solve the equations.

2.22 Write a set of node equations for the circuit of Fig. 2.39, plus any additional equations necessary for calculation of I_5. It is not asked that you solve the equations.

*2.23 Consider the circuit of Fig. 2.40.

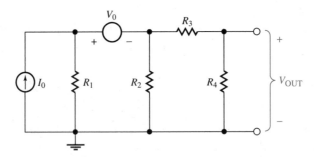

FIGURE 2.40

  a. Using the node method, write equations sufficient for a solution for V_{OUT}.
  b. Solve the equations obtained in part (a).

*2.24 Using the loop method, calculate the current I_1 in Fig. 2.41. (The box through which I_1 flows is an ideal ammeter. The voltage drop across it is zero.)

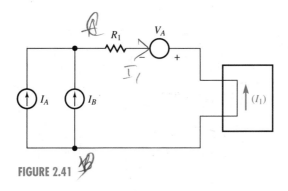

FIGURE 2.41

*2.25 Calculate the current I_1 in Fig. 2.42. Let I_0 = 20 mA, R_2 = 1,600 Ω, V_0 = 80 V, and R_1 = 1,500 Ω.

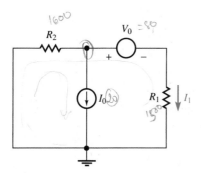

FIGURE 2.42

Section 2.5

2.26 The circuit of Fig. 2.43 is known as a *bridge circuit.* When the voltage V_1 equals zero, the bridge is said to be *balanced.* Suppose R_1 = 27 kΩ, R_2 = 16 kΩ, and R_3 = 10 kΩ. What must R_4 be in order to balance the bridge?

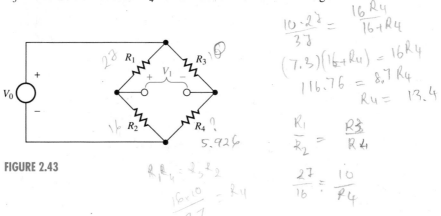

FIGURE 2.43

2.27 Refer to the voltage-divider circuit of Fig. 2.22. What are the limits of V_{OUT} in the following situations?
a. As R_1 approaches infinity?
b. As R_1 approaches zero?
c. As R_2 approaches infinity?
d. As R_2 approaches zero?
In each case give a physical explanation for your result.

2.28 Use the voltage-divider formula to find the voltage V_1 in Fig. 2.44(a) and (b).

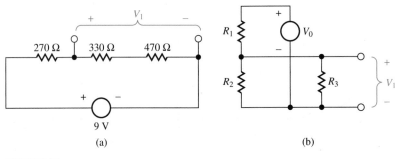

FIGURE 2.44

2.29 Solve Problem 2.16 using the current-divider formula.

*2.30 Figure 2.45(a) shows an infinitely long chain of resistors, all of the same value R. Find the resistance one would measure between its input terminals A and B. *Suggestion:* Note that if two additional resistors are connected as shown in Fig. 2.45(b), the resistance between terminals C and D will be the same as that formerly measured between A and B.

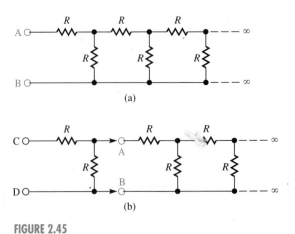

FIGURE 2.45

Section 2.6

2.31 It would in theory be possible to construct a voltmeter by taking an ideal ammeter and a 1,000 Ω resistor and placing them in series.

    a. Suppose the full-scale current of the ideal ammeter is 1 mA. What voltage applied across the series combination gives a full-scale reading?

    b. What is the internal resistance of the real voltmeter made in this way?

2.32 Suppose we wish to construct a voltmeter by the method described in Problem 2.31. However, since ideal ammeters do not really exist, we must use a real physical ammeter, which can be modeled as an ideal ammeter (full-scale current of 1 mA) in series with a 15 Ω resistance. The resulting voltmeter circuit is as shown in Fig. 2.46.

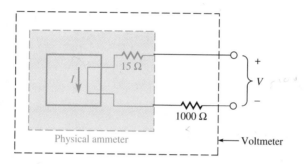

FIGURE 2.46

    a. What voltage *V* gives a full-scale reading? 1.105 V

    b. What is the internal resistance of the voltmeter? ∞

    c. To what value should the 1,000 Ω resistor be changed in order to obtain a full-scale reading when $V = 10.0$ volts? 9.985 k

2.33 We wish to measure the value of the ideal current source I_0 in Fig. 2.47. To do this we connect a voltmeter between points A and B, and we obtain a measurement of 4.63 V. The internal resistance of the voltmeter is 10,000 Ω. What is I_0?

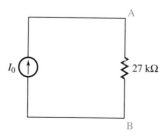

FIGURE 2.47

2.34 Suppose in Problem 2.33 the voltmeter is an electronic voltmeter. Its internal resistance is not known, but is guaranteed to be greater than 1 MΩ. What is the range of values within which I_0 must lie?

2.35 The display of an oscilloscope is shown in Fig. 2.28(c). The horizontal scale is set to 50 msec per division (500 msec for the entire screen width), and the vertical scale is 5 mV per division. Zero voltage is at the center.

  a. What is the maximum value of the voltage?

  b. What is the repetition frequency?

2.36 Suppose the horizontal scale of the oscilloscope in Problem 2.35 were switched to 100 msec per division while the voltage being measured remained the same. Sketch the new appearance of the oscilloscope display.

TECHNIQUES
OF DC
ANALYSIS

I n Chapters 1 and 2 the principles of electric circuits were discussed. Here we consider some special techniques by which circuit operation can be analyzed. Engineering work would go very slowly if every problem had to be attacked from first principles. However, good engineers rely on experience. Being familiar with the common problems, they can solve them quickly and effectively. Thus one needs not only basic knowledge but also practical and efficient techniques for solving problems.

One simplifying technique that is almost always used in complex circuit problems is that of breaking the circuit into pieces of manageable size and analyzing the pieces individually. When the circuit has been broken down, some of the circuit's subunits may be recognized as already familiar. The voltage divider, which was introduced in Chapter 2, is an example of a circuit subunit that occurs very often. Experienced circuit engineers can recognize dozens of common circuits at sight and are thus able to understand complex circuits very quickly. On the other hand, if a subunit is not familiar, it often can be reduced to a simpler form, thus making its function easier to see. Methods for doing this are the subject of Section 3.1.

Circuits containing *nonlinear* circuit elements are quite important. Diodes and transistors, for example, are nonlinear circuit elements. The analysis of circuits containing nonlinear elements requires special methods. Section 3.2 presents a simple graphical method, handy for getting a quick idea of how a nonlinear circuit will operate.

The subject of power transfer and power dissipation arises very often in electrical engineering. Once the currents and voltages in a circuit have been found, it is important to know how power is transferred from one part of the circuit to another. In Section 3.3 we discuss methods for calculating power transfer in various kinds of circuits.

The "application" at the end of this chapter has to do with the storage battery used to start your car. It is a large, heavy lead-acid battery, probably rated (unless your car is very old) at 12 volts. Now it would also be possible to obtain a much smaller 12 V battery by taking eight 1.5 V flashlight batteries and connecting them in series. If you think the flashlight batteries will not work to start your car, you're right. But why? This mystery is explained in Section 3.4.

3.1 Equivalent Circuits

Often one is interested in a subunit of a circuit only to the extent that it affects the rest of the circuit. The internal workings of the subunit do not matter in such a case; we need determine only the relationships between voltage and current at its terminals, where it is connected to the circuit of interest. One is then said to be concerned only with the *terminal properties* of the subunit.

As a simple example, consider a radio receiver plugged into a standard ac power line. The radio and the power line together form a complete circuit; however, let us suppose that we are interested mainly in the operation of the radio. The power line, with its connecting wires, transformers, and generator, may then be regarded as an uninteresting subunit of the complete circuit. Fortunately, we do not need to concern ourselves with the detailed structure of the power line. All we must know are its terminal properties: what voltage is available at the power plug and what voltage variations (if any) will occur as current is drawn. A convenient way to handle such a situation is to imagine that the power line has been replaced by an *equivalent circuit,* a simpler entity that has nearly the same terminal properties as the original subunit. Once the power line has been replaced on the circuit diagram by a much simpler equivalent circuit, analysis of the remainder of the circuit (the radio) can proceed. Of course, when such a substitution is made, nothing further can be learned about the internal operation of the replaced subunit; by means of the equivalent circuit, the features of the original subunit have been eliminated from the problem.

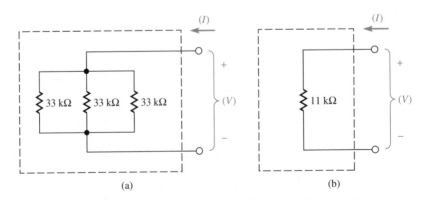

FIGURE 3.1 An example of an equivalent circuit. The replacement subunit (b) has an *I-V* characteristic identical to that of the original subunit (a). The two cannot be distinguished by measurements made at their respective terminals.

We shall now state (without formal proof) a principle that will allow equivalent circuits to be found. Provided that the original subunit contains only resistances, voltage sources, and current sources, *an adequate equivalent circuit will be one which has an I-V relationship identical to that of the original subunit.* As a simple example, consider the subunit of Fig. 3.1(a). Using the formula for parallel resistances, the *I-V* characteristic of this subunit is seen to be given by the equation $V = (1.1 \times 10^4)I$. From the parallel-resistance formula Eq. (2.5) it is clear that the circuit of Fig. 3.1(b) has an identical *I-V* characteristic, and therefore is an equivalent circuit for that of Fig. 3.1(a). If subunit (*a*) appears anywhere within a larger circuit, it clearly may be replaced by (*b*) without changing the operation of the rest of the circuit. Nor can (*b*) be distinguished from (*a*) by external measurements made at their respective terminals. We note, however, that in internal behavior the two circuits are different; in (*b*) the current flows through a single element, while in (*a*) it divides and a part flows through each of three elements.

In the example just given, the simplest equivalent was obvious, and could be found by inspection. However, for a large class of circuit subunits, there exist two simplest equivalent circuits. These are known as the *Thévenin* and *Norton equivalents.* They are found through application of the Thévenin and Norton theorems. The technique is based on identity of *I-V* characteristics for the original subcircuit and its equivalent.

Let us consider a circuit consisting of an arbitrary number of resistances and voltage and current sources having constant values. Let the circuit have two terminals for connection to an external circuit. Then it can be shown that its *I-V* relationship has the form $I = aV + b$. That is, when graphed on the *I-V* plane the *I-V* relationship is a straight line.

EXAMPLE 3.1

Consider the composite circuit element in Fig. 3.2(a), which has two terminals for external connection. Let *I* and *V* be defined at those terminals as shown. Find the *I-V* relationship.

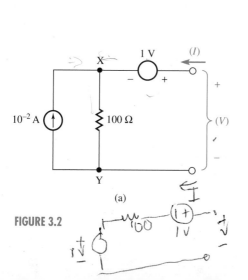

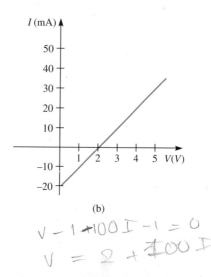

FIGURE 3.2

$V - 1 + 100\,I - 1 = 0$

$V = 2 + 100\,I$

SOLUTION

Designating node voltages V_X and V_Y as shown, we can write a node equation for node X,

$$10^{-2} + \frac{V_Y - V_X}{100} + I = 0$$

which becomes

$$V_X = V_Y + 100I + 1$$

The potential difference V is given by

$$V = (V_X - V_Y) + 1$$

Thus we find that

$$V = (V_Y + 100I + 1 + 1) - V_Y = 100I + 2$$

By algebraic rearrangement we have

$$I = 10^{-2}V - 2 \times 10^{-2}$$

which is of the linear form $I = aV + b$, as stated above. The I-V characteristic, plotted on the I-V graph in Fig. 3.2(b), is a straight line.

EXAMPLE 3.2

Calculate the I-V relationship of the circuit in Fig. 3.3. Show that it is the same as that of the circuit in Example 3.1.

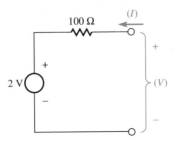

FIGURE 3.3

SOLUTION

Writing a loop equation, we have

$$V - 2 - 100I = 0$$

Manipulating algebraically, we get

$$I = 10^{-2}V - 2 \times 10^{-2}$$

which is identical to the *I-V* relationship of Example 3.1. This circuit is therefore an equivalent for the previous circuit.

EXAMPLE 3.3

The two circuits in Fig. 3.4(a) and (b) have been shown (in the two previous examples) to be equivalents. Let each of these be connected to an external circuit, as shown in Fig. 3.4(c), where the value of *R* may be freely chosen. Show that the current *I* indicated by the ideal ammeter is the same whether subcircuit (*a*) or its equivalent (*b*) is used, and that this is true regardless of the value of *R*.

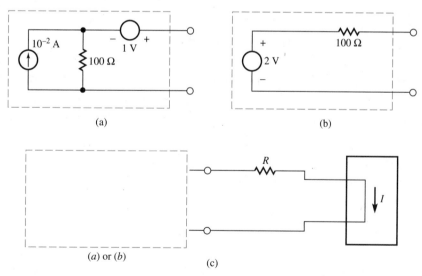

FIGURE 3.4(a), (b), and (c)

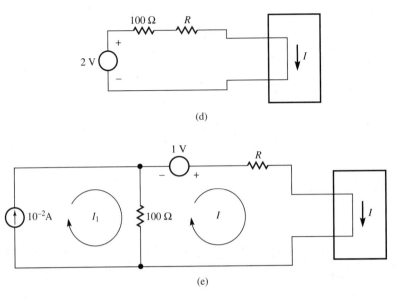

FIGURE 3.4(d) and (e)

SOLUTION

When (*b*) is used, the entire circuit is as shown in Fig. 3.4(d). From Ohm's law we immediately find that

$$I = \frac{2}{100 + R} \quad \text{[subcircuit (*b*)]}$$

When (*a*) is used, the entire circuit is as shown in Fig. 3.4(e). For the sake of variety, let us use the loop method to calculate *I*. Two equations containing the two unknowns I_1 and *I* are obtained. The first is trivial, since I_1 is controlled by the current source:

$$I_1 = 10^{-2} \text{ A}$$

The other loop equation is

$$100(I - I_1) - 1 + IR = 0$$

Eliminating I_1, we have

$$I = \frac{2}{100 + R} \quad \text{[subcircuit (*a*)]}$$

which is identical to what was obtained using subcircuit (*b*). This example illustrates the statement that the operation of the remainder of a circuit is unaffected when a subcircuit is replaced by its equivalent.

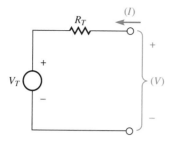

FIGURE 3.5 The general form of a Thévenin equivalent circuit. V_T is the Thévenin voltage, and R_T is the Thévenin resistance.

We shall now discuss the methods whereby two types of equivalent circuits—Thévenin equivalents and Norton equivalents—may be found.

THE THÉVENIN EQUIVALENT

The general form of the Thévenin equivalent circuit is shown in Fig. 3.5. The values V_T and R_T may be called the *Thévenin voltage* and *Thévenin resistance,* respectively.

In order to discuss the *I-V* characteristic, the reference directions for *I* and *V* are arbitrarily chosen to be as indicated in Fig. 3.5. (These conventions are the same as previously used in connection with *I-V* characteristics. They must be adhered to when making use of equations derived in this section.*) Using Ohm's law, it is easily seen that the *I-V* characteristic of the Thévenin circuit is

$$I = \frac{V}{R_T} - \frac{V_T}{R_T} \tag{3.1}$$

The original circuit had an *I-V* relationship of the form $I = aV + b$. In order that the Thévenin circuit have the same *I-V* characteristic as the original circuit, it is necessary only to choose the quantities V_T and R_T so that $1/R_T = a$ and $-(V_T/R_T) = b$. The problem of finding the Thévenin equivalent circuit thus amounts to evaluating V_T and R_T.

Clearly, one way to obtain the values of V_T and R_T is simply to obtain the algebraic form of the *I-V* characteristic of the original circuit, thus finding b and a. However, since only two points are needed to determine completely a straight line, finding the value of *I* for two different values of *V* amounts to finding all the information present in the *I-V* characteristic, which thus should be sufficient for finding b and a. Suppose

---

* We have throughout this book used the convention that the reference direction of current through a set of terminals is directed *into* the (+) terminal. This is consistent with the convention normally used for *I-V* relationships. Furthermore, in more complex circuits it avoids the confusion of having different conventions for different terminal pairs. However, many books do use the opposite convention for *I*. In that case Eq. (3.5) becomes $R_T = +V_{OC}/I_{SC}$, and Eq. (3.8) becomes $I_N = +I_{SC}$.

we obtain, by study of the original circuit, the value of V at the terminals when I is zero. (Note that this is the voltage that would be measured by an ideal voltmeter connected to the terminals.) Let us call this value V_{OC} (for "open-circuit" V). From Eq. (3.1) we then have

$$0 = \frac{V_{OC}}{R_T} - \frac{V_T}{R_T} \qquad (3.2)$$

or

$$V_T = V_{OC} \qquad (3.3)$$

Thus if we can calculate V_{OC}, the value of V_T has been obtained. One may then proceed by obtaining, by study of the original circuit, the value of I when V is zero. (Note that this is the current that would be measured by an ideal ammeter connected between the terminals.) Let us call this value I_{SC} (short-circuit current). From Eq. (3.1) we then have

$$I_{SC} = (0) - \frac{V_T}{R_T} \qquad (3.4)$$

Inserting the value of V_T obtained in Eq. (3.3), we have*

$$R_T = -\frac{V_{OC}}{I_{SC}} \qquad (3.5)$$

Then when V_{OC} and I_{SC} have been computed, both parameters of the Thévenin equivalent can be found.

EXAMPLE 3.4

Find the Thévenin equivalent of the circuit given in Fig. 3.6(a).

SOLUTION

We proceed in two steps, finding first the Thévenin voltage from Eq. (3.3) and then the Thévenin resistance from Eq. (3.5). To obtain the former, it is necessary to calculate the open-circuit voltage, which is the voltage that would be measured by an ideal voltmeter at the terminals. In this case V_{OC} can be found by inspection, since the circuit is a voltage divider (Section 2.3). From Eqs. (3.3) and (2.28) we have

$$V_T = V_{OC} = V_0 \frac{R_2}{R_1 + R_2}$$

---

*In using this formula, it should be remembered that the reference direction for I_{SC} is directed *into* the (+) terminal. Except in very unusual cases, the value of R_T is always positive.

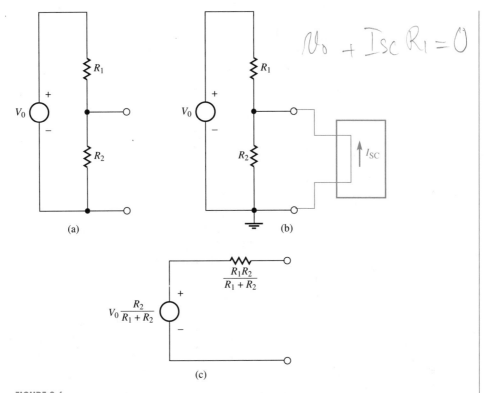

$$V_0 + I_{SC} R_1 = 0$$

FIGURE 3.6

Next we calculate I_{SC}, which is the current measured by an ideal ammeter connected between the terminals, shown in Fig. 3.6(b). (Note that the convention for the positive direction of I must be that used in Fig. 3.1.) To simplify the calculation we choose an arbitrary point of zero potential, as shown. We then write the node equation for the node between R_1 and R_2, noting that the potential of this node is zero:

$$\frac{V_0 - (0)}{R_1} + \frac{(0) - (0)}{R_2} + I_{SC} = 0$$

from which we have

$$I_{SC} = -\frac{V_0}{R_1}$$

From Eq. (3.5) we then obtain the Thévenin resistance:

$$R_T = \frac{V_{OC} R_1}{V_0} = \frac{R_1 R_2}{R_1 + R_2}$$

The complete Thévenin equivalent circuit is shown in Fig. 3.6(c).

• EXERCISE 3.1

Find the Thévenin voltage and the Thévenin resistance for the subcircuit in Fig. 3.7. **Answer:** $V_T = V_0$; $R_T = R$.

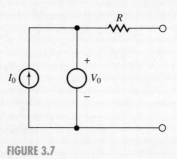

FIGURE 3.7

ALTERNATIVE METHOD

There is an alternative method for finding the Thévenin resistance, which sometimes is easier to use than the one given above. The steps in the alternate method are as follows:

1. Locate all independent voltage and current sources inside the subcircuit whose equivalent is to be found.
2. Replace all independent voltage sources by short circuits.
3. Replace all independent current sources by open circuits.
4. The remaining circuit now contains only resistances, in series and parallel combinations. Now determine what resistance now exists between the subcircuit's two terminals. This resistance is equal to the Thévenin resistance.

EXAMPLE 3.5

Find the Thévenin resistance for the subcircuit given in Fig. 3.8(a) (which is the same as that of Example 3.4) by means of the alternative method.

SOLUTION

The first step is to replace the voltage source by a short circuit. The connection obtained is that shown in Fig. 3.8(b). We observe that in this modified subcircuit, the parallel combination of R_1

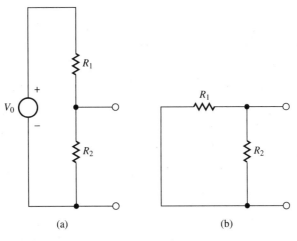

FIGURE 3.8

and R_2, which is called $R_1 \| R_2$, appears across the terminals. Thus $R_T = R_1 \| R_2 = R_1 R_2/(R_1 + R_2)$, in agreement with Example 3.4.

In some cases it may not be possible to see the value of resistance between the terminals simply by inspection. In such a case the following procedure should be used: Connect a voltage source, supplying a constant test voltage V_{TEST}, across the terminals of the subcircuit. Then a current flows through the subcircuit; the value of the current is determined by the effective resistance of the subcircuit. Let us call this current I_{TEST}. The Thévenin resistance is found from Ohm's law: $R_T = V_{TEST}/I_{TEST}$.

EXAMPLE 3.6

Find the Thévenin resistance for the subcircuit of Fig. 3.8(a) by means of a test voltage V_{TEST} applied between the terminals.

SOLUTION

After the subcircuit's internal voltage source has been replaced by a short circuit, the modified subcircuit is as shown in Fig. 3.9(a). We connect a voltage source V_{TEST} across the terminals. Now we have the circuit shown in Fig. 3.9(b). Next we calculate the value of I_{TEST}, the location and

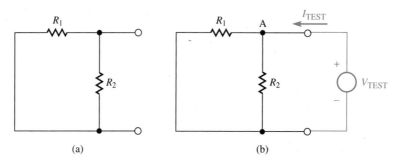

FIGURE 3.9

direction of which are shown on the diagram. We can write a node equation for the node designated as A. Setting to zero the sum of the currents leaving the node, we have

$$\frac{V_{\text{TEST}}}{R_1} + \frac{V_{\text{TEST}}}{R_2} - I_{\text{TEST}} = 0$$

Solving, we get

$$I_{\text{TEST}} = V_{\text{TEST}} \frac{R_1 + R_2}{R_1 R_2}$$

The Thévenin resistance is therefore

$$R_T = \frac{V_{\text{TEST}}}{I_{\text{TEST}}} = \frac{R_1 R_2}{R_1 + R_2}$$

as we found before.

It may seem that the method used in this example is unnecessarily elaborate compared to that of Example 3.5. However, cases where the use of V_{TEST} is necessary will later be encountered.

● EXERCISE 3.2

Repeat Exercise 3.1 using the alternative method.

THÉVENIN EQUIVALENT OF A CIRCUIT WITH TIME-VARYING SOURCES

The method of the previous section may be used unaltered when the original circuit contains time-varying sources. The value of the Thévenin voltage will then vary with time.

EXAMPLE 3.7

Find the Thévenin equivalent of the circuit shown in Fig. 3.10.

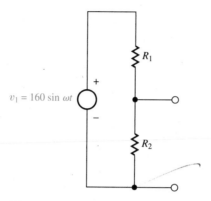

$v_1 = 160 \sin \omega t$

FIGURE 3.10

SOLUTION

The circuit is identical to that of Example 3.4 except that the constant voltage V_0 is replaced by the time-varying voltage 160 sin ωt. Except for this replacement, the steps of the solution are identical. The Thévenin parameters are

$$v_T = v_1 \frac{R_2}{R_1 + R_2} = 160 \sin(\omega t) \frac{R_2}{R_1 + R_2}$$

$$R_T = \frac{R_1 R_2}{R_1 + R_2}$$

Finding the Thévenin equivalent of a circuit containing several dc sources presents no special difficulties. However, if more than one time-varying source is present, some care must be employed in the algebra. For instance, the sum of two sinusoidal voltages of the same frequency, each with amplitude 1 V, is always a sinusoidal voltage, but the amplitude of the sum may not be 2 V. Thus caution must be used.

EXAMPLE 3.8

Find the Thévenin equivalent of the circuit shown in Fig. 3.11.

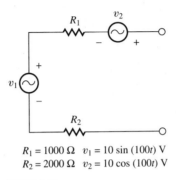

$R_1 = 1000 \ \Omega \quad v_1 = 10 \sin(100t) \ V$
$R_2 = 2000 \ \Omega \quad v_2 = 10 \cos(100t) \ V$

FIGURE 3.11

SOLUTION

We proceed by finding v_{OC} and i_{SC}:

$$v_{OC} = v_1 + v_2 = 10[\sin(100t) + \cos(100t)] \ V$$

$$i_{SC} = -\frac{v_1 + v_2}{R_1 + R_2}$$

Now we find the Thévenin parameters

$$v_T = v_{OC}$$

and

$$R_T = \frac{-v_{OC}}{i_{SC}} = R_1 + R_2 = 3000 \ \Omega$$

Note that v_T is a sinusoid, but its amplitude is not 20. There is a trigonometric identity $\sin x + \cos x = \sqrt{2} \sin(x + \pi/4)$, which is easily verified by expanding the right-hand side. Thus the amplitude of the sinusoid v_T in this case is equal to $10 \sqrt{2}$.

THÉVENIN EQUIVALENT OF A PHYSICAL ELEMENT

In the previous sections attention has been directed to the replacement of a composite subcircuit containing several ideal elements by a Thévenin equivalent. The same principles may be used, however, to obtain an equivalent for a two-terminal *physical* element. Such an element is not a combination of ideal elements, but rather is a physical "black box" whose innards may not even be known. However, for a Thévenin equivalent to exist, the physical element must have a linear *I-V* characteristic. In fact, many physical elements do not have linear *I-V* characteristics. Even then, because it is so convenient to have a Thévenin equivalent available, one may wish to evade the restriction by assuming that the *I-V* characteristic of the physical element is at least approximately linear. In many cases the results obtained by such an approximation are accurate enough.

Replacement of the original element by a Thévenin equivalent depends on identity of the *I-V* characteristics. Two points defining the linear *I-V* characteristic of a physical element can be obtained by actual measurement. That is, a voltmeter that has a very high internal resistance (and therefore is a good approximation to an ideal voltmeter) is connected to the terminals of the physical element to measure V_{OC}, and a low-resistance ammeter is used to measure I_{SC}. Equations (3.3) and (3.5) then give the Thévenin parameters.

One example of a physical element to which this procedure might be applied is a battery. Another is a phonograph pickup.

EXAMPLE 3.9

We wish to find the Thévenin equivalent of a flashlight cell. Measurement of V_{OC} with a high-resistance voltmeter gives 1.50 V, and a (very brief*) measurement of I_{SC} with a low-resistance ammeter gives -0.5 A.

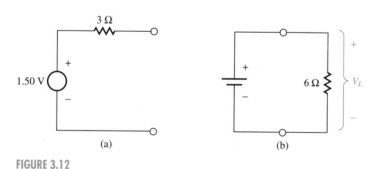

FIGURE 3.12

---

*Drawing such a large current as I_{SC} for even as short a time as 1 sec would probably cause damage to the battery. A better method is described in Example 3.10.

SOLUTION

Using Eqs. (3.3) and (3.5), we find that the equivalent is as shown in Fig. 3.12(a). This example illustrates the fact that the equivalent of a battery is not simply an ideal voltage source, but rather a voltage source in series with a resistance. Practically, this makes an important difference. Even though the battery is nominally a 1.5 V cell, the voltage across its load will not, in general, be 1.5 V, because of voltage drop in the 3 Ω resistance. For instance, let a 6 Ω load be connected to the cell, as shown in Fig. 3.12(b). Replacing the cell with its Thévenin equivalent and using the voltage-divider formula, Eq. (2.28), we find that the voltage across the load V_L is given by

$$V_L = \frac{6}{6 + 3}\ 1.5 = 1.0\ \text{V}$$

EXAMPLE 3.10

We wish to find the Thévenin equivalent of a battery while avoiding the necessity of a short-circuit measurement, which may damage it. Show that this may be done by measuring the open-circuit voltage of the battery, and then measuring the voltage when a known external resistance R is connected across the terminals.

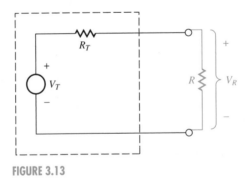

FIGURE 3.13

SOLUTION

The Thévenin voltage is obtained in the usual way, by measuring V_{OC}. Replacing the battery with its assumed Thévenin equivalent, the situation when R is connected across its terminals is as given in Fig. 3.13. The voltage V_R across the known resistance is now measured. By use of the voltage-divider formula we see that the Thévenin voltage is related to the measured V_R by

$$V_R = V_T \frac{R}{R + R_T}$$

Rearranging, we have

$$R_T = R\left(\frac{V_T}{V_R} - 1\right)$$

Since $V_T = V_{OC}$, which, like V_R, is obtained by measurement, all quantities on the right are known, and R_T can be found.

NORTON EQUIVALENT

The principle of the Norton equivalent circuit is similar to that of the Thévenin equivalent. However, the equivalent circuit takes a different form, shown in Fig. 3.14. As in the case of the Thévenin equivalent, the values of the source (now a current source) and resistance are to be chosen so as to give the Norton equivalent the same *I-V* characteristic as the circuit to be replaced. The *I-V* characteristic of the Norton circuit (Fig. 3.14) is found (by writing a node equation) to be

$$I = \frac{V}{R_N} - I_N \qquad\qquad \textbf{(3.6)}$$

Proceeding as in the Thévenin case, we calculate I_{SC} for the original circuit (I_{SC} being what would be measured by an ideal ammeter connected to the terminals). Inserting this value and $V = 0$ in Eq. (3.6), we get

$$I_{SC} = \frac{(0)}{R_N} - I_N \qquad\qquad \textbf{(3.7)}$$

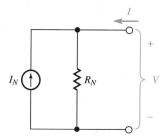

FIGURE 3.14 The general form of a Norton equivalent circuit. I_N is the Norton current, and R_N is the Norton resistance.

Hence,

$$I_N = -I_{SC} \tag{3.8}^*$$

Next we calculate V_{OC} (which is what would be measured by an ideal voltmeter at the terminals of the original circuit). Inserting this value and $I = 0$ into Eq. (3.6), we get

$$(0) = \frac{V_{OC}}{R_N} - I_N \tag{3.9}$$

or, using Eq. (3.8),

$$R_N = -\frac{V_{OC}}{I_{SC}} \tag{3.10}$$

By comparing Eq. (3.10) with Eq. (3.5), we see that the Norton resistance has the same value as the Thévenin resistance.

EXAMPLE 3.11

Find the Norton equivalent of the subcircuit in Fig. 3.15(a) whose two terminals are A and B.

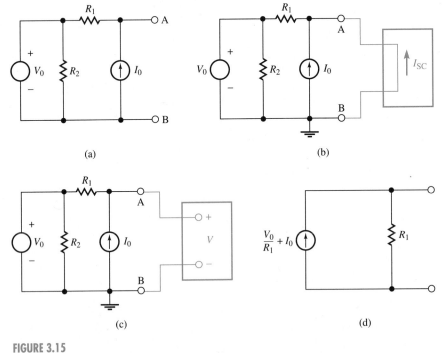

FIGURE 3.15

---

*Remember that the reference direction of I_{SC} is directed *into* the (+) terminal.

SOLUTION

First we find I_{SC}, which is the current that is indicated by an ideal ammeter connected to the terminals, shown in Fig. 3.15(b). To obtain the current through the ammeter, we write the node equation for node A. For convenience, let us define the potential of B as the zero potential. Remember that since there is no voltage drop across an ideal ammeter, the potential at A is zero. We have

$$\frac{V_0 - 0}{R_1} + I_0 + I_{SC} = 0$$

or

$$I_{SC} = -\frac{V_0}{R_1} - I_0$$

Next, to obtain V_{OC} we find the reading of an ideal voltmeter connected to the terminals, shown in Fig. 3.15(c). Writing a node equation for node A (at which point the voltage is V_{OC}), we have

$$\frac{V_0 - V_{OC}}{R_1} + I_0 = 0$$

or

$$V_{OC} = I_0 R_1 + V_0$$

Inserting into Eqs. (3.8) and (3.10) the values we have found for V_{OC} and I_{SC} gives

$$I_N = \frac{V_0}{R_1} + I_0$$

and

$$R_N = R_1$$

The Norton equivalent circuit is as shown in Fig. 3.15(d).

It is interesting to note that the value of R_2 in the original circuit has no effect on the Norton parameters. This reflects the fact that R_2 plays no part in determining the *I-V* characteristics of the original circuit.

3.2 Nonlinear Circuit Elements

All the circuit elements discussed up to now belong to the class known as linear circuit elements. The *I-V* relationships of linear elements are represented mathematically by linear algebraic equations or linear differential equations. However, other circuit elements exist that do not exhibit linear properties. In this section we shall be concerned with the important class of elements for which a graphical *I-V* characteristic exists but does not have linear form. In general, the *I-V* characteristic of such an element is a curve of any form, which can be determined only by measurements on the nonlinear

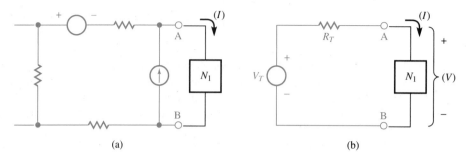

(a)   (b)

FIGURE 3.16   (a) An arbitrary circuit containing a single nonlinear element N_1, as well as voltage and current sources and resistances. (b) Simplified circuit obtained by replacing all of the circuit except N_1 by its Thévenin equivalent.

element. Our concern here is with methods for analyzing circuits containing such non-linear elements.

Let us consider a circuit consisting of voltage and current sources, resistances, and a single nonlinear element, which we shall call N_1. Such a circuit is shown in Fig. 3.16(a). We are interested in calculating the voltage across the nonlinear element $V_A - V_B$ and the current through it, designated in the figure as I. As a first step, the problem may be simplified by breaking the circuit at points A and B and replacing everything to the left of A,B by its Thévenin equivalent. Thus to find I and $V_A - V_B$ it will be sufficient to analyze the circuit of Fig. 3.16(b). To simplify the notation, we shall define $V_A - V_B \equiv V$.

In order to demonstrate the graphical method, let us assume that $V_T = 5$ V and $R_T = 1,000 \ \Omega$. We must also specify the I-V characteristic of N_1. Adopting the sign conventions given in Fig. 3.17(a), let us assume that the I-V characteristic of N_1 is as

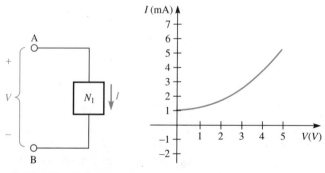

FIGURE 3.17   The I-V characteristic of the nonlinear element N_1. (a) Sign conventions. (b) I-V characteristic.

shown in Fig. 3.17(b). We shall give the mathematical function depicted in Fig. 3.17(b) the name $I_N(V)$.

Our next step is to obtain the *I-V* characteristic for the remainder of the circuit of Fig. 3.16(b) exclusive of N_1. This subcircuit is shown in Fig. 3.18(a). We shall need an *I-V* characteristic expressed in terms of the same variables as were used in the *I-V* characteristic of the other part of the circuit. Therefore our variables must be chosen as shown in Fig. 3.18(a). [Note the reference direction for *I*, chosen to agree with Fig. 3.17(a).] We shall obtain *I* as a function of *V* for this subcircuit. Referring to Fig. 3.18(a), we add voltage drops around the loop A → B → V_T → R_T → A:

$$V - V_T + IR_T = 0 \tag{3.11}$$

Rearranging, we find the relationship

$$I = \frac{V_T}{R_T} - \frac{V}{R_T} \tag{3.12}$$

Clearly, the graph representing this equation is a straight line. To locate the line, two points are sufficient. It is easiest to locate the points where the line crosses the $I = 0$ axis and where it crosses the $V = 0$ axis. We see that when $I = 0$, $V = V_T$ and when $V = 0$, $I = V_T/R_T$. Using the given values $V_T = 5$ V and $R_T = 1{,}000\ \Omega$, we see that the *I-V* characteristic passes through the points $I = 0$, $V = 5$ V and $V = 0$, $I = 5$ mA. The two points and the line through them, which is the *I-V* characteristic, are shown in Fig. 3.18(b). This line, the *I-V* characteristic of all of the circuit except the nonlinear element, is known as the *load line*. Let us refer to the mathematical function shown in Fig. 3.18(b) as $I_L(V)$.

Now the dependence of *I* on *V*, as specified by the nonlinear element, is as shown in Fig. 3.17(b). The dependence of *I* on *V*, as specified by the remainder of the circuit, is as shown in Fig. 3.18(b). The point representing the circuit's operation therefore must lie on both of these curves. Thus to find the solution all we need to do is superimpose the two curves; the point of crossing is the solution. The two curves are shown

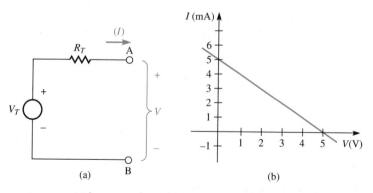

(a)   (b)

FIGURE 3.18   *I-V characteristic of part of the circuit of Fig. 3.16(b) to the left of A, B. The sign conventions (a) are chosen to agree with those of Fig. 3.17(a). The I-V characteristic (b) is shown for $V_T = 5$ V, $R_T = 1000\ \Omega$.*

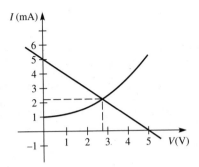

FIGURE 3.19 Graphical solution of the circuit of Fig. 3.16(b). The point where the load line and the *I-V* characteristic of the nonlinear element cross is the operating point.

superimposed in Fig. 3.19. We see that the values that satisfy both parts of the circuit are $I \cong 2.2$ mA and $V \cong 2.6$ V. This point on the *I-V* graph is known as the *operating point* or *bias point*.

• EXERCISE 3.3

The *I-V* characteristic of the nonlinear element N is shown in Fig. 3.20(b). The sign conventions for I and V are given in Fig. 3.20(a). [Note that these conventions are different from those of Fig. 3.17(a).] The nonlinear element is used in the circuit shown in Fig. 3.20(c). Find V_N. **Answer:** Approximately 0.7 V.

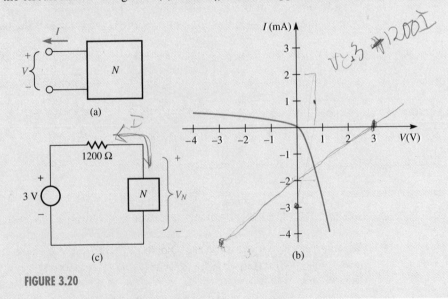

FIGURE 3.20

• EXERCISE 3.4

A large dc generator has the *I-V* characteristics shown in Fig. 3.21. It supplies power to a load that can be modeled as an adjustable resistance, as shown in the inset. Find the power delivered to the load (a) when the generator speed is 1,000 rpm and $R = 100 \ \Omega$; (b) when the generator speed is 1,100 rpm and $R = 200$ Ω. **Answers:** (a) About 97 kW; (b) about 88 kW.

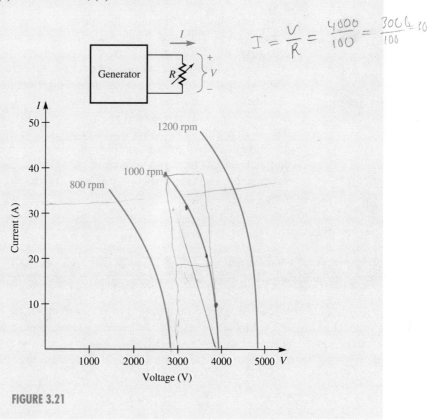

$$I = \frac{V}{R} = \frac{4000}{100} = \frac{3000}{100} \, \text{x0}$$

FIGURE 3.21

3.3 Power Calculations

The operation of any circuit involves the flow of power from one part of the circuit to another. Very often one needs to calculate the amount of power entering or leaving some part of the circuit. This calculation might be made in order to find how much useful power is being delivered to some place where it is needed, such as a loudspeaker. On the other hand, transfer to some parts of the circuit may be undesirable but also unavoidable. For example, any time current flows through a resistor, electrical power must enter the resistor and be converted to heat. Each circuit component is capable of

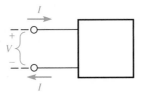

FIGURE 3.22   A box (which can have anything inside it) has two wires for external connection. We define the voltage *V* across the two wires as shown. The positive direction of the current *I* is then chosen to be the direction *into* the terminal marked (+). The choice of which terminal to mark (+) is arbitrary.

converting only a certain amount of power into heat; if this limit is exceeded, the component will be damaged. Thus a circuit designer must be able to calculate the power dissipation of all vulnerable parts.

The fundamental ideas of electrical power were introduced in Chapter 1. At this point we shall discuss power calculations in greater detail.

Let us first consider power flow into or out of a two-terminal circuit element, as shown in Fig. 3.22. As usual, the sign conventions for voltage and current are important. With regard to voltage either terminal may be designated as (+), but the reference direction of current must be defined as *into* the terminal chosen as (+). Since the circuit element has only two terminals, an equal current must flow *out* of the terminal designated (−). Remember that the actual numerical values of *V* and *I* can have either sign.

EXAMPLE 3.12

We are about to calculate the power flow into the current source in Fig. 3.23(a). Following the model of Fig. 3.22, draw a box around the current source, choose suitable sign conventions for *V* and *I*, and find their values.

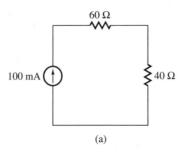

(a)

FIGURE 3.23(a)

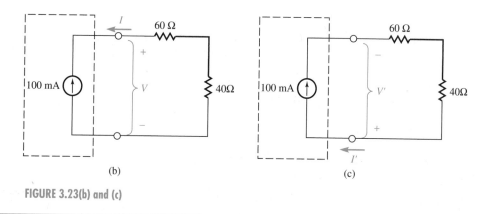

FIGURE 3.23(b) and (c)

SOLUTION

The box and its sign conventions may be drawn as shown in Fig. 3.23(b). With this choice we find that $I = -100$ mA and $V = +10$ V.

However, the choice of which terminal is (+) is arbitrary; thus it is just as good to choose the conventions shown in Fig. 3.23(c). With this choice we find $I' = +100$ mA and $V' = -10$ V. Note that whichever choice is made, we always take the reference direction of I as being *into* the terminal designated (+).

Once the circuit element under consideration has been represented in the form of Fig. 3.22 and V and I calculated (with their correct signs), we use the following rule: The power *entering* the box shown in Fig. 3.22 is given by

$$P = VI \quad \text{(power entering)} \tag{3.13}$$

Note that it is quite possible for P to have a negative numerical value. Such a result just means that power is leaving the box, not entering it.

EXAMPLE 3.13

Find the power entering the current source in Example 3.12.

SOLUTION

Let us adopt the sign conventions shown in Fig. 3.23(b). There it was found that $I = -100$ mA and $V = +10$ V; thus the power entering the current source is -1 W. This is the same as saying that $+1$ W is coming *out* of the current source.

If we instead use the sign conventions of Fig. 3.23(c), we have $I' = +100$ mA and $V' = -10$ V. The power entering the current source is $V'I' = -1$ W. As expected, either choice of sign convention gives the same result, provided that the reference direction of current is correctly chosen — that is, *into* the (+) terminal.

EXAMPLE 3.14

Find the power entering the 40 Ω resistor in Example 3.12.

SOLUTION

In this case we must draw the box around the resistor, in the form of Fig. 3.22; see Fig. 3.24.

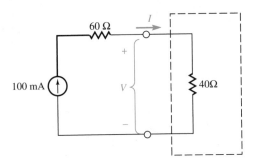

FIGURE 3.24

We now find that $I = 100$ mA and $V = 4$ V; thus the power entering the resistor is 0.4 W. The fact that the result is positive means that electrical power is *entering* the resistor, where it is turned into heat.

A similar calculation performed on the 60 Ω resistor gives the result that 0.6 W is entering it. The sum of the powers entering all three circuit elements is zero, a result which is true in general.

In the preceding examples we found V and I by analyzing the circuit. It should be noted, however, that the calculation based on Eq. (3.13) and Fig. 3.22 is valid regardless of what is inside the box, or what method we use to find V and I.

EXAMPLE 3.15

The box on the right of Fig. 3.25(a) contains an unknown circuit element. Experimentally we connect a voltage source V and measure I with an ammeter as shown. The table (below) gives the measurements of I; find the power entering the nonlinear element for each voltage.

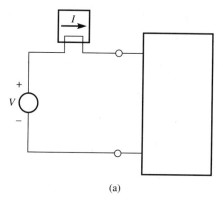

(a)

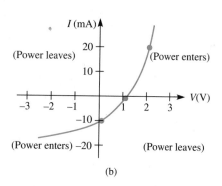

(b)

FIGURE 3.25

SOLUTION

V	I	P
0	-10 mA	0
1 V	-2 mA	-2 mW
2 V	$+20$ mA	$+40$ mW
3 V	$+400$ mA	$+1.2$ W

An interesting sidelight of this example is the following: The unknown box is characterized by some I-V characteristic, which can be graphed. Four points on this I-V graph are listed in the table. Let us suppose that the complete graph is as shown in Fig. 3.25(b). Note that any time the operating point is in the first quadrant both I and V are positive, and power is *entering* the box. Th esame is true when the operating point is in the third quadrant, since the product VI is then also positive. If the operating point is in the second or fourth quadrants, the product VI is negative, and power comes *out* of the box. This particular box has the property that power can either come out of it or go into it, depending on what voltage exists across its terminals. From inspection of Fig. 3.25(b) we see that if $V < 0$ or $V > 1.13$ V, power goes *into* the box. But if $0 < V < 1.13$ V, power comes *out of* the box.

• EXERCISE 3.5

The voltage applied to the dc motor of a streetcar is constant at 4800 V. However, the current drawn by the motor depends on the steepness of the roadbed, as shown in Fig. 3.26. (a) Find the power consumption when the streetcar is traveling level at 20 mph. (b) When the streetcar is going downhill the motor can generate power and return it to the power line, a process known as "regenerative braking." Find the power generated when the streetcar is going downhill at a 2° angle at 30 mph. **Answers:** (a) 48 kW; (b) 122 kW.

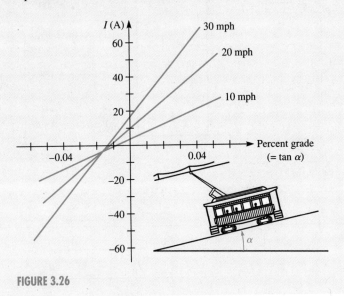

FIGURE 3.26

• EXERCISE 3.6

Refer to Fig. 3.27 and find (a) P_V, the power entering the voltage source; (b) P_I, the power entering the current source; and (c) P_R, the power entering the resistor. **Answers:** (a) $P_V = 18$ mW; (b) $P_I = -31.5$ mW; (c) $P_R = 13.5$ mW.

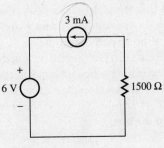

FIGURE 3.27

TWO IMPORTANT SPECIAL CASES

If a voltage V exists across any resistor R, it follows from the preceding discussion that the power dissipated in the resistor is $VI = V(V/R) = V^2/R$. This result holds regardless of what is in the rest of the circuit, so long as V is the voltage that actually exists across R.

Similarly, if a current I passes through any resistor R, the power dissipated in the resistor is I^2R.

EXAMPLE 3.16

In the circuit shown in Fig. 3.28, N_1 and N_2 are nonlinear circuit elements. The voltages at points A and B are measured experimentally and found to be 6.3 V and 3.8 V, respectively. What power is dissipated in R?

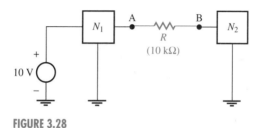

FIGURE 3.28

SOLUTION

The voltage across R is $6.3 - 3.8 = 2.5$ V. The power dissipated in R is $V^2/R = 0.625$ mW.

INSTANTANEOUS POWER AND AVERAGE POWER

It often happens that V and I are functions of time, in which case we write $v(t)$ and $i(t)$. The *instantaneous* power entering the box in Fig. 3.22 is then given by

$$P(t) = v(t)i(t) \quad \text{(instantaneous power)} \tag{3.14}$$

Instantaneous power may or may not be important in a given case; it depends on what one is trying to calculate. Suppose we are attempting to protect a delicate circuit component, which would overheat if excessive power entered it for even an extremely short time. In that case it would be sensible to design the circuit in such a way that instantaneous power is never allowed to exceed the safe value. On the other hand, very often one is more concerned with the *time-averaged power* being transferred. For example, consider an electric light bulb operating on 60 Hz ac voltage. We know that the sinusoidal voltage passes through zero 120 times per second, and the instantaneous power $v(t)i(t)$ must do so as well. For calculating our electric bill, however, we are not interested in such precise details of the time-varying power consumption; we need to know the time-averaged power being consumed. This is given by the formula

$$P_{AV} = \frac{1}{T} \int_0^T v(t)i(t)\, dt \quad \text{(time-averaged power)} \tag{3.15}$$

Here the time T is the time over which the average is taken. It can be an arbitrarily long time, or, if $v(t)$ and $i(t)$ are periodic, it can be one period of their variation.

EXAMPLE 3.17

The voltage source shown in Fig. 3.29(a) produces the repetitive time-varying voltage shown in Fig. 3.29(b). Find the time-average power transferred into R.

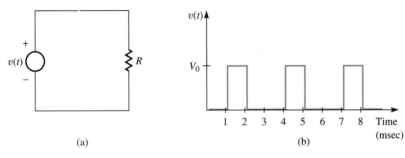

(a)   (b)

FIGURE 3.29

SOLUTION

We note that $v(t)$ is either 0 or V_0; the value of $i(t)$ is 0 or V_0/R amperes, respectively. [Verify that $i(t)$ is in fact $+ V_0/R$ and not $-V_0/R$.] The instantaneous power is either 0 or V_0^2/R. The period of the variation is 3 msec, and so we choose this value for T. Thus

$$P_{AV} = \frac{1}{0.003} \int_0^{0.003} v(t)i(t)\, dt$$

$$= \frac{1}{0.003} \left(\int_0^{0.001} (0)\, dt + \int_{0.001}^{0.002} \frac{V_0^2}{R}\, dt + \int_{0.002}^{0.003} (0)\, dt \right)$$

$$= \frac{1}{0.003} \frac{V_0^2}{R} 0.001 = \frac{V_0^2}{3R}$$

This result might have been guessed quickly by noting that the power flow is turned on one-third of the time; one says that the *duty cycle* is 1/3. Thus the average power is 1/3 of the "peak" power or 1/3 of V_0^2/R. This method of guessing the average power is useful, but it works only when the instantaneous power switches periodically between two steady values. In general, one has to perform the integration required by Eq. (3.15).

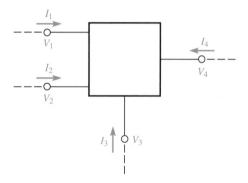

FIGURE 3.30

MULTITERMINAL ELEMENTS

The preceding discussion dealt with finding power flow into two-terminal circuit elements. Some important circuit elements have more than two terminals, however; transistors, for example, usually have three. Thus we need a method more general than that of Fig. 3.22 and Eq. (3.13). Let us consider a circuit element with four terminals, as shown in Fig. 3.30. (The case of four terminals is used as an illustration. The method works with any number of terminals.) The voltages at the four terminals are V_1, V_2, V_3, and V_4, and currents I_1, I_2, I_3, and I_4 are defined with reference directions *into* the terminals. With these definitions, the power flowing into the circuit element is

$$P = V_1 I_1 + V_2 I_2 + V_3 I_3 + V_4 I_4 \qquad (3.16)$$

EXAMPLE 3.18

Show that the method of Fig. 3.22 and Eq. (3.13) can be obtained as a special case of Fig. 3.30 and Eq. (3.16).

SOLUTION

The general formula, Eq. (3.16), gives $P = V_1 I_1 + V_2 I_2$. In Fig. 3.22 I_1 is called I and I_2 is called $-I$. Thus $P = V_1 I - V_2 I = (V_1 - V_2)I$. But in the notation of Fig. 3.22, $V_1 - V_2$ is called V. Thus $P = VI$, in agreement with Eq. (3.13).

EXAMPLE 3.19

The transistor circuit in Fig. 3.31 has the measured values of the voltage and currents as shown. Find the rate at which heat is being produced in the transistor.

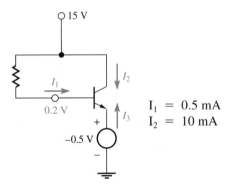

FIGURE 3.31

SOLUTION

We wish to use Fig. 3.30 and Eq. (3.16). The three terminal voltages are $V_1 = 0.2$ V, $V_2 = 15$ V, and $V_3 = -0.5$ V. We note that the current I_3 is not given in the diagram, but this can be found from Kirchhoff's current law: The sum of all currents entering the transistor must be zero. Thus $I_1 + I_2 + I_3 = 0$, $I_3 = -(I_1 + I_2) = -10.5$ mA. Converting all numbers to MKS units, we have

$$P = V_1 I_1 + V_2 I_2 + V_3 I_3$$
$$= (0.2)(5 \times 10^{-4}) + (15)(10^{-2}) + (-0.5)(-1.05 \times 10^{-2})$$

3.4 Starting Your Volkswagen

The engine of an automobile needs to be mechanically turned by an external force in order to start. Normally one uses an electric motor known as the *starter motor*. This is a rather powerful little motor, which may itself produce one-half horsepower or thereabouts. Thus it draws quite a large current from the car battery. If the latter is rated at 12 V, the starter motor current will be in the range of 20 to 40 amperes. The starter motor is a complex electromechanical device, but one can easily construct a model of it. If we take 30 amperes as a typical value for its current, then we can model the motor as a resistance whose value is $R_M = 12/30 = 0.4$ Ω. The simplified circuit diagram of the starter circuit is then as shown in Fig. 3.32. Here the symbol on the left represents a 12 V battery, which is connected to the motor when the starting switch is closed.

The car battery is usually a large, heavy, lead-acid storage battery. Much smaller batteries with 12 V ratings, however, can be imagined. For instance, one could take eight standard flashlight batteries, each rated 1.5 V, and connect them in series. This

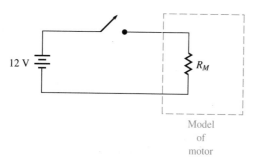

Model
of
motor

FIGURE 3.32

would result in a 12 V battery much smaller and more convenient than a large car battery. The circuit diagram would still look the same as in Fig. 3.32. Intuition tells us
that the smaller battery will not work. But why? The problem here is that the "battery"
symbol in the figure is not very useful. The symbols for both kinds of battery look the
same. We need to replace the batteries by suitable models, so we can see how they
really act.

A first idea might be to model each battery as an ideal 12 V voltage source. However, this would not help us to understand the difference, since the models of the two
kinds of batteries would still be alike! At any rate, doing this would not be correct,
because, as we saw in Section 3.1, the correct model of a two-terminal subcircuit is not
just a bare voltage source, but a Thévenin equivalent circuit consisting of a voltage
source plus a series resistance. The value of the Thévenin voltage source is the same as
the open-circuit voltage, which is 12 V for both batteries. Thus the difference in the
performance of the two kinds of batteries must result from their having different values
of the Thévenin resistance.

Proceeding from Fig. 3.32, we replace the battery symbol with its Thévenin equivalent, obtaining Fig. 3.33. Now we are in a position to actually calculate V_M, the volt-

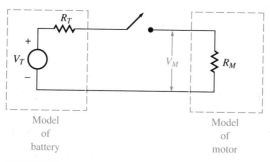

Model
of
battery

Model
of
motor

FIGURE 3.33

age that appears across the motor when the switch is closed. Using the voltage-divider formula we have

$$V_M = V_T \frac{R_M}{R_T + R_M} \qquad (3.17)$$

From this we see that the voltage across the motor is actually never equal to 12 V. Even though V_T, the battery's open-circuit voltage, is 12 V, the voltage divider arising from R_M and R_T reduces the voltage across the motor to some lower value. Presumably the motor will still operate satisfactorily, provided that V_M is at least somewhere *near* 12 V.

For a typical car battery, the Thévenin resistance will probably be on the order of 0.05 Ω. In that case the voltage across the motor will be approximately 12 · 0.4/ (0.4 + 0.05) = 10.67 V, which will probably be sufficient to run the motor. The flashlight batteries, on the other hand, will have a much larger Thévenin resistance, on the order of 20 Ω. Eq. 3.17 then tells us that the voltage across the motor will be only 12 · 0.4/(0.4 + 20) = 0.235 V. Obviously this will be too small to operate the 12 V motor.

The result we have obtained is certainly no surprise. We would not expect that a tiny flashlight battery would provide enough current or power to run the starter motor. The point to be noticed here is the usefulness of the circuit model. Before we replaced the battery symbol with the model, there was no way we could tell how the circuit would operate. But once we insert the models and determine the values of V_T and R_T (by measurements, or from the manufacturer's specifications), the problem disappears. All we have to do is analyze the resulting simple circuit and calculate whatever it is we want to know.

POINTS TO REMEMBER

- Equivalent circuits are circuits that cannot be distinguished from each other by measurements at their terminals. Often circuit analysis can be simplified if a portion of the circuit is replaced by a simpler equivalent. Two general families of equivalents exist for linear circuits: Thévenin equivalents and Norton equivalents.

- Nonlinear circuit elements are those whose *I-V* relationships cannot be expressed as linear equations. Special methods must be used to analyze circuits containing nonlinear elements. Graphical methods are convenient for quickly obtaining insight into circuit operation.

- Power flow can be calculated from the expressions $P = VI$ for two-terminal circuit elements and $P = \sum_N V_N I_N$ for multiterminal circuit elements. However, it is essential that the signs of the various voltages and currents be stated correctly.

- If voltage and current vary, the quantity $v(t)i(t)$ is known as the *instantaneous power*. The *time-averaged power* is the average over time of the instantaneous power.

PROBLEMS

Section 3.1

3.1 Show that the subcircuit of Fig. 3.34(a) is approximately equivalent to that of Fig. 3.34(b).

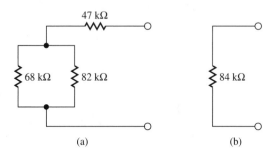

FIGURE 3.34

*3.2 Obtain and graph the *I-V* characteristic of the subcircuit shown in Fig. 3.35(a).

*3.3 Obtain and graph the *I-V* characteristic of the subcircuit shown in Fig. 3.35(b). Is this subcircuit equivalent to that of Fig. 3.35(a)?

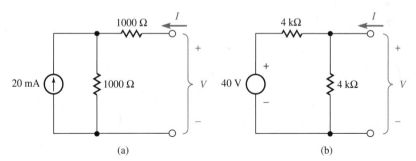

FIGURE 3.35

3.4 Find the open-circuit voltage of the subcircuit shown in Fig. 3.36.

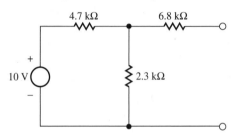

FIGURE 3.36

*3.5 Find the short-circuit current of the subcircuit shown in Fig. 3.36.

*3.6 Find the Thévenin and Norton equivalents of the subcircuit shown in Fig. 3.36.

*3.7 Obtain the Thévenin equivalent of the subcircuit shown in Fig. 3.35(a).

*3.8 Obtain the Thévenin equivalent of the subcircuit shown in Fig. 3.35(b).

*3.9 Find the Norton equivalents of the subcircuits shown in Fig. 3.35(a) and (b).

*3.10 Consider the subcircuit shown in Fig. 3.37. Suppose V and R are adjustable but that the ratio V/R always is kept equal to I_1, a constant. By constructing a Norton equivalent, show that this subcircuit approaches equivalence to an ideal current source when $V \rightarrow \infty$.

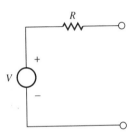

FIGURE 3.37

3.11 Find the Thévenin resistance of the subcircuit of Fig. 3.35(a) using the alternate method in which sources are set equal to zero.

*3.12 Find the Thévenin equivalent of the subcircuit shown in Fig. 3.38.

*3.13 Find the Thévenin equivalent of the subcircuit shown in Fig. 3.38, if the 2 A current source is changed to 1 A. (Can the formula $R_T = -V_{\mathrm{OC}}/I_{\mathrm{SC}}$ be used in this case?)

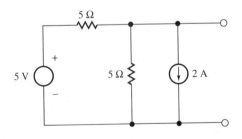

FIGURE 3.38

*3.14 A 10 Ω resistance is connected across the terminals of Fig. 3.38. Calculate the current through this resistance in the following ways:
   a. Directly, by means of the node method.
   b. First construct the Thévenin equivalent of Fig. 3.38. Then replace that part of the circuit by the equivalent and find the current.

**3.15 A battery whose nominal voltage is 6 V is found by experiment to have the *I-V* characteristic shown in Fig. 3.39. Suppose that we attempt to construct a Thévenin equivalent for it by measuring V_{OC} and I_{SC}. Graph the *I-V* characteristic of the equivalent circuit that is obtained and compare it with the actual *I-V* characteristic of the battery. Why do they not agree?

*3.16 The battery whose *I-V* characteristic is shown in Fig. 3.39 is connected across a 40 Ω resistor.
   a. Use a graphical method to find the voltage across the resistor and the current flowing through it. Use the sign conventions of Fig. 3.39.
   b. Let the battery be replaced by the Thévenin equivalent found in Problem 3.15. Find the voltage across the resistor and the current through it in this case.

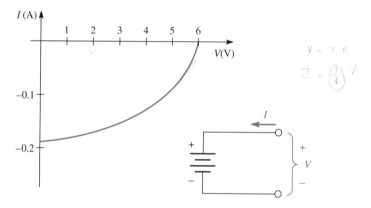

FIGURE 3.39

**3.17 We wish to find the approximate equivalent of a certain battery without measuring the short-circuit current (which is likely to damage the battery). Instead, a 40 Ω resistor is connected across it, and the current is measured to be −0.1125 A. The open-circuit voltage is found to be 6 V. From these measurements find V_T and R_T.

3.18 Show that if the Norton equivalent of any subcircuit is found and then the Thévenin equivalent of the Norton equivalent is found, the result is the same as the Thévenin equivalent of the original subcircuit.

Section 3.2

*3.19 For the circuit shown in Fig. 3.40 the nonlinear element N_1 has the *I-V* characteristic shown in Fig. 3.17(b). Find the current that flows through N_1.

*3.20 For the subcircuit of Fig. 3.41, N_1 has the *I-V* characteristic shown in Fig. 3.17(b). Find the open-circuit voltage and short-circuit current of this subcircuit.

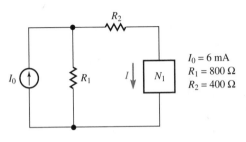

$$I_0 = 6 \text{ mA}$$
$$R_1 = 800 \ \Omega$$
$$R_2 = 400 \ \Omega$$

FIGURE 3.40

*3.21 Consider the circuit of Fig. 3.16(b). Let $V_T = 10$ V and $R_T = 100 \ \Omega$. Furthermore, let the *I-V* characteristic of the nonlinear element N_1 be given by $I = 0.002 \ V^2$, where I is given in amperes and V in volts.
  a. Write a node equation and solve for the voltage across N_1.
  b. Find the voltage across N_1 graphically.

Section 3.3

3.22 Assume that the circuit element of Fig. 3.22 can absorb electrical power but cannot generate electrical power. If the actual value of V is -0.6 V, what is the sign of I?

3.23 Show that if a current I passes through any resistor R, the power dissipated in that resistor is $I^2 R$.

**3.24 Suppose that a current I_1 passes through a resistor R_1 and I_2 passes through another resistor R_2. We make the rule that $I_1 + I_2 = I_0$. I_0 is a constant, but I_1 and I_2 can be varied, subject to their sum remaining I_0. We desire to choose I_1 and I_2 so that the *total* power dissipated in both resistors is minimized. Find I_1 and I_2. What are the resulting voltages across R_1 and R_2?

3.25 The circuit of Fig. 3.16(b) is analyzed in the discussion accompanying Figs. 3.16 to 3.19. Find the power dissipated in N_1.

*3.26 Assume that box N_1 in Fig. 3.40 actually contains some resistor. We wish to find the power *entering* the current source. In order to do this, draw a box around the current source and choose sign conventions for V and I, consistent with Fig. 3.22, such that the meeting point of R_1 and R_2 is the *minus* terminal for measuring V. Without actually calculating, answer the following:
  a. Do you expect that with these conventions I will be a positive or a negative number?
  b. Do you expect V to have a positive or a negative value?
  c. Do you expect that the power entering the current source will be a positive or a negative number?
  d. Is your answer to part (c) consistent with your answers to parts (a) and (b)?

*3.27 The *I-V* characteristic of N_1 for Fig. 3.40 is given in Fig. 3.17(b). Find the power entering N_1.

*3.28 Let the two terminals on the right side of Fig. 3.36 be connected. Find the power entering each of the four circuit elements. Show that the sum is zero.

*3.29 The time-varying voltage shown in Fig. 3.42 is connected across a 470 Ω resistor.
  a. Find and graph the instantaneous power as a function of time.
  b. What is the duty cycle?
  c. Using your answer to parts (a) and (b), find the time-averaged power.

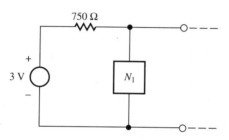

FIGURE 3.41

*3.30 Find the time-averaged power for Problem 3.29 by means of Eq. (3.15).

*3.31 Refer to the discussion associated with Figs. 3.16 to 3.19. In Fig. 3.16(b) the constant voltage source is replaced by a time-varying source whose value is shown in Fig. 3.42. Find the time-averaged power entering N_1.

**3.32 The time-varying voltage shown in Fig. 3.43 is applied across a 50 Ω resistor. Find the time-averaged power.

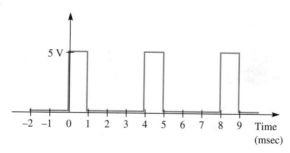

FIGURE 3.42

*3.33 An operational amplifier has voltages and currents at its terminals as shown in Fig. 3.44. (The voltages are produced by external voltage sources omitted from the diagram.) Find the power dissipated in the amplifier.

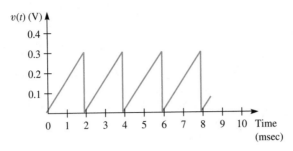

FIGURE 3.43

*3.34 A part of a *pnp* transistor circuit is shown in Fig. 3.45, with measured voltages given at several points of the circuit. Find the power dissipated *in the transistor*.

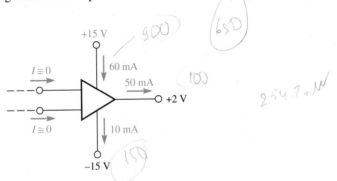

FIGURE 3.44

*3.35 Consider the three-terminal network enclosed by the dashed line in Fig. 3.46.
   a. Find the power dissipated in the network by means of Eq. (3.16).
   b. Check your answer by adding together the values of V^2/R for each resistor.

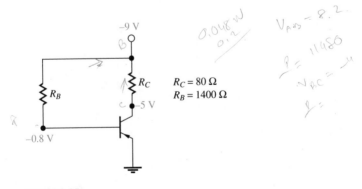

$R_C = 80\ \Omega$
$R_B = 1400\ \Omega$

FIGURE 3.45

Section 3.4

*3.36 Someone has placed an automobile battery on your workbench and asked you to measure the approximate value of its Thévenin resistance. Explain how you would do this. (*Caution:* It is never a good idea to connect a short circuit, or an ammeter, which behaves like a short circuit, across a battery, as the battery will usually be damaged. Furthermore, doing this with a large battery like a car battery would be dangerous! You have to find another method.)

**3.37 a. In Fig. 3.33, find the *power* delivered to R_M when the switch is closed, in terms of V_T, R_T, and R_M.

   b. Assume that V_T and R_T, being properties of a given battery, are fixed. However, you are allowed to choose different motors, so you can select the value of R_M. Show, by differentiating the result of (a) with respect to R_M, that power to the motor is maximized when R_M is made equal to R_T.

   c. When R_M has been optimized, as in part (b), what is the resulting maximum power delivered to the motor?

   d. One might expect that smaller batteries would be capable of delivering less power. Using the results of this problem, can you explain why flashlight batteries have a higher Thévenin resistance than car batteries?

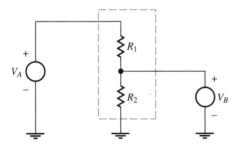

FIGURE 3.46

CHAPTER 4

ACTIVE CIRCUIT ELEMENTS

Electronics is an exciting field because it provides a wide range of powerful techniques. Electronic building blocks such as amplifiers, switches, logic gates, and digital arrays are convenient units, from which large electronic systems can be built. The internal operation of these blocks is based on the use of *active circuit elements*, such as transistors, and they themselves are considered to be active building blocks. The active building blocks can then be combined with each other to build up still larger structures.

In general, a circuit element or building block is called "active" if it is capable of power amplification. Power amplification occurs when a low-powered input signal controls an output signal at a higher power level. The blocks known as *amplifiers* of course have this property. For example, consider a phonograph amplifier. A small signal from a phonograph pickup or other signal source (tape deck, CD player) is applied to the amplifier's input terminals. In the case of a phonograph pickup, this input signal is truly small. It comes from a tiny dynamo actuated by the microscopic motion of the pickup stylus in the record groove. It does contain all the information involved in the music you want to hear, but in the form of extremely small voltages and currents. Conceivably one might connect the phonograph pickup directly to a loudspeaker, and some sound would be produced; this sound would be the desired music, but it would be so weak that it could scarcely be heard at all. What is needed is an amplified version of the input signal. A phonograph amplifier is a device that listens to the instructions contained in the input signal and generates a much larger output signal, which is a magnified version of the input signal and which contains the same information. The large output signal can then be applied to the loudspeaker, as shown in Fig. 4.1, and music is then produced at the desired volume level.

For good-quality music reproduction, one needs a *linear* amplifier, that is, an amplifier for which the output voltage is linearly proportional to the input voltage. But not all active devices are of this type. The ones used in digital technology, for example, are more like externally controlled switches. In these active blocks, a small input signal is used simply to turn a large output signal on or off. The block is still active, because a weak input signal is being made to control the much larger output power. Devices that are not active are referred to as *passive* devices.

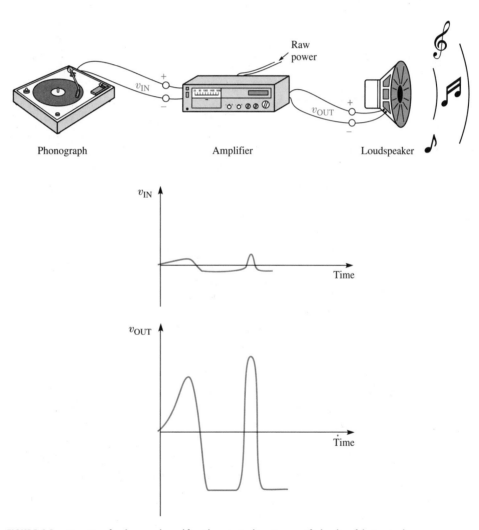

FIGURE 4.1   Operation of a phonograph amplifier. The output voltage is a magnified replica of the input voltage.

A reasonable question that might be asked at this point is whether active devices somehow defy the principle of conservation of energy. That is, where does the amplifier *get* the power that it supplies to the loudspeaker? Not from the phonograph, surely, since that can supply practically no power at all. The answer is that each active device or block requires some kind of raw power from the outside; a phonograph amplifier, for example, will usually be connected not only to the phonograph and the loudspeaker, but also to a power plug in the wall. As another example, imagine a man giving verbal instructions to a trained elephant. The input signal is the man's voice, which contains very little power. The output signal is the work done by the elephant, which can involve a great deal of power. Where does the extra power come from? From the elephant's food, of course. The elephant has all the characteristics of an active device.

Active devices and blocks are comparatively complex, both in their internal construction (which involves transistors or other active circuit elements) and in their behavior. In order to figure out what an active block will do in a circuit, it is useful to represent it by a model. The model usually consists of a few very simple circuit elements, and is capable of imitating the operation of the active block. The idea is the same as appeared in Section 3.4, where we obtained a model for a battery by constructing a Thévenin "equivalent." The word "equivalent" here is enclosed in doubtful quotation marks because the model is never truly equivalent to the block it imitates; the equivalence is only approximate, and applies only over a certain range of operating conditions. Nonetheless, models are extremely useful, and the modeling technique is one of the electronic engineer's most important tools.

The circuit elements so far introduced are not quite enough for the construction of models. In order to model active devices and blocks, we shall require an additional family of voltage and current sources known as *dependent* sources. These will be introduced in Section 4.1. In Section 4.2 we shall introduce our first active block, the amplifier block. Amplifiers are widely used in all kinds of analog circuits.

A special amplifier of great practical interest is the *operational amplifier*, or "op-amp," which we shall discuss in Section 4.3. Op-amps are available at low cost in integrated-circuit form, and are workhorses in analog technology; they can perform a large number of useful operations (hence their name). The Application in Section 4.5 is taken from the music recording studio. We shall see how with a few microphones and an op-amp you might set up shop as a recording engineer!

4.1 Dependent Sources

In order to develop models for active devices, we shall first introduce a new family of ideal circuit elements known as dependent voltage and current sources.* Since dependent sources are quite important, it is worthwhile to discuss the concept in detail.

The sources encountered until now in this book have been *independent* sources. An independent source is one whose value can be independently specified. In contrast,

---

* Also known as "controlled sources."

a dependent source obeys the rule that its value depends on a voltage or current some-where else in the circuit. To be sure, such devices are imaginary; we introduce them in order to use them in models for things that actually exist. From the student's point of view, the vital thing is to understand the rules that these circuit elements always obey. The fact that they do obey simple rules means that circuit analysis can proceed in an orderly way.

Let us begin with the dependent current source shown in Fig. 4.2. In this figure all the components are part of some large circuit, only a piece of which is shown. We see that the value of the current source is βi_1. The number β is a given constant. In order to find out the value of the current source we have to first locate the current i_1, which is the *reference* current for the dependent source. The position of i_1 is indicated on the diagram. We then calculate the value of i_1, and from that we can find the value of the dependent source.

Note the following points:

1. Except that its value depends on its reference, a dependent source functions like any other ideal source. For instance, in Fig. 4.2, the current through R *must* equal βi_1. The voltage across R will be $\beta i_1 R$.

2. The relationship between the ideal dependent source and its reference *cannot* be broken. No matter what the value of i_1, the value of the dependent source is β times as large.

3. In the case of independent sources, we were free to turn sources off for various purposes (for instance, in finding Thévenin resistance by the alternate method.) However, a dependent source *cannot* be turned off. Its value must depend on its

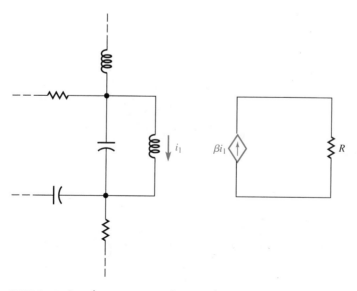

FIGURE 4.2 Dependent current source with current reference.

reference. In Fig. 4.2, if the value of i_1 is not zero, it would violate the rules to set the dependent source to zero.

4. The reference current i_1 in Fig. 4.2 is *not* a given quantity. In general, its value must be found by calculation and will depend on the inputs applied to the circuit. Thus the reference current should not appear in the result of a circuit calculation. If a reference current should appear in an answer, it must be eliminated by solving for it in terms of given quantities. The student should watch out for this; it is often a troublesome point.

EXAMPLE 4.1

Find the voltage at A in Fig. 4.3.

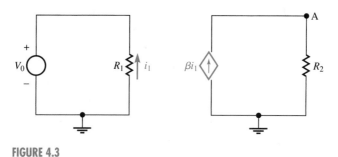

FIGURE 4.3

SOLUTION

From Ohm's law it is clear that $v_A = \beta i_1 R_2$. However, we are not yet finished, because i_1 is not a given quantity. To find its value we must refer to the rest of the circuit. In this case we see that $i_1 = -V_0/R_1$. (Note the minus sign, which comes from the given sign convention for i_1.) The voltage source V_0 is independent because its value does not depend on something else; V_0 is a given quantity and can appear in the answer. Substituting $i_1 = -V_0/R_1$ into $v_A = \beta i_1 R_2$, we have $v_A = -\beta V_0 R_2/R_1$.

A good way to solve circuit problems when dependent sources are present is to write an equation for the unknown to be calculated and then see if this equation also contains an unknown reference current. If it does, we write another equation to obtain the reference current. If the reference current itself depends on something else, still another equation is needed; we work backward through the problem until we arrive at the input, which is usually given. In Example 4.1, for instance, we first noted that we could find v_A if i_1 were known; then we found i_1 from the equation $i_1 = -V_0/R_1$ and eliminated it from the expression for v_A. In less straightforward problems one can usually write enough node (or loop) equations to eliminate the reference currents.

EXAMPLE 4.2

Find v_{OUT} in the circuit shown in Fig. 4.4.

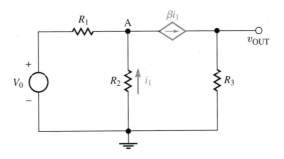

FIGURE 4.4

SOLUTION

Clearly, $v_{OUT} = \beta i_1 R_3$, but this is not sufficient; the unknown reference current must be eliminated. In order to find i_1 we write a node equation for node A:

$$\frac{V_0 - v_A}{R_1} - \frac{v_A}{R_2} - \beta i_1 = 0$$

In writing this equation, however, we have introduced a new unknown, v_A. It can be eliminated by means of the Ohm's law equation $v_A = -i_1 R_2$. Substituting this expression for v_A, we can solve for i_1. Thus we obtain

$$v_{OUT} = \beta R_3 i_1 = \frac{\beta R_3 V_0}{R_1(\beta - 1 - R_2/R_1)}$$

The reader should remember that dependent sources cannot arbitrarily be turned off. This is illustrated by the following example.

EXAMPLE 4.3

Find the Thévenin resistance of the subcircuit shown in Fig. 4.5(a), using the alternative method in which independent sources are set to zero.

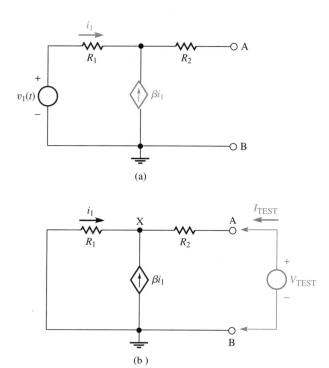

FIGURE 4.5

SOLUTION

In this method we replace the independent source $v_1(t)$ by a short circuit and find the resistance seen looking into terminals A,B. We do *not* replace the dependent source by anything; it continues to function as a dependent source. The resistance looking into terminals A,B cannot be seen by inspection. Thus we apply a test voltage V_{TEST} as shown in Fig. 4.5(b). We then find the resulting current I_{TEST}, and the desired Thévenin resistance is given by $R_T = V_{TEST}/I_{TEST}$.

Writing a node equation at node X, we have

$$i_1 + \beta i_1 + I_{TEST} = 0$$

$$I_{TEST} = -(\beta + 1)i_1$$

We still have to eliminate i_1. Note that $v_X = -i_1 R_1$ and $V_{\text{TEST}} = v_X + I_{\text{TEST}} R_2$. Thus

$$i_1 = \frac{I_{\text{TEST}} R_2 - V_{\text{TEST}}}{R_1}$$

$$I_{\text{TEST}} = -(\beta + 1)i_1 = \frac{\beta + 1}{R_1}(V_{\text{TEST}} - I_{\text{TEST}} R_2)$$

Solving for I_{TEST}, we have

$$R_T = \frac{V_{\text{TEST}}}{I_{\text{TEST}}} = \frac{R_1 + (\beta + 1)R_2}{\beta + 1}$$

In the preceding examples the dependent source was a current source, and the reference was a current. However, there are also dependent voltage sources, and for either kind of source the reference can be either a voltage or a current. Thus there are actually four kinds of dependent sources. Techniques for solving these circuit problems resemble those already seen.

EXAMPLE 4.4

Find v_{OUT} in the circuit shown in Fig. 4.6.

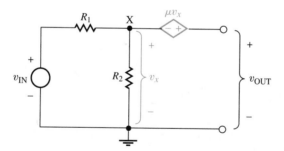

FIGURE 4.6

SOLUTION

In this circuit we have a dependent voltage source whose reference is a voltage. It is easily seen that $v_{\text{OUT}} = v_X + \mu v_X = (1 + \mu)v_X$. However, we must still eliminate the unknown reference voltage v_X. To do this we write a node equation for node X:

$$\frac{v_{\text{IN}} - v_X}{R_1} - \frac{v_X}{R_2} = 0$$

from which we have

$$v_X = v_{IN} \frac{R_2}{R_1 + R_2}$$

$$v_{OUT} = (1 + \mu)v_X = \frac{(1 + \mu)R_2}{R_1 + R_2} v_{IN}$$

● **EXERCISE 4.1**

In the circuit shown in Fig. 4.7 the current source is dependent on the voltage at node X. Find v_{OUT}. **Answer:** $v_{OUT} = R_1 I_0/(1 + gR_1)$.

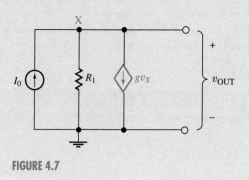

FIGURE 4.7

4.2 Amplifiers

We are now ready to introduce our first active circuit block, the amplifier. As we remarked earlier, the basic function of this block is to multiply an input signal by some constant factor, or, to put it another way, to create an output signal using instructions supplied by an input signal, which is usually smaller. Amplifiers are familiar to most of us from their use in phonographs, public-address systems, and so forth.

When a box marked "amplifier" is seen in a circuit diagram, it needs somehow to be interpreted. There is no way to analyze the operation of the circuit unless we are somehow told what the mysterious amplifier box actually *does*. For this purpose it is convenient to imagine that the amplifier can be replaced by its model. Once the amplifier box is replaced by a suitable model, circuit analysis can proceed in the usual way.

The question of which model is "suitable" is an interesting one. In fact, no model is a perfect imitation of the thing it supposedly represents. The equivalence is always approximate and usually is limited to a certain range of operating conditions. More complicated models, containing many ideal circuit elements, can generally be made more faithful than simple models. On the other hand, a complicated model adds complexity, and the extra accuracy may turn out not to be needed. The trick is to choose just the right model to do the job at hand.

A very simple amplifier model is shown in Fig. 4.8(a). When an amplifier is assumed to be described by this block, it is known as an *ideal voltage amplifier*. It is easy to see the reason for this name. The effect of the dependent voltage source is to make the output voltage larger than the input voltage by the constant factor A. The block simply multiplies the input voltage by A. Note that the two input wires do not go anywhere inside the amplifier block; they simply stop. This means that no current can flow through the input wires of the ideal amplifier. The input current of the ideal amplifier is zero.

Like all "ideal" devices, the ideal voltage amplifier does not really exist. However, the concept of an ideal voltage amplifier is nevertheless widely used in work with larger systems. Often the triangular symbol shown in Fig. 4.8(b) is used to represent it. The meaning is the same as for the model in Fig. 4.8(a): The input voltage is simply multiplied by the constant factor A. Note that only a single input and output connection is shown, as compared with two of each in Fig. 4.8(a). It is assumed that both v_{IN} and v_{OUT} are measured with respect to ground.

Although the model of Fig. 4.8 is convenient and easy to understand, it is too simple for most purposes. It is easy to see that it cannot be correct. For one thing, since the input current is zero, the input *power* (= voltage × current) has to be zero. That would mean that the *power amplification* of the block (output power divided by input power) would be infinitely large, which seems too good to be true. Furthermore, the output end of the block consists only of a bare voltage source. If this were really true, we could extract an infinitely large output power from the block. All we would have to do would be to connect a load resistance R across the output terminals. The power delivered to this load would be v_{OUT}^2/R, and by making R arbitrarily small, we could

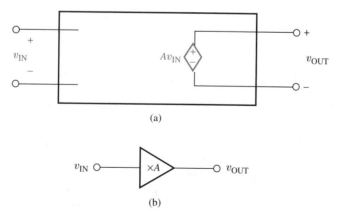

(a)

(b)

FIGURE 4.8 (a) Model for an ideal voltage amplifier. (b) Simplified symbol representing the same block.

make the output power arbitrarily large. Again, this seems a bit too good to be true. We need an improved model, one that leads to more reasonable results.

Amplifiers belong to the family known as *linear* devices. Such devices are described by linear equations, and in general have the property that voltages and currents are linearly proportional to each other. In keeping with this property, it is reasonable to describe the input part of the block by placing a resistor between the input terminals. Once this resistor, known as the *input resistance* of the block, has been introduced, it is no longer true that the input current of the block is zero. In fact, the input current is now given by Ohm's law: $I_{IN} = V_{IN}/R_i$, as shown in Fig. 4.9(a).

We also need a more realistic form for the output end of the block. Relying on our earlier experience with equivalent circuits, we can represent the output end by a Thévenin equivalent, as shown in Fig. 4.9(b). The resistor thus introduced is known as the block's *output resistance R_o*. The Thévenin voltage source is a dependent source controlled by the voltage across R_i. The constant factor A in this figure is now referred to as the *open-circuit voltage amplification*. The reason for this name is that when the output terminals of the block are open-circuited, (i.e., no load is connected), then the ratio of the output voltage to the input voltage is A. Note, however, that when a load is connected, this is no longer true because of the voltage-divider effect. For example, if a load resistance R_L is connected across the output terminals, and we choose $R_L = R_o$, then the ratio of the output voltage to the input voltage is no longer A, but only $A/2$.

We notice that the amplifier is completely characterized by the three parameters R_i, R_o, and A. If the values of these three constants are given (perhaps by the manufacturer of the amplifier), we are in a position to predict its behavior in various systems. The great advantage of the model is that all the complexities of the amplifier's internal

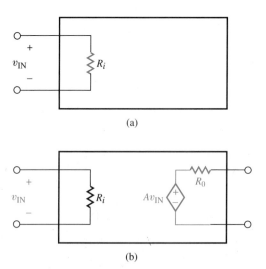

(a)

(b)

FIGURE 4.9   Evolution of the improved amplifier model.

construction become irrelevant. If the three parameters are known, that is enough to predict the behavior of the block. In many amplifiers, the values of the three parameters are truly fixed numbers, constants that always remain the same for that particular amplifier. However, the reader must be cautioned that there are cases in which this is not true. For example, there are times when the value of R_i is altered by the presence of a load resistance connected across the output terminals. This kind of behavior does not invalidate the model; one merely must be careful to use the right R_i, corresponding to the load that is connected.

To use the model, we simply replace the amplifier block with the model and apply conventional circuit analysis. Of course, one must be careful to connect each terminal of the model to the same place as the corresponding terminal of the original block. It should be noted that real amplifiers have other connections to the outside world besides the four shown in Fig. 4.9. In particular, there are always connections to some power source in order to bring in the power necessary to run the amplifier. Power-supply connections are usually not shown in circuit diagrams because it is assumed that they will always be made; showing these extra wires would only clutter up the diagram.

EXAMPLE 4.5

In Figure 4.10(a) the amplifier block can be represented by the model of Fig. 4.9(b). Find v_{OUT}.

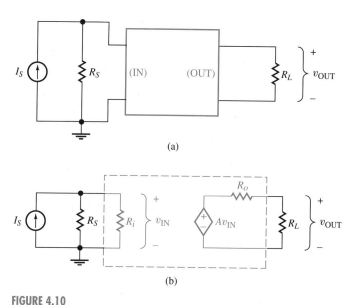

(a)

(b)

FIGURE 4.10

SOLUTION

When the model is substituted, the circuit is as shown in Fig. 4.10(b). Using the voltage-divider formula, we have

$$v_{OUT} = Av_{IN} \frac{R_L}{(R_o + R_L)}$$

However, we must still eliminate the unknown reference voltage v_{IN}. Since R_i and R_s are connected in parallel,

$$v_{IN} = I_s(R_s \parallel R_i) = I_s \frac{R_s R_i}{R_s + R_i}$$

Eliminating v_{IN}, we have

$$v_{OUT} = \frac{AR_L R_s R_i I_s}{(R_o + R_L)(R_s + R_i)}$$

VOLTAGE GAIN

The terms "voltage amplification" and "voltage gain" are often encountered; however, their meanings depend on how they are defined. One must take pains to define them carefully for the specific situation. For example, consider the circuit shown in Fig. 4.11(a), which, with the amplifier model substituted, becomes Fig. 4.11(b). In this circuit a signal voltage v_s is applied to the input. The output terminals are connected to a load resistance R_L. This is reasonable because in order to be useful the output of the amplifier needs to go into something else. (For example, in a music amplifier R_L might represent a loudspeaker.) We may now ask, what is the voltage gain* of this circuit? This question cannot be answered until we choose a definition of voltage gain. It seems reasonable to define a voltage gain G_V as the ratio of the voltage across the load to the signal voltage: $G_V \equiv v_L/v_s$. This is not a general definition of voltage gain; it is just a definition we choose to suit our needs in this particular case.

From Fig. 4.11(b) we see that with G_V defined in this way,

$$G_V = \frac{Av_{IN}R_L}{(R_0 + R_L)v_s} = \frac{AR_L}{R_0 + R_L}$$

It is interesting to note that $G_V \neq A$. The voltage gain is reduced by the voltage divider composed of R_L and R_o. This reduction of the load voltage is known as an output *loading effect*. One can remove the loading effect by taking away the load, as shown in Fig. 4.11(c). This is equivalent to setting $R_L = \infty$, in which case $G_V = A$. We have

---

* The terms "gain" and "amplification" are usually interchangeable.

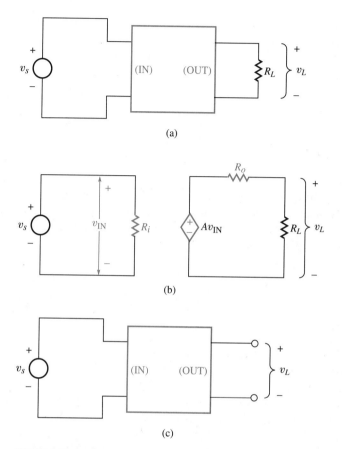

FIGURE 4.11

already mentioned that A is called the "open-circuit voltage amplification," and now we see why: When the output is open-circuited, the value of v_L/v_s is A. In practical cases, however, there is always a load connected, and if through bad planning we make $R_L \ll R_o$, the loss of load voltage can be quite severe. At any rate, the reader should note that the voltage amplification of the entire circuit is not necessarily the same as the amplification of the amplifier by itself.

CURRENT GAIN AND POWER GAIN

As with voltage gain, the terms "current gain" and "power gain" are ones that must be defined to fit the particular case. In Fig. 4.11(a) it is reasonable to define a current gain G_I to be the ratio of the current through R_L to the current through v_s. In that case

$$G_I = \frac{Av_{IN}}{R_o + R_L} \bigg/ \frac{v_{IN}}{R_i} = \frac{AR_i}{R_o + R_L}$$

The power gain G_p for this circuit can be defined (if we wish) as the ratio of the power delivered to the load to the power given out by the signal source v_s. The power delivered to the load is v_L^2/R_L; the power given out by v_s is the same as the power consumed in R_i, which is v_s^2/R_i. Thus

$$G_p = \frac{v_L^2/R_L}{v_s^2/R_i} = \frac{G_V^2 R_i}{R_L}$$

$$= \frac{A^2 R_L R_i}{(R_o + R_L)^2}$$

Whether one is interested most in voltage, current, or power gain depends on the particular application. It should be noted that the existence of power gain does not violate the law of energy conservation. The added power emerging from the output comes from the power source that powers the amplifier, even though the power-supply connections are usually not shown on the circuit diagram.

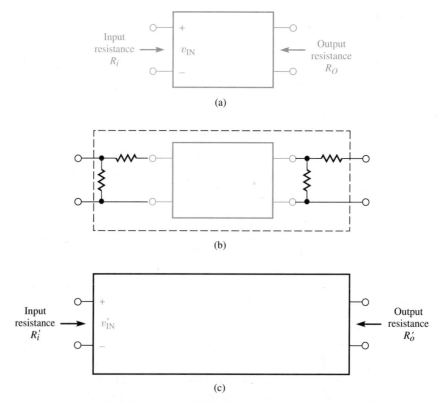

(a)

(b)

(c)

FIGURE 4.12   An amplifier block, with R_i, R_o, and A, is shown in (a). The block is used as part of a larger amplifier circuit, enclosed by the dashed line. (c) The larger circuit is now considered to be a *new* block, with R_i', R_o', and A_i'. Note that the primed quantities have values different from R_i, R_o, and A.

AMPLIFIERS AS SYSTEM COMPONENTS

The beauty of building blocks and their models is their convenience when they are used as parts of larger circuits or systems. Building blocks can be replaced by simple models, which makes analysis of the rest of the circuit easy. Moreover, for the designer, the use of building blocks is a powerful tool. One does not have to begin with individual resistors, transistors, and so forth; one starts with a predesigned and prefabricated building block. Much of the design work has already been done, and furthermore, by combining blocks that are already complex subsystems, one can build up larger and more powerful systems.

Often it happens that a block, such as an amplifier block, is used as part of a larger circuit, and then *that larger circuit is also considered to be a block*. This process is illustrated in Fig. 4.12. In Fig. 4.12(a) we have an amplifier block, whose three parameters are R_i, R_o, and A. In 4.12(b), the block is being used as part of a larger circuit. The larger circuit, enclosed by a dashed line, also has two input terminals and two output terminals. Therefore we can think of the larger circuit as a block in its own right. The new block is shown in 4.12(c). The new block also has its three parameters, but they will be different from those of the original block. Therefore we put "primes" on them, and call them R_i', R_o', and A'. Although these numbers are different from R_i, R_o, and A, they can be calculated through circuit analysis. Note that the dependent source of the larger block is controlled by the voltage between the input terminals of the larger block. This voltage, called V'_{IN}, is not the same as V_{IN}.

As an example, consider the circuit of Figure 4.13(a), where the amplifier block is that of Fig. 4.10(b). Here the dashed box is an amplifier *circuit*, which contains an amplifier block as an internal component. The input resistance of the circuit, as seen

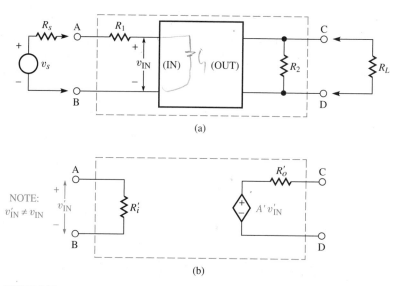

(a)

(b)

FIGURE 4.13

looking into terminals A,B is R_i', and so forth. We are interested in finding R_i', R_o', and A', when R_i, R_o, and A are given. For purposes of calculation, we may imagine a signal source v_s, a resistance R_s connected to the input, and a load resistance R_L to the output, as shown. The reader should remember that in general R_i' can depend on R_L, and R_o' can depend on R_s. It happens that they do not, in this simple example, but cases in which they do will be encountered.

It is easily seen that the input resistance looking into terminals A,B is

$$R_i' = R_1 + R_i \tag{4.1}$$

To find the output resistance, recall that it is the same as the Thévenin resistance for terminals C,D. Turning off independent source v_s, we see that v_{IN} is zero, and thus the dependent source in the amplifier block has the value zero, which means it acts as a short circuit. Looking to the left from C,D we see R_2 and R_o in parallel, so that

$$R_o' = \frac{R_2 R_o}{R_2 + R_o} \tag{4.2}$$

The open-circuit voltage gain of the larger circuit is $A' = v_{CD}/v_{AB}$ with no load connected. (Note that it is *not* v_{CD}/v_s. The input voltage for the larger circuit is located between its input terminals A,B.) In this case we have

$$v_{CD} = \frac{R_2}{R_o + R_2} A v_{IN}$$

$$v_{IN} = \frac{R_i}{R_i + R_1} v_{AB}$$

Thus

$$A' = \frac{v_{CD}}{v_{AB}} = \frac{A R_2 R_i}{(R_o + R_2)(R_i + R_1)} \tag{4.3}$$

Now we can represent the circuit in the dashed box by its own model, shown in Fig. 4.13(b). The parameters R_i', R_o', and A' are as found in Eqs. (4.1) to (4.3). In any further calculations the dashed box can simply be replaced by the model shown in Fig. 4.13(b).

Let us summarize the procedures for finding R_i', R_o', and A'. (These procedures apply to any two-port linear circuit.) Let the input terminals be A,B and the output terminals be C,D. Then

1. The input resistance is found by connecting a load resistance R_L between C and D. Then R_i' is the resistance that is seen looking into terminals A,B. (Note that the value of R_i' may be affected by R_L.)

2. The output resistance is found by first connecting a Thévenin signal source v_s, R_s to terminals A,B. We then find the Thévenin resistance at terminals C,D, which is equal to R_o'. We can evaluate R_o' either by leaving v_s turned on and using $R_o' = -v_{OC}/i_{SC}$, or by turning off v_s and finding the resistance seen looking into C,D. If v_s is turned off, all dependent sources still continue to function. The value of R_o' may be affected by R_s.

3. To find the open-circuit voltage amplification, we apply an ideal voltage source v_s to A,B, find v_{OC}, the voltage at C,D with no load connected, and use $A' = v_{OC}/v_s$.

EXAMPLE 4.6

Figure 4.14(a) shows a circuit containing an amplifier block. The block is represented by the model of Fig. 4.10(b). Find the open-circuit voltage amplification of the circuit.

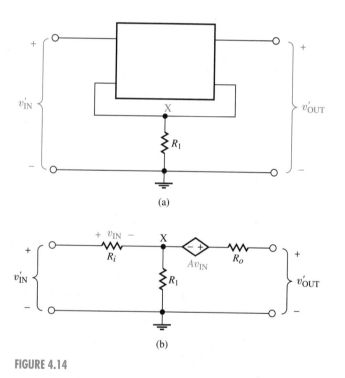

(a)

(b)

FIGURE 4.14

SOLUTION·

After substitution of the model, the resulting circuit is shown in Fig. 4.14(b). (Note the difference between v'_{IN} and v_{IN}: v_{IN} is the input voltage of the amplifier block, and v'_{IN} is the input voltage of the larger circuit.) Since the output of the circuit is open-circuited, there is no current through R_o, and v_x can be found from the voltage-divider formula:

$$v_x = \frac{R_1}{R_1 + R_i}v'_{IN}$$

Also, we see that

$$v'_{OUT} = v_x + Av_{IN}$$

$$= v_x + A(v'_{IN} - v_x)$$

Solving, we have

$$A' = \frac{v'_{OUT}}{v'_{IN}} = \frac{R_1 + AR_i}{R_1 + R_i}$$

• EXERCISE 4.2

For the circuit in Example 4.6, find the input resistance R'_i. The output terminals are connected to a load resistance R_L. **Answer:**

$$R'_i = \frac{(R_o + R_L)(R_1 + R_i) - (A - 1)R_1R_i}{R_o + R_L + R_1}$$

4.3 The Operational Amplifier

In analog work, the most common and widely useful building block is the one known as the *operational amplifier,* or "op-amp." Op-amps are useful because they are inexpensive, in the form of integrated circuits, and also versatile: They can perform many useful operations.

An operational amplifier is exactly the same sort of block as described in Section 4.2. What then makes it special? An op-amp is distinguished from other amplifiers by the values of the three parameters of its model. For an op-amp, the input resistance R_i is always very large, at least 100,000 Ω and often more. The open-circuit voltage amplification A is also very large, on the order of 10^5. And the output resistance R_o is always small, on the order of 30 Ω. Regardless of how it is constructed, any amplifier with parameters of this order can be regarded as an op-amp.

In circuit diagrams, the op-amp is usually represented by a special triangular symbol, as shown in Fig. 4.15(a). This symbol has two input terminals, marked ($-$) and ($+$) and known as the "inverting" and "noninverting" inputs. (If the ($-$) and ($+$) signs are not shown on the diagram, the ($-$), or inverting, input is conventionally the one on top.) The single output terminal is at the pointed right side of the block. We shall refer to the voltage at the inverting input as $v_{(-)}$ and that at the noninverting input as $v_{(+)}$. Both of these voltages are measured with respect to ground. In addition to the two input connections and the output connection, the op-amp will normally have at least two additional connections to the outside, through which power is supplied from an external voltage source. Since the power connections are required for operation, they

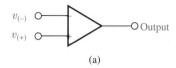

(a)

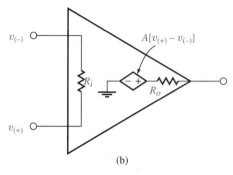

(b)

FIGURE 4.15 Operational amplifier. (a) Circuit symbol; (b) Model.

are always assumed to be there, and thus do not have to be shown in the circuit diagram.

The model of the op-amp is shown in Fig. 4.15(b). In general, one can analyze the operation of any circuit containing an op-amp by substituting the model for the op-amp, and then performing conventional circuit analysis. There is, however, a shortcut method that can be used to understand most op-amp circuits. This approach takes advantage of the fact that, for op-amps, R_i and A are always very large, and R_o is always very small. In essence, the shortcut method makes the approximation that R_i and A are *infinitely* large and that R_o is *infinitely* small. An op-amp with those properties is referred to as an *ideal op-amp*, and the simplified method is called the *ideal op-amp technique*. Of course, like all ideal circuit elements the ideal op-amp does not really exist. But provided that the op-amp parameters are in the expected range, the predictions of the ideal op-amp technique are usually close enough. More will be said about the bases for this technique when we discuss feedback systems in Chapter 9.

THE IDEAL OP-AMP TECHNIQUE

In the ideal op-amp technique, the circuit is analyzed under the following two assumptions:

ASSUMPTION 1

The potential difference between the op-amp input terminals, $v_{(+)} - v_{(-)}$, equals zero.

ASSUMPTION 2

The currents flowing into the op-amp's two input terminals both equal zero.

Neither of these statements is precisely true. For instance, if Assumption 1 were strictly correct, one would expect, according to Fig. 4.15(b), that the op-amp output voltage would always be zero. However, these two statements are *almost* true; hence, if the rest of the circuit is analyzed using these two assumptions, accurate results are usually obtained.

Before proceeding to the use of these assumptions, let us discuss their origins. First, in a correctly designed op-amp circuit, the output voltage of the op-amp will always lie within a certain range, defined by the capabilities of the amplifier. As a practical matter, the absolute values of the output voltage cannot become larger than that of the power supply voltage; that is, $|v_{OUT}| < V_{SUPPLY}$. The value of $|v_{OUT}|$, however, is given by $A|v_{(+)} - v_{(-)}|$. Thus we have

$$|v_{(+)} - v_{(-)}| < \frac{V_{SUPPLY}}{A} \tag{4.4}$$

If A is large, then $|v_{(+)} - v_{(-)}|$ must, according to Eq. (4.4), be small; hence the first assumption.

The second assumption follows from the first, coupled with the assumption that R_i is large. We see that the input current $(v_{(+)} - v_{(-)})/R_i$ will be a very small quantity; hence the approximation that it is equal to zero.

Let us now demonstrate the use of the ideal op-amp technique. Figure 4.16 shows an op-amp *circuit*, that is, a circuit containing an op-amp and a few other components. We wish to find the output voltage v_{OUT}. We observe that $v_{OUT} = v_{(-)}$. According to Assumption 1, $v_{(-)} = v_{(+)}$. According to Assumption 2, there is no current flowing into the (+) input terminal; thus there is no voltage drop in R_S, and $v_{(+)} = v_{IN}$. Thus $v_{OUT} = v_{IN}$. Note that in this case the approximate technique gives an estimate of the output voltage with no algebraic computation at all!

In order to verify that this result, obtained with almost no effort, is close to correct, we can calculate v_{OUT} a second time by substituting the model. This should be a less approximate calculation, since we shall use finite values for the op-amp parameters, rather than assuming they are infinite or zero, as is effectively done in the ideal op-amp technique. However, we will then need the values of the op-amp parameters. Suppose the op-amp manufacturer tells us that $A = 10^5$ and $R_i = 10^4\ \Omega$. To simplify the calculations, we shall assume that $R_o = 0$.

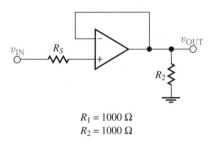

$R_1 = 1000\ \Omega$
$R_2 = 1000\ \Omega$

FIGURE 4.16

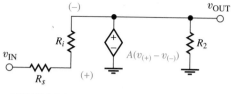

FIGURE 4.17

We now proceed by substituting the model where the op-amp appears in the circuit. The result is as shown in Fig. 4.17. We wish to solve for v_{OUT} as a function of v_{IN}. To proceed, we write a node equation at the node labeled $(+)$:

$$\frac{v_{IN} - v_{(+)}}{R_S} + \frac{v_{OUT} - v_{(+)}}{R_i} = 0$$

However, v_{OUT} is also related to $v_{(+)}$ and $v_{(-)}$ by the relationship $A(v_{(+)} - v_{(-)}) = v_{OUT}$; furthermore, in this circuit $v_{(-)} = v_{OUT}$. Thus

$$v_{OUT} = A(v_{(+)} - v_{OUT})$$

Solving these two equations simultaneously, we obtain

$$v_{OUT} = v_{IN} \frac{A}{A + 1} \frac{R_i}{R_i + R_S/(1 + A)}$$

Evaluating v_{OUT}/v_{IN} according to this result, we obtain $v_{OUT}/v_{IN} = 0.999989$.

The prediction of the ideal op-amp technique was $v_{OUT}/v_{IN} = 1$. Clearly, the two methods are in excellent agreement, but the ideal op-amp calculation is much easier!

EXAMPLE 4.7

Use the ideal op-amp technique to find the closed-loop voltage amplification A' for the circuit shown in Fig. 4.18.

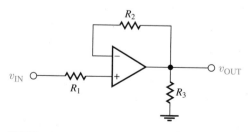

FIGURE 4.18

SOLUTION

No current flows through either resistor R_1 or R_2 because of Assumption 2. Therefore $v_{\text{IN}} = v_{(+)}$ and $v_{\text{OUT}} = v_{(-)}$. However, $v_{(+)} = v_{(-)}$ because of Assumption 1. Thus $v_{\text{OUT}} = v_{\text{IN}}$ and $A' = 1$.

● EXERCISE 4.3

Use the ideal op-amp technique to find v_{OUT} in the circuit shown in Fig. 4.19.
Answer: $v_{\text{OUT}} = -I_0 R_F$.

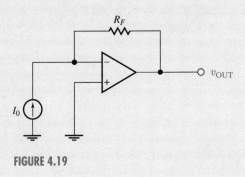

FIGURE 4.19

LIMITATIONS OF THE IDEAL OP-AMP TECHNIQUE

The ideal op-amp technique is based on the approximations $A \cong \infty$, $R_i \cong \infty$, and $R_o \cong 0$. It works whenever the precise values of A, R_i, and R_o have negligible influence on the answer. However, if one tries to calculate something that *is* influenced by the op-amp parameters, the ideal op-amp technique will give a result that is inexact. For instance, suppose we wish to find the input resistance R_i' for the circuit shown in Fig. 4.20, bearing in mind that R_i', the input resistance of the circuit, is in general different from R_i, the input resistance of the op-amp by itself. The value of R_i' can be calculated by substituting the op-amp model; the calculation will be performed below [Eqs. (4.7) to (4.10)]. The result is

$$R_i' = R_i \left(\frac{(A + 1)R_L + R_o}{R_o + R_L} \right) + \frac{R_o R_L}{R_o + R_L} \tag{4.5}$$

On the other hand, if we apply the ideal op-amp technique, we simply conclude that $R_i' = v_{\text{IN}}/i_{\text{IN}} = \infty$, since $i_{\text{IN}} = 0$ according to Assumption 2. The answer $R_i' = \infty$

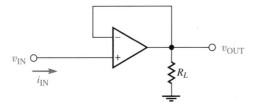

FIGURE 4.20

does not agree with Eq. (4.5) and is therefore inexact. It does correctly indicate the order of magnitude of R_i'; R_i' is in fact very large. This can be seen if we use typical values for R_i, A, and R_o in Eq. (4.5) and let $R_L \cong 1,000\ \Omega$. In that case Eq. (4.5) reduces to $R_i' \cong AR_i \cong 10^{10}\ \Omega$, which, though not infinite, is certainly very large. For most purposes, the knowledge that R_i' is very large is sufficient; if it is not, one must use Eq. (4.5).

How can one know whether the ideal op-amp technique is suitable for a calculation? A good method is to simply apply it on a trial basis. If it leads to a result of infinity (or zero) we can conclude only that the quantity being calculated is very large (or very small); to get the exact value, a calculation must be based on substitution of the op-amp model. On the other hand, if the ideal op-amp technique gives a finite answer, then the quantity being calculated does not depend on the op-amp parameters; in that case answers from the ideal op-amp technique will be quite accurate for practical purposes. A useful tip: The *output voltage* can always be found from the ideal-op-amp technique in the usual op-amp feedback circuits.

The ideal op-amp technique also assumes that R_i is very large, and R_o very small, compared to other resistances in the circuit. Thus in order to be sure that the ideal op-amp technique is accurate, it is necessary that all resistors in the op-amp circuit be in the range $R_o << R << R_i$. With high-quality op-amps the range $500\ \Omega < R < 0.5\ \mathrm{M}\Omega$ should give high accuracy. The need for this restriction can be seen from Fig. 4.20. For this circuit the ideal op-amp technique predicts $A' = 1$, a result which is very accurate so long as $R_L > 500\ \Omega$. On the other hand, it is clear that A' decreases to zero if R_L approaches zero, since in this case the output terminals become short-circuited.

Various practical limitations also affect op-amp performance. These limitations will be discussed later (Section 10.4).

4.4 Operational Amplifier Circuits

In this section several common op-amp circuits are described. In general, the function of each of these circuits is to perform a specific operation on the signal presented to its input. By using these relatively simple circuits, a designer can set up large signal-processing systems quickly and easily.

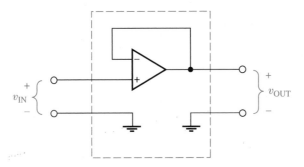

FIGURE 4.21

The op-amp circuits themselves are building blocks. For example, consider the very simple circuit shown in Fig. 4.21, known as the *voltage follower*. A dashed line has been drawn around the circuit to remind us that the voltage follower is itself an amplifier block. Thus it can be represented by our standard amplifier model, Fig. 4.22. The parameters of the model representing the op-amp circuit will be called R_i', R_o', and A'. (The primes are used to separate these quantities from R_i, R_o, and A, which are the parameters of the op-amp by itself.) When we have found R_i', R_o', and A', the op-amp circuit will be fully characterized; we can then predict its behavior using the model of Fig. 4.22.

HOW TO FIND THE CIRCUIT PARAMETERS (R_i', R_o', AND A')

The significance of the circuit parameters R_i', R_o', A' and the way these differ from the op-amp parameters R_i, R_o, A have already been explained. However, it is important to understand how these numbers can be calculated. The steps in the calculation will be summarized here.

The input resistance R_i' is the resistance that would be measured from outside the block, across the input terminals. One way to calculate R_i' is to imagine a "test voltage" v_{TEST} applied to the input terminals, as shown in Fig. 4.23(a). If we designate as i_{TEST} the current that flows in response to v_{TEST} in the direction shown, and calculate its value, we then obtain R_i' from $R_i' = v_{TEST}/i_{TEST}$. Since the value of R_i' may be affected by a load resistance R_L connected across the output terminals, the calculation is made

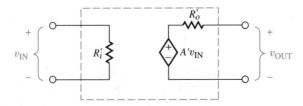

FIGURE 4.22

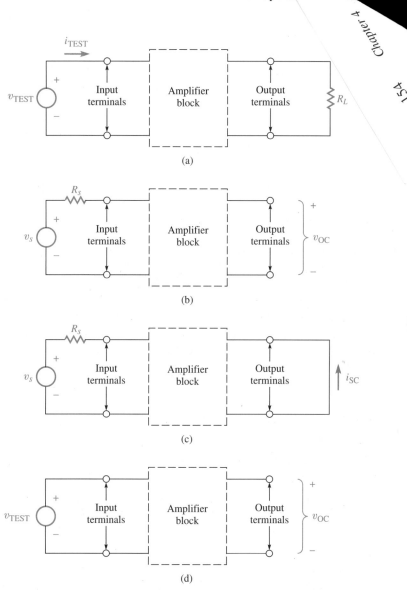

FIGURE 4.23 Test circuits for determining the parameters of the amplifier model. (a) Circuit for finding R_i': $R_i' = v_{TEST}/i_{TEST}$. (b, c) Circuits for finding R_o'. Values of v_{OC} and i_{SC} are found using the circuits of parts (b) and (c), respectively; then $R_o' = -v_{OC}/i_{SC}$. (d) Circuit for determining the voltage amplification; $A' = v_{OC}/v_{TEST}$.

with R_L present, as shown. The value of R_i' will then, in general, be a function of R_L, although the effect of the latter is usually negligible.

Figures 4.23(b) and (c) show circuits that may be used for calculating R_o'. The latter is equal to the Thévenin resistance of the amplifier block as seen from its output terminals. Thus $R_o' = -v_{OC}/i_{SC}$, where v_{OC} and i_{SC} are, respectively, the open-circuit voltage and the short-circuit current at the output terminals, as indicated in Fig. 4.23(b) and (c). In calculating v_{OC} and i_{SC}, a signal source including a source resistance R_S is assumed to be connected at the input terminals, as shown. The value of R'_o that is calculated will in general be a function of R_S, although the effect of R_S (like that of R_L upon R'_i) is usually negligible.

Finally, the open-circuit voltage amplification is obtained using the test circuit of Fig. 4.23(d). Here v_{OC} is the output voltage with no load, when the voltage across the input terminals is v_{TEST}; then $A' = v_{OC}/v_{TEST}$. The ideal op-amp technique can be used.

We shall now proceed to discuss the properties of some important op-amp circuits.

THE VOLTAGE FOLLOWER

The voltage-follower circuit, which has already been considered in various examples, is shown in Fig. 4.24. To find v_{OUT} using the ideal op-amp technique, we note that $v_{OUT} = v_{(-)}$ and that $v_{IN} = v_{(+)}$. But since, according to Assumption 1, $v_{(-)} = v_{(+)}$, we know that $v_{OUT} = v_{IN}$. Thus we have $A' \cong 1$. The "exact" value of A' can be found by substituting the op-amp model. The result is

$$A' = \frac{R_o + AR_i}{R_o + (A + 1)R_i} \tag{4.6}$$

This is clearly very close to unity when the op-amp parameters have their usual values.

The values of R'_i and R'_o for the voltage follower cannot be found by the ideal op-amp technique because R'_i and R'_o are not independent of the op-amp parameters. (Usually only the voltages in an op-amp circuit have this convenient property.) We may, however, fall back on the use of the op-amp model. To find R'_i, we use the method of Fig. 4.23(a). The voltage-follower circuit, Fig. 4.24, is inserted for the amplifier block in Fig. 4.23(a), and the op-amp model of Fig. 4.15(b) is inserted in place of the op-amp itself. The resulting circuit is as shown in Fig. 4.25. We may find the value of i_{TEST} by loop analysis. Let us call the two mesh currents i_1 and i_2; then it is clear that $i_{TEST} = i_1$. The two loop equations are

$$v_{TEST} - i_{TEST}R_i - R_o(i_{TEST} - i_2) - A(v_{(+)} - v_{(-)}) = 0 \tag{4.7}$$

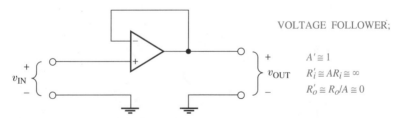

VOLTAGE FOLLOWER;

$$A' \cong 1$$
$$R'_i \cong AR_i \cong \infty$$
$$R'_o \cong R_o/A \cong 0$$

FIGURE 4.24 The voltage follower circuit.

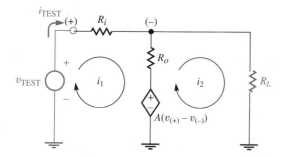

FIGURE 4.25

and

$$A(v_{(+)} - v_{(-)}) - R_o(i_2 - i_{\text{TEST}}) - i_2 R_L = 0 \tag{4.8}$$

The quantities $v_{(+)}$ and $v_{(-)}$ may be eliminated by $v_{(+)} = v_{\text{TEST}}$ and $v_{(-)} = v_{\text{TEST}} - i_{\text{TEST}} R_i$. Solving, we find

$$i_{\text{TEST}} = \frac{v_{\text{TEST}}}{R_i(1 + A) + R_o\left(1 - \dfrac{R_o + AR_i}{R_o + R_L}\right)} \tag{4.9}$$

so that

$$R_i' = \frac{v_{\text{TEST}}}{i_{\text{TEST}}} = R_i \frac{(A + 1)R_L + R_o}{R_o + R_L} + \frac{R_o R_L}{R_o + R_L} \tag{4.10}$$

From Eq. (4.10) it is readily seen that R_i' is always larger than R_i and can be very much larger. For example, in the usual case of $R_L \gg R_o$

$$R_i' \cong AR_i \quad \text{(voltage follower)} \tag{4.11}$$

The voltage follower therefore can have a very large input resistance, typically on the order of $10^{10}\ \Omega$. Note that the value of R_i' is dependent on R_i; thus the ideal op-amp technique cannot be used to find R_i'.

A simple physical argument is helpful in understanding why R_i' is so much larger than R_i. The current i_{TEST} is equal to $(v_{(+)} - v_{(-)})/R_i$. If the potential at the $(-)$ terminal were zero, then we would have $i_{\text{TEST}} = v_{(+)}/R_i = v_{\text{TEST}}/R_i$, and R_i' would equal R_i. This is not the case, however; because of the near-unity voltage gain of the circuit, $v_{(-)}$ $(= v_{\text{OUT}})$ is nearly equal to $v_{(+)}$ $(= v_{\text{IN}})$. Thus $v_{(+)} - v_{(-)}$ always has a very small value, and i_{TEST}, which equals $(v_{(+)} - v_{(-)})/R_i$, is much smaller than $v_{(+)}/R_i$. As a consequence of the small value of i_{TEST}, $R_i' = v_{\text{TEST}}/i_{\text{TEST}}$ has a very large value.

We next calculate R_o' using $R_o' = -v_{\text{OC}}/i_{\text{SC}}$, where v_{OC} is obtained using the circuit of Figs. 4.23(b) and i_{SC} is obtained using the circuit of Fig. 4.23(c). After substituting the voltage-follower circuit into Fig. 4.23(b) and (c), using the op-amp model, and solving for v_{OC} and i_{SC}, we obtain

$$R_o' = \frac{R_o(R_i + R_S)}{R_o + (1 + A)R_i + R_S} \tag{4.12}$$

We note that in the usual case of $AR_i \gg R_o$, $R_i \gg R_S$, this equation reduces to the approximate result

$$R_o' \cong \frac{R_o}{A} \quad \text{(voltage follower)} \tag{4.13}$$

Thus the voltage follower typically has an extremely low output resistance, on the order of 10^{-3} Ω! This remarkably low output resistance is caused by the feedback in the circuit.*

Since $A' \cong 1$ for the voltage follower, it evidently cannot be used to amplify voltages. However, because of its high input resistance and low output resistance, the circuit is useful as a separator, or *buffer*, that can be interposed between two parts of the circuit, when it is desired that the parts not interact. This technique is particularly useful when a low-resistance load is to be driven by a source with high source resistance.

EXAMPLE 4.8

This example demonstrates the use of the voltage follower as a buffer. Let us consider a signal source, represented by a Thévenin equivalent with Thévenin voltage v_S and Thévenin resistance $R_S = 1,000$ Ω. Find the voltage v_L that appears across a load resistor $R_L = 10$ Ω as follows:
(a) When R_L is connected directly to the source.
(b) When a voltage follower is interposed between the source and R_L.
Assume that the properties of the op-amp are $R_i = 10$ kΩ, $R_o = 100$ Ω, and $A = 10^4$.

SOLUTION

(a) The situation is as shown in Fig. 4.26(a). Using the voltage-divider formula, we see immediately that $v_L = R_L v_S/(R_S + R_L) \cong v_S/100$.

---

* One should not conclude that since the output resistance is so low, thousands of amperes could be drawn from the output terminals. Independent of its other characteristics, a given op-amp will have a maximum output current, typically 10 to 50 mA.

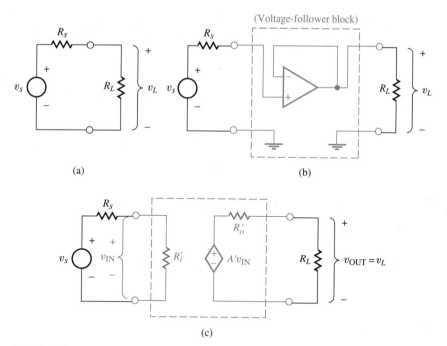

FIGURE 4.26

(b) The circuit is as shown in Fig. 4.26(b). The voltage-follower block may be represented by the amplifier circuit model of Fig. 4.22. Substituting the model for the block, we have Fig. 4.26(c). It is now readily seen, by means of the voltage-divider formula, that $v_{IN} = v_S R_i'/(R_i' + R_S)$ and that $v_L = A' v_{IN} R_L/(R_L + R_o')$. Therefore

$$v_L = \frac{A' R_L}{R_L + R_o'} \frac{v_S R_i'}{R_i' + R_S}$$

We recall that for the voltage follower, $A' \cong 1$. If $R_i' \gg R_S$ and $R_L \gg R_o'$ (which will be verified shortly), then $v_L \cong v_S$. In part (a) the result was $v_L = v_S/100$; v_L was reduced by a factor of 100 due to the loading effect of R_L upon the voltage source. Part (b) shows that through the interposition of the buffer stage, the loading effect is removed, and the load voltage is as high as the voltage of the source when the latter is unloaded. This can be a very useful technique!

It still remains to be verified that $R_i' \gg R_S$ and $R_L \gg R_o'$ for this particular case. Substituting the given parameter values into Eq. (4.10), we find that $R_i' \cong R_i(AR_L/R_o) = 10^3 R_i = 10$ MΩ. Similarly, from Eq. (4.12), we have $R_o' \cong R_o/A = 10^{-2}$ Ω. Thus indeed we have $R_i' \gg R_S$ and $R_L \gg R_o'$, so that $v_L = v_S$ almost precisely. In this case the voltage follower is a nearly ideal buffer.

EXAMPLE 4.9

Find the power amplification of the voltage-follower block in Fig. 4.26(b).

SOLUTION

The power entering the input terminals of the voltage-follower block is the product of the voltage across its input terminals times the current flowing into them. The former is $v_{IN} \cong v_S$ (since $R_i' \gg R_S$), and the latter is $i_{IN} \cong v_S/R_i'$. Thus

$$P_{IN} \cong \frac{v_S^2}{R_i'}$$

The power leaving the amplifier is the same as the power dissipated in R_L, which is v_L^2/R_L. Since $v_L \cong v_S$,

$$P_{OUT} = \frac{v_S^2}{R_L}$$

The power amplification is therefore given by

$$A_P = \frac{P_{OUT}}{P_{IN}} = \frac{R_i'}{R_L} = \frac{10^7}{10} = 10^6$$

Thus we see that notwithstanding its voltage gain of unity, the power gain of the voltage follower can be very large.

The benefits of the voltage-follower circuit are that (1) input resistance has a very high value, preventing loading of the preceding stage; (2) output resistance is very low, preventing loading by the following stage; (3) power gain is obtained.

THE NON-INVERTING AMPLIFIER

In cases in which voltage gain is desired, the *noninverting amplifier* of Fig. 4.27 may be used. This circuit resembles the voltage follower, except that a voltage divider is inserted in the feedback path. Using the ideal op-amp technique, estimation of A' is quite simple. From Assumption 1 we postulate $v_{(-)} = v_{IN}$; from Assumption 2 we postulate that no current flows from the node between R_1 and R_F into the amplifier terminal. Thus we can write a node equation for $v_{(-)}$ ($= v_{IN}$):

$$\frac{v_{IN}}{R_1} + \frac{v_{IN} - v_{OUT}}{R_F} = 0 \tag{4.14}$$

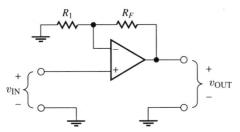

Noninverting amplifier:

$$A' \cong \frac{R_1 + R_F}{R_1}$$

$$R_i' \cong AR_i \frac{R_1}{R_1 + R_F} \cong \infty$$

$$R_o' \cong \frac{R_o}{A} \frac{R_1 + R_F}{R_1} \cong 0$$

FIGURE 4.27

Solving, we have

$$v_{\text{OUT}} = \frac{R_1 + R_F}{R_1} v_{\text{IN}} \tag{4.15}$$

so that

$$A' = \frac{R_1 + R_F}{R_1} \quad \text{(non-inverting amplifier)} \tag{4.16}$$

The accuracy of this result can be verified by the op-amp model; however, since voltages are predicted accurately by the ideal op-amp technique, there is not much to be gained by doing this.

The input resistance may be calculated by the op-amp model and Fig. 4.23(a), just as was done for the voltage follower:

$$R_i' = R_i \left(1 + \frac{AR_1R_L}{R_L(R_1 + R_F) + R_o(R_L + R_1 + R_F)} \right)$$

$$+ R_1 \frac{R_F + (R_o \| R_L)}{R_1 + R_F + (R_o \| R_L)} \tag{4.17}$$

It is easily seen that $R_i' > R_i$. Essentially the same physical explanation applies here as was used to explain the large R_i' of the voltage follower. For the important case of $R_i \to \infty$, $R_o \to 0$, Eq. (4.17) implies that

$$R_i' \cong AR_i \frac{R_1}{R_1 + R_F} \quad \text{(non-inverting amplifier)} \tag{4.18}$$

which is smaller than the value of R_i' for the voltage follower by the factor $(R_1 + R_F)/R_1 = A'$. The output resistance, obtained using the op-amp model, is given by the formidable expression

$$R_o' = R_o \frac{R_F[R_1R_F + (R_1 + R_F)(R_i + R_S)]}{(R_o + R_F)[R_1R_F + (R_1 + R_F)(R_i + R_S)] + R_1[AR_iR_F - R_o(R_i + R_S)]} \tag{4.19}$$

Upon inspection it can be seen that $R_o' < R_o$, and that in the important case $R_i \to \infty$, we have

$$R_o' \cong \frac{R_o}{A} \frac{R_1 + R_F}{R_1} \quad \text{(non-inverting amplifier)} \qquad (4.20)$$

Thus this circuit offers nearly the same advantages of high input resistance and low output resistance as the voltage follower.

EXAMPLE 4.10

A multirange electronic voltmeter is to have ranges of 0 to 0.01, 0 to 0.1, and 0 to 1 V. A precision 0 to 1 V panel-mounting voltmeter and an op-amp are to be used. Design a suitable circuit.

SOLUTION

To obtain a voltmeter circuit that covers the range 0 to 0.01 V, we shall need to insert an amplifier with a voltage gain of 100 between the measuring terminals and the 0 to 1 V voltmeter. Then voltages in the range 0 to 0.01 V at the measuring terminals will be converted to voltages in the 0 to 1 V range at the meter. A reading of 0.6 V on the meter, for example, would then be interpreted as a measurement of 0.006 V. To cover the other two ranges, amplifiers with gains of 10 and unity, respectively, should be used.

Because of its high input resistance, the non-inverting amplifier circuit is well-suited to this application. A possible arrangement is given in Fig. 4.28.

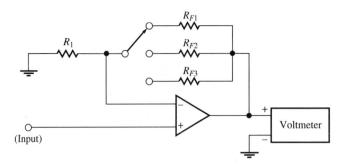

FIGURE 4.28

Let us arbitrarily choose $R_1 = 1,000 \ \Omega$. Then for a gain of 100 we need

$$\frac{R_1 + R_{F1}}{R_1} = 100$$

$$R_{F1} = 99R_1 = 99,000 \ \Omega$$

Similarly, for gains of 10 and unity, we find $R_{F2} = 9,000\ \Omega$ and $R_{F3} = 0$, respectively.

The input resistance of the circuit we have designed will be very high—perhaps 10^9 to 10^{10} Ω. Thus it will function as an almost ideal voltmeter.

THE INVERTING AMPLIFIER

Another circuit capable of providing voltage amplification is shown in Fig. 4.29. This circuit is known as an *inverting amplifier* because the output has the opposite sign from the input. The voltage amplification is easily found with the ideal op-amp technique. From Assumption 1 the voltage at the $(-)$ input terminal is taken to be zero. We write a node equation for this point, postulating, from Assumption 2, that no current enters the amplifier terminal. This equation is

$$\frac{v_{\text{IN}}}{R_1} + \frac{v_{\text{OUT}}}{R_F} = 0 \tag{4.21}$$

from which we have

$$A' = \frac{v_{\text{OUT}}}{v_{\text{IN}}} = -\frac{R_F}{R_1} \quad \text{(inverting amplifier)} \tag{4.22}$$

Unlike the voltage follower and noninverting amplifier, the input resistance of the inverting amplifier can be accurately estimated using the ideal op-amp technique. This is because the input resistance of this circuit is not determined by R_i but rather by R_1. From Assumption 1 the voltage at the $(-)$ input terminal of the op-amp is always nearly zero. Thus when v_{TEST} is applied, i_{TEST} is simply v_{TEST}/R_1. Consequently $R_i' \cong R_1$.

The output resistance, calculated using the op-amp model, is

$$R_o' = R_o \frac{R_F[R_iR_F + (R_i + R_F)(R_1 + R_S)]}{(R_F + R_o)[R_iR_F + (R_i + R_F)(R_1 + R_S)] + R_i(R_1 + R_S)(AR_F - R_o)} \tag{4.23}$$

Inspection of this result indicates that $R_o' < R_o$. In the limit $R_i \to \infty$, $A \to \infty$, it becomes

$$R_o' \cong \frac{R_o}{A} \frac{(R_F + R_1 + R_S)}{(R_1 + R_S)} \tag{4.24}$$

(The simpler result in Fig. 4.29 is for the case $R_S = 0$.)

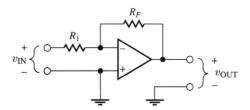

Inverting amplifier:

$$A' \cong -\frac{R_F}{R_1}$$

$$R_i' \cong R_1$$

$$R_o' \cong \frac{R_o}{A} \frac{R_1 + R_F}{R_1} \cong 0$$

FIGURE 4.29

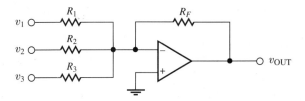

Summing amplifier:

$$v_{OUT} = -\left(\frac{R_F}{R_1}\right)v_1 - \left(\frac{R_F}{R_2}\right)v_2 - \left(\frac{R_F}{R_3}\right)v_3$$

FIGURE 4.30

The inverting amplifier is not as useful a buffer as the noninverting amplifier because of its lower input resistance. However, it is the basis of the *summing amplifier*, a very useful circuit.

THE SUMMING AMPLIFIER

The circuit known as the summing amplifier is shown in Fig. 4.30. This circuit, which is essentially an inverting amplifier with multiple inputs, is capable of adding several input voltages together, each, if desired, with a different scale factor. The operation of the circuit can be demonstrated easily with the ideal op-amp technique. Writing a node equation for the $(-)$ input terminal, we have

$$\frac{v_1}{R_1} + \frac{v_2}{R_2} + \frac{v_3}{R_3} + \frac{v_{OUT}}{R_F} = 0 \tag{4.25}$$

Solving for v_{OUT}, we have

$$v_{OUT} = -\frac{R_F}{R_1}v_1 - \frac{R_F}{R_2}v_2 - \frac{R_F}{R_3}v_3 \tag{4.26}$$

Any number of inputs, of course, may be used. In other respects this circuit resembles the inverting amplifier. However, in the summing amplifier each input has its own input resistance, equal to the resistance through which it is connected to the op-amp (that is, R_1, R_2, or R_3 in Fig. 4.30). The output resistance is identical to the output resistance of the inverting amplifier except that $R_1 + R_S$ in Eq. (4.23) must now be replaced by the parallel combination of all the input resistances. That is, if in Fig. 4.30 the three inputs are fed by sources with Thévenin resistances R_{S1}, R_{S2}, and R_{S3}, then in Eq. (4.23) $R_1 + R_S$ must be replaced by the parallel combination of $(R_1 + R_{S1})$, $(R_2 + R_{S2})$, and $(R_3 + R_{S3})$.

EXAMPLE 4.11

Two sources of signals may be represented by Thévenin equivalents with Thévenin voltages v_{S1} and v_{S2} and Thévenin resistances R_{S1} and R_{S2}, both of which are somewhat variable but in the vicinity of 100 kΩ. We desire to apply the voltage $v_{S1} + 2v_{S2}$ to a load resistance of 100 Ω. Design a circuit to accomplish this. Available op-amps have $R_i = 100$ kΩ, $R_o = 100$ Ω, and $A = 10^5$.

SOLUTION

It is not desirable to connect the voltage sources directly to the inputs of the summing amplifier. The two input resistances of the summing amplifier would be on the order of R_1 and R_2 in Fig. 4.30, and in this case would be less than the source resistances R_{S1} and R_{S2}. The two resulting voltage dividers, composed, respectively, of R_1, R_{S1} and R_2, R_{S2}, would cause the output voltage to depend on the source resistances. Figure 4.31(a) illustrates this effect for the case of the v_{S1} input.

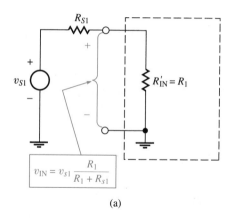

(a)

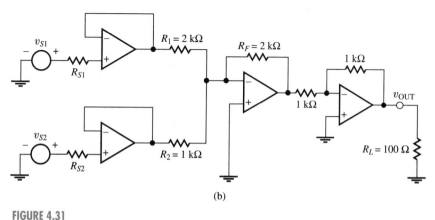

(b)

FIGURE 4.31

Since R_{S1} and R_{S2} have been stated to be variable, this effect would make the amplification of the circuit variable, which is highly undesirable.

A better way to handle this problem is to use buffer stages before the summing amplifier. A good circuit would be that shown in Fig. 4.31(b). In this circuit the load resistances presented to the outputs of the two voltage followers are the two input resistances of the summer, R_1 and R_2, respectively. Since the two R_L's, R_1 and R_2, are much larger than R_o, the input resistances of the two voltage followers are, from Eq. (4.11), on the order of 10^{10} Ω. Thus the inputs to the voltage followers load the 10^5 Ω sources practically not at all; that is, for each voltage follower, $v_{IN} = v_S R_i'/(R_i' + R_S) \cong v_S$. From Eq. (4.13) we find that the output resistances of the voltage followers are

$$R_o' \cong \frac{10^2}{10^5} = 10^{-3} \, \Omega$$

This in turn is very low compared with R_1 and R_2; thus the summing amplifier does not significantly load the voltage followers. From Eq. (4.24) we can estimate the output resistance of the summing amplifier at approximately 2×10^{-3} Ω. The final inverting-amplifier stage, with unity gain, is provided to reinvert the sign of the output voltage. Its input resistance of 1 kΩ does not load the summing amplifier. Like the summing amplifier, the inverting amplifier has a low output resistance and thus is not significantly loaded by R_L.

In addition to the voltage-amplifier variations already discussed, op-amp circuits may be devised in which the input or output signals are currents. Two circuits of this type follow.

CURRENT-TO-VOLTAGE CONVERTER

The circuit shown in Fig. 4.32, which is identical to the inverting amplifier, is also useful for applications requiring an output voltage proportional to the current flowing into the input. To demonstrate this, we use the ideal op-amp technique. No appreciable current flows into the $(-)$ input; thus the source current i_S, which flows out through R_1,

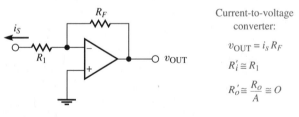

Current-to-voltage converter:

$$v_{OUT} = i_S R_F$$

$$R_i' \cong R_1$$

$$R_o' \cong \frac{R_O}{A} \cong O$$

FIGURE 4.32

must flow in through R_F. Since $v_{(-)} \cong 0$, the output voltage is just the voltage across R_F, which equals $i_S R_F$:

$$v_{\text{OUT}} = i_S R_F \tag{4.27}$$

The input resistance of the circuit is R_1, as already shown. Since R_1 plays no role in the operation of the converter, we may let $R_1 = 0$, in which case the input resistance $\cong 0$.

EXAMPLE 4.12

Design a precision electronic ammeter, using an available 0 to 10 V voltmeter movement with 20,000 Ω resistance. The full-scale reading of the ammeter should be 1 mA.

SOLUTION

We may connect the voltmeter to the output of a simple current-to-voltage converter. The voltmeter's full-scale voltage, 10 V, should correspond to the maximum current 1 mA. Thus from Eq. (4.27),

$$R_F = \frac{v_{\text{OUT}}}{i_S} = \frac{10}{10^{-3}} = 10^4 \ \Omega$$

A suitable circuit is shown in Fig. 4.33. Note that the 20 kΩ resistance of the voltmeter will not load the op-amp output appreciably.

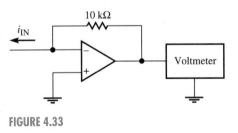

FIGURE 4.33

Op-amps may also be used to construct current amplifiers or voltage-to-current converters with nearly ideal characteristics.

EXAMPLE 4.13

Show that the magnitude of the output current I_{OUT} in the circuit of Fig. 4.34 is 100 times the input current i_{IN} for any load that satisfies $R_L \ll R_X$. Use the ideal op-amp technique.

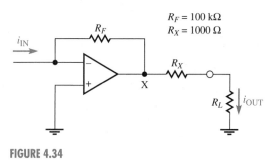

$R_F = 100 \text{ k}\Omega$
$R_X = 1000 \text{ } \Omega$

FIGURE 4.34

SOLUTION

Because the input current flowing into the ideal op-amp is zero, the current i_{IN} flows through R_F. Further, since $v_{(-)} \approx 0$, $v_X = -i_{IN} \cdot R_F$. It is clear that

$$i_{OUT} = \frac{v_X}{R_X + R_L}$$

Thus if $R_L \ll R_X$,

$$i_{OUT} \cong -i_{IN} \frac{R_F}{R_X} = -100 i_{IN}$$

The reader should note that our discussion of op-amps and op-amp circuits is still far from complete. Up to now we have described op-amps' general properties and have shown how they can be used as building blocks in analog systems. However, real op-amps have a number of practical complexities and limitations, about which one should know before one starts to build! More will be said about the theory of op-amp circuits when we discuss feedback in Chapter 9, and practical aspects of amplifier circuits will be considered in Chapter 10.

4.5 Application: Microphone Mixers

A group of musicians has hired you to record their sessions. You have an excellent tape recorder and several good microphones. However, you need a way to adjust the balance between the various instruments. Suppose that somewhere in the middle of a piece, the

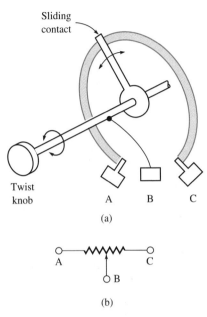

Sliding
contact

Twist
knob   A   B   C

(a)

A   C

B

(b)

FIGURE 4.35 Potentiometer. (a) Physical structure;
(b) Circuit symbol.

pianist wishes to do a solo. Then it is necessary to boost his volume compared to that
of the other instruments, so the piano line stands out. Conceivably, you could do this
mechanically, by moving one microphone closer to the piano at the proper moment,
and having the other musicians back away from the other microphones; but this would
be a very clumsy method! If you know how to use op-amps, you can do your mixing in
a much more ingenious way.

The summing amplifier circuit is ideal for solving this problem. By means of this
circuit you can add together the signals from several microphones; moreover, you can
multiply each signal by a different weighting factor. (See Fig. 4.30). The weighting factors can be made adjustable, by using adjustable resistors in the circuit. The usual way
of obtaining an adjustable resistance is by using what is known as a *potentiometer*. This
consists of a strip of resistive material with a sliding contact that can be moved from
one end of the resistance to the other by twisting a knob, as shown in Fig. 4.35. (The
volume control on your radio or TV is made in this way.)

A possible circuit for the microphone mixer is shown in Fig. 4.36. The values indicated for the potentiometers are their maximum resistances; each channel can thus be
set for any amplification between one-half and (theoretically, at least) infinity. There
is, however, a possible problem with this circuit because the effective impedance of
each microphone (that is, its source or Thévenin resistance) appears in series with the
potentiometer, and thus affects the gain of that channel. Since the microphone is really
a fairly complicated electromechanical device, it is quite possible that its source resis-

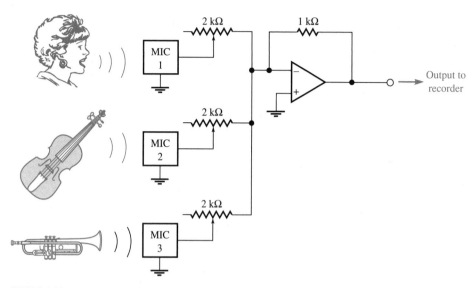

FIGURE 4.36

tance will be a function of frequency. That would not be good: It would mean that the amplification would vary with frequency, which might spoil the quality of the sound. It would be better to make the circuit independent of the microphones' properties. This can easily be done by adding a voltage follower as a buffer between each microphone and the summing amp, as shown in Fig. 4.37. The input resistance of the voltage fol-

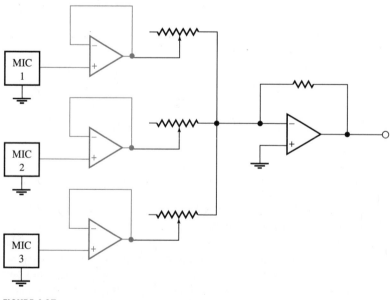

FIGURE 4.37

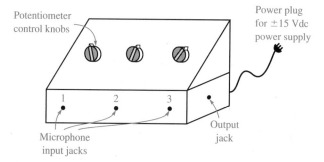

FIGURE 4.38

lowers is so extremely high that the voltage-divider effect between it and the microphone is negligible; thus the Thévenin resistance of the microphone no longer matters. One might at first think that, by adding three additional op-amps, we have quadrupled the complexity and cost of the circuit, but this is not a serious problem. Integrated circuit op-amps are wonderfully cheap; they only cost about one dollar. They require an inexpensive external power supply circuit (not shown on our diagrams) to provide their required operating voltages — often +15 and −15 volts dc. However, the same power supply can provide power for many op-amps, so once a single op-amp is being used, it is easy to add more.

The voltage amplification of the summing amplifier is not only adjustable, but also negative, as we see from Eq. (4.26). Does this minus sign matter to us? Does it mean that the signal is made smaller, instead of larger? No, not at all. The minus sign means that when the input signal voltage is a negative voltage, the output voltage is positive, and vice versa. The ear cannot tell the difference between a signal that goes negative-positive-negative and one that goes positive-negative-positive, so in this case the minus sign has no important effect. If for some reason we wanted to get rid of it, we could always put another inverting amplifier, with unity voltage gain, after the summing amplifier; this would invert the sign of the output signal again, so it would be the same as that of the input.

In your role as recording engineer, you will probably want to assemble your mixer in a neat electronic cabinet. The finished product might look as shown in Fig. 4.38. We hope you hit the top of the charts!

POINTS TO REMEMBER

- Dependent voltage and current sources are sources whose values cannot be specified independently. Instead, the value of such a source is determined by the voltage or current at some other point in the circuit.

- The most common analog building block is the amplifier. Amplifier blocks are conveniently described by a model. When the model is substituted for the amplifier in a larger system, it imitates the action of the amplifier. Using the model makes it easier to analyze the operation of the larger system.

- The typical amplifier model contains a dependent voltage source and two resistances. Each of these components is associated with a number: the dependent source by the open-circuit voltage amplification A, the input resistance by its value R_i, and the output resistance by its value R_o. When these three values have been specified, we have most of the information needed to use the amplifier in a larger system.

- *Voltage gain*, *current gain*, and *power gain* are useful descriptions of amplifier performance. However, the terms are somewhat vague and must be carefully specified for the specific circuit under consideration. One must state which voltages, currents, or powers are being compared.

- Often an amplifier is used as a part of a larger circuit, which may also have input and output terminals. In that case we may wish to describe the larger circuit in terms of its own model. This will have its own parameters, R_i', R_o', and A'. The primed quantities are usually different from the properties of the amplifier block used as a component. However, the primed values can be calculated by analyzing the larger circuit.

- Operational amplifiers are inexpensive, versatile amplifier blocks. Available in the form of integrated circuits, they are widely used as components in analog systems.

- Circuits containing op-amps can be analyzed by substituting the model for the op-amp and performing node or loop analysis. Alternatively, it is often possible to calculate *the output voltage* of an op-amp circuit by means of the *ideal op-amp technique*. Using the ideal op-amp technique is usually much quicker and easier than conventional circuit analysis, and although the method is approximate, it is usually highly accurate. However, it is usually not possible to calculate the input or output resistance of the an op-amp circuit in this way.

- Some common and useful op-amp circuits are the voltage follower (with unity voltage gain); the noninverting amplifier (with high input resistance and positive voltage gain); the inverting amplifier (with negative voltage gain and low input resistance — which is usually a drawback); the summing amplifier, the current-to-voltage converter, and the current amplifier.

PROBLEMS

The Modeling Concept

4.1 We wish to construct an approximate model for a flashlight battery by finding its Thévenin equivalent. The open-circuit voltage of the battery is 1.493 V. When a 100 Ω resistor is connected across it, its measured voltage is 1.356 V. Find its approximate Thévenin resistance and draw a model for the battery.

**4.2 The resistance of a certain resistor depends on its temperature according to

$$R = R_0[1 + \alpha(T - 300° \text{ K})]$$

where α, the *temperature coefficient of resistivity*, is $3.3 \times 10^{-3}/°K$, and R_0 is 10 Ω. When the ambient temperature is $300°$ K, current passing through the resistor causes its temperature to rise according to

$$(T - 300° \text{ K}) = I^2R/G$$

where G, called the *thermal conductance*, is 0.19 W/°K.

a. Sketch a graph of the voltage across the resistor as a function of current, over the range -1 A $< I < 1$ A.

b. In most ordinary work the resistor would be represented by a linear model in which $R = R_0 =$ constant. Sketch $V(I)$ for this model on the graph of part (a).

c. Compare the actual $V(I)$ with that of the linear model and find the maximum percentage error introduced in V if $|I| < 100$ mA. What is the maximum percentage error if $|I| < 1$ A?

Section 4.1

4.3 In the circuit of Fig. 4.39, the dependent current source depends on the current i_1, as shown. Let the value of β be 100. Find the current i_2. Let $R_1 = 10$ kΩ and $R_2 = 1,000$ Ω.

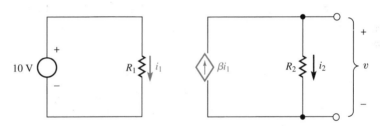

FIGURE 4.39

4.4 For Problem 4.3 find the voltage v.

4.5 For the circuit of Fig. 4.39, let $\beta = 100$.
   a. Sketch a graph of i_2 as a function of R_2, with $R_1 = 10$ kΩ.
   b. Sketch i_2 as a function of R_1 when $R_2 = 1,000$ Ω.

4.6 In Fig. 4.39, let $\beta = 100$.
   a. Sketch v as a function of R_2 when $R_1 = 10$ kΩ.
   b. Sketch v as a function of R_1 when $R_2 = 1,000$ Ω.

4.7 In Fig. 4.40, let $R_1 = 1,000$ Ω, $R_2 = 2,000$ Ω, and $\mu = 3$.
   a. Find i_2.
   b. Find i_2 if R_1 is changed to 500 Ω.

4.8 In Fig. 4.40 let $R_1 = 1,000$ Ω; $\mu = 3$.
   a. Sketch v_2 as a function of R_2.
   b. Sketch i_2 as a function of R_2.

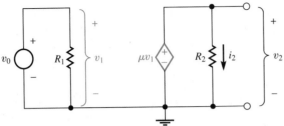

FIGURE 4.40

*4.9 In Fig. 4.41 let $V_1 = 2$ V, $r_\pi = 2,500$ Ω, $R_L = 5$ kΩ, and $\beta = 100$. Find i_1.

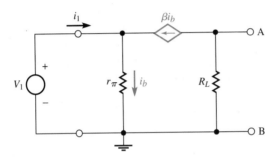

FIGURE 4.41

*4.10 Let the circuit of Fig. 4.41 be regarded as a subcircuit, with terminals A,B for external connection. Obtain the Norton equivalent of the subcircuit. It is not required that numerical values be substituted for the symbols.

*4.11 Find the Thévenin resistance of the subcircuit shown in Fig. 4.41, using the alternative method in which *independent* sources are set to zero. (Use a test voltage at terminals A,B and find I_{TEST}. See Examples 3.6 and 4.3.)

*4.12 Find the Thévenin equivalent of the subcircuit shown in Fig. 4.42. The factor r_m is a constant.

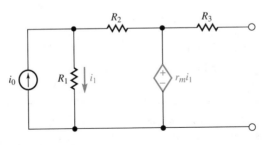

FIGURE 4.42

*4.13 Find the Thévenin resistance for Fig. 4.42 by the alternative method in which independent sources are set to zero. (See Examples 3.6 and 4.3.)

*4.14 Suppose that the device shown in Fig. 4.43(a) is represented by the model shown in Fig. 4.43(b). It is connected in a circuit as shown in Fig. 4.43(c). Calculate v_L. The quantity μ is a constant.

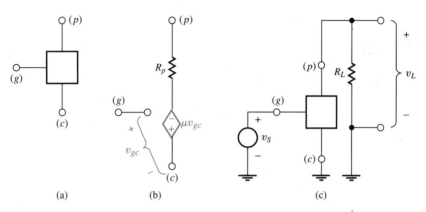

FIGURE 4.43

*4.15 The device shown in Figs. 4.43(a) and (b) is connected as shown in Fig. 4.44. Note that the orientation of the device is upside down compared with Fig. 4.43(a) and (b).

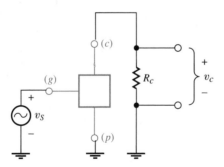

FIGURE 4.44

a. Substitute the device model into the circuit of Fig. 4.44, thus obtaining a circuit model.

b. Use the circuit model to calculate v_c. *Caution:* This problem is slightly tricky. Be careful in locating the reference voltage v_{gc}.

Section 4.2

*4.16 The block shown in Fig. 4.45 is known as a *two-port*. It can be shown that a general linear two-port is described by the matrix equation

$$\begin{pmatrix} v_1 \\ v_2 \end{pmatrix} = \begin{pmatrix} z_{11}\, z_{12} \\ z_{21}\, z_{22} \end{pmatrix} \begin{pmatrix} i_1 \\ i_2 \end{pmatrix} \tag{4.28}$$

Where the four constants z_{11}, z_{12}, z_{21}, z_{22} are known as the "z-parameters" of the two-port.

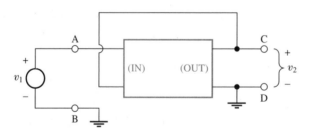

FIGURE 4.45   A two-port block.

a. Show that the amplifier model of Fig. 4.9(b) is a special case of the general two-port, and find z_{11}, z_{12}, z_{21}, z_{22} in terms of R_i, R_o, and A.
b. Devise a model that corresponds to the most general linear two-port, with all four z-parameters different from zero.

*4.17 In Fig. 4.46 the amplifier is represented by the model of Fig. 4.9(b). To simplify the problem, assume $R_o = 0$. Taking v_1 as given, find v_2. (*Suggestion:* Use care in locating the reference voltage v_{IN}.)

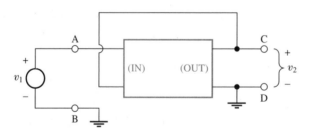

FIGURE 4.46

*4.18 Repeat Problem 4.17 with $R_o \neq 0$.

*4.19 Find the input resistance of the circuit shown in Fig. 4.46, as seen looking into terminals A,B. Output terminals C,D are open-circuited. Use the amplifier model of Fig. 4.9(b). ($R_o \neq 0$.)

*4.20 Find the output resistance of the circuit of Fig. 4.46, using the amplifier model of Fig. 4.9(b).

*4.21 In the circuit of Fig. 4.47 the amplifier block is modeled as in Fig. 4.9(b). Find A', R_o', and R_i' for the circuit. Let $R_i = R_o = 1,000\ \Omega$ and $A = 100$. When finding R_i', assume a load R_L connected across the output. When finding R_o', assume a Thévenin source v_s, R_s at the input.

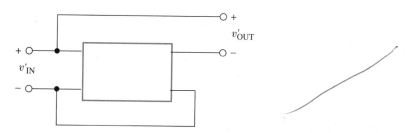

FIGURE 4.47

4.22 Two amplifiers that can be represented by the model of Fig. 4.9(b) are connected in *cascade* (that is, head-to-tail) as shown in Fig. 4.48. Their parameters are A_1, R_{i1}, R_{o1} and A_2, R_{i2}, R_{o2}, respectively. Find v_x. Discuss how your answer is influenced by the ratio R_{i2}/R_{o1}, and explain what happens in the limits as this ratio approaches zero and as it approaches infinity.

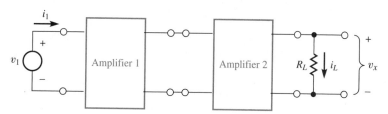

FIGURE 4.48

4.23 For Fig. 4.48 let us define a voltage gain $G_V \equiv v_x/v_1$ and a current gain $G_I \equiv i_L/i_1$. Find G_V and G_I. The amplifiers are those of Fig. 4.9(b).

4.24 In Fig. 4.48 the amplifiers are represented by the model of Fig. 4.9(b). Let us define a power gain G_p as the ratio of the power dissipated in R_L to the power produced by the source v_1. Find G_p.

*4.25 Suppose a load resistor R_L is connected across terminals C,D in Fig. 4.46. Find the power gain (defining G_p as in Problem 4.21). The amplifier is that of Fig. 4.9(b). Let $R_i = 100\ \Omega$, $R_o = 0$, $R_L = 100\ \Omega$, and $A = 5$.

4.26 Figure 4.49(a) shows an amplifier block. An experiment is performed in which known input voltages are applied and the resulting output voltages are measured. The experimental measurements are shown in Fig. 4.49(b).

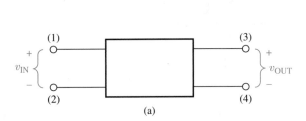

v_{IN}	v_{OUT}
−3	−6.640
−2	−4.320
−1	−1.990
0	+0.080
1	+2.130
2	+4.480
3	+6.800

(a)   (b)

FIGURE 4.49

a. Sketch a graph of v_{OUT} versus v_{IN}. Also construct a straight-line approximation that agrees well with the actual curve for values of $|v_{IN}|$ less than 1 V.

b. Based on your straight-line approximation, what is the open-circuit voltage amplification of this block?

c. Compare the open-circuit v_{OUT} of your model with that of the actual block. What is the maximum percentage error that will be introduced if we keep $|v_{IN}| < 1$ V? 2 V? 3 V?

4.27 In order to find the output resistance of the amplifier block shown in Fig. 4.49, a 50 Ω resistance is connected across output terminals (3), (4) while an input voltage of 1.000 V is applied to terminals (1), (2). The measured value of v_{OUT} is then 2.010 V. What is the output resistance?

4.28 In order to find the input resistance of the amplifier of Fig. 4.49, a Thévenin source is connected to input terminals (1), (2), with $v_T = 1.000$ V and $R_T = 10,000$ Ω. The voltage measured at the open-circuited output terminals is 1.564 V. What is the block's input resistance?

Sections 4.3 and 4.4

Assume that op-amps used in the following problems have $R_i \cong 10^5$ Ω, $R_o \cong 50$ Ω, $A \cong 10^5$. Use the ideal op-amp technique and other simplifying approximations whenever possible.

4.29 Design an op-amp circuit with $R_i' = 1,000$ Ω, $A' = -10$, $R_o' < 10$ Ω.

4.30 Design an op-amp circuit with $R_i' > 10$ MΩ, $A' = -10$, $R_o' < 10$ Ω. Use more than one op-amp if necessary.

4.31 Design an op-amp circuit with $R'_i > 10 \text{ M}\Omega$, $A' = +10$, $R'_o < 10 \Omega$.

4.32 Two input voltages are $v_1(t)$ and $v_2(t)$. Design an op-amp circuit that will generate the voltage $3v_1(t) - 2v_2(t)$. Its input resistances must exceed 1 kΩ and its output resistance must be less than 10 Ω. Use more than one op-amp if necessary.

4.33 A certain signal source has a Thévenin voltage $v_s(t)$ and a Thévenin resistance of 1 MΩ. Design an op-amp circuit that produces the voltage $-10v_s$.

*4.34 Verify Eq. (4.17).

**4.35 Consider the inverting amplifier circuit shown in Fig. 4.29. A signal source with Thévenin resistance R_s is to be connected to the input. Calculate the output voltage by two different methods:

  a. By combining R_s and R_1 into a single resistance R'_1 and using Eq. (4.22).
  b. By finding the input resistance R'_i for the inverting amplifier block and regarding R_s and R'_i as a voltage divider.

Show that the results obtained via parts (a) and (b) are in agreement. (*Caution:* be sure you have used two *different* methods.)

*4.36 Find the open-circuit output voltage of the system shown in Fig. 4.50 as a function of the input voltage v_s.

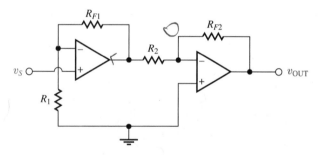

FIGURE 4.50

4.37 In the circuit of Fig. 4.51 find i_L in terms of v_{IN}.

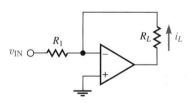

FIGURE 4.51

4.38 In Fig. 4.52, use the ideal-op-amp technique.

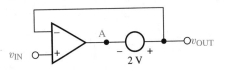

FIGURE 4.52

a. Find v_{OUT} as a function of v_{IN}.
b. What is the voltage at A?

4.39 In the circuit of Fig. 4.53(a), the diode has the approximate I-V relationship $I_D = I_0 \exp[qV_D/kT]$. The sign conventions for I_D and V_D are shown in Fig. 4.53(b). I_0, q, kT, R_1, and R_2 are given constants. Assume that v_{IN} is negative with respect to ground. Use the ideal-op-amp technique to find v_{OUT} as a function of v_{IN}.

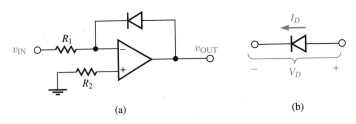

(a)   (b)

FIGURE 4.53

*4.40 In the circuit of Fig. 4.54, the box is a nonlinear circuit element with the I-V relationship $V_{NL} = 2\,I_{NL}^3$, where V_{NL} is given in volts and I_{NL} in mA, and the sign conventions are as shown. Let $R_1 = R_F = R_L = 10\text{ k}\Omega$, $R_2 = 2\text{ k}\Omega$, and use the ideal-op-amp technique. Find V_{NL}.

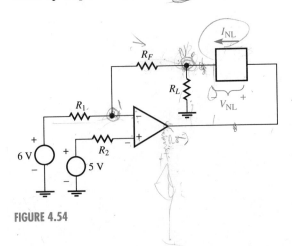

FIGURE 4.54

Section 4.5

4.41 As a recording engineer, you are using the microphone mixer of Section 4.5 to record musical performances. In practice, you find that some additional difficulties appear. One of these is that when the volume controls of the mixer are set too high, it is possible to overload the tape recorder that follows the mixer; this results in unpleasant distortion of the music. Unfortunately you don't discover this until the recording is finished and you try to play it back; then you have to ask the musicians to perform all over again. What might you add to your system to avoid this difficulty?

Another problem that turns up is that you really can't tell, while you are recording, exactly what the gain settings of the various channels should be. When you play the recording back, you see that one channel should have been louder, and another one should have been quieter; but the musicians will be annoyed if you ask them to play the piece all over again. Assuming that you have a good-sized budget to work with, how would you get around this problem?

TIME-VARYING SIGNALS IN CIRCUIT ANALYSIS

CHAPTER 5

INDUCTANCE AND CAPACITANCE

U ntil this point we have dealt with "dc" circuit analysis, that is, with voltages and currents that are assumed not to vary with time. We began with this case for the sake of simplicity. However, it is easy to see that cases of real practical interest will involve time variations. That is because we will be concerned with *signals*. Signals are time-varying voltages or currents that represent information. Very little information is contained in a non–time-varying signal. If you stand at one end of a pair of wires and observe a constant voltage between them, all you learn is that the constant voltage source at the other end is still connected. On the other hand, a time-varying voltage arriving through the wires can represent, for example, a series of numbers. It could be agreed that a voltage of 10 volts represents the binary digit "one" and that a voltage of zero volts represents the binary digit "zero". Then a rapidly time-varying voltage could represent an intense stream of ones and zeros, and could thus transmit a large amount of information. We note that the faster the voltages can change, the more numbers per second can be sent. In the dc case, the voltage cannot change at all, and thus the rate of information is almost zero. Since modern electrical engineering deals primarily with the storage, transmission, and processing of information, we see that time-varying signals are all-important.

To be sure, everything we have learned up to now about dc analysis will be found to remain true in the "ac", or time-varying, case. The circuit elements we have seen — ideal voltage sources, current sources, and resistances — behave in the same way whether or not the voltages vary in time; Kirchhoff's laws remain true as well. However, there are some additional circuit elements that are useful only in ac circuits.

In this chapter we introduce two of the most important such elements, the capacitor and the inductor. The remaining chapters in this section then deal with techniques for ac circuit analysis when inductances and capacitances are present.

The interaction of time-varying signals with inductors and capacitances can produce some fairly striking results. For instance, the reader may have wondered how very large voltages — tens of thousands of volts — can appear in the sparkplugs of a car when the electricity is supplied by a battery of only 12 volts. This amazing voltage transformation is usually obtained through the use of a single inductor, as the Application in Section 5.4 explains.

5.1 Ideal Capacitors and Inductors

Physically, a capacitor consists of two metallic conductors that are close enough together to interact, but are not in actual contact. An electric charge Q can be applied to one plate of the capacitor; in that case a charge $-Q$ appears on the other plate. The presence of this charge gives rise to a voltage V across the capacitor. This voltage is linearly proportional to the stored charge, according to the equation $Q = CV$. The quantity C is a constant, known as the *capacitance*, whose value depends on the physical structure of the capacitor. The MKS unit of capacitance is the *farad*. However, the farad is an extremely large unit for electronic purposes; capacitors as large as 1 farad are almost never encountered. Typical capacitances are usually described in *microfarads* ($1 \ \mu F = 10^{-6}$ farad) or *picofarads* ($1 \ pF = 10^{-12}$ farad).

One might consider that the basic physics of the capacitor lies in the fact that the voltage across the capacitor is linearly proportional to the charge on its plates. However, circuits are not usually described in terms of charge, but rather in terms of current. It is easy to shift from charge to current by simply differentiating the equation $Q = CV$. This results in $dQ/dt = C \ dV/dt$. However, the rate at which the charge on the capacitor increases, dQ/dt, is just equal to the current flowing into the capacitor, I. Thus we have $I = C \ dV/dt$, which is the usual equation used in circuit analysis.

● EXERCISE 5.1

A capacitor with value 5 microfarads is initially uncharged. Beginning at time $t = 0$, a 1-milliampere current flows into its terminals. Find the charge on the capacitor after 10 seconds have elapsed. **Answer:** 0.01 Coulomb.

Although we have established the physically motivated equation $I = C \ dV/dt$, we cannot make use of it until we have specified algebraic signs. Let us adopt the same nomenclature used earlier for resistances. The voltage at terminal A is V_A, and the potential difference $V_A - V_B$ is V_{AB}. The current through the capacitor, defined with reference direction from A toward B, is $I_{A \to B}$. The circuit symbol for the capacitor and

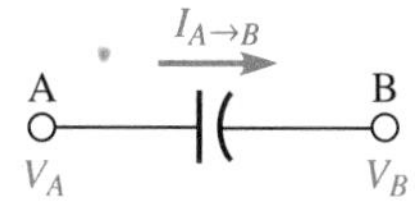

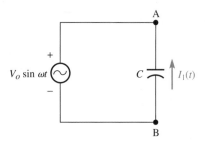

FIGURE 5.1

its associated voltage and current are then as shown in Fig. 5.1. Now the *I-V* relationship for the capacitor can be stated without ambiguity as to signs:

$$I_{A \to B} = C \frac{d}{dt} V_{AB} \tag{5.1}$$

We observe that the relationship in Eq. (5.1) is fundamentally more complicated than Ohm's law, because the latter is only an algebraic relationship. Eq. (5.1), on the other hand, involves differentiation, an operation belonging to calculus. As a result, circuit equations involving time-varying currents and capacitors become differential equations instead of the merely algebraic equations characteristic of dc circuits. Historically this higher level of mathematical sophistication was a roadblock, interfering with the understanding and use of ac circuits. Eventually powerful techniques were developed for dealing with ac circuit equations; some of these techniques are presented in the following chapters.

We see, from Eq. (5.1), that if the voltage across a capacitor is *constant*, then the current through the capacitor will be zero. Thus a capacitor that happened to be present in a dc circuit would have no current flowing through it; it would behave just like an open circuit. Another point to observe is the sequence of the subscripts in Eq. (5.1), which is the same as that previously used in writing Ohm's law. As an aid to memory, remember that the current in the direction from A to B depends on the voltage at A minus the voltage at B. Just as with Ohm's law, the algebraic signs of the current and voltage are of the greatest importance.

EXAMPLE 5.1

Find the current $I_1(t)$ that passes through the capacitor shown in Fig. 5.2. The voltage produced by the ideal voltage source is the sinusoid $V_o \sin \omega t$, where V_o and ω are given constants and t is time.

FIGURE 5.2

SOLUTION

We note that the voltage across the capacitor is time-varying. Thus we expect the current through it to be something other than zero, and to be given by Eq. (5.1). From the diagram we see that

$$V_A - V_B = V_o \sin \omega t$$

Thus from Eq. (5.1) we have

$$I_{A \to B} = C \frac{d}{dt}(V_o \sin \omega t)$$

$$= C V_o \omega \cos \omega t$$

However, in this case the reference direction for the current I_1 is from B to A. Thus

$$I_1 = -C V_o \omega \cos \omega t$$

EXAMPLE 5.2

Find the voltage $v_C(t)$ across the capacitor shown in Fig. 5.3. The current I_o through the current source is constant.

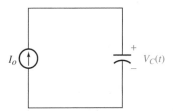

FIGURE 5.3

SOLUTION

In this case we have, from Eq. (5.1),

$$I_o = C \frac{d}{dt} v_C(t)$$

or equivalently,

$$dv_C = \frac{I_o}{C} dt$$

Integrating, we have

$$v_C = \frac{I_o}{C} \int dt$$

$$= \frac{I_o t}{C} + K$$

where K is an arbitrary constant of integration. From the form of our answer we see that K represents the value of v_C at time $t = 0$. Physically, what happens in this example is that the flow of current from the current source charges up the capacitor, causing its voltage to increase linearly with time. It is possible, however, that the capacitor already has some charge at time $t = 0$. In that case the voltage resulting from this initial charge is given by the constant K.

In passing we note that an ideal capacitor, once given a charge, can retain that charge (and its associated voltage) indefinitely. Suppose a capacitor is charged up to some voltage by the method of Fig. 5.3, and the current source is then disconnected. Disconnecting the current source is equivalent to setting $I_o = 0$. This implies that $C \, dv_C/dt = 0$; therefore v_C does not vary in time. As a result, the disconnected capacitor retains the charge and voltage it had at the moment it was disconnected. One should think twice before picking up a capacitor found lying on a workbench. If someone charged it up yesterday, a large voltage might still be lurking between its terminals!

CAPACITORS IN SERIES AND PARALLEL

In Chapter 2 we saw that two resistors, R_1 and R_2, connected in series, are equivalent to a single resistor of value $R_1 + R_2$. We also found that, when connected in parallel, the combination is equivalent to a single resistor of value $R_1 R_2/(R_1 + R_2)$. Let us now enquire as to whether some similar results apply to capacitors in series and in parallel.

When two capacitors, C_1 and C_2, are connected in parallel, the voltage across both capacitors is identical. Thus the total current entering the combination is

$$I = C_1 \frac{dV}{dt} + C_2 \frac{dV}{dt}$$

hence

$$I = (C_1 + C_2) \frac{dV}{dt}$$

This is the same equation that we would have if there were only a single capacitor of value $C_1 + C_2$. Thus two capacitors connected in parallel are equivalent to a single capacitor of this value.

On the other hand, when two capacitors (or two anythings, for that matter) are connected in series, the *current* through both of them is the same. Thus

$$I = C_1 \frac{dV_1}{dt} = C_2 \frac{dV_2}{dt}$$

where V_1 is the voltage across C_1 and V_2 is the voltage across C_2. Let the total voltage across the series combination be $V = V_1 + V_2$. Then

$$\frac{dV}{dt} = \frac{dV_1}{dt} + \frac{dV_2}{dt} = \frac{dV_1}{dt} + \frac{C_1}{C_2}\frac{dV_1}{dt}$$

$$= \frac{dV_1}{dt}\left(\frac{C_2 + C_1}{C_2}\right)$$

However, we know that

$$I = C_1\frac{dV_1}{dt}$$

Substituting for $\dfrac{dV_1}{dt}$ from the previous equation, we find that

$$I = C_{\text{SER}}\frac{dV}{dt}$$

where

$$C_{\text{SER}} = \frac{C_1 C_2}{C_1 + C_2}$$

Note the resemblance of this formula for capacitors in series to the formula for resistors connected in *parallel*.

THE IDEAL INDUCTOR

Like the ideal capacitor, the ideal inductor is a two-terminal circuit element. Its circuit symbol is shown in Fig. 5.4. Adopting the sign conventions of that figure, the *I-V* relationship of the ideal inductor is

$$V_{AB} = L\frac{d}{dt}I \tag{5.2}$$

Here L is a constant known as the *inductance*. The MKS unit of inductance is the henry. Large iron-core inductors used in electric-power applications may have inductances larger than 1 henry, but in electronics, values on the order of millihenries (1 mH $= 10^{-3}$ H), microhenries (1 μH $= 10^{-6}$ H), and nanohenries (1 nH $= 10^{-9}$ H) are more common.

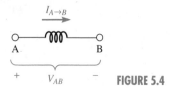

FIGURE 5.4

Comparing Eq. (5.2) with Eq. (5.1), which describes the ideal capacitor, we observe a certain similarity. If one begins with Eq. (5.1) and replaces current by voltage, and voltage by current, one arrives at Eq. (5.2). Thus the inductor behaves much like a capacitor, with the roles of voltage and current reversed. We see that if the *current* through an inductor is constant, the *voltage* across the inductor will be zero. Moreover, if the two terminals of an inductor are short-circuited (i.e., connected together), the voltage across the inductor will be zero, and from Eq. (5.2), we see that the current through the inductor cannot change. This is a slightly weird result, analogous to the ability of an open-circuited capacitor to retain a charge: If a current is established through an inductor and its terminals are connected together by an ideal wire, the current should continue to flow through the inductor forever! We should remember, however, that all this only applies to *ideal* inductors. Real inductors (and real wires) usually possess some resistance, in which case they are no longer ideal. But, in fact, something like the ideal behavior really can be obtained, if one is prepared to use superconducting material for the inductor and shorting wire. In that case a current, once started, really will go on forever.

EXAMPLE 5.3

Initially the current through an inductor L is zero. At time $t = 0$ a time-varying voltage $v(t)$ is applied across its terminals. Find the current through the inductor as a function of time.

SOLUTION

From Eq. (5.2) we know that the current $i(t)$ obeys

$$v(t) = L\frac{d}{dt}i(t)$$

Thus

$$i(t) = \frac{1}{L}\int_0^t v(t)dt$$

As a special case, we observe that if $v(t)$ happens to be a constant, the current $i(t)$ increases linearly with time.

Very often one needs to analyze circuits containing inductors and capacitors, as well as resistors, in order to find the unknown voltages and currents. This is done through the use of Kirchhoff's laws, just as in the purely resistive case. However, because of Eqs. (5.1) and (5.2), the resulting node or loop equations now contain derivatives of the voltages and currents and thus become differential equations. Some powerful techniques for dealing with these equations will be discussed in the following chapters.

● **EXERCISE 5.2**

Write a loop equation for the loop current $i(t)$ in Fig. 5.5. **Answer:**

$$L\frac{di}{dt} + i(t)R = 0$$

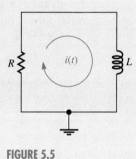

FIGURE 5.5

It is readily shown that inductors L_1 and L_2 in parallel and in series can be replaced by single equivalent inductors according to the formulas

$$L_{SER} = L_1 + L_2$$

and

$$L_{PAR} = \frac{L_1L_2}{L + L_2}$$

The proofs are similar to those used in the case of capacitors.

5.2 Energy Storage

Capacitors and inductors differ from resistors in an interesting and important way, having to do with how they handle energy. As we have seen, resistors *dissipate* energy; that is, they convert energy from electrical form into heat. When a current I passes through a resistance R, the power entering the resistor is $VI = I^2R = V^2/R$. This power is con-

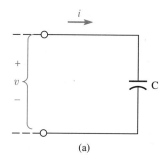

(a)

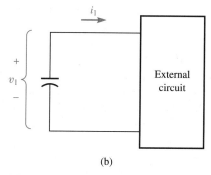

(b)

FIGURE 5.6

verted by the resistor into heat. However, according to the laws of thermodynamics, there is no way to reverse the process: No one has ever seen a resistor become colder and produce electrical power. Once electrical power is converted into heat in a resistor, the energy is lost. Thus resistors are referred to as *dissipative circuit elements*.

For capacitors and inductors, the situation is quite different. When a capacitor is charged, energy is stored in the electric fields inside the capacitor. Energy is *not* dissipated; it is simply stored. When the capacitor is eventually discharged, it does work on the external circuit that discharges it, and all the energy that was used to charge it is regained. Inductors also have this property. Thus capacitors and inductors are known as *energy-storage* circuit elements. Neither capacitors nor inductors are physically capable of turning energy into heat. Thus if a circuit contains only capacitors and inductors, no heat will be produced in it, and any energy that enters the circuit can later be recovered.

Let us demonstrate the phenomenon of energy storage using the capacitor of Fig. 5.6. Suppose the capacitor is initially uncharged, and beginning at time $t = 0$ a voltage $v(t)$ is applied. The instantaneous power entering the capacitor is then $p(t) = v(t)i(t)$. From time $t = 0$ to time t, the energy entering the capacitor is

$$E = \int_0^t p(t)\, dt = \int_0^t v(t)i(t)\, dt$$

Using Eq. (5.1), this becomes

$$E = \int_0^t v \, C \frac{dv}{dt} \, dt$$

$$= \int_0^t \frac{1}{2} C \frac{d}{dt}(v^2)$$

$$= \frac{1}{2} C[v(t)]^2 - \frac{1}{2} C[v(t = 0)]^2 = \frac{1}{2} C[v(t)]^2$$

[Here we have used the fact that the capacitor was initially uncharged, which tells us that $v(t = 0) = 0.$] We notice that it does not matter how $v(t)$ goes from 0 to $v(t)$; only its final value is important. The conclusion is that, in order to charge a capacitor to a voltage V, we must store an energy equal to $\frac{1}{2}CV^2$.

Let us now verify that the stored energy can be regained. Suppose the capacitor, initially charged to voltage V, is allowed to discharge through some arbitrary external circuit, as shown in Fig. 5.6(b). The instantaneous power entering the external circuit will be $p(t) = v_1(t)i_1(t)$, and the energy recovered from the capacitor, after an infinite length of time, will be

$$E = \int_0^\infty v(t)i(t) \, dt$$

$$= \int_0^\infty v(t)\left[-C\frac{dv}{dt}\right] dt$$

(Note the minus sign that appears in this equation, resulting from the reference direction of i in Fig. 5.6(b). Compare with Fig. 5.1.)

$$= -C \int_0^\infty \frac{d}{dt}\left(\frac{1}{2}v^2\right) dt$$

$$= -\frac{1}{2} C\left[v(\infty)^2 - v(0)^2\right]$$

Since the capacitor is assumed to be fully discharged at $t = \infty$, and the initial voltage $v(t = 0)$ is called V, we find that the total energy supplied to the external circuit is

$$E = \frac{1}{2} C V^2 \tag{5.3}$$

Clearly, this is the same energy that was required to charge the capacitor. Thus the capacitor stores all the energy used to charge it, with no energy being lost in the form of heat.

A similar result can be shown to apply to inductors. However, in this case it is the final *current* through the inductor (rather than the voltage across it) that determines the energy that is stored. For an inductor,

$$E = \int_0^t v(t)\,i(t)\,dt = \int_0^t L\frac{di}{dt}\,i(t)\,dt \qquad (5.4)$$

$$= \frac{1}{2}L\,I^2$$

where I is the final current at time t.

• EXERCISE 5.3

The current through an inductor is initially zero. At time $t = 0$ a constant voltage V_o is applied across its terminals. Find the energy stored in the inductor after a time T has elapsed. **Answer:** $\frac{1}{2}V_o^2 T^2/L$.

5.3 Practical Capacitors and Inductors

Our discussion so far of capacitors and inductors has assumed that they are ideal. However, the circuit elements used to build real circuits inevitably differ from ideal elements to a greater or lesser degree. In practice, real capacitors often approximate ideal elements fairly well, but inductors seldom do.

Capacitors for use below 1 GHz come in several standard types, such as mica, ceramic, and tantalum. They are specified in terms of (a) their capacitance, (b) the working voltage (maximum voltage that can be continuously applied without damaging the capacitor), and (c) their tolerance (i.e., how far the actual capacitance can differ from the stated value). In addition, one should be aware that the materials of a real capacitor also contain some resistance, which may influence the way the capacitor performs in a circuit. When this resistance is important, the capacitor should be represented by an appropriate model, as shown in Fig. 5.7(b). This model consists of an ideal

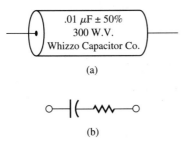

(a)

(b)

FIGURE 5.7 (a) A real capacitor; (b) A suitable model.

FIGURE 5.8   A possible model for a real inductor.

capacitance plus a series resistance. When the circuit is analyzed, one treats the real capacitor as though it were the combination of ideal elements shown in Fig. 5.7(b). Thus the effects of the resistance can be included in the circuit analysis. The value R of this resistance often can be found in the manufacturer's literature. Usually it is not specified directly, but in terms of the parameter Q. Q is related to R by the formula $Q = (2\pi f CR)^{-1}$ where f is the frequency of interest. Values of Q around 100 are typical.

Special capacitors are used for certain applications. Electrolytic capacitors are a special type used when comparatively large values of capacitance (say 10 to 1,000 μF) are required in a reasonably small space. These are used in the power supplies of most electronic equipment, where they assist in converting power-line ac to the dc required by electronic equipment. They have the unusual characteristic of being *polarity-dependent;* that is, they require that one end of the capacitor (which is marked on the outside) be at a positive voltage with respect to the other end. (Ideal capacitors, of course, do not have this property; voltages of either sign can be applied.) The price of obtaining their large values of capacitance is that electrolytic capacitors tend to be failure-prone. A loud humming sound from a radio or television set often announces that an electrolytic capacitor needs to be replaced. Another special type of capacitor is the "variable" capacitor, whose capacitance value can be changed by mechanical adjustment. This is accomplished, for example, by moving one capacitor plate so that it is nearer to, or farther from, the other. Variable capacitors are often used in radio receivers to select the station being received.

Practical inductors for use below 1 GHz usually consist of coils of wire. Since the wire may be long, its resistance may be quite large, and thus is more likely to be important than the smaller resistance of a capacitor. In addition, there is also some capacitance between the turns of the inductor, and this can also be included in a model, as shown in Fig. 5.8. The additional circuit elements that appear in the model are known as *parasitic* elements. Because of the complexity of the inductor model, the values of its constituent elements are most often determined by measurement. However, one may find that the values of the circuit elements in the model actually vary with frequency. This is because the model is really only a simplified representation of a fairly complex structure.

At frequencies above 1 GHz, modern planar integrated-circuit technology is likely to be used. In integrated circuit technology, all structures lie on the surface of a flat piece of semiconductor, and thus must be nearly two-dimensional. Capacitors lend themselves fairly well to planar technology; two types of planar capacitor, the "interdigital" and "overlay" capacitors, are shown in Fig. 5.9. However, inductors, if they are coils of wire, are difficult to construct in planar form. One type of inductor that has evolved for use in high-frequency planar circuits is the "spiral inductor" sketched in Fig. 5.10. (Note that "bridges" of some sort are required to permit one end of the

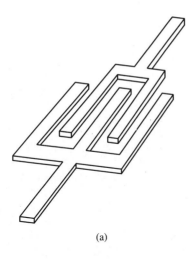

(a)

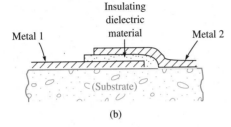

(b)

FIGURE 5.9 Two types of planar capacitor. (a) Inter-digital; (b) Overlay.

inductor to escape from the center of the spiral.) Inductors like this take up a great deal of space in an integrated circuit, which is highly undesirable; moreover, the maximum inductance that can be obtained is not very large (on the order of 50 nF), the parasitic capacitances tend to be very significant (so that at high frequencies the inductor may even behave more like a capacitor!), and the inductors are difficult to design and

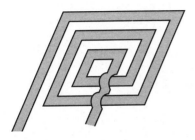

FIGURE 5.10 Spiral inductor, a planar inductor for use in integrated circuits.

model. Similar disadvantages often apply to conventional inductors in low-frequency circuits. As a result, there is a tendency to avoid the use of inductors when some alternative is possible.

5.4 Application: Automobile Ignition Systems

The fuel–air mixture in each cylinder of a gasoline engine must be ignited at the proper times. This is done by means of a spark that is formed across the terminals of a sparkplug, as shown in Fig. 5.11. The sparkplug is nothing more than a pair of metal electrodes, separated by an adjustable distance. (The adjustment is made by simply bending one of the electrodes.) When a large voltage, perhaps 10,000 volts, is applied between these electrodes, a spark—a miniature lightning bolt—jumps across the gap, igniting the fuel.

Electricity in cars is usually supplied by a 12 volt electrical system. If the car battery supplies only 12 volts, how can 10,000 volts for the sparkplugs be obtained? The usual way is by means of an inductor, known as the *spark coil*. To obtain the high voltage, we take advantage of the inductor equation $v = L \, di/dt$. The quantity di/dt is the rate of change of the current through the inductor. If a large change in the current happens in a very small time, then di/dt can be very large. In fact, if the current could be turned off instantaneously, in principle an infinitely large voltage could be generated.

A typical circuit for this purpose is shown in Fig. 5.12. Initially the switch is closed, and current flows from the battery through the resistor and through the inductor. At this time the current is nearly constant, and so there is no voltage drop across the inductor (since di/dt is zero.) The current through the inductor is thus approximately $12/R$ amperes. At the proper time the switch (known in cars as the *breaker points*) is suddenly opened. This causes the current through the inductor to decrease suddenly,

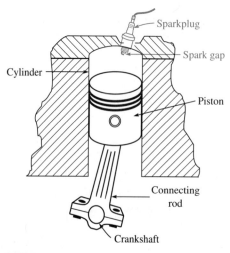

FIGURE 5.11

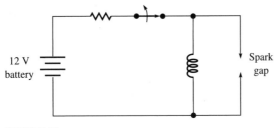

FIGURE 5.12

and a large voltage is developed across the inductor. When this voltage becomes large enough, a spark forms across the spark gap. The current through the inductor then can continue to flow, but now, instead of flowing through the switch, it is flowing through the spark. The spark now continues until all the energy stored in the inductor ($\frac{1}{2}LI^2$, where $I = 12/R$) has been dissipated in the spark discharge.

This same effect is often rediscovered accidentally in student laboratories. Suppose one wishes to find the resistance of an inductor; one would naturally connect an ohmmeter to the inductor and make the measurement. An ohmmeter works by passing a known small current through the element and measuring the resulting voltage across it. (The resistance is then equal to the measured voltage divided by the known current.) Since the current from the ohmmeter is quite small, one feels safe. But when the ohmmeter is disconnected, the current changes very suddenly from its small value to zero, and di/dt can be quite large. If one's fingers are in the wrong place, one gets a very nasty shock!

POINTS TO REMEMBER

- The voltage across an ideal capacitor is related to the charge on its plates according to $Q = CV$, where C is the capacitance.

- By differentiating $Q = CV$ with respect to time, we arrive at the I-V relationship for the ideal capacitor, which is $I = CdV/dt$. Note that the current is not determined by the voltage, as in a resistor, but by *the rate of change* of the voltage.

- In a static or dc situation—that is, when the voltage is not changing with time—the current through a capacitor is zero. In static situations a capacitor behaves just like an open circuit.

- The I-V relationship of an ideal inductor is $V = LdI/dt$. Note that this resembles the I-V relationship of the capacitor, with V and I interchanged.

- In a static, or dc, situation, the current through an inductor will not be varying in time. Thus the voltage across the inductor will be zero. In dc situations an inductor behaves just like a short circuit.

- Inductors and capacitors both have the ability to store energy. This happens whenever a voltage is applied to a capacitor; the stored energy is $\frac{1}{2}CV^2$. The energy stored in an inductor depends on the current that is established in it, and is equal to $\frac{1}{2}LI^2$.

- Ideal inductors and capacitors are incapable of dissipating energy (that is, turning electrical energy into heat). Any energy that flows into these elements is recovered when the capacitor is discharged or when the current through the inductor returns to zero. (Note that resistors are different: They cannot store energy, and do convert it into heat.)

- Practical capacitors and inductors differ from the ideal elements. Real capacitors generally contain some resistance, and real inductors contain not only resistance but also significant capacitance. When analyzing circuits containing real circuit elements, one should proceed by first substituting suitable models for the real circuit elements.

- Surprisingly large voltages can be obtained using a low-voltage battery and an inductance. This is done by producing a sudden change in the current through an inductor and taking advantage of the relationship $V = L dI/dt$. The high sparkplug voltage in automobile engines is usually obtained in this way.

PROBLEMS

Section 5.1

5.1 A constant current of 1 mA flows into a 1 μF capacitor that is initially uncharged. After what time is the voltage across the capacitor equal to 2 V?

5.2 Prove that two inductors in series are equivalent to a single inductor of value $(L_1 + L_2)$.

*5.3 Prove that two inductors in parallel are equivalent to a single inductor of value

$$\frac{L_1 L_2}{L + L_2}$$

5.4 Write a node equation for the circuit of Fig. 5.13, stating that the sum of the currents entering node A is zero. This should be a differential equation in the variable V_A.

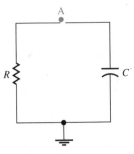

FIGURE 5.13

**5.5 Write a loop equation for the circuit of Fig. 5.14. This will be a differential equation; it should contain I and its time derivative but not V_C. The latter should be eliminated by substitution of Eq. (5.1). You will have to differentiate the loop equation before you can substitute.

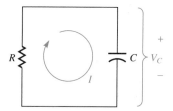

FIGURE 5.14

*5.6 In the circuit of Fig. 5.15, the capacitor is uncharged at time $t = 0$. Use the ideal op-amp technique to show that

$$V_{OUT}(t) = -\frac{1}{RC} \int_0^t V_{IN}(t') \, dt'$$

This useful op-amp circuit is known as an *integrator*. It is widely used in specialized machines used for solving differential equations, known as *analog computers*.

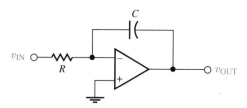

FIGURE 5.15

5.7 In the circuit shown in Fig. 5.16, the voltages and currents do not vary in time. Let $V_0 = 50$ V, $R_1 = 2$ kΩ, $R_2 = 4$ kΩ, $R_3 = 3$ kΩ, $R_4 = 2,500$ Ω. Find the current I through R_4.

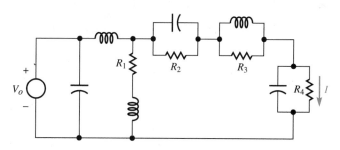

FIGURE 5.16

5.8 **a.** Find a single capacitor equivalent to the connection of Fig. 5.17(a).
  b. Find a single inductor equivalent to the connection of Fig. 5.17(b).

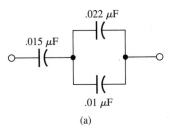

(a)

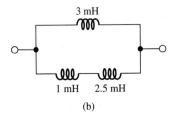

(b)

FIGURE 5.17

**5.9 The ideal current source in Fig. 5.18 produces a sinusoidal current $I_o \cos \omega t$.

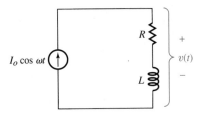

FIGURE 5.18

  a. Obtain an expression for the time-varying voltage $v(t)$ indicated in the figure.
  b. Show that this voltage can be expressed in the form

$$v(t) = A \cos(\omega t + \phi)$$

where

$$A = I_o \sqrt{R^2 + (\omega L)^2}$$

and

$$\phi = \tan^{-1}\left(\frac{\omega L}{R}\right)$$

5.10 The triangular voltage waveform shown in Fig. 5.19 is applied across a 1 µF capacitor. Make a sketch of the current through the capacitor as a function of time.

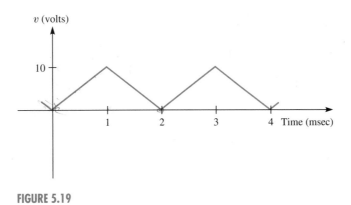

FIGURE 5.19

**5.11 The voltage of Fig. 5.19 is applied across a 5 mH inductor. Make a sketch of the current through the inductor as a function of time.

Section 5.2

5.12 In Fig. 5.20, the capacitor is initially charged to a voltage V_o. At time $t = 0$ the switch is closed. The charge stored on the capacitor then flows out through the inductor. At some later time the charge on the capacitor has fallen to zero. Find the current through the inductor at that time.

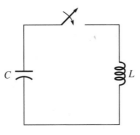

FIGURE 5.20

*5.13 A sinusoidal voltage source, producing a voltage $v(t) = V_o \cos \omega t$, is connected across a capacitor C.
  a. Find the instantaneous power entering the capacitor, as a function of time.
  b. Find the time-averaged power entering the capacitor. Perform the time average over one period T of the sinusoidal voltage. (*Note:* $T = 2\pi/\omega$.)

*5.14 Repeat Problem 5.13 with the capacitor replaced by an inductor *L*.

*5.15 An ideal sinusoidal voltage source produces a voltage $V_o \cos 2\pi ft$, where *f* is the frequency of the voltage in Hz. The voltage source is connected across a 1 μF capacitor.

    a. Show that the current through the capacitor is a sinusoidal function of the same frequency, and find I_m, the maximum instantaneous current.

    b. Make a sketch of I_m as a function of the frequency *f*.

CHAPTER 6

SINUSOIDAL SIGNALS

In this chapter and the next we shall consider the special case in which voltages and currents are sinusoidal functions of time. The special case of sinusoidal signals is of particular importance. The alternating currents and voltages that appear in electrical power systems are usually sinusoidal; so are the high-frequency currents used in radio communications. Furthermore, sinusoidal signals have particular mathematical properties that make them particularly convenient in circuit analysis.

Our interest lies in what are known as *linear* circuits, that is, in circuits containing only resistors, capacitors, inductors, and ideal voltage and current sources. Such circuits are referred to as "linear" because the node and loop equations arising in their analysis are always found to be linear differential equations. The particular property of sinusoidal signals that makes them interesting to us is this: When all the voltage and current sources that drive a linear circuit are sinusoidal with a certain frequency, then all the voltages and currents everywhere in the circuit turn out to also be sinusoidal, with the same frequency. This situation, in which all the currents are sinusoids that do not begin or end, but go on forever, is known as the *sinusoidal steady state*. Special techniques are available for calculating voltages and currents in the sinusoidal steady state, and these are among the engineer's most useful tools.

We shall begin in Section 6.1 by reviewing the general properties of sinusoidal signals. Then we shall be ready to begin our study of circuit analysis when such signals are present. As we have seen, the node and loop equations for circuits containing inductors and capacitors usually turn out to be differential equations, which in general are difficult to solve. However, we shall see that these differential equations can be converted to ordinary algebraic equations, which are much easier to handle. This is done through a special technique, known as the *phasor* technique, in which sinusoidal signals are represented by special numbers known as phasors. These numbers are, however, not ordinary real numbers but complex numbers, containing both real and imaginary parts. It is assumed that the reader has already had some exposure to the algebra of complex numbers, but this knowledge will need to be reviewed.

Problems involving the sinusoidal steady state arise in many other fields besides electrical engineering. For instance, the vibrations that arise in mechanical systems are also generally sinusoidal. In the Application in Section 6.5 we shall see how the phasor technique can be a useful tool in mechanical problems.

6.1 Properties of Sinusoids

We use the word "sinusoid" to mean functions of time having the form

$$f(t) = A \cos (\omega t + \phi) \tag{6.1}$$

Functions of this mathematical form are all similar to the well-known sine and cosine functions. However, each function of the form of Eq. (6.1) contains the three constant parameters A, ω, and ϕ. The presence of these three adjustable parameters means that there are a very great number of different sinusoidal functions possible. In sinusoidal steady-state analysis, one knows from the beginning that the voltage or current being calculated is a sinusoid. One only needs to calculate the three numbers A, ω, and ϕ. Actually one of these calculations (ω) is usually trivial, so the problem boils down to calculating the parameters A and ϕ.

It is important to have a good understanding of what the three parameters A, ω, and ϕ actually mean. The quantity A is known as the *amplitude* of the sinusoid. It is equal to the maximum value that the sinusoid can reach. (This maximum occurs at those instants of time that make $\cos (\omega t + \phi)$ equal to one.) Clearly the minimum value reached by the sinusoid is $-A$. Two sinusoids having different values of A (but the same values of ω and ϕ) are illustrated in Fig. 6.1.

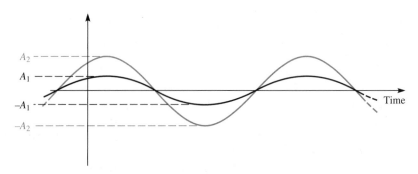

FIGURE 6.1   Two sinusoids having different amplitudes A_1 and A_2, but the same ω and ϕ.

The parameter ω is known as the *angular frequency* of the sinusoid. It specifies how often the maxima occur. To see this, we note that when the argument of the cosine function in Eq. (6.1) is zero, that is, at an instant of time $t = t_0$ such that

$$(\omega t_0 + \phi) = 0 \tag{6.2}$$

the sinusoid will be at its maximum. At times slightly later than t_0 the sinusoid will have decreased; it goes to its minimum value and then comes back to its maximum when the argument of the cosine function reaches 2π. Let us call the time when this happens t_1; then t_1 is defined by

$$(\omega t_1 + \phi) = 2\pi \tag{6.3}$$

Subtracting Eq. (6.2) from Eq. (6.3), we have

$$\omega(t_1 - t_0) = 2\pi \tag{6.4}$$

Let us define $\tau \equiv t_1 - t_0$. Then τ is the time that elapses between maxima of the sinusoid; τ is known as the *period*, as illustrated in Fig. 6.2. From Eq. (6.4) we have

$$\tau = \frac{2\pi}{\omega} \quad \text{(period of a sinusoid)} \tag{6.5}$$

The *ordinary frequency* f is equal to the number of repetitions of the sinusoid per second; it is given by

$$f = 1/\tau \tag{6.6}$$

It is not the same as the angular frequency; however, it is related to the latter by the formula

$$\omega = 2\pi f \tag{6.7}$$

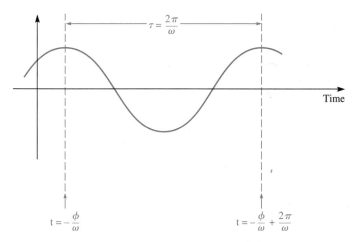

FIGURE 6.2

Regrettably, in casual usage both ordinary frequency and angular frequency are often referred to simply as "frequency." When the word "frequency" appears, one must make certain which of the two is meant. Angular frequency is given in units of radians per second. The units of ordinary frequency are hertz (Hz); units such as kilohertz (kHz) and megahertz (MHz) are also used.*

EXAMPLE 6.1

The transmitter of a certain radio station produces a sinusoidal voltage at its antenna. The frequency of this sinusoid is 98.7 MHz. What is the time that elapses between voltage maxima?

SOLUTION

Since the frequency is stated in units of MHz, we know that ordinary frequency f (rather than angular frequency ω) is being stated. The time that elapses between maxima is the period of the sinusoid τ. From Eq. (6.6),

$$\tau = 1/f = (9.87 \times 10^7)^{-1} \text{ sec}$$

$$= 1.013 \times 10^{-8} \text{ sec} = 10.13 \text{ nsec}$$

The *phase angle* ϕ specifies at what instants of time the sinusoid reaches its maxima. For instance, we know there must be a maximum of the sinusoid when the argument of the cosine, $\omega t + \phi$, equals zero. Thus there must be a maximum when $t = -\phi/\omega$, and also, because of the periodicity of the sinusoid, at $t = -\phi/\omega \pm 2n\pi/\omega$, where n is any integer. The phase angle ϕ may be thought of as specifying the right and left position of the sinusoid with respect to the time axis. Increasing ϕ moves the sinusoid to the left, that is, toward earlier times, as shown in Fig. 6.3. Since it is an angle, ϕ can be specified either in degrees or in radians. However, note that ω is usually stated in radians per second. If a sum such as $\omega t + \phi$ is formed, one must take care that ωt and ϕ are stated in the same units.

The sinusoid of Eq. (6.1) happens to be expressed in terms of a cosine function. However, the use of the cosine function instead of the sine is an arbitrary choice; $f(t)$ could just as well be specified in terms of a sine function. There is a trigonometric identity that states that $\cos \alpha = \sin (\alpha + \pi/2 \text{ radians})$, where α is any angle. Apply-

---

*Until the 1960s ordinary frequencies were stated in cycles per second. This fine, self-descriptive old unit has now been dropped from standard usage.

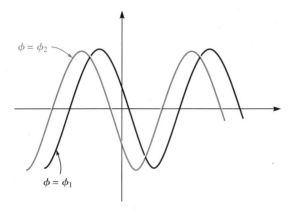

FIGURE 6.3 Two sinusoids having the same frequency and amplitude but different phase angles. Note that ϕ_2 is *larger* than ϕ_1.

ing this identity to Eq. (6.1), we can find an alternative expression for the same sinusoid in terms of the sine function:

$$f(t) = A \cos(\omega t + \phi) = A \sin(\omega t + \phi + \pi/2) \qquad \textbf{(6.8)}$$

EXAMPLE 6.2

We wish to convert the function $g(t) = B \sin(\omega t - 37°)$ to the form $g(t) = B \cos(\omega t + \phi)$, where ϕ is given in radians. Evaluate ϕ.

SOLUTION

We can use the trigonometric identity $\sin x = \cos(x - \pi/2 \text{ radians})$. Thus $g(t) = B \cos(\omega t - 37° - \pi/2 \text{ radians})$. The angle $37°$ must now be expressed in radians. One radian equals approximately $57.3°$; therefore $37° \cong 0.65$ radians. Thus $\phi = -0.65 - \pi/2 \cong -0.65 - 1.57 \cong -2.22$ radians.

Besides the form $f(t) = A \cos(\omega t + \phi)$, there is another alternative way in which the same sinusoid can be described: We may write $f(t) = B \cos \omega t + C \sin \omega t$. Of course there is a relationship between the quantities A and ϕ, on the one hand, and B and C, on the other, used to describe the same sinusoid. If A and ϕ are given, it is easy to find B and C using trigonometric identities.

EXAMPLE 6.3

Let $v(t) = V_0 \cos(\omega t + \phi)$. We wish to express $v(t)$ in the form $V_1 \cos \omega t + V_2 \sin \omega t$. Find V_1 and V_2.

SOLUTION

We must have

$$V_0 \cos(\omega t + \phi) = V_1 \cos \omega t + V_2 \sin \omega t$$

Expanding the left side,

$$V_0(\cos \omega t \cos \phi - \sin \omega t \sin \phi) = V_1 \cos \omega t + V_2 \sin \omega t$$

This equation must be true at all times. This can be the case only if the coefficient of $\cos \omega t$ on the left is equal to the coefficient of $\cos \omega t$ on the right, and similarly for $\sin \omega t$. This situation requires

$$V_1 = V_0 \cos \phi$$
$$V_2 = -V_0 \sin \phi$$

One can also express a sinusoid in the form $A \cos(\omega t + \phi)$ if it is given in the form $B \cos \omega t + C \sin \omega t$. This is done, as in Example 6.3, by equating coefficients of $\cos \omega t$ and $\sin \omega t$.

EXAMPLE 6.4

Let $i(t) = I_1 \cos \omega t + I_2 \sin \omega t$. The same current is to be expressed in the form $i(t) = I_0 \cos(\omega t + \phi)$. Find I_0 and ϕ.

SOLUTION

We must have

$$I_1 \cos \omega t + I_2 \sin \omega t = I_0 \cos(\omega t + \phi)$$
$$= I_0(\cos \omega t \cos \phi - \sin \omega t \sin \phi)$$

Equating coefficients of $\cos \omega t$

$$I_1 = I_0 \cos \phi \tag{A}$$

Equating coefficients of sin ωt

$$I_2 = -I_0 \sin \phi \qquad \text{(B)}$$

If we square Eqs. (A) and (B) and add them, we find

$$I_1^2 + I_2^2 = I_0^2 (\cos^2 \phi + \sin^2 \phi) = I_0^2$$

Thus

$$I_0 = \sqrt{I_1^2 + I_2^2}$$

To find ϕ, we divide Eq. (B) by Eq. (A), obtaining

$$\tan \phi = -\frac{I_2}{I_1}$$

The formulas derived in this example are useful. It may be helpful in remembering them if we imagine that I_1 and I_2 are sides of a right triangle, as shown in Fig. 6.4.

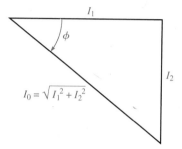

FIGURE 6.4

● EXERCISE 6.1

The sinusoid $2 \cos \omega t + 3 \cos (\omega t - 45°)$ is to be expressed in the form $V_0 \cos (\omega t + \phi)$. Find V_0 and ϕ. **Answer:** $V_0 = 4.635$; $\phi = -27.24°$.

DESCRIPTION IN TERMS OF AMPLITUDE AND PHASE

There is a commonly used "shorthand method" for the description of sinusoids. In this method one simply specifies the amplitude A and the phase ϕ, using the notation $A \angle \phi$. For example, instead of stating $v(t) = 27$ V $\cdot \cos (\omega t + 60°)$, in this notation we would say $v(t) = 27$ V $\angle 60°$. The frequency of the sinusoid is not described by this notation and must be specified separately. This is not much of a shortcoming, as in sinusoidal steady-state situations all signals in the circuit usually have the same frequency. The amplitude-and-phase description is the basis of the important phasor notation, to be introduced in Section 6.3.

EXAMPLE 6.5

A voltage $v_1(t)$ is described by the sinusoid 160 V $\angle 37°$. The ordinary frequency is 60 Hz. What is $v_1(t)$ at $t = 34$ msec?

SOLUTION

The desired voltage is

$$v_1(t = 34 \text{ msec}) = 160 \text{ V} \cdot \cos [2\pi(60)(0.034) \text{ radians} + 37°]$$

where the factor 2π has been inserted to convert ordinary frequency to angular frequency in radians per second. In computing the argument of the cosine, it must be remembered that the first term is now in radians and the second in degrees. Converting the latter to radians, we find

$$v_1(t = 34 \text{ msec}) = 160 \text{ V} \cdot \cos [12.82 \text{ radians} + 0.65 \text{ radians}]$$

$$= 160 \text{ V} \cdot \cos [13.46 \text{ radians}]$$

$$= 160 \text{ V} \cdot \cos [(13.46 - 4\pi) \text{ radians}]$$

$$= 160 \text{ V} \cdot \cos [0.90 \text{ radians}] = 99 \text{ V}$$

6.2 Review of Complex Numbers

In Section 6.3 we shall introduce special numbers, known as *phasors*, which are used to describe sinusoids. It is assumed that the frequency of a sinusoid is already known. Thus to describe it we need only specify its amplitude and its phase. This seems to indicate that we need *two* numbers to describe the sinusoid; how can it be described with only one? The trick is to describe the sinusoid by a single *complex number*. Each complex number is really composed of two separate real numbers, known as its real and imaginary parts. Thus a single complex number conveys two separate pieces of information and can be used to represent a sinusoid. Exactly how this is done will be discussed in Section 6.3. First let us briefly review the arithmetic of complex numbers.

Ordinary numbers, such as 2.3, $-3/4$, or π, are known as *real numbers*. The square of any real number is always a positive real number. A second family of numbers includes those whose squares are negative real numbers. Such numbers are called *imaginary numbers*. The particular imaginary number which is the square root of -1 is given the symbol j. Other imaginary numbers are expressed as multiples of j. For instance, $\sqrt{-4} = \sqrt{(4)(-1)} = (\sqrt{4})(\sqrt{-1}) = 2j$. Since $j^2 = -1$, we note that $1/j = -j$.

A number that is the sum of a real number and an imaginary number is called a *complex number*. For instance, if x and y are any two real numbers, $z = x + jy$ is a complex number. In this book the symbols for complex numbers are printed in bold-face type—for example, $\mathbf{z}$.

If $\mathbf{z} = x + jy$, we call x the *real part* of $\mathbf{z}$, abbreviated Re ($\mathbf{z}$). We shall refer to jy as the *imaginary part* of $\mathbf{z}$, abbreviated Im ($\mathbf{z}$). To every complex number there corresponds a second complex number known as its *complex conjugate*. If $\mathbf{z} = x + jy$, the complex conjugate of $\mathbf{z}$, whose symbol is $\mathbf{z}^*$, is by definition equal to $x - jy$. As a practical matter, the complex conjugate of any complex number can be found by reversing the algebraic sign before every term in which j appears. For example, the complex conjugate of $(1 + 2j)/(3 - 4j)$ is equal to $(1 - 2j)/(3 + 4j)$. When a number is multiplied by its own complex conjugate, the product is a real number. This may be proven as follows: $\mathbf{z}\mathbf{z}^* = (x + jy)(x - jy) = x^2 + jxy - jxy - j^2y^2 = x^2 + y^2$. Since x and y are both real, $\mathbf{z}\mathbf{z}^*$ is also real. The number $\mathbf{z}\mathbf{z}^*$ is given the symbol $|\mathbf{z}|^2$ and is called the *absolute square* of $\mathbf{z}$.

EXAMPLE 6.6

Find the real and imaginary parts of the complex number

$$\mathbf{z} = \frac{2 + 3j}{4 + 5j}$$

What is its complex conjugate? Its absolute square?

SOLUTION

To find the real and imaginary parts, it is necessary to express the number in the form $x + jy$. This can be done by multiplying the numerator and denominator of $\mathbf{z}$ by the complex conjugate of the denominator. Doing this, we obtain

$$\mathbf{z} = \frac{(2 + 3j)(4 - 5j)}{(4 + 5j)(4 - 5j)} = \frac{8 + 12j - 10j + 15}{16 + 25} = \frac{23 + 2j}{41} = \frac{23}{41} + \frac{2}{41}j$$

The real part of $\mathbf{z}$ is therefore 23/41, and its imaginary part is $2j/41$. The complex conjugate is $\mathbf{z}^* = 23/41 - 2j/41$. The absolute square is

$$|\mathbf{z}|^2 = \mathbf{z}\mathbf{z}^* = \left(\frac{23}{41} + \frac{2}{41}j\right)\left(\frac{23}{41} - \frac{2}{41}j\right) = \left(\frac{23}{41}\right)^2 + \left(\frac{2}{41}\right)^2 = \frac{533}{1681}$$

An alternative way to obtain the complex conjugate (although in a less simple form) is to reverse the sign of all terms containing j in the original expression for $\mathbf{z}$. Thus

$$\mathbf{z}^* = \frac{2 - 3j}{4 - 5j}$$

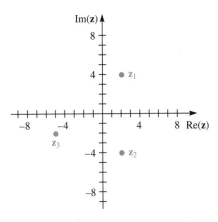

FIGURE 6.5 **FIGURE 6.5** Locations of three complex numbers in the complex plane: $z_1 = 2 + 4j$; $z_2 = 2 - 4j$; $z_3 = -5 - 1.5j$. Note that $z_2 = z_1^*$.

The value of a complex number can be depicted graphically as a point in the *x-y* plane. One simply plots the value of *x* as the *x*-coordinate of the point and the value of *y* as the *y*-coordinate. One speaks of the *x-y* plane as the *complex plane*. Several examples of numbers in the complex plane are shown in Fig. 6.5.

When the number **z** is expressed in the form $x + jy$, it is said to be expressed in *rectangular form* because it is specified in terms of its rectangular coordinates in the complex plane. It is also possible to specify the same number by means of its polar coordinates. The relationship between the two descriptions is shown in Fig. 6.6. From trigonometry it is immediately clear that the length of the radius vector *M* is given by

$$M = \sqrt{x^2 + y^2} \tag{6.9}$$

and that the polar angle θ is given by*

$$\theta = \tan^{-1}\left(\frac{y}{x}\right) \tag{6.10}$$

Conversely, it is also evident from Fig. 6.6 that

$$x = M \cos \theta \tag{6.11}$$

and

$$y = M \sin \theta \tag{6.12}$$

Thus the complex number $z = x + jy$ may be expressed in the form $z = M \cos \theta + jM \sin \theta$, or

$$z = M(\cos \theta + j \sin \theta) \tag{6.13}$$

---

* It should be noted that Eq. (6.10) contains an ambiguity with regard to the quadrant in which θ lies. For example, if y/x is a positive number, θ could, according to Eq. (6.10), lie in either the first or third quadrants. The ambiguity is removed by noting the signs of *y* and *x* individually. For instance, if *y* is a positive number we know θ must lie in either the first quadrant or the second.

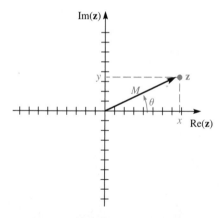

FIGURE 6.6 Relationship between rectangular and polar representations for the complex number **z**. By trigonometry we see that $z = x + jy = M(\cos \theta + j \sin \theta)$. The radius vector *M* is called the absolute value of the complex number, and is given the symbol $|z|$.

When expressed as in Eq. (6.13), **z** is said to be expressed in its *polar form*.

The radius vector *M* is known as the *absolute value* of the complex number **z**, and it is given the symbol $|z|$. We have already seen that $zz^* = x^2 + y^2$. Thus from Eq. (6.9)

$$|z| = M = \sqrt{zz^*} \tag{6.14}$$

Equation (6.14) is a very useful formula. The polar angle θ is called the *argument* of the complex number **z**. Its value is given by Eq. (6.10).

EXAMPLE 6.7

What is the absolute value of the number $z = 6.4 - 5.6j$?

SOLUTION

The absolute value $|z|$ is obtained from Eq. (6.14). Here $z^* = 6.4 + 5.6j$. The product zz^* is given by

$$zz^* = (6.4 - 5.6j)(6.4 + 5.6j) = (6.4)^2 - (5.6j)(6.4) + (5.6j)(6.4) + (5.6)^2$$

$$= (6.4)^2 + (5.6)^2 \cong 72$$

Therefore

$$|z| = \sqrt{zz^*} = \sqrt{72} \cong 8.5$$

Still a third way of expressing a complex number is by a mathematical identity known as *Euler's formula*. This identity states that if θ is any angle, then

$$e^{j\theta} \equiv \cos\theta + j\sin\theta \tag{6.15}$$

[Euler's formula may be proved by showing that the power series expansions for the left and right sides of Eq. (6.15) are identical.] As important special cases of Eq. (6.15), note that $e^{j(0)} = 1$, $e^{j(\pi/2)} = j$, $e^{j(\pi)} = -1$, and $e^{j(3\pi/2)} = e^{-j(\pi/2)} = -j$. Comparing Eq. (6.15) with Eq. (6.13), we now find that

$$\mathbf{z} = Me^{j\theta} \tag{6.16}$$

where the values of M and θ are still as given by Eqs. (6.9) and (6.10). When expressed as in Eq. (6.16), $\mathbf{z}$ is said to be expressed in its *exponential form*.

We can show that if $\mathbf{z} = Me^{j\theta}$, then $\mathbf{z}^* = Me^{-j\theta}$. Referring to Eq. (6.15), $\mathbf{z} = M\cos\theta + jM\sin\theta$. Therefore $\mathbf{z}^* = M\cos\theta - jM\sin\theta = M\cos(-\theta) + jM\sin(-\theta)$. Now comparing again with Eq. (6.15), we see that $\mathbf{z}^* = Me^{j(-\theta)} = Me^{-j\theta}$.

The various definitions and relationships derived above are summarized in Table 6.1.

TABLE 6.1   Properties of Complex Numbers

Alternative Expressions for the Complex Number $\mathbf{z}$:
   Rectangular: $\mathbf{z} = x + jy$
   Polar: $\mathbf{z} = M(\cos\theta + j\sin\theta)$
   Exponential: $\mathbf{z} = Me^{j\theta}$

Relationships Between Expressions:
   To convert from rectangular to polar or exponential forms:
   $$M = \sqrt{x^2 + y^2}$$

   $$\theta = \tan^{-1}\frac{y}{x}$$

To convert from polar or exponential to rectangular form:
   $x = M\cos\theta$
   $y = M\sin\theta$

Complex Conjugate of $\mathbf{z}$:

$\mathbf{z} = x + jy$	$\mathbf{z}^* = x - jy$
$\mathbf{z} = \dfrac{A + jB}{C + jD}$	$\mathbf{z}^* = \dfrac{A - jB}{C - jD}$
$\mathbf{z} = M(\cos\theta + j\sin\theta)$	$\mathbf{z}^* = M(\cos\theta - j\sin\theta)$
$\mathbf{z} = Me^{j\theta}$	$\mathbf{z}^* = Me^{-j\theta}$

Absolute Value of $\mathbf{z}$:
   $|\mathbf{z}| = M$
   $|\mathbf{z}| = \sqrt{\mathbf{z}\mathbf{z}^*}$

The manipulation of complex numbers follows the rules of ordinary algebra, with the added rule that $j^2 = -1$. Note that when two numbers expressed in exponential form are multiplied, the absolute value of the product is the product of the absolute values, but the argument of the product is the *sum* of the arguments. For example, if $z_1 = M_1 e^{j\theta_1}$, and $z_2 = M_2 e^{j\theta_2}$, then, following the usual rules of algebra, $z_1 z_2 = M_1 M_2 e^{j\theta_1 + j\theta_2} = M_1 M_2 e^{j(\theta_1 + \theta_2)}$.

EXAMPLE 6.8

Divide $3.1 e^{j(1.8)}$ by $[-3.6 + 2.9j]$ and express the quotient in exponential form.

SOLUTION

Let us first convert $-3.6 + 2.9j$ to exponential form. We use Eq. (6.16), where M and θ are given by Eqs. (6.9) and (6.10). Thus

$$M = \sqrt{(3.6)^2 + (2.9)^2} = 4.6$$

$$\theta = \tan^{-1} \frac{2.9}{-3.6}$$

We must be careful to select the proper quadrant for θ. Inasmuch as the real part of the number is negative and the imaginary part positive, θ must lie in the second quadrant. Thus we have

$$\theta = 141° = 2.46 \text{ radians}$$

and therefore

$$-3.6 + 2.9j = 4.6 e^{2.46j}$$

The required quotient is

$$Q = \frac{3.1 e^{1.8j}}{4.6 e^{2.46j}}$$

$$= 0.67 e^{j(1.8 - 2.46)} = 0.67 e^{-0.66j}$$

EXAMPLE 6.9

Let $z_1 = 3.9 e^{j(4.2)}$ and $z_2 = 0.63 e^{-j(1.8)}$. Calculate $\text{Re}(z_1 z_2^*)$.

SOLUTION

The complex conjugate of z_2 is $z_2^* = 0.63 e^{+j(1.8)}$. The product of z_1 and z_2^* is

$$\mathbf{z_1 z_2}^* \ = \ 3.9 e^{j(4.2)} \cdot 0.63 e^{j(1.8)} \ = \ (3.9)(0.63) e^{j(4.2 \ + \ 1.8)}$$

$$= \ 2.46 e^{j(6.0)}$$

To find the real part of this number, we can use Eq. (6.11):

$$\text{Re} \ (\mathbf{z_1 z_2}^*) \ = \ 2.46 \cos{(6.0)} \ = \ 2.46 \cos{(-16°)}$$

$$= \ (2.46)(0.96) \ = \ 2.36$$

● **EXERCISE 6.2**

The complex number

$$\frac{3 \ + \ 2j}{2 \ - \ 3j} \ + \ 3 e^{j(-50°)}$$

is to be expressed in the exponential form $A e^{j\theta}$. Find A and θ. **Answer:** $A \ = \ 2.324$; $\theta \ = \ -33.95°$.

6.3 Phasors

We shall now introduce those particular complex numbers known as phasors. A phasor is a complex number used to represent a sinusoid. We shall be considering the steady-state forced response of passive circuits. It is a characteristic of linear circuits that if the sinusoidal source driving such a circuit has frequency ω, all voltages and currents in the circuit will be sinusoids with this same frequency. It hence is not necessary for the phasor to contain information about the frequency of whatever sinusoid it represents, as this is already known to be the driving frequency. The other two parameters necessary to describe a sinusoid are its amplitude and phase. Thus the phasor that represents a particular sinusoid is a complex number containing information about the sinusoid's amplitude and phase.

It should be carefully noted that phasors are *not* time-varying quantities themselves. A phasor is a *constant* complex number that tells us the amplitude and phase of a certain sinusoid. To each sinusoidal function there corresponds a phasor, and to each phasor there corresponds a particular sinusoid.

Let us consider a time-varying voltage $v(t)$ given by $v(t) \ = \ V_0 \cos{(\omega t \ + \ \phi)}$. We shall denote the phasor representing this sinusoid by the symbol **v**. We use the same symbol for the phasor—small letter "v"—as for the quantity it represents, but the phasor is a complex number and hence is set in boldface type. We now must ask how to find the value of the complex number **v** so that it will contain the desired amplitude and phase information. This is done according to the following rule:

RULE 1

If a sinusoid is described by the formula $v(t) = V_0 \cos(\omega t + \phi)$, the phasor representing the sinusoid is $\mathbf{v} = V_0 e^{j\phi}$.

Clearly the phasor defined in this way contains both amplitude and phase information. The amplitude of the sinusoid is equal to the absolute value of the phasor, and the phase of the sinusoid is equal to the argument of the phasor.

EXAMPLE 6.10

A certain current is given by the formula $i(t) = I_0 \sin(\omega t + \theta)$, where θ is a constant angle. What is the corresponding phasor $\mathbf{i}$?

SOLUTION

The function $i(t)$ must first be put into the form of a cosine function. This may be done using the trigonometric identity $\sin \alpha = \cos(\alpha - \pi/2)$. Thus

$$i(t) = I_0 \cos[\omega t + \theta - \pi/2]$$
$$= I_0 \cos[\omega t + (\theta - \pi/2)]$$

which now has the desired form. The required phasor is therefore

$$\mathbf{i} = I_0 e^{j(\theta - \pi/2)} = I_0 e^{j\theta - j\pi/2} = -jI_0 e^{j\theta}$$

$$\frac{I e^{j\theta}}{e^{j\pi/2}} = \frac{I e^{j\theta}}{j}$$

We also need the converse rule, which tells us how to find the sinusoid corresponding to a given phasor. This is done as follows:

RULE 2

To obtain the sinusoid corresponding to a given phasor, multiply the phasor by $e^{j\omega t}$ and take the real part. Thus the sinusoid corresponding to the phasor $\mathbf{v}$ is $\mathrm{Re}\,(\mathbf{v}e^{j\omega t})$.

It is of course necessary that this rule be consistent with the first one. Let us make sure that this is true. Beginning with a sinusoid $v(t) = V_0 \cos(\omega t + \phi)$, Rule 1 tells us that the corresponding phasor is $V_0 e^{j\phi}$. Now, according to Rule 2, the sinusoid corresponding to this phasor is $\mathrm{Re}\,(V_0 e^{j\phi} e^{j\omega t}) = \mathrm{Re}\,(V_0 e^{j(\omega t + \phi)}) = \mathrm{Re}\,\{V_0[\cos(\omega t + \phi) + j\sin(\omega t + \phi)]\} = V_0 \cos(\omega t + \phi)$. This is the same as the original sinusoid, and thus the two rules are consistent.

Often we need to know the *amplitude* of a sinusoid when we know its phasor. This is easily found if we remember that the amplitude of the sinusoid is equal to the absolute value of its phasor.

EXAMPLE 6.11

A sinusoid is represented by the phasor $\mathbf{f} = 2 - 3j$. Find A, the amplitude of the sinusoid.

SOLUTION

The amplitude is equal to the absolute value of f. Thus

$$A = |\mathbf{f}| = \sqrt{\mathbf{f}\mathbf{f}^*}$$
$$= \sqrt{(2 - 3j)(2 + 3j)} = \sqrt{4 + 9} = \sqrt{13}$$

We may also wish to know the phase of a sinusoid given the phasor that represents it. This is done by recalling that the phase of the sinusoid is equal to the argument of the phasor.

EXAMPLE 6.12

A sinusoid is represented by the phasor $\mathbf{f} = 2 - 3j$. Find ϕ, the phase angle of the sinusoid.

SOLUTION

The phase angle ϕ is equal to the argument of the phasor $\mathbf{f}$. According to Eq. (6.10), this is given by $\phi = \tan^{-1}(y/x)$, where $jy = \text{Im}(\mathbf{f})$ and $x = \text{Re}(\mathbf{f})$. In this case $y = -3$ and $x = 2$. Thus $\phi = \tan^{-1}(-3/2) \cong -56.3°$.

We note that Examples 6.11 and 6.12 provide an alternative way of finding the sinusoid when the phasor is known:

RULE 3

    a. The amplitude of the sinusoid corresponding to a phasor $\mathbf{v}$ is $|\mathbf{v}|$.

    b. The phase of the sinusoid corresponding to a phasor $\mathbf{v}$ is arg $\mathbf{v}$. [That is, if $\mathbf{v} = A + jB$, the phase angle is $\tan^{-1}(B/A)$.]

This rule will in many cases be more convenient to use than Rule 2.

It is quite easy to show that *the phasor representing the sum of two sinusoids of the same frequency is the sum of the phasors representing the individual sinusoids.* This is important for the calculations we shall wish to perform later on.

EXAMPLE 6.13

Find the phasor corresponding to the sinusoid $V_0 \cos \omega t + V_0 \sin \omega t$.

SOLUTION

The phasor representing $V_0 \cos \omega t$ is $V_0 e^{j(0)} = V_0$. To find the phasor representing $V_0 \sin \omega t$ we note that $\sin \omega t = \cos(\omega t - \pi/2)$. Thus the phasor for $V_0 \sin \omega t$ is $V_0 e^{-j\pi/2} = -jV_0$. The phasor representing the sum of the sinusoids is thus $V_0 - jV_0 = V_0(1 - j)$.

EXAMPLE 6.14

Let $v_1(t) = 2.3 \cos(\omega t + 0.6)$ and $v_2(t) = 1.9 \cos(\omega t - 1.0)$. Find the phasor representing $v_3(t) = v_1(t) + v_2(t)$. Express the answer in rectangular and in exponential form.

SOLUTION

$$\mathbf{v}_3 = \mathbf{v}_1 + \mathbf{v}_2 = 2.3e^{j(0.6)} + 1.9e^{j(-1.0)}$$

One way to simplify this expression is to convert the complex numbers to rectangular form by means of Eqs. (6.16), (6.14), and (6.15):

$$\mathbf{v}_3 = 2.3[\cos(0.6) + j \sin(0.6)] + 1.9[\cos(-1.0) + j \sin(-1.0)]$$

$$= [2.3 \cos(0.6) + 1.9 \cos(-1.0)] + j[2.3 \sin(0.6) + 1.9 \sin(-1.0)]$$

$$\cong 2.9 - j(0.3)$$

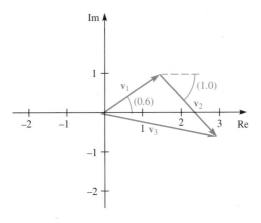

FIGURE 6.7

The addition just performed appears graphically in Fig. 6.7 (the numbers in parentheses are angles in radians). The complex number $v_3 = 2.9 - j(0.3)$ is the sum, expressed simply in rectangular form. It may be desired to express it in exponential form, so that the amplitude or the phase of the sinusoid $v_3(t)$ can be found. This is done by means of Eq. (6.9):

$$M = \sqrt{(2.9)^2 + (-0.3)^2} = 2.92$$

$$\phi = \tan^{-1}(-0.3/2.9) = -0.10 \text{ radian}$$

Therefore $v_3 = 2.92e^{j(-0.10)}$. The amplitude of the sinusoid, which is the sum of v_1 and v_2, is 2.92, and its phase is -0.10 radian.

It is also easy to show that if several sinusoids add to zero, the sum of their phasors is also zero.

The properties of phasors are summarized in Table 6.2.

It is important to become proficient with calculations involving phasors. This skill is best developed through practice, which can be obtained by working as many as possible of the relevant problems at the end of this chapter.

TABLE 6.2   Properties of Phasors

Finding Phasor When Sinusoid Is Known:

Express sinusoid in the form $f(t) = M \cos(\omega t + \theta)$. The phasor representing $f(t)$ is then equal to $Me^{j\theta}$.

Finding Sinusoid When Phasor Is Known:

1. The sinusoid corresponding to the phasor f is $f(t) = \text{Re}(fe^{j\omega t})$.
2. If $f = Me^{j\theta}$, $f(t) = M \cos(\omega t + \theta)$.
3. The amplitude of the sinusoid $f(t)$ is equal to $|f|$.
   a. If $f = x + jy$, the amplitude is $\sqrt{x^2 + y^2}$.
   b. If $f = Me^{j\theta}$, the amplitude is M.
4. The phase angle of the sinusoid is equal to the argument of f.
   a. If $f = x + jy$, $\theta = \tan^{-1}(y/x)$.
   b. If $f = Me^{j\theta}$, θ is the phase angle.

Sum of Sinusoids of the Same Frequency:

1. If $f(t)$ and $g(t)$ are sinusoids of the same frequency whose phasors are f and g, respectively, the phasor representing the sinusoid $f(t) + g(t)$ is $f + g$.
2. If the sum of several sinusoids of the same frequency is zero, the sum of their phasors is zero.

You may wish to test your skills with the examples that follow.

EXAMPLE 6.15

Find the amplitude and phase of the sinusoid obtained by adding $3\cos(\omega t + 20°)$ to $4\cos(\omega t - 100°)$.

SOLUTION

It will be sufficient to find the phasor $\mathbf{v}_s$ representing the sum of these sinusoids, expressed in the exponential form $\mathbf{v}_s = Me^{j\theta}$. The answer to the problem will then be amplitude $= M$, phase $= \theta$. Proceeding, we have

$$\mathbf{v}_s = 3e^{j20°} + 4e^{-j100°}$$

$$= 3(0.94 + 0.34j) + 4(-0.17 - 0.98j) = 2.11 - 2.90j$$

$$M = \sqrt{(2.11)^2 + (2.90)^2} = 3.59$$

$$\theta = \tan^{-1}(-2.90/2.11) = -53.96°$$

[handwritten: $3.59\ \cos(\omega t - 54.96°)$]

EXAMPLE 6.16

Find the amplitude M and phase θ of the sinusoid whose phasor is $\mathbf{v} = (4 + 6j)/(-3 + 4j)$.

SOLUTION

First the phasor can be converted to simple rectangular form by multiplying numerator and denominator by the complex conjugate of the denominator:

$$\mathbf{v} = \frac{(4 + 6j)(-3 - 4j)}{(-3 + 4j)(-3 - 4j)} = \frac{12 - 34j}{25} = 0.48 - 1.36j$$

Hence

$$M = \sqrt{(0.48)^2 + (1.36)^2} = 1.44$$

$$\theta = \tan^{-1}(-1.36/0.48) = -71°$$

EXAMPLE 6.17

Find the real part of $\mathbf{z} = (26 - 17j)e^{j\omega t}$. Express the result in the form $M \cos(\omega t + \phi)$.

SOLUTION

It is convenient to first express $(26 - 17j)$ in the form $Ae^{j\theta}$:

$$A = \sqrt{(26)^2 + (17)^2} = 31.1$$

$$\theta = \tan^{-1}\frac{-17}{26} = -33.2°$$

Thus

$$\mathbf{z} = (31.1)e^{j(-33.2°)}e^{j\omega t} = 31.1e^{j(\omega t - 33.2°)}$$

$$= 31.1[\cos(\omega t - 33.2°) + j\sin(\omega t - 33.2°)]$$

Hence $\mathrm{Re}(\mathbf{z}) = 31.1\cos(\omega t - 33.2°)$.

EXAMPLE 6.18

Let $\mathbf{v} = 1 + 2j$ and $\mathbf{i} = 2 - 3j$. Find $\mathrm{Re}(\mathbf{vi}^*)$.

SOLUTION

$$\mathbf{vi}^* = (1 + 2j)(2 + 3j) = -4 + 7j$$

Thus $\mathrm{Re}(\mathbf{vi}^*) = -4$.

• EXERCISE 6.3

The output sinusoid of a circuit is represented by the phasor

$$\mathbf{v} = \frac{4e^{j(3/2)}(2 - 3j)}{(-1 + 0.5j)} \quad \text{volts}$$

Find the amplitude A and phase angle ϕ of the sinusoid. (The exponent 3/2 is in radians.) **Answer:** $A = 12.9$ V; $\phi = -2.16$ radians.

EXAMPLE 6.19

Let $\mathbf{v} = 2e^{j15°}$ and $\mathbf{i} = 5e^{j30°}$. Find Re $(\mathbf{vi}^*)$.

SOLUTION

$$\mathbf{vi}^* = (2e^{j15°})(5e^{-j30°}) = 10e^{-j15°}$$
$$= 10(\cos 15° - j \sin 15°)$$
$$\text{Re}(\mathbf{vi}^*) = 10 \cos(-15°) = 9.66$$

EXAMPLE 6.20

Show that the real part of the sum of two complex numbers is equal to the sum of their real parts.

SOLUTION

Let the two numbers be $\mathbf{z}_1 = A + jB$ and $\mathbf{z}_2 = C + jD$. The sum of their real parts is $A + C$. The sum of the two numbers is $\mathbf{z}_1 + \mathbf{z}_2 = (A + C) + j(B + D)$. Thus Re $(\mathbf{z}_1 + \mathbf{z}_2)$ is $A + C$, which was to be shown.

EXAMPLE 6.21

Prove that if $\mathbf{v}_1$ is the phasor representing the sinusoid $v_1(t)$ and $\mathbf{v}_2$ represents the sinusoid $v_2(t)$ with the same frequency, then $\mathbf{v}_3$, the phasor representing the sinusoid $v_3(t) = v_1(t) + v_2(t)$, is obtained from the formula $\mathbf{v}_3 = \mathbf{v}_1 + \mathbf{v}_2$.

SOLUTION

Let us see what sinusoid v_3 represents. From Rule 2 this sinusoid is
$$f(t) = \text{Re}(\mathbf{v}_3 e^{j\omega t}) = \text{Re}[(\mathbf{v}_1 + \mathbf{v}_2)e^{j\omega t}]$$
Using the result of the previous example,
$$f(t) = \text{Re}(\mathbf{v}_1 e^{j\omega t}) + \text{Re}(\mathbf{v}_2 e^{j\omega t})$$
$$= v_1(t) + v_2(t)$$
which was to be shown.

PHASOR REPRESENTING THE DERIVATIVE

We now arrive at the property of phasors that accounts for their usefulness in solving differential equations:

RULE 4

If $\mathbf{v}$ is the phasor representing the sinusoid $v(t)$, then the phasor representing the sinusoid dv/dt is $j\omega\mathbf{v}$.

This rule is so important that it seems worthwhile to give its proof. Suppose $\mathbf{v} = V_0 e^{j\theta}$. Then $v(t) = V_0 \cos(\omega t + \theta)$. Differentiating, we find

$$\frac{dv}{dt} = -V_0\omega \sin(\omega t + \theta)$$

$$= -V_0\omega \cos(\omega t + \theta - \pi/2)$$

The phasor representing this sinusoid is

$$-V_0\omega e^{j(\theta - \pi/2)} = -V_0\omega e^{j\theta} e^{-j(\pi/2)} = -V_0\omega e^{j\theta}(-j)$$

$$= j\omega V_0 e^{j\theta} = j\omega\mathbf{v}$$

which proves the theorem. Rule 4 is useful because it allows us to replace differentiation with simple multiplication by the constant $j\omega$, as will now be seen.

USE OF PHASORS IN CIRCUIT ANALYSIS

We are now ready to demonstrate the use of phasors representing sinusoidal voltages and currents. As a first example let us use the circuit of Fig. 6.8.

We wish to find the amplitude and phase of the voltage v, indicated in the figure, as a function of the angular frequency ω. To do this we write a node equation for node x, where the voltage is v:

$$\frac{v}{R} - \frac{V_1 \cos \omega t}{R} + C\frac{dv}{dt} = 0 \tag{6.17}$$

Since the sum of the three sinusoidal functions on the left side of Eq. (6.17) is zero, the sum of the phasors representing them must also be zero. The phasor representing

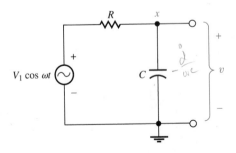

FIGURE 6.8

the sinusoid $V_1 \cos \omega t$ is $V_1 e^{j(0)} = V_1$. We denote the (as yet unknown) phasor representing $v(t)$ by **v**. Then the phasor form of Eq. (6.17) is

$$\frac{1}{R}(\mathbf{v} - V_1) + Cj\omega\mathbf{v} = 0 \qquad (6.18)$$

Solving Eq. (6.18) algebraically for **v**, we have

$$\mathbf{v} = \frac{V_1}{1 + j\omega RC} = V_1 \frac{1 - j\omega RC}{1 + \omega^2(RC)^2} \qquad (6.19)$$

The amplitude of the sinusoid $v(t)$ is equal to the absolute value of the phasor **v**:

$$|\mathbf{v}| = (\mathbf{vv}^*)^{1/2} \qquad (6.20)$$

$$= \frac{V_1}{\sqrt{1 + \omega^2(RC)^2}}$$

The phase angle of $v(t)$ is equal to the argument of **v**, which is obtained from Eq. (6.19) by means of Rule 3(b):

$$\phi = \tan^{-1}(-\omega RC) \qquad (6.21)$$

Using the results obtained in Eqs. (6.20) and (6.21), graphs can be constructed of the amplitude and phase of $v(t)$ as a function of frequency, as in Fig. 6.9. As is custom-

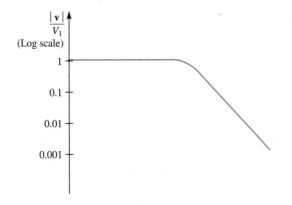

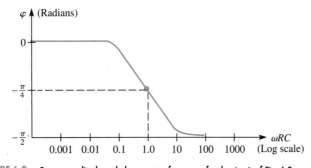

FIGURE 6.9   Output amplitude and phase versus frequency for the circuit of Fig. 6.8.

ary, logarithmic scales are used for amplitude and frequency (but not phase). We observe that, thanks to the phasor method, the results have been obtained without the need for solving any differential equations, and thus there is a considerable saving of computational effort. When the circuit to be analyzed is more complicated, perhaps containing several capacitors and inductors, the simplification afforded by the phasor method is practically indispensable.

EXAMPLE 6.22

Find the current $i(t)$ that flows in the circuit shown in Fig. 6.10.

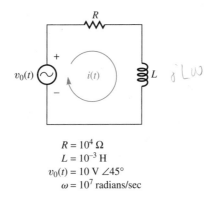

$$R = 10^4\ \Omega$$
$$L = 10^{-3}\ H$$
$$v_0(t) = 10\ V\ \angle 45°$$
$$\omega = 10^7\ \text{radians/sec}$$

FIGURE 6.10

SOLUTION

Writing a loop equation, we have

$$v_0(t)\ -\ i(t)R\ -\ L\frac{d}{dt}i(t)\ =\ 0$$

The phasor representing $v_0(t)$ is

$$\mathbf{v}_0\ =\ 10e^{j(\pi/4)}\ =\ \frac{10}{\sqrt{2}}\,(1\ +\ j)$$

The phasor form of the loop equation is therefore

$$\frac{10}{\sqrt{2}}(1\ +\ j)\ -\ \mathbf{i}R\ -\ Lj\omega\mathbf{i}\ =\ 0$$

Solving for **i**,

$$\mathbf{i} = \frac{10}{\sqrt{2}} \frac{1 + j}{(R + j\omega L)} = \frac{10}{\sqrt{2}} \frac{(R + \omega L) + j(R - \omega L)}{(R^2 + \omega^2 L^2)}$$

Using Rule 3 (compare Examples 6.11 and 6.12), the amplitude and phase of the current are

$$M = \frac{10}{\sqrt{2}} \frac{[(R + \omega L)^2 + (R - \omega L)^2]^{1/2}}{R^2 + \omega^2 L^2} = \frac{10}{\sqrt{R^2 + \omega^2 L^2}} = \frac{10^{-3}}{\sqrt{2}} \quad \text{amperes}$$

$$\phi = \tan^1 \frac{(R - \omega L)}{(R + \omega L)} = \tan^1 (0) = 0$$

The current $i(t) = 7.07 \times 10^{-4} A \angle 0°$.

6.4 Power in Sinusoidal Signals

The terms "instantaneous power" and "time-averaged power" have already been mentioned. Instantaneous power is the instantaneous product of the voltage and the current: $p(t) = v(t)i(t)$. The instantaneous power is the power being transferred from one part of the circuit to another at a particular moment in time. If the voltage and current are time-varying, then the instantaneous power will also be time-varying. It is even possible for the instantaneous power to be positive at some times and negative at others. This would simply mean that at some moments energy flows from part *A* of the circuit to part *B*, but that at other moments the direction of power is reversed, and energy moves from part *B* of the circuit to part *A*. For example, if a battery were alternately connected to a light bulb, and then to a battery charger, one would see such an effect.

In the case of ac circuits, the voltage and current both are sinusoids, and change signs rapidly. The instantaneous power in such circuits also varies rapidly in time. In a great many cases, the quantity of interest is not the instantaneous power, but rather the time-averaged power. For instance, if we are using an electric light bulb, we might wish to know how much energy it consumes per unit time. The energy *E* consumed in time *T* is equal to the time integral of the instantaneous power:

$$E = \int_0^T p(t)dt \tag{6.22}$$

This can be rewritten in the form

$$E = \left[\frac{1}{T} \int_0^T p(t)dt \right] T$$

The quantity in brackets is the time-averaged power *P* over the time interval 0 to *T*. Thus the energy consumed in time *T* is simply *PT*, the product of the time-averaged power and the time. At present we are interested in the sinusoidal steady state, in which all voltages and currents are sinusoidal and have the same period τ. In this case it is convenient to assume that the time integral in Eq. (6.22) is taken over a time of duration τ. Then it will not matter at what precise point in the cycle one begins to integrate, and the time-averaged power P is well-defined.

The time-averaged power in ac circuits can always be found by direct integration if the voltage and current are known. Suppose we have

$$v(t) = V_0 \cos(\omega t + \phi_1) \tag{6.23}$$

$$i(t) = I_0 \cos(\omega t + \phi_2)$$

Then the time-averaged power over one period is

$$P = \frac{1}{\tau} \int_0^\tau V_0 I_0 \cos(\omega t + \phi_1) \cos(\omega t + \phi_2)\, dt$$

Using the triginometric identity $\cos A \cos B = 1/2\,[\cos(A+B) + \cos(A-B)]$ this becomes

$$P = \frac{V_0 I_0}{2\tau} \int_0^\tau \left[\cos(2\omega t + \phi_1 + \phi_2) + \cos(\phi_1 - \phi_2) \right] dt$$

The integral of the first term in the integrand over a complete period vanishes. This leaves

$$P = \frac{V_0 I_0}{2} \cos(\phi_1 - \phi_2) \tag{6.24}$$

EXAMPLE 6.23

Find the time-averaged power entering the capacitor in Fig. 6.11. The voltage amplitude V_0, the angular frequency ω, and the capacitance C are given constants.

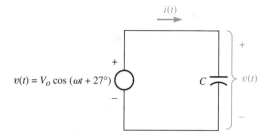

FIGURE 6.11

SOLUTION

The voltage across the capacitor is already given. The current $i(t)$ is given by

$$i(t) = C\frac{dv}{dt} = -CV_0\omega \sin(\omega t + 27°)$$

$$= CV_0\omega \cos(\omega t + 117°)$$

(where the trigonometric identity $\cos(A + 90°) = -\sin A$ has been used.) Using Eq. (6.24), we now have

$$P = \frac{1}{2}\omega V_0^2 C \cos(117° - 27°) = \frac{1}{2}\omega V_0^2 C \cos 90° = 0$$

This result should not be surprising, when we recall that the capacitor is an energy-storage element. It is incapable of converting energy into heat; whatever energy is stored in it ultimately flows out again. In the sinusoidal case, energy flows in during half of the sinusoidal cycle and flows out during the other half. Thus the time-averaged power flowing into the capacitor is zero. This is a general result: It holds for *any* capacitor in *any* sinusoidal steady-state circuit. It is also true for ideal inductors in the sinusoidal steady state.

An important special case is that in which a sinusoidal voltage $V_0 \cos(\omega t + \phi)$ is applied across a pure resistance R. In this case we have

$$i(t) = \frac{V_0}{R} \cos(\omega t + \phi)$$

and, using Eq. (6.24), we find that

$$P = \frac{1}{2}V_0\frac{V_0}{R} \cos(\phi - \phi)$$

For the special case of the pure resistance, the phase of the current is the same as that of the voltage. Thus the cosine factor becomes unity, and

$$P = \frac{V_0^2}{2R} \tag{6.25}$$

The factor of 1/2 appearing in Eq. (6.25) should be noted. In the dc case, the power dissipated in a resistance is simply V_0^2/R. However, if the voltage is sinusoidal, the instantaneous power varies with time between this value and zero. When the average over time is taken, it results in the factor of 1/2.

In order to avoid the appearance of the factor of 1/2, some workers introduce quantities known as the *rms voltage* and *rms current*. For sinusoidal signals these are simply related to the amplitudes by

$$V_{RMS} = \frac{V_0}{\sqrt{2}} \tag{6.26}$$

$$I_{RMS} = \frac{I_0}{\sqrt{2}}$$

In terms of the new parameters, we see immediately that

$$P_{AV} = \frac{V_{RMS}^2}{R} = I_{RMS}^2 R \tag{6.27}$$

The initials *rms* stand for *root-mean-square*, because the *general* formula for rms voltage (valid for all periodic signals, not just sinusoids) is

$$V_{RMS} = \sqrt{\frac{1}{T}\int_0^T v^2(t)\, dt} \tag{6.28}$$

where the integral is taken over one period. (A similar expression is used to obtain I_{RMS}.) Clearly this expression amounts to taking the square *root* of the *mean* (that is, the average) of the *square* of v, which accounts for the name. It can easily be shown that Eq. (6.27) is the correct expression for average power even when $v(t)$ is not sinusoidal, provided that V_{RMS} is calculated by means of Eq. (6.28).

TIME-AVERAGED POWER USING PHASORS

In work with sinusoidal signals, one often works with the phasors representing the voltage and current, and it is convenient to find the power in terms of the phasors. There is an important formula for finding the time-averaged power when the *phasors* representing $v(t)$ and $i(t)$ are known:

RULE 5

If $v(t)$ and $i(t)$ are represented by the phasors **v** and **i,** the time average of the product vi is given by

$$\text{Avg } (vi) = \frac{1}{2}\text{Re } (\mathbf{vi}^*)$$

Rule 5 may be proved as follows. Let $v(t)$ and $i(t)$ be as given in Eq. (6.23). Then $\mathbf{v} = V_0 e^{j\phi_1}$ and $\mathbf{i} = I_0 e^{j\phi_2}$. Thus (noting that the complex conjugate of $e^{j\phi_2}$ is $e^{-j\phi_2}$), we have

$$\frac{1}{2}\text{Re } (\mathbf{vi}^*) = \frac{V_0 I_0}{2}\text{Re } e^{j(\phi_1 - \phi_2)}$$

$$= \frac{V_0 I_0}{2}\cos (\phi_1 - \phi_2)$$

This agrees with Eq. (6.24), and thus Rule 5 has been proved.

EXAMPLE 6.24

Repeat Example 6.23, using the phasor method (Rule 5) to perform the calculation.

SOLUTION

In this case we have $\mathbf{v} = V_0 e^{j27°}$, and from Rule 4, $\mathbf{i} = Cj\omega\mathbf{v} = Ce^{j90°}\omega V_0 e^{j27°}$ (where we have used the fact that $j = e^{j90°}$.) Thus

$$P = \frac{1}{2}\,\mathrm{Re}\,[V_0 e^{j27°}\,V_0 C\,\omega e^{-j117°}]$$

$$= \frac{1}{2}\,V_0^2 C\,\omega\,\cos\,(-90°) = 0$$

in agreement with Example 6.23.

● **EXERCISE 6.4**

Suppose the phasor representing the voltage across a circuit element is $\mathbf{v} = 6 - 2j$ volts. The current through the element is $\mathbf{i} = (2 + 4j)\mathbf{v}$ amperes. What is the power entering the circuit element? **Answer:** 40 watts.

6.5 Application: Mechanical Vibrations

Sinusoidal steady-state analysis is useful not only in electronics, but in many other branches of engineering and science. Whenever one has a linear system driven by something, at some particular frequency, the techniques of these chapters can be applied. A good example is the theory of mechanical vibrations, which is of considerable importance to mechanical engineers.

Suppose, for example, that a certain airplane engine runs at 2,000 rpm. Because it is not perfectly balanced, the engine produces vibrations; every time it turns over it shakes slightly, exerting forces on the rest of the airplane. The frequency of this shaking is well-defined: There are 2,000 shakes per minute, or 33.3 per second; in other words, the frequency of the shaking is 33.3 Hz. This shaking drives the rest of the structure mechanically, just as a sinusoidal voltage source drives the rest of a circuit. The effect is that every part of the airplane vibrates, and all parts of the plane vibrate at this same frequency, the frequency of the driving force. The aeronautical engineer will be interested in these vibrations. They will be larger at some places, smaller at others. If the design of the plane is not a good one, some of these vibrations may turn out to be resonant with the driving force, in which case they might become very large. In that case the safety of the aircraft would be threatened.

Let us see how mechanical vibrations can be studied using sinusoidal analysis. Rather than talk about a structure as complicated as an airplane, we can illustrate the analysis with a much simpler system. We shall see that the mechanical system is typically composed of certain components—masses, springs, and dashpots—that behave much like capacitances, inductances, and resistances in electrical circuits.

Three typical mechanical elements are shown in Fig. 6.12. The first, the mass, obeys Newton's law $F = MA$, where F is the force, M is the mass, and A, the acceleration, is equal to the time derivative of the velocity, or, equivalently, the second derivative of the position. Thus the mass obeys

$$F = M\frac{d^2x}{dt^2}$$

Physically, we expect that we can do work on the mass by pulling on it until it starts to move; then the mass will have kinetic energy. However, we could recover the energy of

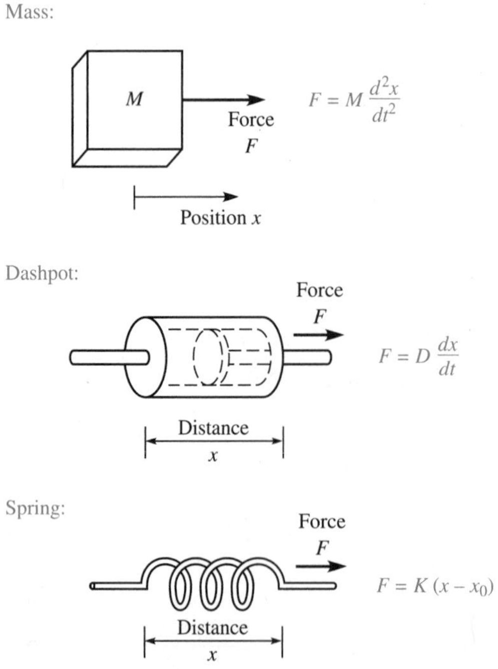

Mass:

$$F = M\frac{d^2x}{dt^2}$$

Dashpot:

$$F = D\frac{dx}{dt}$$

Spring:

$$F = K(x - x_0)$$

FIGURE 6.12 Three elements of mechanical systems. In mechanics, these play roles similar to those of *L, R,* and *C* in electric circuit theory.

the moving mass by making it turn a machine as it stops. Thus the mass will be an energy-storage element.

The second element, the dashpot, is a friction device much like the shock absorber in a car; it might consist of a plunger moving through a viscous fluid. The damping force is proportional to the velocity, and thus it obeys

$$F = D\frac{dx}{dt}$$

where D is a constant describing the drag. Physically it seems clear that we can do work on this element. If the plunger is moved back and forth, energy will be converted by friction into heat. Thus the dashpot is a dissipative element, like the resistance.

The third mechanical element is the spring. When no force is applied, the spring has its natural length x_0. When a stretching force F is applied, the length of the spring increases to a value x. The force required to produce this stretch is given by

$$F = k(x - x_0)$$

where k is the spring constant. Work that is done in stretching the spring can be recovered when the spring contracts. Thus the spring, like the mass, is an energy-storage element.

Now let us consider a mechanical system, for instance the one shown in Fig. 6.13. In this example a force $F_0 \cos \omega t$ pulls on the mass toward the right; it is resisted by the forces of the spring and the dashpot. Writing Newton's law for the mass, we have

$$M\frac{d^2x}{dt^2} = F_0 \cos \omega t - kx - D\frac{dx}{dt} \tag{6.29}$$

Here we have a linear differential equation that can readily be solved by the phasor technique. The position $x(t)$ will vary sinusoidally in time, at the driving frequency ω. Let $\mathbf{x}$ be the phasor representing $x(t)$. The derivatives are converted into multiplication by $j\omega$ according to Rule 4. Then Eq. (6.29) is turned into a phasor equation:

$$-M\omega^2\mathbf{x} = F_0 - k\mathbf{x} - j\omega D\mathbf{x} \tag{6.30}$$

The point to notice here is that we no longer have a differential equation, but only an ordinary algebraic equation. The ordinary algebraic equation is of course much easier to solve. Doing the algebra, we obtain

$$\mathbf{x} = \frac{F_0}{k + j\omega D - M\omega^2} \tag{6.31}$$

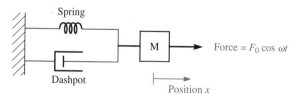

FIGURE 6.13 A mechanical system with sinusoidal excitation.

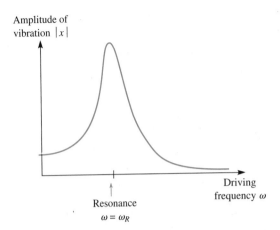

FIGURE 6.14   Vibrational amplitude versus drive frequency for the mechanical system of Fig. 6.13. Note the mechanical resonance at the frequency ω_R.

The motion of the mass will be oscillatory; it will vibrate sinusoidally back and forth at the frequency ω. The maximum displacement that occurs in this motion is the amplitude of the sinusoid represented by **x;** this amplitude is equal to the absolute value of the phasor, that is, to $|\mathbf{x}|$. Using the formula $|\mathbf{x}| = \sqrt{\mathbf{xx}^*}$, we obtain

$$|\mathbf{x}| = \frac{F_0}{\sqrt{(k - M\omega^2)^2 + (\omega D)^2}} \qquad (6.32)$$

This interesting result shows that mechanical resonance can indeed occur. The denominator of Eq. (6.32) is minimized, and thus the amplitude of the motion is maximized, when the frequency is chosen to make $(k - M\omega^2) = 0$. Note that this happens even though the magnitude of the driving force, F_0, is held constant. The resonant behavior of the vibration amplitude as a function of frequency is shown in Fig. 6.14. If the damping coefficient D happens to be small, very large vibrations may set in, when the driving frequency is close to the resonant frequency $\omega_R = \sqrt{k/M}$. In an airplane this would be very undesirable, and might cause the plane to fall apart.

There are said to have been occasions in history when a bridge has collapsed as an army marched over it. This was a result of the army's marching in step, at a speed that made the rhythm of the pounding feet come close to a resonant frequency of the structure. Even today, armies traditionally break step and walk in random time when passing over a bridge.

POINTS TO REMEMBER

- A sinusoidal signal is characterized by three parameters: amplitude, frequency, and phase.

- When a linear circuit is driven by a sinusoidal voltage or current source, all voltages and currents in the circuit are sinusoids with the same frequency as that of the source. This condition is known as the *sinusoidal steady state*.
- A phasor is a complex number that describes the amplitude and phase of a sinusoid.
- If the sinusoid is described by the formula $v(t) = V_0 \cos(\omega t + \theta)$, the phasor representing the sinusoid is $\mathbf{v} = V_0 e^{j\theta}$.
- The sinusoid corresponding to the phasor $\mathbf{v}$ is $\mathrm{Re}\,(\mathbf{v}e^{j\omega t})$.
- The amplitude of the sinusoid represented by the phasor $\mathbf{v}$ is equal to $|\mathbf{v}|$. Its phase is equal to the argument of $\mathbf{v}$.
- The phasor representing the sum of two sinusoids of the same frequency is the sum of their individual phasors.
- If the sinusoid $v(t)$ is represented by the phasor $\mathbf{v}$, then the sinusoid dv/dt is represented by the phasor $j\omega\mathbf{v}$.
- If the voltage across a resistance R is sinusoidal with amplitude V_0, the time-averaged power dissipated in the resistance is $V_0^2/2R$. Similarly, if the current through a resistance R is sinusoidal with amplitude I_0, the time-averaged power dissipated in the resistance is $I_0^2/2R$.
- The rms value of a voltage (current) is the square root of the time average of its square. If the voltage (current) varies sinusoidally in time and has amplitude V_0 (I_0), its rms value is $V_0/\sqrt{2}$ ($I_0/\sqrt{2}$).
- If a voltage with rms value V_{rms} is applied across a resistance R, the power dissipated in R is V^2_{rms}/R. We note that this is the same formula as would apply for a dc voltage.
- If $v(t)$ and $i(t)$ are a sinusoidal voltage and current with the same frequency, the instantaneous power is $i(t)v(t)$. The time-averaged power is given by Avg $[i(t)v(t)] = 1/2\,\mathrm{Re}\,(\mathbf{i}\mathbf{v}^*)$, where $\mathbf{i}$ and $\mathbf{v}$ are the phasors representing $i(t)$ and $v(t)$.
- Mechanical vibrations exhibit sinusoidal motions and often are found in the sinusoidal steady state. Thus the same mathematical methods (such as phasors) useful in electric circuit analysis can be used in problems involving mechanical vibration.

PROBLEMS

In the following problems angles (such as $377t$) are given in radians, and t is in seconds, unless otherwise indicated.

Section 6.1

6.1 The ordinary frequency of a certain power-line voltage is 60 Hz.
    a. Find the angular frequency and the period.
    b. If a second sinusoidal voltage has the same frequency as the first but a phase angle 18° larger, how much earlier will the maxima of the second occur than the maxima of the first?

6.2   A sinusoid described by $f(t) = A \cos(\omega t + \phi)$ has two maxima, at $t = 0.014$ sec and at $t = 0.018$ sec, with no other maxima between these two. Evaluate ω and ϕ. Is there more than one possible answer for ϕ?

6.3   A sinusoid with ordinary frequency 10 kHz has a maximum at $t = 0$. What is the least number of radians its phase angle ϕ can be increased to have a zero at $t = 0$?

*6.4   It is desired to express the function $f(t) = 27 \sin(18t + 47°)$ in the form $A \cos(\omega t + \phi)$. Find A, ω, and ϕ.

6.5   A sinusoid with ordinary frequency 1500 Hz is described by the notation 27 V $\angle 60°$. Express this sinusoid in the form $A \cos(\omega t + \phi)$, where ϕ is given in radians and ω in radians per second.

*6.6   Express the sinusoid $A(\sin \omega t + \cos \omega t)$ in the form $B \cos(\omega t + \phi)$. (That is, find B and ϕ.)

**6.7   The sinusoid $A \sin \omega t + B \cos \omega t$ is to be written in the form $C \sin(\omega t + \phi)$. Find C and ϕ.

*6.8   Let $v(t) = (27 \text{ V}) \cos(376t - 80°)$. Express this voltage in the form $V_1 \cos \omega t + V_2 \sin \omega t$.

Section 6.2

6.9   Consider the complex number $z = 12 + 17j$.
   a. What is $z*$?
   b. Express z in exponential form.
   c. Express $z*$ in exponential form.
   d. What is $|z|$?

6.10   Let $z_1 = 0.6e^{-0.8j}$.
   a. Find $z*_1$.
   b. Express z_1 in rectangular form.
   c. Express $z*_1$ in rectangular form.
   d. What is $|z_1|$?

6.11   Prove that $\exp[j(2.23 \text{ radians})] = j \exp[j(0.66 \text{ radian})]$.

6.12   Let $z_1 = 3 - 2j$ and $z_2 = -1 + 6j$. Find $z_1 + z_2$. Illustrate the summation by means of vectors in the complex plane.

6.13   Let $z_1 = 2 + 3j$ and $z_2 = 4 - 2j$.
   a. Calculate their product $z_1 z_2$ in rectangular form.
   b. Convert the result to exponential form.
   c. Convert z_1 and z_2 to their exponential forms.
   d. Multiply together the exponential forms obtained in part (c). The result should agree with that of part (b).

6.14   Let $z_3 = 0.2 - 0.3j$ and $z_4 = 6 + 5j$. Consider their quotient $z_4/z_3 = (6 + 5j)/(0.2 - 0.3j)$.
   a. Convert this fraction to a form in which the denominator is real. *Suggestion:* Multiply denominator and numerator by the complex conjugate of the denominator.

b. Find the real and imaginary parts of the quotient z_4/z_3.

c. Convert the result obtained in part (b) to exponential form.

d. Convert z_3 and z_4 to exponential form.

e. Compute z_4/z_3 using the exponential forms obtained in part (d). The result should agree with part (c).

6.15 If $z_1 = a + bj$ and $z_2 = c + dj$, find $|z_1/z_2|$.

6.16 Use the power series expansions of the sine, cosine, and exponential functions to verify that $e^{j\theta} = \cos\theta + j\sin\theta$.

*6.17 Prove that if z_1 and z_2 are any two complex numbers, Re $(z_1 z^*_2)$ = Re $(z_2 z^*_1)$.

Section 6.3

6.18 Find the phasors that represent the following sinusoidal functions and express them in exponential form:

a. $v_1(t) = \cos(\omega t + 10°)$ V

b. $v_2(t) = 27.6 \sin(\omega t - 54°)$ V

c. $i_1(t) = 3.4 \cos(\omega t - 0.42 \text{ radian})$ mA

6.19 Obtain the sinusoids corresponding to the following phasors and express them in the form $A \cos(\omega t + \phi)$:

a. $\mathbf{v}_1 = 12e^{j(34°)}$ V

b. $\mathbf{v}_2 = 7 + 9j$ V

c. $\mathbf{i}_1 = 6 - 2j$ mA

6.20 Find the amplitudes and phases of the sinusoids whose phasors are as follows:

a. $3 + 2j$

b. $15e^{-j(24°)}$

c. $(1.3 - 2.1j)/(0.3 - 0.8j)$

*6.21 The sinusoid $v_1(t)$ is represented by the phasor $\mathbf{v}_1 = 10 + 12j$, and another sinusoid of the same frequency $v_2(t)$ is represented by $\mathbf{v}_2 = -7 - 9j$.

a. Find the phasor corresponding to the sinusoid $v_1(t) + v_2(t)$.

b. Use the phasor found in part (a) to express $v_1(t) + v_2(t)$ in the form $A \cos(\omega t + \phi)$ (that is, find A and ϕ).

*6.22 Find the sum of $3\sin(\omega t + 28°)$ and $4\cos(\omega t - 71°)$. Express your answer in the form $A \cos(\omega t + \theta)$.

*6.23 Prove that if the sum of several sinusoids of the same frequency is zero, the sum of their phasors must be zero.

6.24 Find the phases of the sinusoids whose phasors are as follows:

a. $12 + 14j$

b. $12/15 + (14/15)j$

c. $12/15j + (14/15j)j$

*6.25 Find the amplitude of the sinusoid whose phasor is

$$\frac{(6 + 3j)e^{-j(27°)}}{2 - 6j}$$

*6.26 Find the amplitude and phase of the sinusoid $v_3(t)$ if $v_3(t) = v_2(t) - v_1(t)$, $v_2 = 48 \sin(\omega t + 230°)$, and $v_1(t) = 17 \cos(\omega t - 8°)$.

*6.27 Find the real part of

$$\frac{19e^{j(\omega t + 100°)}}{2 + 3j}$$

and express the result in the form $A \cos(\omega t + \theta)$.

6.28 Let $\mathbf{v} = 6 - 4j$ and $\mathbf{i} = 2 + 4j$.
  a. Find Re $(\mathbf{vi}^*)$.
  b. Find Re $(\mathbf{iv}^*)$.

6.29 Let $\mathbf{v} = 16e^{j(208°)}$ and $\mathbf{i} = 11e^{-j(43°)}$.
  a. Find Re $(\mathbf{vi}^*)$.
  b. Find Re $(\mathbf{iv}^*)$.

6.30 Find the amplitude and phase of the sinusoid whose phasor is

$$\frac{V_0 \omega RC}{1 + j\omega RC}$$

where V_0, R, C, and ω are all given constants.

*6.31 Find the amplitude and phase of the sinusoid represented by the phasor

$$\frac{V_0(R + j\omega L)}{(1 - \omega^2 LC) + jR\omega C}$$

where V_0, R, L, C, and ω are given constants.

6.32 The voltage across a 20 μF capacitor is $v(t) = 160 \cos(377t \text{ radians})$ V, where t is given in seconds.
  a. What is the phasor representing $v(t)$?
  b. What is the phasor representing $i(t)$, the current through the capacitor?
  c. Express $i(t)$ in the form $A \cos(\omega t + \phi)$.
  d. Do the maxima of $i(t)$ come earlier in time or later than the maxima of $v(t)$? (If the former is true, one says the current "leads" the voltage; if the latter, one says the current "lags" the voltage.)

Section 6.4

6.33 The sinusoidal voltage $v(t) = (15 \text{ V}) \cos(376t + 30°)$ is applied across a 50 Ω resistor.
  a. What is the *instantaneous* power dissipated in the resistor?
  b. What is the *time-averaged* power dissipated in the resistor?

6.34 A voltage $v(t) = (10 \text{ V}) \cos 376t - (12 \text{ V}) \sin 376t$ is applied across a 4700 Ω resistor. What is the time-averaged power dissipated in the resistor?

*6.35 A current $i(t) = (15 \text{ mA}) \cos 100t + (20 \text{ mA}) \cos(100t + 0.7 \text{ radians})$ passes through a 10,000 Ω resistor. What is the time-averaged power dissipated in the resistor?

6.36 A voltage $v(t) = (24 \text{ V}) \cos (376t - 20°)$ is applied across a 3,000 Ω resistor.

    a. What is the rms voltage across the resistor?

    b. What is the rms current through the resistor?

    c. What is the power dissipated in the resistor?

    d. Would the answer to part (c) change if the phase angle were changed?

*6.37 Let $v(t) = V_0 \cos \omega t$. Use Eq. (6.28) to find the rms value of v. The result should agree with Eq. (6.26).

*6.38 Repeat Problem 6.37 for the case $v(t) = V_0 \cos (\omega t + \phi)$.

*6.39 Find the rms value of the voltage shown in Fig. 6.15(a).

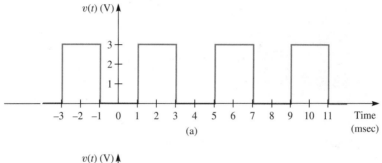

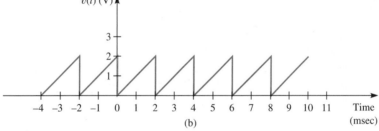

FIGURE 6.15

**6.40 Find the rms value of the voltage shown in Fig. 6.15(b).

*6.41 The current through a circuit element is $i(t) = I_1 \cos \omega t$, and the voltage across it is $v(t) = V_1 \cos (\omega t + \phi)$. Calculate the average power entering the element as a function of ϕ. What does negative power mean, physically?

6.42 A sinusoidal voltage represented by the phasor $\mathbf{v}_1$ is applied to the terminals of a resistor with value R. How much time-averaged power enters the resistor?

*6.43 Repeat the preceding problem with a capacitor C replacing the resistor.

CHAPTER 7

PHASOR ANALYSIS

P roblems involving sinusoidal voltages and currents arise so often that efficient techniques are needed. Engineers like to have powerful tools at hand, so that the expected problems can be handled quickly and effectively. When it comes to circuit problems with sinusoidal signals, the appropriate technique is *phasor analysis*. Phasor analysis is a classic tool of electrical engineers, dating back to the introduction of alternating current in the nineteenth century. Similar techniques are used in mechanical engineering, acoustics, and electromagnetics—anywhere that vibrations of one sort or another are present. Phasor analysis is a powerful and versatile method for many kinds of important problems.

We have already begun the study of phasor analysis in the last chapter, where we introduced the complex numbers known as phasors. As we have seen, a phasor is simply a complex number used to describe a sinusoid. The frequency of the sinusoid is regarded as given (because in linear circuits all voltages and currents have the same frequency). The other two parameters of the sinusoid, its amplitude and phase, are described by the phasor, according to Rules 1 and 2 in Section 6.3.

In this chapter we shall show how phasors can be used to simplify the analysis of circuits with sinusoidal voltages and currents. In fact, we shall reduce this class of problems to a form very similar to dc circuit analysis. In order to do this, we shall introduce the concept of *impedance*. As we shall see, impedance, in sinusoidal analysis, plays a role similar to that of resistance in dc analysis.

7.1 Impedance

In the case of a resistance, if the voltage across the terminals is given, the current is determined by Ohm's law. The ratio of voltage to current is equal to the constant R. That is, $v(t)/i(t) = R$. The existence of this fixed ratio of v to i gives the *I-V* relationship of the resistor a very simple form. Through the use of phasors, the *I-V* equations of inductors and capacitors can be reduced to an equally simple form.

Suppose that the phasor **v** represents the sinusoidal voltage across the terminals of a circuit element, and that the phasor representing the sinusoidal current through the element is **i.** Then *we define the impedance* **Z** *of the circuit as the ratio of the phasor representing the sinusoidal voltage across it to the phasor representing the sinusoidal current flowing through it:*

$$\frac{\mathbf{v}}{\mathbf{i}} = \mathbf{Z} \qquad (7.1)$$

Not every circuit element possesses an impedance; an ideal voltage source does not, since the voltage across a voltage source is not functionally related to the current through it, and the ratio **v/i** has no fixed value. The elements for which the definition of impedance is most useful are resistance, capacitance, and inductance. The impedances of these elements will now be found.*

For resistance, the constitutive equation is $v = iR$. Therefore $\mathbf{v} = \mathbf{i}R$, and the impedance is $\mathbf{v/i} = R$:

$$\boxed{\mathbf{Z}_R = R} \qquad (7.2)$$

The equation describing a capacitor (leaving aside the question of algebraic signs) is, according to Eq. (7.1), $i = C\frac{dv}{dt}$. Therefore the phasor representing the sinusoid i is equal to C times the phasor representing the sinusoid $\frac{dv}{dt}$. In Rule 4, Section 6.3, however, we saw that the phasor representing $\frac{dv}{dt}$ is simply $j\omega\mathbf{v}$, where **v** is the phasor representing the sinusoid v. Thus for a capacitor we have

$$\mathbf{i} = Cj\omega\mathbf{v} \qquad (7.3)$$

Thus the impedance of a capacitor, which we shall call $\mathbf{Z}_C$, is given by

$$\mathbf{Z}_C \equiv \frac{\mathbf{v}}{\mathbf{i}} = \frac{1}{j\omega C} \qquad (7.4)$$

The impedance of an ideal inductor is found similarly, using Eq. (5.2). We have

$$\mathbf{Z}_L = j\omega L \qquad (7.5)$$

Note that the symbol for an impedance (**Z**) is boldface because an impedance is a complex number. However, the use of boldface type does *not* mean that impedance is a

---

*The definition of impedance given in this paragraph is not the most general definition possible. More general definitions of impedance are used in advanced work.

phasor. Impedances describe the properties of circuit elements; phasors describe sinusoidal voltages or currents.

When the phasor voltage across a resistance, capacitance, or inductance is known, the current through it can be found immediately by means of Eqs. (7.2), (7.4), or (7.5).

EXAMPLE 7.1

The sinusoidal voltage across a 10^{-6}F capacitor is represented by the phasor $\mathbf{v} = 6e^{j\phi}$ V, where $\phi = 0.6$ radian. The frequency ω is 10^4 radians/sec. What is the amplitude of the sinusoidal current $i(t)$ through the capacitor?

SOLUTION

According to Eq. (7.4), the impedance of the capacitor is given by

$$\mathbf{Z}_C = \frac{1}{j\omega C}$$

From Eq. (7.1) we have

$$\mathbf{i} = \frac{\mathbf{v}}{\mathbf{Z}_C} = \mathbf{v}j\omega C$$

$$= 6\,\omega C e^{j(\phi + \pi/2)} = 6(10^4)(10^{-6})e^{j(\phi+\pi/2)}$$

$$= 6 \times 10^{-2} \cdot e^{j(\phi+\pi/2)} \text{ A}$$

The absolute value of $\mathbf{i}$ is 6×10^{-2} A; therefore this is the amplitude of the sinusoidal current through C.

When circuit elements are connected in series or in parallel, the impedance of the combination is obtained from rules identical in form to those for series or parallel resistances:

$$\mathbf{Z}_{\text{SER}} = \mathbf{Z}_1 + \mathbf{Z}_2 \quad \text{(impedances in series)} \tag{7.6}$$

$$\mathbf{Z}_{\text{PAR}} = \frac{\mathbf{Z}_1\mathbf{Z}_2}{\mathbf{Z}_1 + \mathbf{Z}_2} \quad \text{(impedances in parallel)} \tag{7.7}$$

(The derivations of these formulas are similar to those for resistances in series and parallel. The proofs are deferred to the problems at the end of the chapter.) Formulas (7.6) and (7.7) are very useful in ac circuit analysis.

EXAMPLE 7.2

Find the impedance of the combination of elements shown in Fig. 7.1.

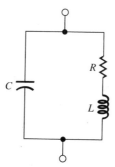

FIGURE 7.1

SOLUTION

The impedance of the combination, using Eqs. (7.6) and (7.7), is

$$\mathbf{Z} = \mathbf{Z}_C \parallel (\mathbf{Z}_R + \mathbf{Z}_L)$$

$$= \frac{\mathbf{Z}_C (\mathbf{Z}_R + \mathbf{Z}_L)}{\mathbf{Z}_C + \mathbf{Z}_R + \mathbf{Z}_L}$$

$$= \frac{(1/j\omega C)(R + j\omega L)}{(1/j\omega C) + R + j\omega L}$$

$$= \frac{R + j\omega L}{1 + j\omega RC - \omega^2 LC}$$

EXAMPLE 7.3

Suppose a current source whose phasor is $\mathbf{i}_0$ is connected to the terminals of the network in Example 7.2, as shown in Fig. 7.2(a).

(1) What is the phasor $\mathbf{v}$ representing the voltage at the terminals indicated in the figure?

(2) What is the amplitude of the sinusoidal voltage $v(t)$ at these terminals?

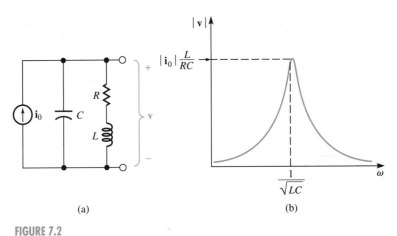

FIGURE 7.2

SOLUTION

(1) By the definition of impedance, $\mathbf{v} = \mathbf{i}_0\mathbf{Z}$, where $\mathbf{Z}$ is the impedance of the network found in Example 7.2.

$$\mathbf{v} = \mathbf{i}_0 \, \frac{R + j\omega L}{(1 - \omega^2 LC) + (j\omega RC)}$$

(2) The amplitude of the sinusoidal voltage $v(t)$ is equal to the absolute value of the phasor **v.** Since the absolute value of the product of two complex numbers is the product of their absolute values, $|\mathbf{v}| = |\mathbf{i}_0| \cdot |\mathbf{Z}|$.

The value of $|\mathbf{Z}|$ can be found most easily by means of the identity $|\mathbf{Z}| = \sqrt{\mathbf{Z}\mathbf{Z}^*}$. Thus

$$|\mathbf{Z}| = \left(\frac{R^2 + (\omega L)^2}{[1 - \omega^2 LC]^2 + [\omega RC]^2} \right)^{1/2}$$

The amplitude of $v(t)$ is then equal to this quantity multiplied by $|\mathbf{i}_0|$. For instance, if $i_0(t)$ were 20 mA $\angle 60°$, $|\mathbf{i}_0|$ would be 20 mA, and $|\mathbf{v}| = (20 \text{ mA}) \cdot |\mathbf{Z}|$ V.

The result obtained in this example illustrates the phenomenon of *resonance* in an RLC circuit. Because of the form of the first term in the denominator of $|\mathbf{Z}|$, the absolute value of the impedance of the network, regarded as a function of frequency, has a maximum at a certain frequency. This frequency is called the *resonant frequency*. At the resonant frequency the voltage across its terminals is a maximum. An example of $|\mathbf{v}|$ versus ω, for the case of $(RC)^2 \ll LC$, is given in Fig. 7.2(b). We see that the value of the resonant frequency is

$$\omega_{\text{RES}} \cong \frac{1}{\sqrt{LC}}$$

RESISTANCE AND REACTANCE

In general, an impedance is a complex number. Like any complex number, an impedance can be decomposed into its real and imaginary parts. It is customary to write $\mathbf{Z} = R + jX$. Here R is called the *resistive part* of the impedance, and X is called the *reactive part*. For brevity one sometimes calls R the *resistance* and X the *reactance* of the subcircuit. The resistance of any RLC subcircuit is always a positive number. (Negative resistances occur only rarely and then only in connection with active circuits.) We note, however, that reactance can be either positive or negative. For example, if a resistance R and a capacitance C are connected in series, the resulting impedance is $R + 1/j\omega C = R - j/\omega C$. Thus the resistive part of the impedance is R and the reactive part is $-1/\omega C$. Reactance, like resistance, is measured in ohms.

EXAMPLE 7.4

A resistance R and an inductance L are connected in parallel. Find the resistive and reactive parts of the resulting impedance.

SOLUTION

The impedance of the combination is

$$\mathbf{Z} = \frac{(R)(j\omega L)}{R + j\omega L}$$

Multiplying top and bottom by the complex conjugate of the denominator, we have

$$\mathbf{Z} = j\omega RL \cdot \frac{R - j\omega L}{R^2 + (\omega L)^2} = \frac{\omega^2 RL^2 + j\omega R^2 L}{R^2 + (\omega L)^2}$$

The resistive part of $\mathbf{Z}$ is

$$R = \text{Re}\,(\mathbf{Z}) = \frac{\omega^2 RL^2}{R^2 + (\omega L)^2}$$

The reactive part of $\mathbf{Z}$ is

$$X = \frac{1}{j}\text{Im}\,(\mathbf{Z}) = \frac{\omega R^2 L}{R^2 + (\omega L)^2}$$

● EXERCISE 7.1

A 1,000 Ω resistance and a 10 μF capacitor are connected in parallel. Find the impedance of the combination when $f = 100$ Hz. **Answer:** $(24.71 - 155.2j)\,\Omega$.

PHASOR FORM OF KIRCHHOFF'S LAWS

In Section 6.3 it was pointed out that if the sum of several sinusoids is zero, the sum of their phasors likewise must be zero. We can make use of this fact to express Kirchhoff's two laws in phasor form. Let $i_1(t)$, $i_2(t)$, . . . , $i_n(t)$ be the sinusoidal currents entering a node. Then Kirchhoff's current law, in phasor form, is

$$\sum_{k=1}^{n} \mathbf{i}_k = 0 \quad \text{(Kirchhoff's current law)} \tag{7.8}$$

Similarly, let $v_1(t)$, $v_2(t)$, . . . , $v_n(t)$ be the sinusoidal voltage drops around a complete loop. Then Kirchhoff's voltage law, in phasor form, is

$$\sum_{k=1}^{n} \mathbf{v}_k = 0 \quad \text{(Kirchhoff's voltage law)} \tag{7.9}$$

We now have at our disposal everything needed to analyze sinusoidal ac circuits in a manner exactly analogous to that for dc circuits. Voltages and currents are represented by their phasors. The node method or the loop method can be used, taking advantage of Eqs. (7.8) and (7.9). Ohm's law is replaced by $\mathbf{v} = \mathbf{iZ}$. Thus—except that complex quantities are used—sinusoidal analysis can now be performed in just the same way as the dc analysis of Chapter 2.

As an example of the use of the node method in phasor analysis, let us consider the circuit of Fig. 7.3(a). This circuit is similar to the voltage divider, except that inductors are present instead of resistors. To obtain $\mathbf{v}_{\text{OUT}}$, we write a phasor node equation for the node between L_1 and L_2. The phasor current through each of the inductors is equal to the phasor voltage across the inductor, divided by its impedance. Thus

$$\frac{\mathbf{v}_{\text{OUT}} - \mathbf{v}_{\text{IN}}}{\mathbf{Z}_{L_2}} + \frac{\mathbf{v}_{\text{OUT}}}{\mathbf{Z}_{L_1}} = 0 \tag{7.10}$$

Since $\mathbf{Z}_{L_1} = j\omega L_1$ and $\mathbf{Z}_{L_2} = j\omega L_2$, we have

$$(\mathbf{v}_{\text{OUT}} - \mathbf{v}_{\text{IN}})L_1 + \mathbf{v}_{\text{OUT}} L_2 = 0 \tag{7.11}$$

$$\mathbf{v}_{\text{OUT}} = \mathbf{v}_{\text{IN}} \frac{L_1}{L_1 + L_2}$$

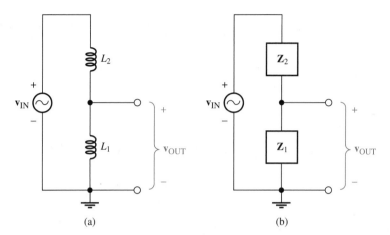

(a)   (b)

FIGURE 7.3

The reader will note that no differential equations are needed for this solution; the algebra also is much like that used in solving the resistive voltage divider, except that complex voltages and currents are used.

We can distinguish a general family of circuits, characterized by the form of Fig. 7.3(b). This general circuit resembles the resistive voltage divider, except that now the boxes designated Z_1 and Z_2 may represent any resistance, capacitance, or inductance, or two-terminal combinations of these elements. The impedance between the terminals of box Z_1 is determined by whatever elements or combination of elements happens to be in the box, and the same is true for Z_2. (For example, if the box marked Z_2 contains a resistor R and inductance L in series, then $Z_2 = R + j\omega L$.) Proceeding as in Eq. (7.10), we find

$$\frac{\mathbf{v}_{OUT} - \mathbf{v}_{IN}}{\mathbf{Z}_2} + \frac{\mathbf{v}_{OUT}}{\mathbf{Z}_1} = 0 \qquad (7.12)$$

$$\mathbf{v}_{OUT} = \mathbf{v}_{IN} \frac{\mathbf{Z}_1}{\mathbf{Z}_1 + \mathbf{Z}_2}$$

This result closely resembles that for the resistive voltage divider, Eq. (2.18), except that resistances are now replaced by impedances. We may call the circuit of Fig. 7.3(b) the *generalized voltage divider*. The case of Fig. 7.3(a) is now seen to be just a specific example of the general case of Fig. 7.3(b), and Eq. (7.11) can be obtained from Eq. (7.12).

EXAMPLE 7.5

Find the phasor $\mathbf{v}_{OUT}$ in the circuit shown in Fig. 7.4. What is the amplitude of $\mathbf{v}_{OUT}$?

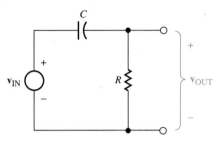

FIGURE 7.4

SOLUTION

This circuit has the form of a generalized voltage divider, as may be seen by comparison with Fig. 7.3(b). Accordingly we may use Eq. (7.12). In this case $\mathbf{Z}_1 = R$ and $\mathbf{Z}_2 = 1/j\omega C$. Thus

$$\mathbf{v}_{OUT} = \mathbf{v}_{IN} \frac{R}{R + 1/j\omega C}$$

$$= \mathbf{v}_{IN} \frac{j\omega RC}{1 + j\omega RC}$$

The amplitude of the output sinusoid is given by

$$|\mathbf{v}_{OUT}| = |\mathbf{v}_{IN}| \cdot \left| \frac{j\omega RC}{1 + j\omega RC} \right|$$

$$= |\mathbf{v}_{IN}| \sqrt{\frac{j\omega RC}{1 + j\omega RC} \left(\frac{j\omega RC}{1 + j\omega RC} \right)^*}$$

$$= |\mathbf{v}_{IN}| \frac{\omega RC}{\sqrt{1 + (\omega RC)^2}} \quad ?$$

EXAMPLE 7.6

In Fig. 7.5, find the phasor $\mathbf{v}$ representing the voltage at the output terminals.

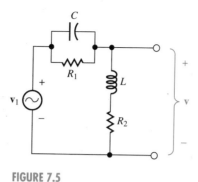

FIGURE 7.5

SOLUTION

This is a special case of the generalized voltage divider of Fig. 7.3(b), with

$$\mathbf{Z}_2 = \mathbf{Z}_C \parallel \mathbf{Z}_{R_1} = \frac{R_1/j\omega C}{R_1 + 1/j\omega C} = \frac{R_1}{1 + j\omega R_1 C}$$

$$\mathbf{Z}_1 = \mathbf{Z}_{R_2} + \mathbf{Z}_L = R_2 + j\omega L$$

Using Eq. (7.12), we have

$$\mathbf{v} = \mathbf{v}_1 \frac{R_2 + j\omega L}{R_2 + j\omega L + R_1/(1 + j\omega R_1 C)}$$

The following example demonstrates the use of the loop method in phasor analysis:

EXAMPLE 7.7

In Fig. 7.6, find the phasor representing the current flowing through R (as shown) in terms of the input current phasor $\mathbf{i}_1$.

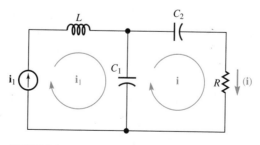

FIGURE 7.6

SOLUTION

We introduce two phasor loop currents as shown in the figure. The left loop current is controlled by the current source and must have the value i_1. The second phasor loop current equals the unknown i. Writing a loop equation for the right-hand loop yields

$$(i - i_1)Z_{C1} + iZ_{C2} + iZ_R = 0$$

Inserting the values of the three impedances, we find

$$\frac{(i - i_1)}{j\omega C_1} + \frac{i}{j\omega C_2} + iR = 0$$

$$i = i_1 \frac{1}{1 + C_1/C_2 + jR\omega C_1}$$

● **EXERCISE 7.2**

In Example 7.7, find the maximum instantaneous voltage that occurs across C_1. Let $i_1 = 10$ mA $\angle 30°$, $R = 100$ Ω, $C_1 = 2$ μF, $C_2 = 3$ μF, and $f = 650$ Hz. **Answer:** 0.69 V.

With phasor techniques, generalized Thévenin and Norton equivalent circuits can be found for subcircuits containing voltage and current sources, resistance, capacitance, and inductance. The procedure is the same as that described in Section 3.1 for purely resistive circuits, with the following changes: (1) Phasors representing voltages and currents replace the voltages and currents themselves, and (2) the Thévenin and Norton resistances are replaced by Thévenin and Norton impedances. The Thévenin and Norton parameters thus are found by using the following rules:

$$\mathbf{v}_T = \mathbf{v}_{OC} \qquad \text{(7.13)}$$

(general Thévenin equivalent)

$$\mathbf{Z}_T = -\mathbf{v}_{OC}/\mathbf{i}_{SC}$$

$$\mathbf{i}_N = -\mathbf{i}_{SC} \qquad \text{(7.14)}$$

(general Norton equivalent)

$$\mathbf{Z}_N = -\mathbf{v}_{OC}/\mathbf{i}_{SC}$$

The sign conventions of $\mathbf{v}_{OC}$, $\mathbf{i}_{SC}$, $\mathbf{v}_T$, and $\mathbf{i}_N$ are as shown in Fig. 7.7.

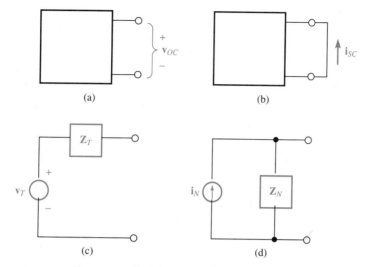

FIGURE 7.7 Sign conventions for finding generalized Thévenin and Norton equivalent circuits. (a) The sign convention for open-circuit voltage. (b) The sign convention for short-circuit current. (c) The generalized Thévenin equivalent. (d) The generalized Norton equivalent.

EXAMPLE 7.8

Find the Thévenin equivalent of the circuit shown in Fig. 7.8(a).

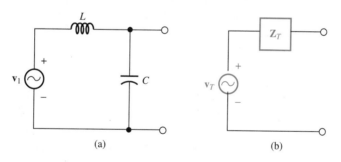

FIGURE 7.8

SOLUTION

This will be recognized as a generalized voltage divider of the type of Fig. 7.3(b). Using Eq. (7.12) we have

$$\mathbf{v}_{OC} = \mathbf{v}_1 \frac{\mathbf{Z}_C}{(\mathbf{Z}_C + \mathbf{Z}_L)} = \mathbf{v}_1 \frac{1/j\omega C}{1/j\omega C + j\omega L} = \frac{\mathbf{v}_1}{1 - \omega^2 LC}$$

The short-circuit phasor current is given by

$$\mathbf{i}_{SC} = -\frac{\mathbf{v}_1}{j\omega L}$$

(the minus sign results from the sign convention for $\mathbf{i}_{SC}$). Thus from Eq. (7.13), we find

$$\mathbf{v}_T = \frac{\mathbf{v}_1}{1 - \omega^2 LC} \qquad \mathbf{Z}_T = \frac{j\omega L}{1 - \omega^2 LC}$$

The Thévenin equivalent circuit is shown in Fig. 7.8(b).

It is interesting to note that the Thévenin parameters take on infinite values when $\omega^2 = 1/LC$. As in Example 7.3, this is again the phenomenon of resonance. When $\omega^2 = 1/LC$, the impedance of the series combination of L and C is

$$\mathbf{Z}_L + \mathbf{Z}_C = j\omega L + \frac{1}{j\omega C}$$

$$= \frac{jL}{\sqrt{LC}} - \frac{j\sqrt{LC}}{C} = j\left(\sqrt{\frac{L}{C}} - \sqrt{\frac{L}{C}}\right) = 0$$

Thus at this frequency the impedance across $\mathbf{v}_1$ would seem to become zero; consequently, the current that flows from $\mathbf{v}_1$ through the series combination would seem to become infinite. If such a circuit were constructed, the current would indeed have a maximum when $\omega = 1/\sqrt{LC}$ but of course could not be infinite. The actual physical circuit would inevitably contain some resistance (especially in the coil windings), and this resistance, if included in the equations, would prevent the impedance from becoming zero for any real value of ω.

Note that the parallel combination of L and C in Example 7.3 had an impedance maximum at the resonant frequency, while in this example, with a series combination of L and C, there is an impedance minimum at the resonant frequency.

LIMITING CASES

Operation of an RLC circuit in either the limit $\omega \to \infty$ or the limit $\omega \to 0$ can usually be found by inspection. The reason for this is that the impedance of each inductor and capacitor takes on a simple value at the extremes of frequency, as follows:

$$\lim_{\omega \to 0} \mathbf{Z}_{capacitor} = \lim_{\omega \to 0} \frac{1}{j\omega C} = \infty \qquad (7.15)$$

$$\lim_{\omega \to 0} \mathbf{Z}_{inductor} = \lim_{\omega \to 0} j\omega L = 0$$

$$\lim_{\omega \to \infty} \mathbf{Z}_{capacitor} = \lim_{\omega \to \infty} \frac{1}{j\omega C} = 0$$

$$\lim_{\omega \to \infty} \mathbf{Z}_{inductor} = \lim_{\omega \to \infty} j\omega L = \infty$$

An infinite impedance allows no current to flow for any finite value of applied voltage and hence is equivalent to an open circuit. An impedance of zero will give rise to zero voltage no matter what current flows through it and thus is equivalent to a short circuit. Thus: (1) At sufficiently low frequencies, capacitors are equivalent to open circuits, and inductors are equivalent to short circuits. (2) In the limit as $\omega \rightarrow \infty$, capacitors become equivalent to short circuits and inductors to open circuits. The operation of a circuit in either limit can usually be guessed rather easily, simply by imagining capacitors and inductors replaced by short or open circuits according to the foregoing rules.

EXAMPLE 7.9

In the circuit of Fig. 7.9(a), find $\mathbf{v}_{OUT}$ in the limit $\omega \rightarrow 0$ and in the limit $\omega \rightarrow \infty$.

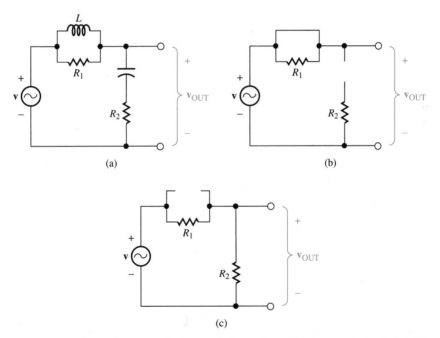

(a) (b) (c)

FIGURE 7.9   Behavior of capacitors and inductors in the limits of low and high frequency. At $f = 0$, circuit (a) behaves like circuit (b). At $f = \infty$, circuit (a) behaves like circuit (c).

SOLUTION

In the limit $\omega \rightarrow 0$, L acts as a short circuit and C as an open circuit, resulting in the circuit shown in Fig. 7.9(b). In this case the output is directly connected to the input; thus $\mathbf{v}_{OUT} \rightarrow \mathbf{v}$.

In the limit $\omega \to \infty$, the circuit behaves as shown in Fig. 7.9(c). In this limit the circuit reduces to a resistance voltage divider, and

$$\lim_{\omega \to \infty} \mathbf{v}_{OUT} = \mathbf{v} \frac{R_2}{R_1 + R_2}$$

It should be noted that Eq. (7.15) applies to *ideal* capacitors and inductors. A practical capacitor is likely to be contaminated with parasitic inductance, and vice versa. In the high-frequency limit these parasitic elements often become important and must be taken into account. The low-frequency limits, however, usually are not affected by parasitics.

The usefulness of these limiting cases will be seen when we look at transistor/ amplifier circuits. There it is often necessary to provide a path of conduction for the signal frequency while blocking dc current through the same path. This can be done quite nicely by means of a "blocking capacitor." Like any capacitor, it prevents the flow of dc; moreover, it can simultaneously act as a short circuit at the signal frequency, provided that frequency is high enough.

7.2 Frequency Response

In phasor problems we generally assume that the frequency of all voltages and currents has the same, given value. This value is determined by the voltage source or current source that drives the circuit. However, the impedances of all the inductors and capacitors in the circuit are functions of frequency. Thus it is clear that the operation of the circuit will change as the driving frequency is changed. We have already seen an ex-treme case of this in the previous section, when we discussed the limiting cases $\omega \to \infty$ and $\omega \to 0$. The way in which circuit operation varies with frequency is known as the *frequency response* of the circuit.

As an example, let us consider the simple passive subcircuit shown in Fig. 7.10. The value of the voltage source varies in time and is equal to $V_1 \cos \omega t$. We shall

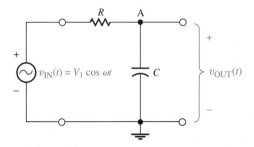

FIGURE 7.10

regard the passive subcircuit (composed of R and C) as having a pair of input terminals across which the voltage is called v_{IN}, as shown in the figure. In this case $v_{IN} = V_1$ cos ωt. The circuit also has a pair of output terminals across which the voltage is v_{OUT}. This is the voltage we wish to calculate. Each pair of terminals is known as a *port*. This particular circuit, with one pair of input terminals and one pair of output terminals, is a *two-port* passive circuit.

We recognize the circuit as a type of voltage divider; the phasor representing v_{OUT} can be found from Eq. (7.12). The result is

$$\mathbf{v}_{OUT} = \mathbf{v}_{IN} \frac{1}{1 + j\omega RC} \tag{7.16}$$

Let us now investigate the way the *amplitude* of the output sinusoid varies if we change the *frequency* of v_{IN} (while keeping the amplitude of v_{IN} constant). This is the sort of question that might arise in connection with a music system, in which low notes or high notes can be applied to the input. It would not be a very good music system if some notes got through and others did not!

We recall that the amplitude of a sinusoid is given by the absolute value of its phasor. Let us designate the amplitude of v_{IN} by V_1 and that of v_{OUT} by V_2. (Note that V_1 and V_2 are constant, non–time-varying numbers.) Thus

$$V_2 = |\mathbf{v}_{OUT}| = \sqrt{\mathbf{v}_{OUT}\mathbf{v}_{OUT}^*} \tag{7.17}$$

$$= \sqrt{\mathbf{v}_{IN}\mathbf{v}_{IN}^*} \left[\frac{1}{1 + j\omega RC} \frac{1}{1 - j\omega RC} \right]^{1/2}$$

$$= \frac{V_1}{\sqrt{1 + \omega^2(RC)^2}}$$

We can now see how the output amplitude V_2 varies with the driving frequency ω, assuming the input amplitude V_1 to be held constant.

To begin with, we can consider the two limiting cases, $\omega \rightarrow \infty$ and $\omega \rightarrow 0$. From Eq. (7.17) we see that when $\omega \rightarrow \infty$ the output amplitude approaches zero; when $\omega \rightarrow 0$ the output amplitude approaches the input amplitude. (The same conclusions would be reached using Eq. (7.15) and the method of Example 7.9. Check this for yourself; it is a useful exercise.) For frequencies that lie between 0 and ∞, the ratio V_2/V_1 is given by Eq. (7.17).

A graph of the ratio V_2/V_1 versus ωRC is shown in Fig. 7.11. Clearly this is a circuit that transmits low-frequency signals well, while high-frequency signals experience a loss of amplitude. In general, a two-port subcircuit used to transmit some frequencies while discriminating against others is called a *filter* circuit. This particular circuit would be called a *low-pass filter*. Such a circuit might be used, for example, to remove "hiss" (high-frequency noise) from the output of a tape player or phonograph, while allowing the music, composed mostly of lower frequencies, to pass without significant loss.

Figure 7.11 is graphed on linear scales. In this kind of graph it is difficult to see how the function varies over a wide range of frequencies. Thus it is customary to give

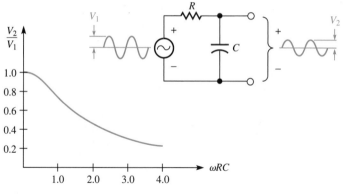

FIGURE 7.11

this type of information in the form of a log-log plot. The same function V_2/V_1, plotted on logarithmic scales, is given in Fig. 7.12. The log-log graph of frequency response is sometimes called a *Bode plot*. A complete Bode plot would also have a second curve showing the phase of the output as a function of frequency.

We note that on the Bode plot the function V_2/V_1 is asymptotic to a set of straight line segments. The frequencies at which these segments meet are known as *break frequencies* or *corner frequencies*. The significance of the break frequencies is that they are the frequencies at which the slope of the Bode plot changes. In the example we have been discussing (Figs. 7.10 to 7.12), the behavior of V_2 [Eq. (7.17)] is different for the two cases $\omega \ll 1/RC$ and $\omega \gg 1/RC$. The frequency range $\omega \ll 1/RC$ is the same as the range $\omega RC \ll 1$. Inspecting Eq. (7.17), we see that in this range

$$V_2 = \frac{V_1}{\sqrt{1 + (\text{small})^2}} \cong V_1 \qquad (7.18)$$

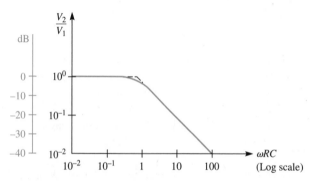

FIGURE 7.12

Note that Eq. (7.18) is an approximate result that is reasonably accurate only when ω is much less than $1/RC$. On the other hand, when $\omega \gg 1/RC$ (i.e., $\omega RC \gg 1$) we can make the approximation

$$1 + (\omega RC)^2 \cong (\omega RC)^2$$

so we have, in this frequency range,

$$V_2 \cong \frac{V_1}{\omega RC} \qquad (7.19)$$

(The reader should make sure he or she understands these approximations. The method is useful in many kinds of problems.) Referring now to the Bode plot, we see that the low-frequency part of the curve is nearly horizontal; that is because according to Eq. (7.18) V_2 is nearly constant in this regime. On the other hand, for $\omega > 1/RC$, Eq. (7.19) applies and V_2 decreases in proportion to $1/\omega$.

Since the high-frequency region is characterized by $\omega RC > 1$, and the low-frequency region by $\omega RC < 1$, it is logical to choose the boundary between the regions to be at $\omega RC = 1$ (or, equivalently, $\omega = 1/RC$). The frequency at this boundary is called the *break frequency*. In general, one finds break frequencies associated with terms like $1 + (\omega RC)^2$. The break frequency occurs when the first term (1) and the second term $(\omega RC)^2$ are the same size. Thus we find the break frequency by writing $(\omega_B RC)^2 = 1$, from which we have $\omega_B = 1/RC$.

On the Bode plot, the curve within each of the two regimes is nearly a straight line. The two straight lines intersect at the break frequency. In Fig. 7.12 the actual V_2/V_1 curve is the solid line, while the high-frequency and low-frequency approximations are shown as dashed lines. Everywhere except near the break frequency the solid and dashed lines nearly coincide, because away from the break frequency, approximations (7.18) and (7.19) are very nearly correct.

Two vertical scales are shown in Fig. 7.12. In addition to the logarithmic scale of V_2/V_1, there is another scale calibrated in *decibels* (dB). A decibel scale is often used as a means of comparing the value of some quantity (in this case V_2/V_1) to some arbitrary reference value (in this case chosen to be unity). In general, if the quantity being described is V and the reference value is V_0, then V is said to be above the reference value by a number of decibels given by*

$$\text{Decibels} = 20 \log_{10} \frac{V}{V_0} \qquad (7.20)$$

It should be noted that the decibel is not quite a unit in the same sense that the kilogram is a unit. The decibel is useful only in comparing one number with another number selected as a reference. It is not meaningful to speak of a sound having a level

---

* The unit being defined here is known more specifically as the *voltage decibel*. There is another unit also in use called the *power decibel,* defined dB $= 10 \log_{10}(P/P_0)$. Power decibels are used when power levels, rather than voltages or currents, are being compared. In this book the word "decibel" always means voltage decibel.

of 96 dB unless you also state to what reference value your sound is being compared. (Nonetheless, one does in fact often hear sound levels described as being so many decibels. This is possible because the acoustics profession has established a reference level of loudness, which is accepted everywhere as the standard.)

EXAMPLE 7.10

By how many decibels does V_2/V_1 lie above the reference level of unity when $V_2/V_1 = 0.5$?

SOLUTION

The number of decibels is given by Eq. (7.20):

$$dB = 20 \log_{10} \frac{0.5}{1.0} = 20 \log_{10} 0.5$$

$$= 20 \cdot -0.301 = -6.02$$

The result is negative because V_2/V_1 is less than the reference level; one can also say V_2/V_1 is 6.02 dB *below the reference level.* In practice one rounds the number off; it is customary, when the numbers differ by a factor of 2, to say that one is 6 dB below the other.

It is left to the student to verify that at the break frequency in Fig. 7.12, the meeting point of the straight-line asymptotes is above the actual curve by almost exactly 3 dB. This separation (3 dB) is the discrepancy at the break frequency between the true curve and the straight-line approximation. At other frequencies the discrepancy is less.

We observe from Fig. 7.12 that above the break frequency the slope of the function V_2/V_1 is -20 dB per decade; that is, every time the frequency increases by a factor of 10, V_2/V_1 drops by another 20 dB.

EXAMPLE 7.11

Analysis of the circuit of Fig. 7.13(a) shows that V_3, the amplitude of the sinusoid v_{OUT}, is given by

$$\frac{V_3}{V_1} = \frac{\omega RC}{\sqrt{1 + (\omega RC)^2}}$$

(The derivation of this result is left as an exercise.) Construct a Bode plot of V_3/V_1 versus frequency. Find the break frequency. What are the slopes above and below the break frequency, expressed in dB per decade?

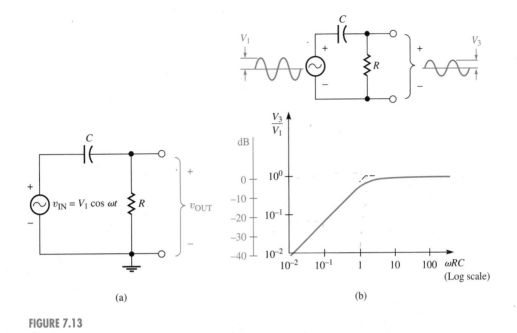

FIGURE 7.13

SOLUTION

By plotting a few points and connecting them with straight line segments, we obtain the Bode plot shown in Fig. 7.13(b). The break frequency is the frequency at which $(\omega RC)^2 = 1$; thus the break frequency is $1/RC$. Above the break frequency the slope is seen to be 0 dB per decade; below it the slope is 20 dB per decade. This circuit acts as a *high-pass filter*.

EXAMPLE 7.12

Suppose in a certain frequency range the ratio of output amplitude to input amplitude is proportional to $1/\omega^2$. What is the slope of the Bode plot in this frequency range, expressed in dB per decade?

SOLUTION

Let us imagine that we keep the input amplitude constant and vary the frequency from some frequency ω_0 (which lies in the frequency range of interest) to the frequency $10\omega_0$ (which is one decade higher). Since the output amplitude is proportional to $1/\omega^2$, the output amplitude $V_{OUT}(\omega)$ must obey

$$\frac{V_{OUT}(10\omega_0)}{V_{OUT}(\omega_0)} = \frac{1/100}{1} = 10^{-2}$$

The number of decibels representing this ratio is given by

$$dB = 20 \log_{10}\left[\frac{V_{OUT}(10\omega_0)}{V_{OUT}(\omega_0)}\right]$$

$$= 20 \log_{10} 10^{-2} = -40$$

Thus the slope of the Bode plot is -40 dB per decade. The fact that the slope is negative indicates that the output amplitude gets *smaller* as frequency is increased.

Similarly, we can show that in general when the output amplitude is proportional to ω^N, the slope of the Bode plot is $20N$ dB per decade.

EXAMPLE 7.13

Suppose that V_4, the amplitude of the output voltage of a circuit, is $V_0(2 + \omega^2/\omega_0^2)^{1/2}$. Find the slope of the Bode plot (a) for frequencies much less than ω_0 and (b) for very high frequencies. What is the break frequency?

SOLUTION

For case (a), $\omega^2/\omega_0^2 \ll 2$, and thus the amplitude is approximately equal to $\sqrt{2}\,V_0$. Since the amplitude is constant in this frequency range, the rate of change is 0 dB per decade.

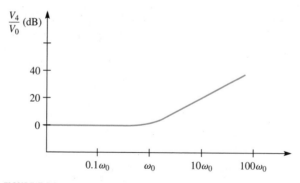

FIGURE 7.14

For case (b), the output voltage is approximately equal to $\omega V_0/\omega_0$. Using the rule that when output is proportional to ω^N the slope is $20N$ dB per decade, we find that the slope in this case is 20 dB per decade.

By setting ω^2/ω_0^2 equal to 2, we find that the break frequency occurs at $\omega_B = \sqrt{2}\omega_0$.

The up-and-down position of the curve is found by noting that in the low-frequency limit $V_4 = V_0$. Taking the reference level of V_4/V_0 to be $\sqrt{2}$, the Bode plot is as shown in Fig. 7.14.

• EXERCISE 7.3

Suppose that the output amplitude of a circuit depends on frequency according to

$$V = \frac{A\omega}{\sqrt{B + C\omega^2}}$$

Find (a) the break frequency; (b) the slope of the Bode plot (in dB per decade) above the break frequency; (c) the slope of the Bode plot below the break frequency; (d) the high-frequency limit of V. **Answer:** (a) $\omega_B = \sqrt{B/C}$; (b) zero; (c) 20 dB per decade; (d) $A/\sqrt{C}$.

EXAMPLE 7.14

The function of a loudspeaker *crossover network* is to channel frequencies higher than a given crossover frequency f_c into the high-frequency speaker ("tweeter") and frequencies below f_c into the low-frequency speaker ("woofer"). The circuit is shown in Fig. 7.15(a). Figure 7.15(b) shows an approximate equivalent circuit, or circuit model. The amplifier is represented by its Thévenin equivalent; for simplicity we shall assume (although it is unlikely to be true in practice) that its Thévenin resistance is vanishingly small. Each speaker acts as an 8 Ω resistance.

(a) If the crossover frequency is chosen to be 2,000 Hz, evaluate C and L.

(b) Make a graph showing the time-averaged power delivered to each speaker as a function of frequency, using log-log scales.

SOLUTION

(a) The part of the circuit containing S_1 is identical with that of Example 7.11. The rms voltage across R_1 is thus

$$V_1 = 10 \frac{2\pi f C R_1}{\sqrt{1 + (2\pi f C R_1)^2}}$$

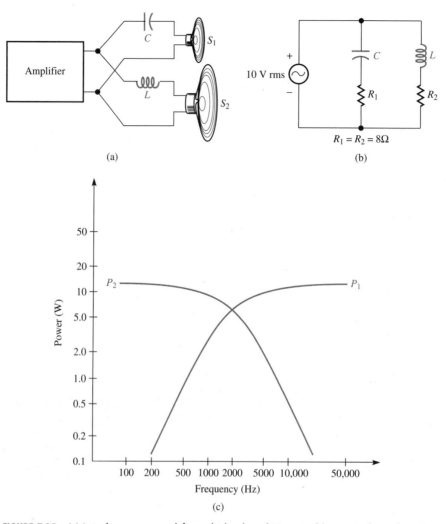

FIGURE 7.15   (a) A simple crossover network for two loudspeakers. (b) Circuit model, assuming, for simplicity, that amplifier source impedance is zero. (c) Output power of low- and high- frequency speakers versus frequency.

where f is the ordinary frequency. It is reasonable to set the crossover frequency equal to the break frequency $(2\pi RC)^{-1}$. Thus we have

$$C = \frac{1}{(2\pi)(8)(2{,}000)} = 9.95\ \mu F$$

The part of the circuit containing L is similar, but here we are interested in the voltage across R_2. This rms voltage is

$$V_2 = \frac{10}{\sqrt{1 + (2\pi f L/R_2)^2}}$$

The break frequency is the value of f that makes $2\pi fL/R_2 = 1$. Thus

$$L = \frac{R_2}{2\pi f_c} = 0.637 \text{ mH}$$

(b) The functions to be graphed are

$$P_1(f) = \frac{V_1^2(f)}{R_1} = \frac{100}{8} \frac{1}{1 + (f_c/f)^2}$$

and

$$P_2(f) = \frac{V_2^2(f)}{R_2} = \frac{100}{8} \frac{1}{1 + (f/f_c)^2}$$

where $f_c = 2,000$ Hz. These functions are graphed in Fig. 7.15(c). Note that at f_c, which is the break frequency for both curves, both functions are reduced from their maximum values by a factor of 2, or 3 dB.

*7.3 Examples and Applications

RESONANCE

As an example of the phasor technique, let us consider the interesting phenomenon known as *resonance*. A circuit is said to be resonant if its impedance shows either a pronounced maximum or a pronounced minimum at one particular frequency, known as the *resonant frequency*. This effect can be quite dramatic. It is accompanied by a maximum of energy stored in the circuit, and large currents or voltages may appear. The analogous mechanical effect is quite familiar; the resonance of a plucked string or tuning fork is essentially the same phenomenon. Another mechanical analogy is the case of a child being pushed in a swing. The swing has its own natural frequency (that at which the child oscillates when not being pushed); this is the swing's resonant frequency. If you give the swing periodic pushes at a frequency not equal to the swing's resonant frequency, not much motion of the swing will occur on the average. This is because your pushes sometimes add to the motion of the swing (if you happen to push when the swing is moving away from you) and sometimes subtract from the motion (if you happen to push when the swing is coming toward you). However, if the frequency of your pushes is equal to the resonant frequency you can contrive to always push when the swing is going away from you. (Imagine yourself pushing a child in a swing. You will see that you do exactly this.) In this case each push adds to the motion, which tends to become larger and larger.

     The most usual type of resonant circuit contains one inductance and one capacitance. (In fact, each circuit must also contain some resistance, but this will be neglected for the moment.) The type known as a *parallel resonant circuit* is shown in Fig.

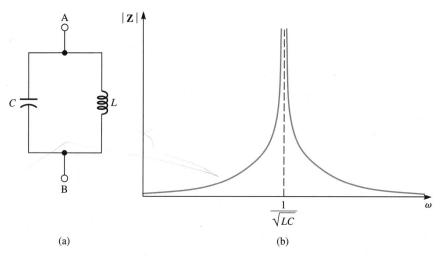

FIGURE 7.16   Lossless parallel-resonant circuit.

7.16(a). The impedance that appears between terminals A and B of this parallel combination is

$$\mathbf{Z} = \frac{j\omega L(1/j\omega C)}{j\omega L + 1/j\omega C} = \frac{j\omega L}{1 - \omega^2 LC} \tag{7.21}$$

Often one is interested in the absolute value $|\mathbf{Z}|$. (For example, if a current source $\mathbf{i}_0$ were connected across terminals A,B, the amplitude of the voltage between the terminals would be $|\mathbf{v}| = |\mathbf{i}_0| \cdot |\mathbf{Z}|$.) The absolute value of the parallel impedance given in Eq. (7.21) is

$$|\mathbf{Z}| = \frac{\omega L}{1 - \omega^2 LC} \quad \text{(parallel resonance)} \tag{7.22}$$

This function of ω is graphed in Fig. 7.16(b). We see that the function $|\mathbf{Z}|$ has a singularity at the frequency $\omega_R = 1/\sqrt{LC}$; its value appears to become infinite when $\omega = \omega_R$. This frequency ω_R given by

$$\omega_R \equiv \frac{1}{\sqrt{LC}} \quad \text{(resonant frequency)} \tag{7.23}$$

is called the *resonant frequency* of the circuit.

In a practical circuit the impedance would not actually become infinite at any frequency: Actual physical circuit elements would contain some resistance in their wires. For example, the windings of the inductor may contain substantial resistance. If this resistance is taken into consideration, the circuit has the form shown in Fig. 7.17(a).

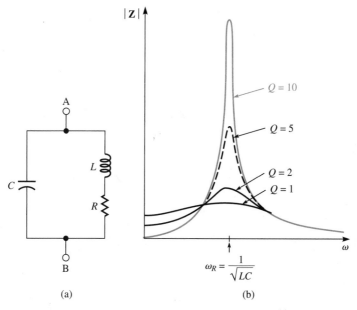

FIGURE 7.17

This circuit has already been considered in Examples 7.2 and 7.3. There it was found that the impedance between terminals A and B is

$$\mathbf{Z} = \frac{R + j\omega L}{1 - \omega^2 LC + j\omega RC} \tag{7.24}$$

and thus the absolute value of the impedance is

$$|\mathbf{Z}| = \left(\frac{R^2 + (\omega L)^2}{(1 - \omega^2 LC)^2 + (\omega RC)^2}\right)^{1/2} \tag{7.25}$$

This quantity is plotted in Fig. 7.17(b) for several cases corresponding to the same resonant frequency but to different values of R. We shall define a dimensionless parameter Q, known as the *quality factor* of the circuit (or, more frequently, simply as its "Q"), according to the formula*

$$Q = \frac{\omega_R L}{R} \tag{7.26}$$

---

* Equation (7.27) gives the value of Q for this particular circuit. Other resonant circuits may require different expressions for Q.

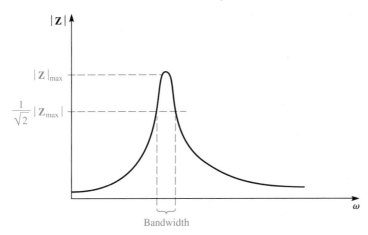

FIGURE 7.18   The bandwidth of a high-Q circuit is given by $BW = \omega_R/Q$.

The curves of Fig. 7.17(b) are designated by their values of Q. We see that when Q is large (that is, when R is small) the curves have a sharply peaked form, which approaches that of Fig. 7.16(b) as Q approaches infinity. On the other hand, when $Q = 1$ the curve becomes broad, and in fact the maximum of $|\mathbf{Z}|$ occurs at a frequency slightly less than ω_R. As the value of Q increases, the width of the peak becomes "narrower." Let us define the *bandwidth* of the resonant circuit as the range of frequencies over which $|\mathbf{Z}|$ is larger than $1/\sqrt{2}$ times its maximum value. This definition is illustrated by Fig. 7.18. Then it can be shown from Eq. (7.25) that if $Q \gg 1$, the bandwidth is given by

$$BW \cong \frac{\omega_R}{Q} \quad \text{(radians per second)} \qquad (7.27)$$

The above definition of bandwidth makes quantitative our previous statement that larger values of Q imply a "narrower" resonance curve.

EXAMPLE 7.15

A parallel-resonant circuit of the type shown in Fig. 7.17(a) is to be used as a "filter" to remove a signal of an unwanted frequency. The circuit contains two current sources of equal magnitude but with different frequencies, 1 MHz and 1.01 MHz, as shown in Fig. 7.19. Assume that ω_R for the resonant circuit equals 1 MHz, that $Q \gg 1$, and that $C = 3 \times 10^{-10}$ F. Find the value of R that will cause the amplitude of the 1.01 MHz voltage at the voltmeter to be 3 dB less than that of the 1 MHz voltage.

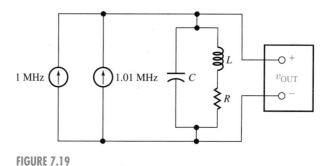

FIGURE 7.19

SOLUTION

As can be seen from Fig. 7.18, we need a circuit for which the bandwidth extends approximately from 0.99 MHz to 1.01 MHz, so that the 1.01 MHz signal will lie at one end of the bandwidth and be attenuated by $1/\sqrt{2}$, or 3 dB. Thus we wish BW = 0.02 MHz. From Eq. (7.27) we have

$$Q \cong \frac{\omega_R}{\text{BW}} = \frac{2\pi(1)}{2\pi(0.02)} = 50$$

The value of Q is related to that of R by Eq. (7.26). However, before Eq. (7.26) can be used we must find L. We can do so by using the formula for resonant frequency, Eq. (7.23):

$$L = \frac{1}{\omega_R^2 C} = \frac{1}{(2\pi)^2(10^6)^2 \cdot 3 \times 10^{-10}} = 8.4 \times 10^{-5}\,\text{H}$$

Now from Eq. (7.26) we get

$$R = \frac{\omega_R L}{Q} = \frac{2\pi \cdot 10^6 \cdot (8.4 \times 10^{-5})}{50} = 10.6\,\Omega$$

Thus we find that an inductor with resistance of 10.6 Ω or less is required to make the filter.

Another important resonant circuit is the *series resonant circuit*, shown in Fig. 7.20(a). Assuming for the moment that $R = 0$, the impedance of this connection is

$$\mathbf{Z} = j\omega L + \frac{1}{j\omega C} = \frac{1 - \omega^2/\omega_R^2}{j\omega C} \tag{7.28}$$

where ω_R, as before, is defined by $\omega_R = 1/\sqrt{LC}$. We see that for the series resonant circuit the impedance decreases to zero when $\omega = \omega_R$. The absolute value of $\mathbf{Z}$, as given by Eq. (7.28), is graphed as the solid curve in Fig. 7.20(b) as a function of frequency. Of course, resistance must also be present in the series resonant circuit, and in practice this prevents the impedance from actually becoming zero. However, provided

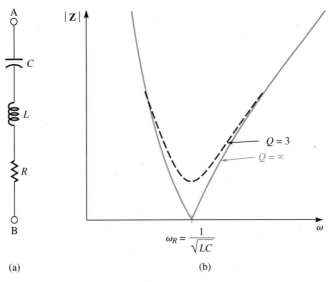

FIGURE 7.20 (a) A series resonant circuit. (b) For the case of $Q = \infty$ ($R = 0$), the impedance vanishes at the resonant frequency, as shown by the solid curve. For comparison the dashed curve shows the case of $Q = 3$.

that Q, as defined in Eq. (7.26), is much greater than unity, the absolute value of the series impedance will still have a well defined minimum at the resonant frequency. For comparison, a curve with $Q = 3$ is also shown in Fig. 7.20(b).

IMPEDANCE MATCHING

For our next example of the use of phasors, let us discuss a useful result known as the *power-transfer theorem*. This theorem arises from the following problem: Suppose we have some signal source represented by its generalized Thévenin equivalent. The Thévenin parameters of this source, $\mathbf{V}_T$ and $\mathbf{Z}_T$, are not adjustable. The problem is to discover what load impedance $\mathbf{Z}_L$ should be connected across the terminals of the source, as shown in Fig. 7.21(a), so that the maximum power will be delivered to the load. In practice, the signal source might be a music amplifier; in that case, we would be asking what impedance a loudspeaker should have in order to take maximum power from the amplifier output. The answer to this problem is found as follows.

When the signal source is replaced by its Thévenin equivalent, the situation is as shown in Fig. 7.21(b). Using the generalized voltage-divider formula, we see that the voltage across the load is given by

$$\mathbf{v}_L = \frac{\mathbf{Z}_L}{\mathbf{Z}_L + \mathbf{Z}_T}\mathbf{v}_T \qquad (7.29)$$

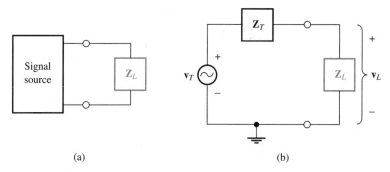

FIGURE 7.21   The power-transfer theorem.

The current flowing through the load is given by

$$\mathbf{i}_L = \frac{\mathbf{v}_T}{\mathbf{Z}_L + \mathbf{Z}_T} \tag{7.30}$$

The time-averaged power is given by

$$\tfrac{1}{2} \operatorname{Re}(\mathbf{v}_L \mathbf{i}_L^*) = \frac{|\mathbf{v}_T|^2}{2} \operatorname{Re}\left(\frac{\mathbf{Z}_L}{|(\mathbf{Z}_L + \mathbf{Z}_T)|^2}\right) \tag{7.31}$$

Since $\mathbf{v}_T$ and $\mathbf{Z}_T$ are not adjustable, our problem is to adjust $\mathbf{Z}_L$ so that the quantity on the right of Eq. (7.31) is maximized. Let us set $\mathbf{Z}_L = R_L + jX_L$ and $\mathbf{Z}_T = R_T + jX_T$. Then we must maximize

$$\operatorname{Re}\left(\frac{\mathbf{Z}_L}{|(\mathbf{Z}_L + \mathbf{Z}_T)|^2}\right) = \frac{R_L}{(R_L + R_T)^2 + (X_L + X_T)^2} \tag{7.32}$$

It is clear by inspection that we should choose $X_L = -X_T$. However, we admit only the possibility of positive resistances, so we are not allowed to choose $R_L = -R_T$. If R_L becomes very large, the power approaches zero; the same is true if R_L is made very small. Somewhere between zero and infinity there is a value of R_L that maximizes the quantity $R_L/(R_L + R_T)^2$. To find it, we differentiate the quantity being maximized with respect to R_L and set the derivative equal to zero:

$$\frac{d}{dR_L} \frac{R_L}{(R_L + R_T)^2} = \frac{(R_L + R_T)^2 - 2R_L(R_L + R_T)}{(R_L + R_T)^4} = 0 \tag{7.33}$$

Solving, we find $R_T = R_L$. Thus for maximum power transfer, we choose $R_L = R_T$ and $X_L = -X_T$; that is, we choose

$$\mathbf{Z}_L = \mathbf{Z}_T^* \qquad \text{(power-transfer theorem)} \tag{7.34}$$

EXAMPLE 7.16

For the circuit of Fig. 7.21(b), $\mathbf{Z}_T$ consists of a 10 Ω resistance and a 20 μH inductance in series. The angular frequency $\omega = 10^6$. Design a load $\mathbf{Z}_L$ that will receive the maximum power obtainable from the source.

SOLUTION

In this case the value of $\mathbf{Z}_T$ is $10 + j(2 \times 10^{-5})(10^6) = 10 + j20 \ \Omega$. Therefore we must have $\mathbf{Z}_L = 10 - j20 \ \Omega$. There are two simple solutions to the problem of designing a load with this impedance: we may use either a series combination or a parallel combination of elements for $\mathbf{Z}_L$. Choosing the former, we see that a series combination of a resistance and an inductance will not work: the series impedance $R + j\omega L$ of such a combination will have a positive imaginary part, and a negative imaginary part is needed. However, a series combination of a resistance and a capacitance can be used. Such a combination has series impedance

$$\mathbf{Z}_L = R + \frac{1}{j\omega C} = R - \frac{j}{\omega C}$$

We set $R = 10 \ \Omega$ and choose C such that

$$\frac{1}{\omega C} = 20$$

$$C = \frac{1}{20\omega} = 5 \times 10^{-8} \ \text{F}$$

The required load is shown in Fig. 7.22.

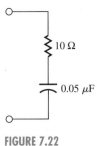

10 Ω

0.05 μF

FIGURE 7.22

7.4 Application: Automobile Suspension System

In Section 6.5 we saw how phasor methods could be used to study mechanical vibrations. These methods are in fact widely used by mechanical and aeronautical engineers, among others. Let us now demonstrate by working out a design for an automobile suspension system.

For simplicity, let us deal with the "unicycle" shown in Fig. 7.23. In this figure h is the height of the wheel; the origin of coordinates is such that when the wheel is rolling on flat, level ground we have $h = 0$. The mass M represents the weight of the car. The height of mass M above its reference level is called y, and the reference level is chosen so that when the car is in level motion, $y = 0$. Under these conditions the length of the spring is L_0.

The spring exerts forces in the y-direction. The total force acting on M depends on the length of the spring, according to

$$\text{Force} = -k\,(\text{Length} - L_0) \tag{7.35}$$

where k is the spring constant. (This total force includes gravity. At equilibrium the spring is slightly compressed from its natural length, making its length equal to L_0. Then the total force acting on M—gravity plus spring force—is zero, and M neither rises nor falls.) If M should move up, the length of the spring would increase to $L_0 + y$; if the wheel moves up, the length of the spring *decreases* to $L_0 - h$. Thus

$$\text{Length} = L_0 + y - h \tag{7.36}$$

Combining Eqs. (7.35) and (7.36), we have

$$M\frac{d^2y}{dt^2} + ky = kh \tag{7.37}$$

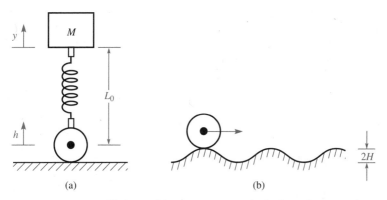

FIGURE 7.23   (a) Simplified automobile suspension system. (b) The driving frequency is determined by the spacing of the bumps and the speed of the car.

Now suppose our vehicle rolls along a bumpy road, as shown in Fig. 7.23(b). If we imagine that the bumps in the road form a sinusoid, then the wheel moves up and down as it travels. Supposing the height of the road varies between $-H$ and $+H$, then

$$h = H \cos \omega t \tag{7.38}$$

The frequency of the wheel's motion, ω, depends on the spacing of the bumps in the road and on the horizontal speed of the vehicle. Inserting Eq. (7.38) into Eq. (7.37) and applying the phasor technique, we have

$$-M \omega^2 \mathbf{y} + k\mathbf{y} = kH \tag{7.39}$$

Now in designing a car suspension system, we are interested in minimizing the amplitude of the up-and-down motion of the car. This amplitude is $|\mathbf{y}|$. Solving Eq. (7.39), we obtain

$$|\mathbf{y}| = \frac{H}{\left| 1 - \dfrac{\omega^2 M}{k} \right|} \tag{7.40}$$

From this result we see that the amplitude of the car's up-and-down motion depends on ω. For very high frequencies (the limit $\omega \to \infty$) the spring does a good job of isolating the car from the bumps in the road, because in that limit $|\mathbf{y}| \ll H$. For very low frequencies ($\omega \to 0$), though, the spring is not helpful; in this limit $|\mathbf{y}| = H$, and the car moves up and down with the road just as if there were a rigid connection between the wheel and M. But worse yet, the system contains a resonance at the frequency $\omega_R = \sqrt{k/M}$. If the applied frequency happens to have this value, the car's motion becomes infinitely large! The passengers will come back to you and complain about a terribly bumpy ride.

Well then, what to do? Probably the value of M cannot be changed much, as the size of the car is determined for us by its use. Conceivably we could, however, make k larger or smaller, which changes the resonant frequency. If we make k large, we just return to the non-useful case $|\mathbf{y}| = H$. But we cannot make k small, because, we remember, the spring supports the car. Making k small means making the spring weak. If k is small, the car will sag down and touch the road.

In practice, what is done is to add another component to the suspension system, as shown in Fig. 7.24. The new component is called (somewhat misleadingly) a *shock absorber*. It is the same as the dashpot described in Section 6.5; it is a viscous-damping device consisting of a plunger moving through a viscous fluid. It adds an additional force $-D\frac{d}{dt}$ (length), where "length" is still given by Eq. (7.28), and D is a constant determined by the size and shape of the shock absorber. Now Eq. (7.37) is changed to

$$M\frac{d^2 y}{dt^2} + ky = kh - D\frac{dy}{dt} + D\frac{dh}{dt} \tag{7.41}$$

and Eq. (7.40) becomes

$$|\mathbf{y}| = H \frac{\left[1 + \left(\dfrac{\omega D}{k}\right)^2\right]^{1/2}}{\left[\left(1 - \dfrac{\omega^2 M}{k}\right)^2 + \left(\dfrac{\omega D}{k}\right)^2\right]^{1/2}} \tag{7.42}$$

This result is sketched in Fig. 7.24(b). Here frequency is plotted on the horizontal axis in multiples of the resonant frequency $\omega_R = \sqrt{k/M}$, and we express D in terms of the dimensionless parameter

$$A \equiv \frac{D^2}{kM} \tag{7.43}$$

We observe that the displacement no longer reaches infinity at resonant frequency, although it still has a maximum near there. For the case of slight damping (small A) the resonant peak is large, but the decrease in vertical motion above the resonant frequency is very good. The case of large damping (large A, corresponding to a very "stiff" shock absorber), on the other hand, almost eliminates the resonance, but performance above the resonant frequency is little better than a rigid connection would be. In practice, the designer will have to balance the advantage of high-frequency performance against the disadvantage of large resonant oscillations, and choose some compromise value of A.

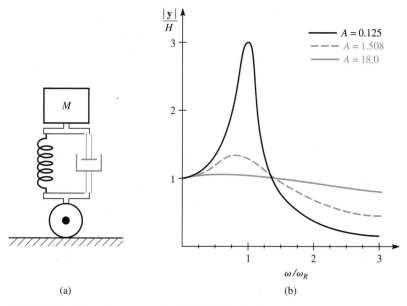

(a)   (b)

FIGURE 7.24 Improved suspension system with "shock absorber."

POINTS TO REMEMBER

- A phasor is a complex number that describes the amplitude and phase of a sinusoid.
- If the sinusoid is described by the formula $v(t) = V_0 \cos(\omega t + \theta)$, the phasor representing the sinusoid is $\mathbf{v} = V_0 e^{j\theta}$.
- The sinusoid corresponding to the phasor $\mathbf{v}$ is Re $(\mathbf{v} e^{j\omega t})$.
- The amplitude of the sinusoid represented by the phasor $\mathbf{v}$ is equal to $|\mathbf{v}|$. Its phase is equal to the argument of $\mathbf{v}$.
- The phasor representing the sum of two sinusoids of the same frequency is the sum of their individual phasors.
- If the sinusoid $v(t)$ is represented by the phasor $\mathbf{v}$, then the sinusoid dv/dt is represented by the phasor $j\omega \mathbf{v}$.
- If $v(t)$ and $i(t)$ are a sinusoidal voltage and current with the same frequency, the instantaneous power is $i(t)v(t)$. The time-averaged power is given by Avg $[i(t)v(t)] = \frac{1}{2}$ Re $(\mathbf{i}\mathbf{v}^*)$, where $\mathbf{i}$ and $\mathbf{v}$ are the phasors representing $i(t)$ and $v(t)$.
- The impedance of a circuit element is the ratio of the phasor for the voltage across it to the phasor for the current flowing through it. Impedances are complex numbers, which play a role similar to that of resistance in dc calculations.
- The impedance of a series or parallel combination of circuit elements is found using formulas similar to those used for series or parallel combinations of resistances.
- The imaginary part of an impedance is known as a reactance.
- The node and loop methods can be used for sinusoidal analysis of ac circuits. Voltages and currents are represented by their phasors in the calculation. The use of phasors eliminates the need for solving differential equations.
- Thévenin and Norton equivalents can be constructed for sinusoidal ac circuits. The Thévenin voltage is represented by its phasor, and the Thévenin resistance is replaced by a Thévenin impedance.
- In the limit of infinitely high frequency, a capacitor behaves like a short circuit and an inductor behaves like an open circuit.
- In the low-frequency limit (that is, for frequencies approaching zero) a capacitor behaves like an open circuit and an inductor like a short circuit.
- Resonant circuits are circuits that have singular behavior at or near a special frequency that is characteristic of the circuit. This frequency is called the resonant frequency. A parallel combination of inductance and capacitance makes a parallel resonant circuit; such a connection has a maximum of impedance at the resonant frequency. A series combination of inductance and capacitance makes a series-resonant circuit; such a connection has a minimum of impedance at the resonant frequency. If there is little resistance present in the circuit, the resonant frequency in either case is approximately $1/\sqrt{LC}$.
- Power transfer is greatest when the load impedance is the complex conjugate of the source impedance. Such a load is said to be "impedance-matched."

PROBLEMS

B B R G

Section 7.1

7.1 **a.** Find $\mathbf{Z}_1$, the impedance of a 1 μF capacitor at 100 MHz.
  b. What is $|\mathbf{Z}_1|$?
  c. What is $\mathbf{Z}_1^*$?

7.2 Find the reactance of a 1 mH inductance at 60 Hz (U.S. power line frequency) and at 1 MHz (in the AM radio broadcast band).

7.3 Find the reactance of a 0.01 μF capacitor at
  a. 1 MHz.
  b. 100 MHz.

7.4 Let $\mathbf{Z}_L$ be the impedance of a 10 μH inductance and $\mathbf{Z}_C$ be the impedance of a 10 μF capacitance.
  a. Over what range of ordinary frequencies is $|\mathbf{Z}_L| > 1{,}000 \ \Omega$?
  b. Over what range of ordinary frequencies is $|\mathbf{Z}_C| > 1{,}000 \ \Omega$?

*7.5 Prove that if two elements with impedances $\mathbf{Z}_1$ and $\mathbf{Z}_2$, respectively, are connected in series, the impedance of the combination is $\mathbf{Z}_1 + \mathbf{Z}_2$.

*7.6 Prove that if two elements with impedances $\mathbf{Z}_1$ and $\mathbf{Z}_2$, respectively, are connected in parallel, the impedance of the combination is $\mathbf{Z}_1\mathbf{Z}_2/(\mathbf{Z}_1 + \mathbf{Z}_2)$.

7.7 A 100 Ω resistor and a 1 μF capacitor are connected in series. What are the resistive and reactive parts of the resulting impedance if the frequency is 1,000 Hz?

7.8 Repeat Problem 7.7 with R and C connected in parallel.

*7.9 Find the impedances between terminals A and B for each of the combinations shown in Fig. 7.25 at an arbitrary frequency ω. Express your answers in their simplest rectangular form (that is, in the form $R + jX$).

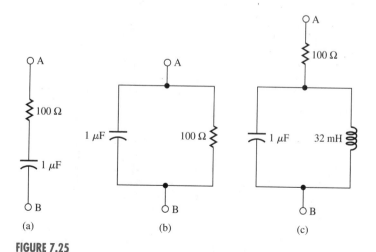

(a)   (b)   (c)

FIGURE 7.25

7.10 Show that if $\mathbf{v} = \mathbf{iZ}$, then $|\mathbf{v}| = |\mathbf{i}|\,|\mathbf{Z}|$.

*7.11 A sinusoidal voltage of amplitude 100 V and ordinary frequency 1,000 Hz is applied across each of the impedances shown in Fig. 7.25. Calculate the amplitude of the current that flows through terminals A,B in each case.

7.12 A current $\mathbf{i}$ passes through an impedance $\mathbf{Z} = A + jB$. Find the average power entering the impedance.

7.13 A voltage $\mathbf{v}$ is applied across the impedance $\mathbf{Z} = A + jB$. Find the average power entering the impedance.

7.14 The reciprocal of impedance is a complex number $\mathbf{Y}$ known as the *admittance*, defined by $\mathbf{Y} = 1/\mathbf{Z}$. We may write $\mathbf{Y} = G + jB$, in which case G is known as the *conductance* and B as the *susceptance*. Find the conductance and susceptance of the following:

a. A parallel combination of R and L at the angular frequency ω.

b. A series combination of R and L.

7.15 For the circuit of Fig. 7.26 write a phasor loop equation and solve for $\mathbf{v}$.

7.16 Write a phasor node equation for the circuit of Fig. 7.26 and solve for $\mathbf{v}$.

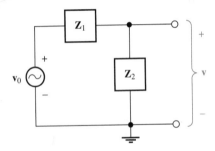

FIGURE 7.26

*7.17 In Fig. 7.26 find the numerical value of $\mathbf{v}$. Let $\mathbf{v}_0 = 10\,e^{j(20°)}$ and $f = 1,000$ Hz, and let $\mathbf{Z}_1$ and $\mathbf{Z}_2$ be the compound elements shown in Figs. 7.25(a) and 7.25(b), respectively. Use Eq. (7.12).

7.18 Write a set of loop equations sufficient to analyze the circuit of Fig. 7.27. It is not asked that you solve the equations.

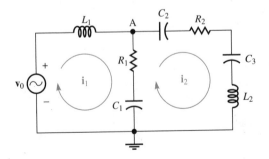

FIGURE 7.27

*7.19 Write a node equation for the circuit of Fig. 7.27 and solve for the voltage at node A.

*7.20 We wish to find a Thévenin equivalent for the subcircuit of Fig. 7.28(a) in the form shown in Fig. 7.28(b). The angular frequency is 230 radians per second.
  a. Find the numerical values of $\mathbf{v}_T$ and $\mathbf{Z}_T$.
  b. Devise a connection of ideal circuit elements with impedance equal to $\mathbf{Z}_T$.

7.21 Find the phasor current through the 1,000 Ω resistor in Fig. 7.28, in the high-frequency limit $f \to \infty$.

7.22 In the circuit of Fig. 7.29(a), find $\mathbf{v}$ for the following cases:
  a. In the limit $\omega \to 0$.
  b. In the limit $\omega \to \infty$.

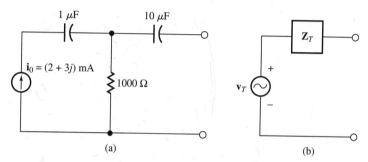

(a)   (b)

FIGURE 7.28

7.23 If the circuit of Fig. 7.29(a) is actually constructed, there will be additional parasitic (unintended) circuit elements, such as the inductances of the wires and capacitances arising from proximity. The resulting circuit is likely to look like that of Fig. 7.29(b).
  a. Find $\mathbf{v}$ in the limits $\omega \to 0$ and $\omega \to \infty$.
  b. In which limit do parasitics seem to have the greatest effect?

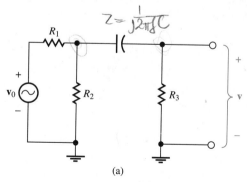

(a)

FIGURE 7.29(a)

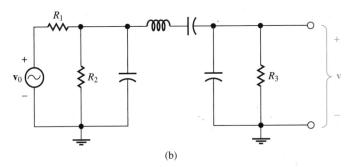

(b)

FIGURE 7.29(b)

7.24 A sinusoidal current source with phasor I_0 (real) and frequency ω is connected in series with a resistor R and capacitor C. Find the time-averaged power produced by the source.

*7.25 Repeat the previous problem with the current source replaced by a voltage source represented by the phasor V_0 (real).

*7.26 The following circuit elements are connected in parallel: An ideal current source $10 \cos (\omega t + 27°)$ mA; a 100 Ω resistor; and a 20 mH inductor. The ordinary frequency f is 400 Hz. Find the power produced by the current source.

*7.27 For the case of the preceding problem, calculate the power consumed by the resistor. Does this answer agree with the answer to Problem 7.26? *Ought* it to agree?

Section 7.2

7.28 A certain dc voltage increases from 6 to 16 V. Express the change in decibels. What is the change, in decibels, if a voltage increases from 60 to 160 V? If it changes from 160 to 60 V?

7.29 A sinusoidal signal with amplitude 14 V is described as being 82.92 dB in amplitude. What is the reference level?

7.30 How many times larger does a quantity become if it increases 1 dB?

7.31 Consider the Bode plot shown in Fig. 7.13(b). What is the slope of the low-frequency portion of this curve, expressed in dB per decade? Repeat for the curve in Fig. 7.30.

*7.32 Just as "decade" refers to an increase in frequency by a factor of 10, the term "octave" refers to an increase in frequency by a factor of 2. (For example, two notes separated by an octave on a piano keyboard differ in frequency by a factor of 2.) Describe the slope of Fig. 7.30 in terms of dB per octave.

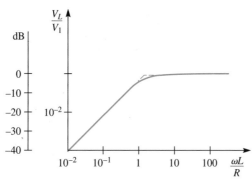

FIGURE 7.30

***7.33** Suppose a certain voltage V_1 is inversely proportional to frequency according to $V_1 = K/\omega$, where K is a constant. How many decibels does V_1 decrease per decade of increase in ω? How many dB per octave?

***7.34** A certain sinusoid has amplitude

$$V_1 = \frac{V_0 \omega RC}{1 + \omega^2 (RC)^2}$$

where V_0, R, and C are constants. What is the limit of V_1 as ω approaches zero? As ω approaches infinity? Sketch a Bode plot of V_1 versus ω. Show the corner frequency on your plot.

***7.35** For the circuit of Fig. 7.13(b), calculate the amplitude of the sinusoid V_{OUT} as a function of frequency.

**7.36 A sinusoid has amplitude

$$V_2(\omega) = \frac{V_0(1 + A\omega)}{(B^2 + C^2\omega^2)^{1/2}}$$

Find the limit of V_2 as $\omega \to 0$ and as $\omega \to \infty$. Sketch the Bode plot, showing important frequencies and magnitudes. Consider the two different cases: (a) $C \ll AB$; (b) $C \gg AB$.

Section 7.3

7.37 A 10 mH inductance, a 3 Ω resistance, and a variable capacitance are connected in series as shown in Fig. 7.31. (The arrow through the capacitor symbol indicates that its value is adjustable.) A voltage source with ordinary frequency $f = 800$ Hz is connected across the combination as shown. At what value of the capacitance is the circuit resonant?

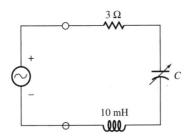

FIGURE 7.31

**7.38 An inductance L, a capacitance C, and a resistance R are connected in parallel. Calculate $|\mathbf{Z}|$, the absolute value of the impedance of this parallel combination.
   a. Make a graph on linear scales of $|\mathbf{Z}|$ as a function of ω over the range $0 < \omega < 25{,}000$ radians/sec. Take $L = 1$ mH, $C = 10$ μF, and $R = 30$ Ω.
   b. Repeat part (a) with $C = 5$ μF.
   c. Repeat part (a) with $L = 1$ mH, $C = 10$ μF, and $R = 60$ Ω.

**7.39 A sinusoidal voltage source with an amplitude V_0, a resistance R, an inductance L, and a capacitance C are all connected in parallel. Assume the frequency of the source is chosen to make the circuit resonant.
   a. What is the amplitude of the current $\mathbf{i}_0$ drawn from the voltage source?
   b. What is the amplitude of $\mathbf{i}_L$, the current through L?
   c. Find the ratio $|\mathbf{i}_L/\mathbf{i}_0|$. How does this ratio vary as R approaches infinity? Explain physically how $|\mathbf{i}_L|$ can be larger than $|\mathbf{i}_0|$.

**7.40 Use Eq. (7.25) to verify Eq. (7.27). Assume $Q \gg 1$.

7.41 Devise a load consisting of two elements in parallel that will accept maximum power from the Thévenin source of Example 7.16.

7.42 A Thévenin source consists of a sinusoidal voltage source with phasor $\mathbf{V}_0$ in series with a pure resistance R_0. It is connected to a load consisting of a pure resistance R. Find and sketch the power P_R delivered to R, versus R, over the range $0.1R_0 < R < 10R_0$.

*7.43 The *available power* of a signal source is defined as the largest power that it can deliver into any load. Suppose we have a Thévenin source consisting of a voltage source with phasor $\mathbf{v}_T$ in series with $\mathbf{Z}_T = \mathbf{R}_T + j\mathbf{X}_T$. Show that its available power is $|\mathbf{v}_T|^2/8R_T$. (Note the important factor 8.)

**7.44 Find the available power (defined in Problem 7.43) of a Norton source consisting of $\mathbf{i}_n$ in parallel with $\mathbf{Z}_n$, where $\mathbf{Z}_N \equiv R_N + jX_N$.

Section 7.4

*7.45 Investigate the behavior of result (7.42) in the limits $D \to 0$ (flabby dashpot) and $D \to \infty$ (rigid dashpot). Show that these limits agree with what one expects from physical reasoning.

7.46 (Computer exercise) Use a suitable software package to automatically graph $|\mathbf{y}|$ vs. ω from result (7.42). For simplicity assume $k = M = 1$ in the system of units being used.

You are now asked to find the value of D that makes the *maximum* value of $|\mathbf{y}|/H$ equal to 1.2. Find the value of D that satisfies this requirement.

CHAPTER 8

TRANSIENT RESPONSE OF PASSIVE CIRCUITS

I n the previous two chapters we have been describing the *forced* response of passive circuits. Here the term "forced" indicates that the circuit is responding to an ongoing stimulus from outside, the so-called forcing function. However, the same passive circuits that we have discussed in connection with forced response are also capable of another kind of behavior, known as their *natural response*. Natural response is the behavior that continues *after* some disturbance has occurred. For example, let us think of a mass hanging from a spring. We can apply a constant sinusoidal up-and-down force to the mass, and it will oscillate up and down at the frequency of the driving force. Then we are observing the forced response of the spring-mass system. But now let us imagine that instead of applying a sinusoidal outside force, we simply push the mass away from its position of rest and then let it move freely. In this case the mass will bob up and down in a motion that is damped by friction, so that after a while it comes to rest. This is not a case of forced response; the force that started the motion is no longer acting. What we are now seeing is the natural response of the system. It is referred to as "natural" response because in this case the oscillation frequency and the time needed for the motion to damp out are not determined by the force that originally started the motion; on the contrary, they are determined by the mass, spring constant, and friction of the system itself. Natural response is sometimes also called *transient response* because the natural response to the initial stimulus tends to eventually die out.

In mechanical systems it is easily seen that the transient response decays due to friction. In electric circuits the decay arises from energy loss in the resistances.

It is possible for a circuit to exhibit both forced and natural response at the same time. This can happen, for example, if an originally quiescent circuit is subjected to a sinusoidal outside force, which is turned on only at time $t = 0$. In this case there will be an initial period in which natural response (as well as forced response) is important, because the system is responding to the change in its environment that occurred when the force was turned on. This part of the response is transient, however, and dies away as the system becomes accustomed to its new environment. Ultimately the "transient" is gone, and what is left is the sinusoidal steady-state response. Most electronic systems do exhibit transients whenever they are turned on.

Fortunately, many problems involving transients are of a particularly simple kind. This is because in many circuits only one energy-storage element (that is, only one capacitor or one inductor) plays a significant role. Such circuits are described by differential equations of only the first order; thus we shall refer to them as "first-order" circuits. In Section 8.1 we show how to find the transient response in these simple cases. This section requires little previous experience with differential equations. For those readers who wish to go more deeply into the subject of transient response, a brief review of linear differential equations with constant coefficients is presented in Section 8.2. Equipped with this mathematical background, we are then ready to proceed to circuits of order higher than the first. The subject of *complete response* (that is, the general case of simultaneous natural and forced response) is considered briefly in Section 8.3.

8.1 Transients in First-Order Circuits

In this section we deal with those very common circuits in which only one capacitance or one inductance plays an important role. The subject of transient response includes the delayed response to any kind of external stimulus, but let us begin by considering a particular kind of starting force. We shall assume that before time $t = 0$ the input voltage to the circuit had some constant value, and that at $t = 0$ the input voltage changes to a different constant value. Somewhere in the circuit is another pair of terminals marked "output," and our goal is to find the voltage at the output terminals as a function of time. We anticipate that for $t < 0$ the output voltage will be constant, since before $t = 0$ the circuit had been undisturbed for a long time. Beginning at $t = 0$ we expect the output voltage to change. After a very long time we expect that the output voltage will again settle down to some constant value. The intermediate period, during which the output voltage is changing, is the transient phase we wish to study. We can see that the transient response tells us something about how quickly a circuit can respond to changes of input. The fact that an instantaneous change of input gives a slower, delayed change of output is a key factor determining the operating speeds of electronic circuits.

STEP RESPONSE

A case that occurs often is illustrated by Fig. 8.1. Here the input voltage $v_1(t)$ is equal to V_1 (a constant) for $t < 0$ and to V_2 (another constant) for $t > 0$. This of course gives a stepwise change of input voltage at $t = 0$, as shown in Fig. 8.1(b). In order to indicate that $v_1(t)$ has the form of a voltage step, the voltage source in Fig. 6.1(a) has a "step" symbol in its center. The transient response resulting from this kind of input is known as the *step response* of the circuit.

We now wish to find the output voltage $v_{OUT}(t)$ when the network contains any combination of resistors and either one capacitor or one inductor. Under these restrictions it can be shown that the output voltage for $t > 0$ always has the following general form:

$$v_{OUT}(t) = A + Be^{-t/\tau} \tag{8.1}$$

In this expression, A, B, and τ are constants to be determined. A and B have the dimensions of voltage, while τ, which has the dimensions of time, is known as the circuit's *time constant*. Note that Eq. (8.1) is valid only for $t > 0$, since it describes the transient response to the change made at $t = 0$. The instant of time $t = 0$ (that is, t equals *exactly* zero) is exceptional, because at $t = 0$ the input voltage is in the process of changing from V_1 to V_2 and, strictly speaking, is not a single-valued function of

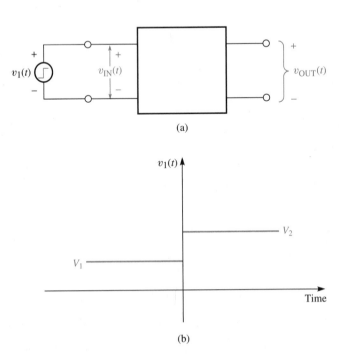

(a)

(b)

FIGURE 8.1

time. To avoid confusion, we shall say that the validity of Eq. (8.1) begins at the time $t = 0+$. Here "0+" is the time just *after* $t = 0$. Similarly we shall use "$t = 0-$" to denote the time just *before* $t = 0$.

After the transient begins, at $t = 0+$, the rate of change of v_{OUT} decreases. This can be seen from Eq. (8.1): dv_{OUT}/dt decreases exponentially with time. After a time τ has passed, dv_{OUT}/dt has decreased to $1/e$ of its value at the time $t = 0+$. As more time passes, the rate of change of v_{OUT} approaches zero, and v_{OUT} approaches some constant final value. Because the rate of change of v_{OUT} decreases to $1/e$ (37%) of its maximum value after a time τ has passed, it is common to refer to τ as the duration of the transient. Of course, strictly speaking, the transient persists for an infinitely long time, since $e^{-t/\tau}$ never quite reaches zero but only approaches zero asymptotically. However, *most* of the change from the output's initial value to its final value (63% of the change, to be precise) will have occurred by the time $t = \tau$ is reached. Thus the idea of τ being the approximate duration of the transient is quite useful.

It should be noted that $v_{OUT}(t)$ may in some circuits be discontinuous at $t = 0$. This would be the case, for example, in a very simple circuit where the input is directly connected to the output, with $v_{OUT} = v_{IN}$. However, for $t = 0+$ and all later times v_{OUT} is the continuous function given in Eq. (8.1).

We shall now state two useful rules concerning *sudden* changes of voltages and currents in circuits. First, let us consider a capacitor C across which the voltage is v. The current through the capacitor is $i = C \, dv/dt$. If the voltage v were to change instantaneously, dv/dt would be infinite, and the current through the capacitor would also be infinite. But infinite current is an impossibility, and thus we have

RULE 1

The voltage across a capacitor cannot change instantaneously.

Similarly, for an inductor we have $v = L \, di/dt$. We see that in order to have a sudden change of current, an infinite voltage would be required. Thus we have

RULE 2

The current through an inductor cannot change instantaneously.*

Two additional rules can be stated concerning circuits that are in a dc *steady state*. Let us assume that no changes have been made for a long time in a circuit that contains

---

*In more advanced work, an idealized input known as an "impulse function" is sometimes used. An impulse is a voltage (or current) that is assumed to be infinitely large for an infinitesimal length of time. When the input to a circuit is assumed to be an impulse, Rules 1 and 2 do not apply. However, impulses do not actually occur in real circuits, because v and i are always finite.

TABLE 8.1

	C	L
Quantity that cannot be discontinuous	Voltage	Current
Quantity that is zero in the dc steady state	Current	Voltage

only *constant* voltage and current sources. In that case all transients will have had time to die away, and all voltages and currents will have become constant. We refer to this condition as the dc steady state. The current through a capacitor is given by $i = C \, dv/dt$. Since in the dc steady state $dv/dt = 0$, we have

RULE 3

In the dc steady state the current through a capacitor is zero.

Similarly, the voltage across an inductor is $v = L \, di/dt$. Thus we have

RULE 4

In the dc steady state the voltage across an inductor is zero.

Rules 1 to 4 are summarized in Table 8.1. Their use will be illustrated by the following examples.

EXAMPLE 8.1

In the circuit shown in Fig. 8.2 the switch is set at position A until time $t = 0$, when it is changed to position B. Find v_C, the voltage across the capacitor, immediately after the switch is changed. Also, find v_R immediately after the switch is changed.

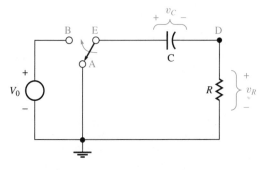

FIGURE 8.2

SOLUTION

Before the switch was changed, the capacitor discharged any charge it may once have had through R. Thus v_C was zero.* Since v_C cannot change suddenly (Rule 1), v_C is still zero immediately after $t = 0$.

Immediately after $t = 0$, the voltage at E is V_0. We have just seen that $v_C(0+)$ is zero; but $v_C = V_E - V_D$, and therefore the potential at D is the same as at E. Thus V_D is V_0; thus $v_R = V_D - (0) = V_0$. We note that before the switch was changed, v_R was zero. Thus v_R *does* change instantaneously when the switch is changed. This does not violate any of the rules, since the rules say nothing about the voltage across a *resistor*.

---

*We could also reason as follows: At $t = 0 -$ the circuit is in its dc steady state. In the dc steady state the current through C must be zero (Rule 3). Thus the current through R is zero. Thus $v_D = 0$ and $v_C = v_E - v_D = 0$.

EXAMPLE 8.2

In the circuit shown in Fig. 8.3 the switch is at position B until time $t = 0$, when it is changed to A. (Assume that the action of the switch is truly instantaneous, so the current through L is not interrupted.)

(a) Find the voltage at A at $t = 0+$.

(b) Find $(d/dt)(v_A)$ at $t = 0+$.

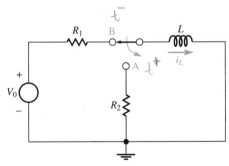

FIGURE 8.3

SOLUTION

(a) At times before $t = 0$, the circuit is in its dc steady state. Therefore the voltage across L is zero (Rule 4). All the applied voltage V_0 thus appears across R_1, and i_L (the current through L) is equal to V_0/R_1.

Immediately after the switch is moved from B to A, i_L must still be V_0/R_1 (Rule 2). However, this current is now moving upward through R_2. Thus by Ohm's law $[(0) - v_A]/R_2 = V_0/R_1$, and

$$v_A = -\frac{R_2}{R_1} V_0$$

(It is rather easy to overlook the important minus sign. The reader should note how it appears in the result.)

(b) We note that at $t = 0+$, $v_A = L\, di_L/dt$. Thus $di_L/dt = v_A/L = -(R_2/R_1 L)V_0$. But from Ohm's law, $v_A = -i_L R_2$. Thus $(d/dt)(v_A) = -R_2\, di_L/dt = (R_2^2/R_1 L)V_0$ at $t = 0+$.

• EXERCISE 8.1

In the circuit shown in Fig. 8.4 the switch is moved from position B to position A at $t = 0$. (a) Find $v_D(t = 0-)$. (b) Find $i_1(t = 0-)$. (c) Find $v_D(t = 0+)$. (d) Find $(d/dt)(v_D)$ at $t = 0+$. **Answers:** (a) Zero. (b) V_0/R_2. (c) $-V_0 R_1/(R_1 + R_2)$. (d) $V_0 R_1^2 R_2/(R_1 + R_2)^2 L$.

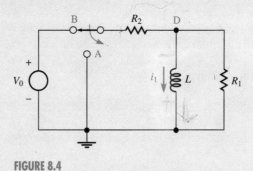

FIGURE 8.4

Let us now find the transient response of the network shown in Fig. 8.5. We shall consider the input $v_1(t) = 0$ for $t < 0$ and $v_1(t) = V$ for $t > 0$. (Thus the input is as shown in Fig. 8.1(b), with $V_1 = 0$ and $V_2 = V$.) We begin by writing a node equation for v_{OUT}:

$$\frac{v_1(t) - v_{OUT}(t)}{R} - C\frac{dv_{OUT}}{dt} = 0 \tag{8.2}$$

which can be rewritten

$$\frac{dv_{OUT}}{dt} + \frac{1}{RC} v_{OUT}(t) = \frac{1}{RC} v_1(t) \tag{8.3}$$

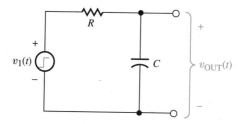

FIGURE 8.5

We know that the solution will have the form of Eq. (8.1); all we need to do is evaluate the unknown constants in Eq. (8.1). Thus we proceed by substituting the general answer, Eq. (8.1), into the equation for this particular problem, Eq. (8.3). The result is

$$-\frac{B}{\tau}e^{-t/\tau} + \frac{A}{RC} + \frac{B}{RC}e^{-t/\tau} = \frac{1}{RC}V \tag{8.4}$$

where the constants A, B, and τ are still to be determined. Note that we have substituted $v_1(t) = V$ on the right side of Eq. (8.4). We do so because we are interested only in the time domain $t > 0$, during which the value of v_1 is V.

We see that Eq. (8.4) can be rewritten

$$\left(\frac{A}{RC} - \frac{V}{RC}\right) + \left(\frac{B}{RC} - \frac{B}{\tau}\right)e^{-t/\tau} = 0 \tag{8.5}$$

Now in Eq. (8.5) the first term is constant in time, but the second varies because of the factor $e^{-t/\tau}$. The only way that Eq. (8.5) can be true for *all values of t* is for *each* of the two quantities in parentheses to separately equal zero. Thus we must have

$$\frac{A}{RC} - \frac{V}{RC} = 0 \tag{8.6}$$

$$\frac{B}{RC} - \frac{B}{\tau} = 0$$

From Eq. (8.6) we immediately obtain the values of two of our unknown constants, finding that $A = V$ and $\tau = RC$. The third constant, B, must still be found by making use of the initial condition.

From Eq. (8.1) we note that at $t = 0+$, $v_{OUT} = A + B$. Thus* if we know v_{OUT} $(t = 0+)$ we have the third equation needed to find B. The value of $v_{OUT}(t = 0+)$ can be found using Rule 1 discussed previously, which tells us that $v_{OUT}(0+)$ must

---

*More precisely, we are taking the limit of Eq. (8.1) as t approaches zero from the positive side. However, let us not obscure the discussion with mathematical refinements. The desired result is simply obtained by substituting $t = 0$ into Eq. (8.1).

equal $v_{OUT}(t = 0-)$. But the value of v_{OUT} must be zero for all times earlier than zero because the input was zero at all times less than zero. Thus

$$v_{OUT}(0+) = A + B = 0 \qquad (8.7)$$

Since $A = V$, we have $B = -V$, and from Eq. (8.1) the solution is

$$v_{OUT} = V(1 - e^{-t/RC}) \qquad (8.8)$$

which is valid for $t > 0$.

The behavior of v_{OUT} as a function of time is shown in Fig. 8.6. We can visualize what happens physically as follows: Initially the capacitor is uncharged, and v_{OUT}, the voltage across it, is zero. When the step input changes to V at $t = 0$, a voltage whose value is $V - v_{OUT}$ is applied across R. This causes a current $(V - v_{OUT})/R$ to flow through R and into C. As C is charged, the voltage across it, v_{OUT}, increases. This in turn causes the charging current $(V - v_{OUT})/R$ to have a smaller value, and the capacitor charges more slowly. As the capacitor continues to charge, the charging current continues to decrease. The charging current never quite falls to zero, but it approaches zero asymptotically; similarly, v_{OUT} never quite reaches V, but it approaches V as a limit. We have already noted that the value of τ provides an order-of-magnitude estimate for the duration of the transient. In the present example $\tau = RC$.

The steps used in finding the step-input response of a first-order circuit are thus as follows:

1. Write a node or loop equation.

2. Substitute the general solution, Eq. (8.1), obtaining two equations in the three unknowns A, B, and τ.

3. Use the initial condition of the circuit (Rules 1 and 2) to obtain the third equation.

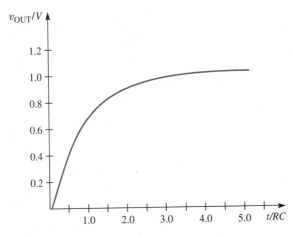

FIGURE 8.6

EXAMPLE 8.3

For the circuit of Fig. 8.5, find $v_{OUT}(t)$ when the input $v_1(t) = V_1$ for $t < 0$ and V for $t > 0$. Assume $V_1 > V$.

SOLUTION

Equations (8.2) through (8.6) are obtained exactly as before. However, the initial condition has been changed, and it is necessary to find $v_{OUT}(0+)$ for the present case.

We note that for all times $t < 0$ the input voltage was constant. Presumably any transients resulting from the original turn-on of v_1 have long since died away, and before $t = 0$ all voltages and currents in the circuit will be *constant*. The current through C, which is $C\,dv_{OUT}/dt$, must thus be zero for $t < 0$. But the current through C is equal to the current through R, which must therefore also be zero. If the current through R is zero for $t < 0$, then the voltage drop across R must be zero, and thus for $t < 0$, v_{OUT} must equal v_1, which equals V_1 in that time period. Thus $v_{OUT}(0-) = V_1$, and from Rule 1 we have $v_{OUT}(0+) = v_{OUT}(0-) = V_1$.

Setting $t = 0$ in Eq. (8.1), we find that $v_{OUT}(0+) = A + B$. Thus

$$A + B = V_1$$

From Eq. (8.6) $A = V$, $\tau = RC$, and hence $B = V_1 - V$. The solution is

$$v_{OUT} = V + (V_1 - V)e^{-t/RC}$$

The duration of the transient is again equal to RC, just as it was in the previous case. This is because the time constant arises from the natural response of the circuit, which depends only on the circuit itself and not on the initial conditions. The initial value of v_{OUT} (found by setting $t = 0$) is V_1, and its final value (found by taking the limit at $t \to \infty$) is V. The function $v_{OUT}(t)$ is as shown in Fig. 8.7.

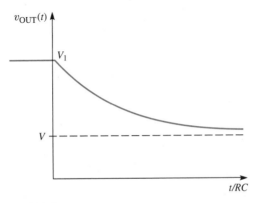

FIGURE 8.7

EXAMPLE 8.4

For the circuit of Fig. 8.8(a) find the output voltage $v_{OUT}(t)$. Sketch $v_{OUT}(t)$ as a function of time, for both $t < 0$ and $t > 0$. The input step voltage $v_1(t)$ is equal to 0 for $t < 0$ and equal to V for $t > 0$.

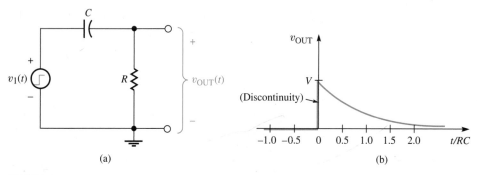

(a)   (b)

FIGURE 8.8

SOLUTION

Since $v_1(t) = 0$ for $t < 0$, v_{OUT} will also be zero until $t = 0$. For $t > 0$, $v_1 = V$. Thus the node equation for v_{OUT} is

$$C\frac{d}{dt}(v_{OUT} - V) + \frac{v_{OUT}}{R} = 0$$

Rearranging and noting that since V is a constant, we have $dV/dt = 0$, we obtain

$$\frac{d}{dt}(v_{OUT}) + \frac{v_{OUT}}{RC} = 0$$

As usual, we substitute the general first-order solution $v_{OUT}(t) = A + Be^{-t/\tau}$, Eq. (8.1), obtaining

$$\left(\frac{A}{RC}\right) + \left(\frac{B}{RC} - \frac{B}{\tau}\right)e^{-t/\tau} = 0$$

This equation can hold at all times only if the quantities enclosed by parentheses separately equal zero. Setting the first term to zero gives us $A = 0$; setting the second to zero, we have $\tau = RC$. To find B we must make use of initial conditions.

We know that at $t = 0-$ (the time just before $t = 0$), $v_{OUT} = 0$. For $t < 0$, $v_1 = 0$; hence the voltage *across the capacitor* is zero at $t = 0-$. From Rule 1 we know that the voltage *across the capacitor* cannot change instantaneously. Therefore $v_{OUT} - v_1$ must be zero at $t = 0+$. But notice that at $t = 0+$ we have $v_1 = V$ (because v_1 changed suddenly from 0 to V at $t = 0$). Therefore $v_{OUT}(t = 0+) = V$. We note that $v_{OUT}(t)$ changes discontinuously from zero to V at $t = 0$. This does not violate Rule 1. Rule 1 does *not* say that v_{OUT} cannot change discontinuously; Rule

1 says that the voltage *across a capacitor* cannot change discontinuously. In this case the voltage across the capacitor is not v_{OUT}; it is $v_{OUT} - v_1$.

Now that we have $v_{OUT}(0+) = V$, we are in a position to find B. This is done by setting t to zero in $v_{OUT} = A + Be^{-t/\tau}$; thus

$$V = A + B$$

However, we found earlier in this example that $A = 0$. Thus we have $B = V$, and

$$v_{OUT} = Ve^{-t/RC}$$

The function $v_{OUT}(t)$ is shown in Fig. 8.8(b). Again we note that it contains a discontinuity at $t = 0$.

• EXERCISE 8.2

In connection with step response one sometimes speaks of the "10 to 90% rise time," T_R. This is defined as the time required for a rising voltage to change from 10% to 90% of its final value. Find T_R for the circuit of Fig. 8.5 with the step input shown in Fig. 8.1(b). Let $V_1 = 0$ and $V_2 = 3$ V. Express your answer in terms of RC. **Answer:** 2.197 (RC).

When a circuit contains a single inductor instead of a capacitor, the procedures to be used are similar. When there is an inductor we may use Rule 2, which states that the *current* through an inductor cannot change instantaneously. Another point to be noted is that, according to Rule 4, in the dc steady state it is the *voltage* across the inductor that is zero (whereas for a capacitor the *current* is zero in the steady state).

EXAMPLE 8.5

Find and sketch the voltage $v_{OUT}(t)$ in the circuit shown in Fig. 8.9(a). The step input voltage is given by $v_1(t < 0) = +1$ V and $v_1(t > 0) = -1$ V. Let $L = 1$ mH and $R = 1,000\ \Omega$.

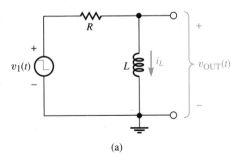

(a)

FIGURE 8.9(a)

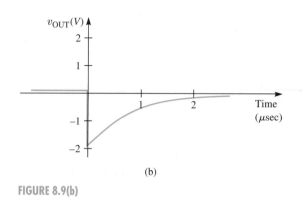

(b)

FIGURE 8.9(b)

SOLUTION

Before $t = 0$ the circuit is in its steady state, and from Rule 4 (or Table 8.1) we see that the voltage across L is zero. In order to find v_{OUT} for $t > 0$ we write a node equation,

$$\frac{v_{OUT} + 1}{R} + i_L = 0$$

[where we have used the fact that $v_1(t > 0) = -1$]. Differentiating this equation with respect to time and noting that $v_{OUT} = L\, di_L/dt$, we have

$$\frac{dv_{OUT}}{dt} + \frac{v_{OUT}}{(L/R)} = 0$$

Substituting the general first-order solution $v_{OUT} = A + Be^{-t/\tau}$, we find that $\tau = L/R$ and $A = 0$. To determine B we then make use of the initial conditions.

As we have already noted, $v_{OUT}(t = 0-) = 0$. Therefore the value of $i_L(0-)$ is given by Ohm's law: $i_L(0-) = [(+1) - (0)]/R = 1$ mA. Since i_L cannot change instantaneously (Rule 2), we find that $i_L(0+)$ also equals 1 mA. Again from Ohm's law, $v_{OUT}(0+) = v_1(0+) - i_L(0+)R = -1 - (1 \text{ mA})(1{,}000\ \Omega) = -2$ V. Substituting $t = 0$ into $v_{OUT} = A + Be^{-t/\tau}$ and remembering that A is zero, we find $B = -2$. The solution for times $t > 0$ is thus

$$v_{OUT} = -2e^{-10^6 t}\ \text{V}$$

where t is given in seconds. The behavior of v_{OUT} for times both earlier and later than zero is shown in Fig. 8.9(b). Note that once again v_{OUT} is discontinuous at $t = 0$.

EXAMPLE 8.6

Find and graph $i_L(t)$ for the circuit of Fig. 8.9(a).

SOLUTION

Since $v_{OUT}(t)$ is already known, i_L can be found using Ohm's law:

$$i_L = \frac{v_1 - v_{OUT}}{R}$$

Thus for $t < 0$ we have $i_L = [1 - (0)]/R = 1$ mA, and for $t > 0$ we have $i_L = [-1 - (-2e^{-10^6 t})] = 2e^{-10^6 t} - 1$. These results are shown in Fig. 8.10. Note that there is *no* discontinuity in i_L at $t = 0$, as we know must be the case. As $t \to +\infty$, i_L approaches -1 mA asymptotically. This represents the approach to the final steady state, in which the voltage across the inductor must approach zero.

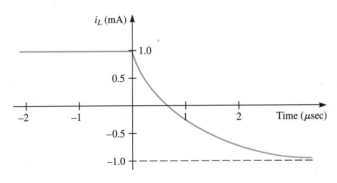

FIGURE 8.10

● EXERCISE 8.3

For the circuit of Example 8.5, let $v_1(t < 0) = +1.7$ V and $v_1(t > 0) = +3.2$ V. (a) Find $v_{OUT}(t)$ for $t > 0$. (b) What is the approximate duration of the transient? **Answers:** (a) $v_{OUT}(t) = 1.5e^{-10^6 t}$ V. (b) 1 μsec.

The examples just presented involved rather simple circuits. However, one may treat more complex circuits with the same methods, provided that they contain only one capacitor or one inductor. The algebraic effort may increase, of course. Alternatively, we note that more complex circuits can often be reduced in complexity through use of Thévenin's or Norton's theorems. This is done by replacing the entire circuit, except for the capacitor or inductor, by an appropriate equivalent.

EXAMPLE 8.7

Simplify the circuit shown in Fig. 8.11(a) by means of Thévenin's theorem.

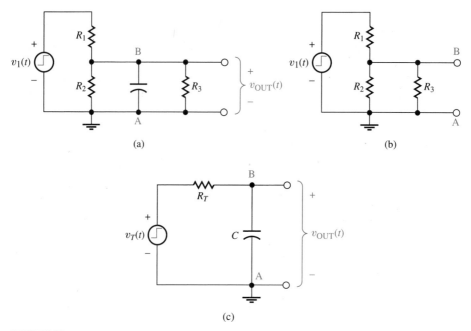

FIGURE 8.11

SOLUTION

We shall regard points A and B as the terminals of a subcircuit containing v_1, R_1, R_2, and R_3. In other words, we need the Thévenin equivalent of the subcircuit that is connected across the capacitor; this subcircuit is shown in Fig. 8.11(b). Using the voltage-divider formula and the methods of Chapter 3, we easily find

$$v_T(t) = v_1(t) \frac{R_2 \| R_3}{R_2 \| R_3 + R_1}$$

$$R_T = R_2 \| R_3 \| R_1$$

Now we can replace the subcircuit by its equivalent, obtaining the simplified circuit shown in Fig. 8.11(c). Note that the simplified circuit is the same as that of Fig. 8.5, which we have already studied. Let $v_1(t) = 0$ for $t < 0$ and V for $t > 0$. Then with no further calculation we have from Eq. (8.8) that

$$v_{\text{OUT}}(t) = V \frac{R_2 \| R_3}{R_2 \| R_3 + R_1} (1 - e^{-t/R_T C})$$

where R_T is as given above.

● **EXERCISE 8.4**

Find the time constant (approximate duration of transients) in the circuit shown in Fig. 8.12 by first replacing v_1, R_1, and R_2 with their Thévenin equivalent, and then using the results of Example 8.5. **Answer:** $\tau = L/R_2$.

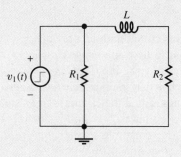

FIGURE 8.12

A slightly different situation can arise when the step input results from the opening or closing of a switch somewhere in the circuit—for example, as shown in Fig. 8.13(a). (Actual mechanical switches do not turn up very often in electronic circuits, but electronic switching circuits, in which a transistor plays the role of a switch, are common.) In this circuit, let the switch be closed for all times $t < 0$ and open for all

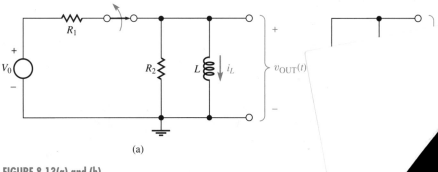

(a)

FIGURE 8.13(a) and (b)

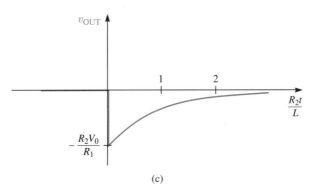

(c)

FIGURE 8.13(c)

times $t > 0$; we wish to find $v_{OUT}(t)$ for all times. Let us note the subtle difference between this circuit and the previous type, as, for example in Fig. 8.1. In Fig. 8.1 v_{IN} is constrained to have the value V_2 when $t > 0$. In the present case, after the switch is opened at $t = 0$ the input *voltage* is not controlled at all; the constraint imposed by the open switch is that the input *current* is zero.

At times $t < 0$ the circuit is in its steady state and (from Rule 4) $v_{OUT} = 0$. Thus at $t = 0-$ we have $i_L = V_0/R_1$. Since i_L cannot change instantaneously, $i_L(0+)$ must also be V_0/R_1. For times greater than zero, the circuit is as shown in Fig. 8.4(b), with the initial condition $i_L(0+) = V_0/R_1$. Since $v_{OUT} = L\, di_L/dt$ and $i_L = -v_{OUT}/R_2$, we have

$$\frac{dv_{OUT}}{dt} + \frac{v_{OUT}}{(L/R_2)} = 0 \qquad (8.9)$$

Substituting the general first-order solution, Eq. (8.1), we have

$$v_{OUT} = Be^{-R_2 t/L} \qquad (8.10)$$

where B is a constant still to be determined. We do this by noting that $v_{OUT}(0+) = -R_2 i_L(0+) = -R_2 V_0/R_1$. Thus

$$v_{OUT} = -\frac{R_2 V_0}{R_1} e^{-R_2 t/L} \qquad (8.11)$$

This output is shown in Fig. 8.13(c). This circuit is interesting because we notice that as R_2 increases, the size of the voltage pulse seems to increase without limit. What would happen, then, if we made R_2 infinitely large by simply omitting it from the circuit? The answer is that the voltage pulse would grow so large that breakdown would take place; a spark would occur somewhere. In fact, this circuit is used in automobile ignition systems for this very purpose. The spark plug is connected in parallel with L, V_0 is provided by the car battery, and the switching action is provided by the breaker points or, in modern systems, by a transistor switch. Using this system, several thousand volts can be obtained although V_0 is only 12 V!

• EXERCISE 8.5

Find the approximate duration of the transient in Fig. 8.13(a), if the switch is initially *open* and is *closed* at $t = 0$. **Answer:** $\tau = L(R_1 + R_2)/R_1 R_2$.

RESPONSE TO A RECTANGULAR PULSE

It often happens (especially in digital systems) that a circuit is excited by a rectangular input voltage pulse, as shown in Fig. 8.14. In this case $v_{\text{IN}}(t)$ is zero for $t < 0$, equals V_0 for $0 < t < T$, and is again zero for $t > T$, as shown in Fig. 8.14(b). This kind of input is equivalent to excitation by *two* step inputs of the kind discussed above. It is easily seen that the input to the block is the same if we redraw the circuit as shown in

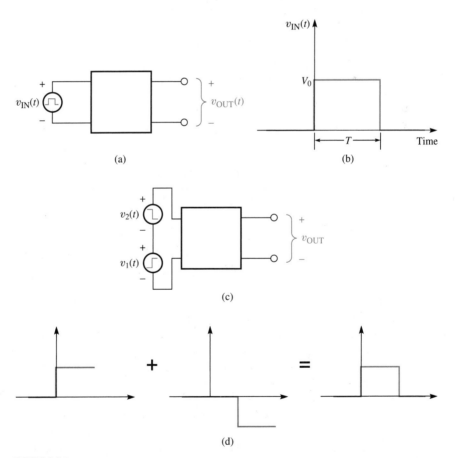

(a)

(b)

(c)

(d)

FIGURE 8.14

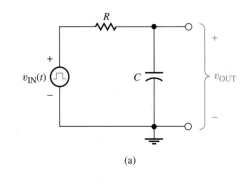

(a)

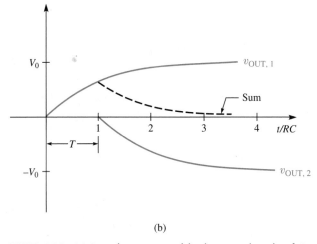

(b)

FIGURE 8.15   (a) First-order circuit excited by the rectangular pulse of Fig. 8.14(b). (b) Output for the special case of $T = RC$. The total output, marked "Sum," is the sum of the outputs for a positive-going step at $T = 0$ and a negative-going step at $t = T$.

Fig. 8.14(c). Here $v_1(t)$ is 0 for $t < 0$ and V_0 for $t > 0$; $v_2(t)$ is 0 for $t < T$ and $-V_0$ for $t > T$. The way that these two steps add up to the original pulse is shown in Fig. 8.14(d).

We have already learned how to find the response to a single step input; the response to two steps can be found by addition. The response to the rectangular pulse will simply be the sum of the response to v_1 and the response to v_2. However, we must remember that the step of v_2 does not occur at $t = 0$, but rather at $t = T$. Thus the response to v_2 is delayed by that amount of time.

Let us see what happens when the rectangular pulse of Fig. 8.14(b) excites the circuit of Fig. 8.15. This circuit is (except for the form of v_{IN}) the same as that of Fig. 8.5; we have already found its step response in Eq. (8.8). The response to v_2 is found by noting (1) that the amplitude of v_2 is *minus* V_0; (2) that the step occurs at $t = T$; and (3) that for any function $f(t)$ the same function, delayed by a time T, is $f(t - T)$. For the three time periods in question the functions are as follows:

	$t < 0$	$0 < t < T$	$T < t$	(8.12)
Response to v_1	0	$V_0(1 - e^{-t/RC})$	$V_0(1 - e^{-t/RC})$	
Response to v_2	0	0	$-V_0(1 - e^{-(t-T)/RC})$	
Sum $= v_{OUT}$	0	$V_0(1 - e^{-t/RC})$	$V_0 e^{-t/RC}(e^{T/RC} - 1)$	

The two step responses and their sum, which is v_{OUT}, are shown in Fig. 8.15(b).

The response of this circuit to the rectangular pulse depends in an important way on the quantity T/RC. From Fig. 8.15(b) we see that v_{OUT} reaches its maximum at $t = T$. If T is short compared with the time constant RC, v_{OUT} does not have time to rise very high before it must start downward. On the other hand, if $T >> RC$, then $v_{OUT,1}$ will have time to increase to almost its final value (V_0) before $v_{OUT,2}$ begins to bring the total down. Figure 8.16 shows the response to a pulse of duration 1 msec for three different values of RC. As expected, we see that the shorter the time constant RC, the higher the output pulse rises. The output pulse also becomes shorter and more rectangular as RC decreases, and in fact in the limit of $RC \rightarrow 0$, v_{OUT} becomes exactly the same rectangular pulse as v_{IN}.

The foregoing example illustrates the phenomenon of *pulse distortion*. It often happens that we wish the output pulses of an electronic circuit to have nearly the same shape as the input pulses. For example, a nearly ideal amplifier might increase the size of incoming pulses but leave their shape unchanged. Unfortunately, capacitance is always present in electronic circuits, and very often its effect is similar to that shown in Fig. 8.16. If we wish the output pulses to be relatively undistorted in shape, we must take care to use pulses that are much longer than the time constant RC. But the longer

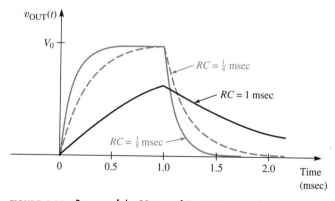

FIGURE 8.16 Response of the RC circuit of Fig. 8.15(a) to a rectangular pulse of amplitude V_0 and duration 1 msec. The three curves are for the cases $RC = \frac{1}{8}$ msec, $\frac{1}{4}$ msec, and 1 msec.

the pulses, the fewer of them per second, and the less information per second the circuit can handle. For this reason a great deal of effort has gone into reducing capacitance and its effects in electronic devices.

EXAMPLE 8.8

For the circuit of Fig. 8.15, the duration of the rectangular input pulse [as shown in Fig. 8.14(b)] is 1 μsec. It is required that v_{OUT} reach at least 0.95 V_0. If $R = 1,000\ \Omega$, what is the largest value that C can be allowed to have?

SOLUTION

Expressions for $v_{OUT}(t)$ in the three time periods are given in the bottom line of Eq. (8.12). From Fig. 8.15(b) we note that the maximum value of v_{OUT} is reached at $t = T$. Thus we can use either the expression in the middle or the one in the right-hand column of (8.12); they should be equal at $t = T$. Choosing the simpler expression valid for $0 < t < T$ and substituting $t = T$, we find

$$v_{OUT}(\text{max}) = V_0(1 - e^{-T/RC})$$

We must find the value of C that makes $v_{OUT}(\text{max}) = 0.95\ V_0$. Thus

$$e^{-T/RC} = 0.05$$

$$C = -\frac{T}{R\ \ln(0.05)} = 3.34 \times 10^{-10}\ \text{F}$$

• EXERCISE 8.6

For the circuit of Fig. 8.15(a), let $R = 1,000\ \Omega$. A rectangular input pulse of duration 1 msec is applied from $t = 0$ to $t = 1$ msec. Find the value of C that maximizes v_{OUT} at $t = 2$ msec. **Answer:** 1.44 μF.

*8.2 Higher-Order Circuits

We now turn our attention to circuits that contain more than one capacitor or inductor. Since we will deal only with circuits containing linear circuit elements, the equations governing them are *linear differential equations with constant coefficients*. Let us first review the properties of these equations.

The general differential equation we shall consider is of the form

$$A_n \frac{d^n v}{dt^n} + A_{n-1} \frac{d^{(n-1)} v}{dt^{(n-1)}} + \cdots + A_1 \frac{dv}{dt} + A_0 v = f(t) \tag{8.13}$$

The A_n are all constants, $v(t)$ is the unknown to be found, and $f(t)$ is a forcing function, which arises from the voltage and current sources that drive the circuit. The order of the equation, n, is less than or equal to the sum of the number of capacitors plus the number of inductors in the circuit. The equations of the preceding section—for example, Eq. (8.3)—all belonged to the special case of $n = 1$.

Equation (8.13) is very general in that it allows any kind of forcing function $f(t)$. In fact, the case of sinusoidal $f(t)$ was the subject of Chapters 6 and 7. [Compare, for example, Eq. (6.17).] Now, however, we are interested in transient response, and so we shall consider forcing functions that are constant except for a sudden change at time $t = 0$. In other words, $f(t)$ will have the general form shown in Fig. 8.1(b).

From the theory of differential equations, it is known that the solution of Eq. (8.13) can be expressed in the form

$$v(t) = h(t) + p(t) \tag{8.14}$$

Here $h(t)$, known as the homogeneous solution, is a solution of the *homogeneous equation*

$$A_n \frac{d^n h}{dt^n} + A_{n-1} \frac{d^{(n-1)} h}{dt^{(n-1)}} + \cdots + A_1 \frac{dh}{dt} + A_0 h = 0 \tag{8.15}$$

The homogeneous equation is just the original equation with the forcing function replaced by zero. The homogeneous solution $h(t)$ contains n arbitrary constants, which can be evaluated by means of the initial conditions. The other part of the solution, $p(t)$, is known as the *particular solution*. It can be any function of time that satisfies Eq. (8.13).

EXAMPLE 8.9

Show that a homogeneous solution of the equation

$$\frac{d^2 v}{dt^2} + \Omega^2 v = K$$

(where Ω and K are constants) is $f(t) = C \sin \Omega t + D \cos \Omega t$ (where C and D are also constants).

SOLUTION

If $f(t)$ is a homogeneous solution, it must satisfy $d^2 f/dt^2 + \Omega^2 f = 0$. Differentiating $f(t)$ twice, we see that this requirement is satisfied *regardless* of the values of C and D.

EXAMPLE 8.10

Show that for the equation of Example 8.9 a particular solution can be found by assuming $p(t)$ is a constant, and find $p(t)$.

SOLUTION

If we assume $p(t) = $ constant, the second derivative vanishes and we have $\Omega^2 p(t) = K$. Thus $p(t) = K/\Omega^2$, consistent with the assumption that $p(t)$ is a constant.

EXAMPLE 8.11

For the case of the two preceding examples, find $v(t)$, given the initial conditions $v(0) = 1$ and $dv/dt\,(t = 0) = 0$.

SOLUTION

We have established that $v(t) = C \sin \Omega t + D \cos \Omega t + K/\Omega^2$, where C and D are still to be determined. From the given initial conditions we have

$$v(0) = D + \frac{K}{\Omega^2} = 1$$

and

$$\frac{dv}{dt}(t = 0) = C\Omega = 0$$

Solving these two equations for C and D, we have $C = 0, D = 1 - K/\Omega^2$, and

$$v(t) = \left(1 - \frac{K}{\Omega^2}\right) \cos \Omega t + \frac{K}{\Omega^2}$$

From the foregoing examples it is easy to see that when the forcing function is a constant, the particular solution is also a constant. It is the natural response of the system, however, which is of most interest in this chapter, and this is contained in the homogeneous solution. Let us now recall how $h(t)$ can be found.

HOMOGENEOUS SOLUTIONS

If $n \geq 1$, Eq. (8.15) is satisfied by $h(t) = e^{st}$. To show this we simply substitute into Eq. (8.15), obtaining

$$A_n s^n + A_{(n-1)} s^{n-1} + \cdots + A_1 s + A_0 = 0 \qquad (8.16)$$

This algebraic equation, known as the *characteristic equation*, has at least one solution and possibly as many as n different solutions. We shall refer to these solutions s_1, $s_2, \ldots, s_m$ (where $m \leq n$) as the *roots* of the characteristic equation. It is easy to show, by substituting into Eq. (8.15), that if $e^{s_1 t}, e^{s_2 t}, \ldots$, etc. satisfy the homogeneous equation, so does the function $C_1 e^{s_1 t} + C_2 e^{s_2 t} + \cdots + C_m e^{s_m t}$, where $C_1, \ldots, C_m$ are any constants. If the characteristic equation has n *different* roots, then the most general homogeneous solution is

$$h(t) = \sum_{i=1}^{n} C_i e^{s_i t} \qquad (8.17)$$

Equation (8.17) contains n constants C_i, which must be found from n equations describing initial conditions. On the other hand, if m, the number of different roots of the characteristic equation, is *less* than n, Eq. (8.17) is not the most general homogeneous solution. We shall consider one case of $m < n$ later in connection with second-order circuits.

EXAMPLE 8.12

Solve the first-order equation (8.3) with $v_1(t) = V$ (a constant) by the methods just described. Show that the result agrees with Eq. (8.1), the general first-order solution, which was stated (but not derived) in Section 8.1.

SOLUTION

Equation (8.3),

$$\frac{dv_{\text{OUT}}}{dt} + \frac{1}{RC} v_{\text{OUT}} = \frac{V}{RC}$$

is of the form of Eq. (8.13) with $n = 1$, $A_1 = 1$, $A_0 = 1/RC$, and $f(t) = V/RC$. We have already noted that since $f(t)$ is a constant, the particular solution $p(t)$ must also be a constant; substituting, we find $p(t) = V$. The characteristic equation, obtained by substituting $v_{\text{OUT}} = e^{st}$ into the homogeneous equation, is

$$s + \frac{1}{RC} = 0$$

from which we have $s = -1/RC$. From Eq. (8.17) the homogeneous solution is thus $C_1e^{-t/RC}$, and

$$v_{OUT}(t) = p(t) + h(t) = V + C_1e^{-t/RC}$$

in agreement with Eq. (8.1). It is not possible to evaluate the constant C_1 until we are given an initial condition. For example, if we are given that $v_{OUT}(t = 0) = V_1$, then

$$v_{OUT}(0) = V_1 = V + C_1e^{(0)} = V + C_1$$

from which we have $C_1 = V_1 - V$. Thus

$$v_{OUT}(t) = V + (V_1 - V)e^{-t/RC}$$

Note that this case is identical with Example 8.3.

ALTERNATIVE METHOD USING IMPEDANCE

An alternative method for obtaining the characteristic equation is often used. First, all voltage and current sources are set to zero. (As in Chapter 3, voltage sources become short circuits and current sources become open circuits.) We then write the circuit equations in terms of impedances, just as was done in Chapter 7. However, we write "s" instead of "$j\omega$" in all impedances. Thus $\mathbf{Z}_c = 1/sC$, $\mathbf{Z}_L = sL$, and $\mathbf{Z}_R = R$. For example, the circuit equation for the circuit in Example 8.4 can immediately be written $v_{OUT}/R + sCv_{OUT} = 0$. If necessary, the circuit equations are then manipulated by substitution to obtain an equation containing only a single voltage or current. This voltage or current can then be divided out of the equation; what remains is the characteristic equation for s. In the case of Example 8.4 we immediately have $s = -1/RC$. This method is mathematically equivalent to the one already outlined, but it provides a very convenient method for writing the characteristic equation and finding its roots. By analogy with $j\omega$, the quantity s is often referred to as the *complex frequency*.

SECOND-ORDER CIRCUITS

When a circuit contains two capacitors, two inductors, or one of each, it is usually described by equations of the second order. This case arises often and exhibits interesting phenomena not found in first-order circuits. One such phenomenon, that of *resonance*, has already been considered in connection with forced response, in Chapter 7. Here we shall see that resonance appears in the natural response as well.

Let us consider the circuit shown in Fig. 8.17(a). The voltage source v_1 has the value V for $t < 0$ and is zero for $t > 0$, as shown in Fig. 8.17(b). We wish to find the current $i(t)$. First, we must write a differential equation, expressing the fact that the sum of the voltage drops across the three passive elements is equal to $v_1(t)$. The voltage drop across R is iR, and the drop across L is $L\,di/dt$. However, it is not quite as straightforward to write v_c, the voltage across C, in terms of i. We can do this by integrating the equation $i = C\,dv_c/dt$, obtaining

$$v_c(t) = \frac{1}{C}\int_0^t i(t')\,dt' + v_c(t = 0) \tag{8.18}$$

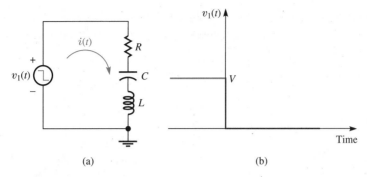

FIGURE 8.17

The term $v_c(t = 0)$ is the constant of integration. [We can see that this constant term is correct by setting $t = 0$ in Eq. (8.18); this gives $v_c(t = 0) = v_c(t = 0)$, which is obviously correct.] Now we can write the equation describing current in the circuit of Fig. 8.17:

$$i(t)R \; + \; \frac{1}{C} \int_0^t i(t') \, dt' \; + \; v_c(t = 0) \; + \; L\frac{di}{dt} \; = \; v_1(t) \tag{8.19}$$

EXAMPLE 8.13

Write an expression for the current through an inductor L as a function of the voltage across it.

SOLUTION

Integrating the expression $v_L = L \, di_L/dt$, we have

$$i_L(t) \; = \; \frac{1}{L} \int_0^t v(t) \, dt \; + \; K$$

where K is a constant of integration. To find K we let $t = 0$, in which case the integral vanishes and we find $i_L(t = 0) = K$. Thus

$$i_L(t) \; = \; \frac{1}{L} \int_0^t v(t) \, dt \; + \; i_L(t = 0)$$

In order to convert Eq. (8.19) to our standard form, Eq. (8.13), we differentiate once, obtaining

$$L\frac{d^2i}{dt^2} \; + \; R\frac{di}{dt} \; + \; \frac{i(t)}{C} \; = \; 0 \tag{8.20}$$

Note that we have set $dv_1/dt = 0$ because at present we are solving only for times $t >$ 0, and for such times v_1 is a constant. The characteristic equation, obtained by substituting $i = e^{st}$ into Eq. (8.20), is

$$Ls^2 + Rs + \frac{1}{C} = 0 \tag{8.21}$$

and its two roots are

$$s_1 = -\frac{R}{2L} + \sqrt{(R/2L)^2 - (1/LC)} \tag{8.22}$$

$$s_2 = -\frac{R}{2L} - \sqrt{(R/2L)^2 - (1/LC)}$$

For the moment we shall assume that s_1 and s_2 are real numbers, with $s_1 \neq s_2$. Noting that $p(t) = 0$ is a suitable particular solution (because $v_1 = 0$ for $t > 0$), we have

$$i(t) = C_1 e^{s_1 t} + C_2 e^{s_2 t} \tag{8.23}$$

where the constants C_1 and C_2 are to be determined from initial conditions.

EXAMPLE 8.14

Evaluate C_1 and C_2 in Eq. (8.23).

SOLUTION

We require two independent initial conditions to evaluate the two constants. Since the current through L cannot change instantaneously, $i(0+) = i(0-)$. But since at $t = 0-$ the circuit was in the dc steady state, the current through C must have been zero; thus $i(0+) = 0$.

For the second initial condition we need the value of di/dt at $t = 0+$. Since the sum of the voltages across the three passive elements is zero at $t = 0+$, we have

$$i(0+)R + \frac{1}{C}\int_0^{0+} i(t')\,dt' + v_c(t = 0+) + L\frac{di}{dt} = 0$$

Since $0+$ is arbitrarily close to 0, the integral vanishes. Since v_c cannot change instantaneously, $v_c(t = 0+) = v_c(t = 0-)$. Now at $t = 0-$ the circuit was in the dc steady state. Therefore $i(0-) = 0$, $v_L(0-) = 0$, and hence $v_c(0-) = v_1(0-) = V$. Thus we have

$$i(0+)R + V + L\frac{di}{dt}(t = 0+) = 0$$

We have already shown that $i(0+) = 0$; thus we have

$$\frac{di}{dt}(t = 0+) = \frac{-V}{L}$$

From $i(t = 0+) = 0$ we have

$$C_1 + C_2 = 0$$

and from $(di/dt)(t = 0+) = -V/L$ we have

$$s_1 C_1 + s_2 C_2 = -V/L$$

Solving simultaneously, we obtain

$$C_1 = -C_2 = \frac{V/L}{s_2 - s_1}$$

Physically, what happens in this example is the following: Before $t = 0$ the circuit is in its dc steady state. The capacitor is charged to a voltage V, but no current is flowing. When v_1 changes to zero, the capacitor begins to discharge through R and L. However, current through L cannot change from zero instantaneously; thus the discharge current has to build up from zero. After a long time we expect that the capacitor will discharge and i_L will again approach zero. Let us choose the values $L = 1$ mH, $C = 250$ pF, $R = 8 \times 10^3$ Ω, and $V = 3$ μV; $i(t)$ is then as shown in Fig. 8.18.

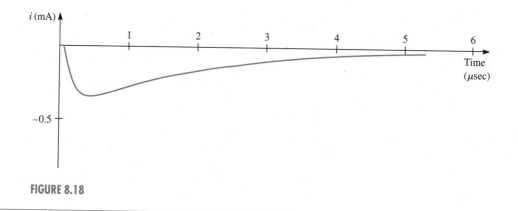

FIGURE 8.18

•EXERCISE 8.7

In the circuit shown in Fig. 8.19, both resistors have the value R and both capacitors have the value C. (a) Find the two roots of the characteristic equation for this circuit. (b) Find $v_{OUT}(t)$ for $t > 0$ if $v_1(t) = 0$ for $t < 0$, $v_1(t) = 1$ for $t > 0$, and $R = 1,000 \ \Omega$ and $C = 1 \ \mu F$. **Answers:** (a) $s_1 = -0.382/RC$; $s_2 = -2.62/RC$. (b) $v_{OUT}(t) = 0.447(e^{-382t} - e^{-2620t})$.

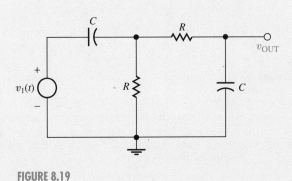

FIGURE 8.19

COMPLEX ROOTS

In general, the roots of the characteristic equation are complex numbers. In first-order circuits the single root is always a real number, but for second- and higher-order circuits there is no such restriction: One must solve the characteristic equation to find out what the roots are. For the circuit we have been considering (Fig. 8.17), the two roots are given in Eq. (8.22). In this case, whether or not the roots are real depends on the quantity beneath the square-root sign, which is known as the *discriminant*. If $[(R/2L)^2 - (1/LC)] > 0$, we see that the two roots are both negative real numbers. (This is what was assumed until now.) However, if $[(R/2L)^2 - (1/LC)] < 0$, the two roots are both complex, with $s_1 = s^*_2$. This is an interesting case because, as we shall see, resonant behavior appears and the solutions become oscillatory in nature.

In order to gain a feeling for what happens in the case of complex s, let us first make the simplifying assumption that $R = 0$. Although $R = 0$ is not entirely realistic (since all circuits do contain some resistance), the assumption causes Eq. (8.22) to simplify to

$$s_1 = j \sqrt{\frac{1}{LC}} \qquad \text{(undamped case)} \quad \textbf{(8.24)}$$

$$s_2 = -j \sqrt{\frac{1}{LC}}$$

As previously noted, a particular solution of the differential equation (8.20) is $p(t) = 0$. Thus we need only the homogeneous solution $h(t)$:

$$i(t) = h(t) = \mathbf{C}_1 e^{j(1/LC)^{1/2}t} + \mathbf{C}_2 e^{-j(1/LC)^{1/2}t} \qquad \textbf{(8.25)}$$

$$= \mathbf{C}_1 \cos \sqrt{\frac{1}{LC}}t + j\mathbf{C}_1 \sin \sqrt{\frac{1}{LC}}t$$

$$+ \mathbf{C}_2 \cos \sqrt{\frac{1}{LC}}t - j\mathbf{C}_2 \sin \sqrt{\frac{1}{LC}}t \tag{8.26}$$

$$i(t) = (\mathbf{C}_1 + \mathbf{C}_2) \cos \sqrt{\frac{1}{LC}}t + j(\mathbf{C}_1 - \mathbf{C}_2) \sin \sqrt{\frac{1}{LC}}t$$

We note that $\mathbf{C}_1$ and $\mathbf{C}_2$ are arbitrary constants, which in general are complex numbers. (This is indicated by the use of boldface type.) However, since $h(t)$ represents a physical quantity (current, in this case), $h(t)$ must be *real*. This requires that $\text{Im}\,(\mathbf{C}_1 + \mathbf{C}_2) = 0$ and $\text{Re}\,(\mathbf{C}_1 - \mathbf{C}_2) = 0$. Thus Eq. (8.26) can be rewritten

$$i(t) = C_3 \cos \omega_R t + C_4 \sin \omega_R t \tag{8.27}$$

where $C_3 = \text{Re}\,(\mathbf{C}_1 + \mathbf{C}_2)$ and $C_4 = j\text{Im}\,(\mathbf{C}_1 - \mathbf{C}_2)$ are arbitrary *real* constants, and where by definition

$$\omega_R \equiv \frac{1}{\sqrt{LC}} \tag{8.28}$$

We shall refer to ω_R as the *undamped resonant frequency* of the circuit.* From Eq. (8.27) we see that the homogeneous solution is now an oscillatory function that goes on forever, never dying out. Of course, in any real circuit, transients are expected to die out; the fact that in this case they do not arises from the unrealistic assumption $R = 0$. In electric circuits R plays a role similar to that of friction in mechanical systems. Resistance provides an energy-loss mechanism, which causes transients to damp out. Thus we refer to a system in which no energy is lost in resistances as *undamped*.

EXAMPLE 8.15

Evaluate the constants C_3 and C_4 in Eq. (8.27) describing the circuit of Fig. 8.17 for the case $R = 0$.

SOLUTION

As shown in Example 8.14, the initial conditions are $i(0+) = 0$ and $(di/dt)(t = 0+) = -V/L$. Thus we have

$$C_3 = 0$$

---

*This name refers to the fact that ω_R is the natural frequency of oscillation in circuits containing no resistance.

and, differentiating Eq. (8.27),

$$C_4 = -\frac{V}{L\omega_R}$$

The behavior of $i(t)$ for the case $L = 1$ mH, $C = 250$ pF, and $V = 3$ μV is as shown by Fig. 8.20.

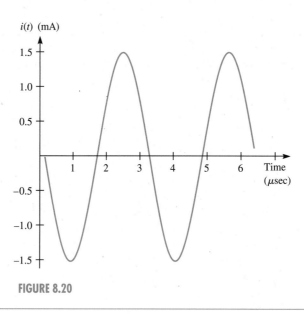

FIGURE 8.20

Let us now consider the more general case in which the discriminant of the circuit is negative but $R \neq 0$. In this case, from Eq. (8.22) we have

$$\mathbf{s}_1 = s' + js'' \tag{8.29}$$

$$\mathbf{s}_2 = \mathbf{s}_1^* = s' - js''$$

where

$$s' = -\frac{R}{2L} \tag{8.30}$$

$$s'' = \sqrt{\frac{1}{LC} - \left(\frac{R}{2L}\right)^2}$$

The solution is

$$i(t) = e^{s't}[\mathbf{C}_1 e^{js''t} + \mathbf{C}_2 e^{-js''t}] \tag{8.31}$$

Again using the fact that $i(t)$ must be real, Eq. (8.31) can be rewritten

$$i(t) = e^{s't}[C_3 \cos s''t + C_4 \sin s''t] \tag{8.32}$$

Clearly Eq. (8.32) reduces to Eq. (8.27) when $R = 0$. When R is not zero, Eq. (8.32) shows that $i(t)$ is an oscillatory function multiplied by a decaying exponential term. The frequency of the oscillations is not in general equal to the resonant frequency, which we have defined as $(LC)^{-1}$. However, if R is small, $s'' \cong \omega_R$, and also the decay time of the transient is comparatively long. We can refer to this as the *lightly damped* case.

An interesting special case is that in which the discriminant $(R/2L)^2 - (1/LC)$ vanishes. Let us imagine that we begin with the undamped case, $R = 0$, and gradually increase R. As this is done the rate of damping increases, and the discriminant approaches zero from the negative direction. When the discriminant passes through zero and becomes positive, the character of the solution changes from oscillatory to nonoscillatory. This change occurs when

$$R = 2\sqrt{\frac{L}{C}} \quad \text{(critical damping)} \tag{8.33}$$

When R has exactly this value, the circuit is said to be *critically damped*. Systems in which the damping is less than critical (and therefore oscillate) are said to be *underdamped*; solutions with greater than critical damping (which therefore decay without oscillation) are said to be *overdamped*.

There is a mathematical peculiarity associated with the case of critical damping. In this special case the two roots of the characteristic equation are equal, and Eq. (8.23), which applies only when all roots are different, is no longer the correct solution. It can be shown (by substitution) that the solution for the critically damped case is

$$i(t) = C_3 e^{-(R/2L)t} + C_4 t e^{-(R/2L)t} \tag{8.34}$$

The undamped, underdamped, critically damped, and overdamped solutions for the circuit of Fig. 8.17 are shown in Fig. 8.21(a) to (d). In this sequence of figures, the

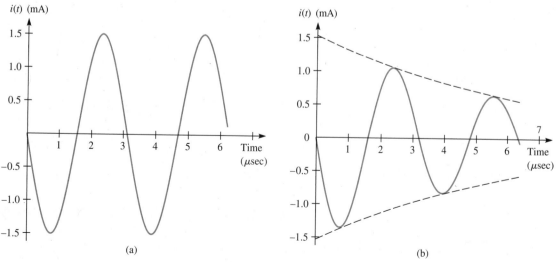

(a)   (b)

FIGURE 8.21(a) and (b)

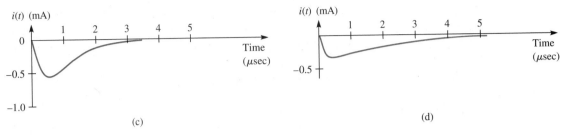

FIGURE 8.21(c) and (d)

forcing function is kept constant (with $V = 3\ \mu$V), and the product LC also is held constant (with $L = 1$ mH and $C = 250$ pF). The values of $i(0+)$ ($=0$) and (di/dt) $(t = 0+)(= -V/L)$ are hence the same in all four curves. In Fig. 8.21(a), $R = 0$ and the oscillations are undamped. In Fig. 8.21(b), $R = 309\ \Omega$ and the circuit is underdamped. The dashed curve is the instantaneous amplitude of the function and is proportional to $e^{s't}$. It is known as the "envelope" of the sinusoid because the sinusoid is enclosed in it. The oscillation frequency is 0.32 MHz, just below the AM radio band. The damping is sufficient to make the amplitude decrease by $1/e$ in roughly two periods of the oscillation. In Fig. 8.21(c) $R = 4,000\ \Omega$ and the damping is critical. For this and larger R, no oscillations appear in the solution. Figure 8.21(d) shows $i(t)$ for the case of $R = 8,000\ \Omega$. It is interesting that, although R is larger, Fig. 8.21(d) shows a long "tail" that decays more slowly than in Fig. 8.21(c). This tail occurs because in Fig. 8.21(d) one of the two roots of the characteristic equation has a very small absolute value. Optimally fast decay behavior is an advantage of the critically damped case.

● EXERCISE 8.8

Consider a resistor, capacitor, and inductor in *parallel.* (a) Find the roots of the characteristic equation. (b) Taking L and C as given, find the value of R that gives critical damping. **Answers:**

$$\text{(a) } s_{1,2} = -\frac{1}{2RC} \pm \sqrt{\left(\frac{1}{2RC}\right)^2 - \frac{1}{LC}} \qquad \text{(b) } R = \frac{1}{2}\sqrt{\frac{L}{C}}$$

Second-order circuits containing only capacitors or only inductors never exhibit oscillatory behavior in their natural response because the roots of the characteristic equation in those cases always turn out to be real. In general, in order to determine the nature of the solutions it is necessary to write the appropriate characteristic equation and find its roots. [The reader should note that Eqs. (8.20) to (8.22) are not general; they were derived specifically for the circuit of Fig. 8.17.] Once the roots have been found, we can determine the form of the solution as follows: (1) both roots imaginary: undamped sinusoids; (2) two complex roots, complex conjugates of each other: damped sinusoids; (3) two equal roots: critical damping; or (4) two real roots: exponential solutions. To review, the procedure for finding the transient response of a second-order circuit is as follows:

1. Write node or loop equations for the circuit.

2. Manipulate the circuit equations to obtain a second-order differential equation containing only one unknown voltage or current. We shall call this unknown $f(t)$.

3. Find a particular solution for $t > 0+$ by assuming $p(t) =$ constant.

4. Substitute $f(t) = e^{st}$ into the homogeneous equation to obtain the characteristic equation.

5. Find the roots of the characteristic equation. From these the form of the solution (oscillatory or nonoscillatory) can be determined.

6. Write the appropriate homogeneous solution $h(t)$ in the form of Eq. (8.23), Eq. (8.32), or Eq. (8.34). The solution for the unknown in question is then $f(t) = h(t) + p(t)$.

7. Using Rules 1 to 4 of Section 8.1, find $f(t = 0+)$ and (df/dt) $(t = 0+)$. (This step can require some ingenuity. It may be helpful to use the circuit equations found in step 1.)

8. Use the initial values of f and df/dt to evaluate the two arbitrary constants in $h(t)$.

*8.3 Complete Response

To conclude this chapter, we note that natural and forced response often occur simultaneously. In fact, many of the examples treated in Sections 8.1 and 8.2 involved forced as well as natural response. Any time that the forcing function on the right side of Eq. (8.13) is not zero, forced response is taking place. Until now, we have assumed the forcing function to be a constant. In such a case there is always a constant particular solution. After the natural response dies away, any nonvanishing steady-state behavior is the ongoing forced response.

The most general case, however, arises when natural response occurs in the presence of a forcing function that is *not* a constant. For instance, let us recall the sinusoidal steady-state solutions found in Chapter 7. There we imagined that the sinusoidal forcing functions had been applied for a very long time, so that any transients had already died out. In reality, however, every forcing function must be turned on at some time. Turning on the forcing function is likely to generate a transient response, which will coexist with the forced response until the transient dies away.

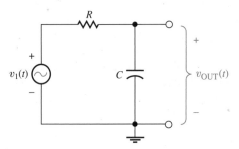

FIGURE 8.22

As an example, consider the circuit of Fig. 8.22. In this circuit the sinusoidal forcing function is "turned on" at $t = 0$; that is, $v_1(t) = 0$ for $t < 0$, and $v_1(t) = V_0 \sin \omega t$ for $t > 0$. Using the methods of Section 8.2, we find easily that the homogeneous solution is $C_1 e^{-t/RC}$, where C_1 is a constant to be determined. Since the particular solution is what remains after the homogeneous solution dies away, the particular solution is identical with the sinusoidal steady-state solutions found in Chapter 7. Using phasors, it is simple to find that the particular solution for v_{out} is

$$p(t) = V' \cos (\omega t + \phi) \tag{8.35}$$

where

$$V' = \frac{V_0}{\sqrt{1 + (\omega RC)^2}} \tag{8.36}$$

$$\phi = \tan^{-1}\left(\frac{1}{\omega RC}\right)$$

(the value to be taken for the arctangent is in the third quadrant). The complete response is

$$v_{\text{OUT}}(t) = h(t) + p(t) = C_1 e^{-t/RC} + V' \cos (\omega t + \phi) \tag{8.37}$$

The initial condition is $v_{\text{OUT}}(t = 0+) = 0$. Hence

$$C_1 = -V' \cos \phi = \frac{\omega RC V_0}{1 + (\omega RC)^2} \tag{8.38}$$

From Eq. (8.37) we see that the complete response is just the sum of the sinusoidal steady-state response plus a transient that decays away in a time on the order of RC. It is interesting to note that the forcing function $v_1(t)$ is *not* discontinuous at $t = 0$; thus we see that simply making the forcing function continuous does not necessarily prevent generation of a transient.

EXAMPLE 8.16

Find the complete response of the circuit of Fig. 8.22 if $v_1(t) = 0$ for $t < 0$ and $v_1(t) = V_0 \cos \omega t + V_1$ for $t > 0$. (V_0 and V_1 are constants.) Determine the value of V_1 for which the transient part of the solution vanishes.

SOLUTION

As before, the homogeneous solution (which does not depend on the forcing function) is $C_1 e^{-t/RC}$. The particular solution can be found by superposition. The sinusoidal steady-state solution is $V' \cos(\omega t + \phi)$, where $V' = V_0[1 + (\omega RC)^2]^{-1/2}$ and $\phi = \tan^{-1}(-\omega RC)$. The part of $p(t)$ due to V_1 is the dc steady-state solution, which is just $v_{OUT} = V_1$. Thus

$$v_{OUT}(t) = C_1 e^{-t/RC} + V' \cos(\omega t + \phi) + V_1$$

Using the initial condition $v_{OUT}(t = 0+) = 0$, we have

$$C_1 + V' \cos \phi + V_1 = 0$$

from which we obtain

$$C_1 = -V' \cos \phi - V_1 = -\frac{V_0}{1 + (\omega RC)^2} - V_1$$

Thus we can make the transient vanish by choosing

$$V_1 = -\frac{V_0}{1 + (\omega RC)^2}$$

● **EXERCISE 8.9**

In the circuit of Fig. 8.22, let $v_1(t) = V_2$ (a constant) for $t < 0$ and $v_1(t) = V_0 \cos \omega t$ when $t > 0$. Taking V_0 as given, what value of V_2 will result in elimination of the transient? **Answer:** $V_2 = V_0/[1 + (\omega RC)^2]$.

8.4 Application: Computer-Aided Design

Computer software packages, such as SPICE, can perform sinusoidal steady-state analysis or analysis of transient response. Just as in the case of dc analysis, the computer allows us to avoid tedious hand labor while obtaining results of high accuracy. This is what is known as computer-aided analysis. However, pure analysis, in practice, is something engineers do not do very often. Usually one practices not analysis, but *design*. The difference is that in analysis one is given a circuit and asked what it does, while in design, one is asked to produce a circuit that will perform as required. Design is a much broader and more challenging activity than simple analysis, although analysis is usually one of the many steps involved in design. Many people believe that design is the one key activity that characterizes engineering, and distinguishes it from the sciences and other professions.

In most cases design problems do not have a single right answer. There may be many ways of achieving the desired result. For example, an engineer asked to provide a road connection between two cities on opposite sides of a river may consider suspension bridges, cantilever bridges, and tunnels. The decision will be based on many factors, such as construction costs, maintenance costs, reliability, aesthetic appearance, and so on. The design problem is complicated because there are so many design choices that can be made. The suspension bridge, for example, might have more towers or fewer, tall towers or short ones, and cables of different sizes. To compare possible designs, we will need to calculate the construction costs for all the different possible choices. Calculating the cost of each may be a large, time-consuming task, and we will be glad to have a computer to do this for us. In that case the computer will still be doing analysis, but we will be using it as part of the design process.

True computer-aided design, or CAD, as it is known, is a different matter again. In CAD, the computer finds it own way to the best design. Not only does it analyze possible designs, but it finds new designs that work better until it finds the one that works best. For example, let us refer back to Figs. 8.15 and 8.16. Suppose that the value of R is given, and we are asked to design C in order that the output pulse (Fig. 8.16) has a desired maximum value, such as $\frac{1}{2}V_0$. For this simple circuit the problem could be solved analytically, but we are only using it as an example; if the circuit had a few extra R's and C's in it, it would be very laborious to solve. In CAD, the computer would vary the value of C, each time performing the circuit analysis, until it finds a value that gives the desired result, or at least comes close enough, according to the accuracy we require.

In selecting different values of C to try, a CAD program might simply use trial and error. Since the computer analysis of each trial design will go very quickly, a great many designs could be tried and the program might eventually find one that would work. But this is a very primitive approach. It is slow, and if the circuit had several capacitors—say, four—that could all be varied, it might take an extremely long time to stumble onto a set of four values that would work. Instead, one can use what are known as "derivative" methods. These are based on the same idea as the well-known Newton's method for solving equations.

In a derivative method, the program would first choose some random initial value of C, and calculate the resulting maximum output voltage. Of course this will be nowhere near the desired value of 0.5 V_0; perhaps it is 0.7 V_0 instead. Then the program would try a slightly larger C and again calculate the maximum output voltage. Suppose this time the result is 0.65 V_0. "Aha," the program says to itself, "I am going in the right direction," and it proceeds to try still larger values of C. If the goal is being approached slowly, the program jumps to a much larger value of C; when it gets close to the goal, it makes only a small change in C. Clearly this "intelligent" way of finding trial values is much better than merely guessing. In effect, trial values are found by using the rate of change of the result in response to the design variable, that is, the *derivative* of the result with respect to the design variable—hence the name derivative method.

Derivative methods can be extended to design problems with many variables. The program can be designed to simultaneously adjust all the variables so as to move toward the best design in an efficient way. For example, let us consider the problem of filter design. A traditional problem in circuit design is that of the *bandpass filter*. The idealized transmission characteristic (amplitude versus frequency) of such a filter is the rectangular shape shown as the solid curve in Fig. 8.23(a). The design objective is that sinusoidal signals with frequency between ω_1 and ω_2 should be equally transmit-

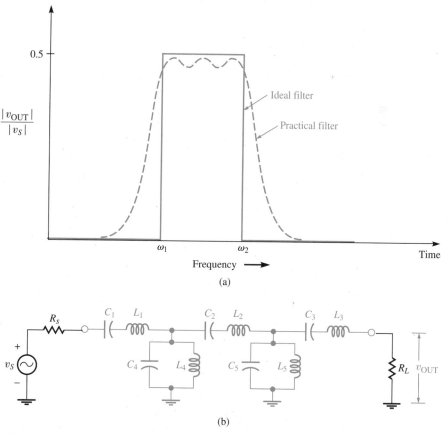

FIGURE 8.23

ted, while signals of other frequencies should be completely blocked. In practice such ideal performance cannot be achieved, and one must put up with performance something like the dashed curve shown in the figure.

Standard circuits exist for filters; one such circuit is shown in Fig. 8.23(b). The source and load resistances R_S and R_L are given, but the designer must find values for each of the five capacitors and five inductors. Analyzing by hand and using trial and error, this would be an impossible job. However, a derivative CAD program can try new sets of values for the ten elements in an orderly and efficient way, until results within the design specifications are obtained. This is an extremely powerful and useful design technique.

As it happens, filter design problems have been around for a long time, and have been studied very extensively. In fact, filter design is an advanced mathematical discipline that is almost a science in itself. As a result, analytic techniques do exist, through which optimal values for the L's and C's in our example can be found. However, as CAD technology advances, we may find that CAD techniques are preferable to analytical methods. For one thing, CAD techniques are very versatile. One need not be restricted to a standard circuit; one can try novel designs, perhaps using nonlinear circuit elements. Nonlinear circuits cannot even be analyzed analytically, much less designed; but CAD doesn't care. With the power of the computer available for the analyses, CAD will go almost as fast as with linear circuits.

Some people feel sorry to see analytic methods replaced by CAD. They regard analytic techniques as sophisticated and elegant, while CAD methods smack of brute force. But the world of engineering is highly competitive, and in order to compete, one must always have the latest and most powerful tools. Engineers can never expect to stand pat with what they learned in school. Better and more powerful techniques constantly appear, and it is vital to grab them and use them, because your competition will!

POINTS TO REMEMBER

- The term *transient response* refers to behavior of a circuit in response to a change of input that has occurred at some time in the past. The transient response dies away in time. What is left after the transient has died away is called the *steady-state response*.

- The *voltage* across a *capacitor* cannot change suddenly. The *current* through an *inductor* cannot change suddenly.

- In the dc steady state, the current through a capacitor and the voltage across an inductor must be zero.

- For *first-order circuits*, transient voltages and currents are of the form $A + Be^{-t/\tau}$. The quantity τ is called the *time constant*. The constants A and τ are found by substituting the preceding expression into the circuit equation. The constant B is found from the initial condition.

- The response to a rectangular pulse is the sum of responses to positive- and nega-tive-going step inputs. The form of the output pulse depends strongly on whether the duration of the input pulse is long or short compared with the time constant of the circuit.

- For a circuit of any order, the solution of the circuit equation is the sum of a particu-lar solution and a solution of the homogeneous equation. The homogeneous solution contains constants, which are evaluated using initial conditions. If the forcing func-tion is constant, the particular solution is a constant. If the forcing function is a sinusoid, a particular solution can be found by using phasors.

- The homogeneous solution for a circuit of second order is found by means of the characteristic equation. The characteristic equation is found by substituting $v(t) = e^{st}$. An alternative method using generalized impedance can also be used.

- If the roots s_1 and s_2 of the characteristic equation are real and unequal, the homo-geneous solution is $C_1 e^{s_1 t} + C_2 e^{s_2 t}$, where C_1 and C_2 are constants.

- If the roots of the characteristic equation are complex numbers $s' \pm js''$, the homo-geneous solution is $e^{s't}(C_3 \cos s''t + C_4 \sin s''t)$, where C_3 and C_4 are constants. In this case the response is a sinusoid whose amplitude decays exponentially.

- The special case in which the two roots of the characteristic equation are equal is called the case of *critical damping*.

- When transient response and forced response occur simultaneously, the combined behavior is known as *complete response*. The complete response is the sum of the homogeneous solution, which is transient, and the particular solution, which gives the steady-state response.

- Computer-aided design (CAD) is a family of techniques for obtaining optimized design. Typically, computer analysis is used to predict performance for different trial choices of the design parameters. "Intelligent" methods are used to find succes-sively better trial choices, until near-optimal values are found.

PROBLEMS

Section 8.1

8.1 Graph the exponential function $v(t) = Ae^{-t/\tau}$, where $A = 3$ V and $\tau = 1$ msec. For what value of t is $v(t) = A/2$?

8.2 Graph the function $v(t) = B(1 - e^{-t/\tau})$, where $B = 2$ V and $\tau = 1$ msec. For what value of t is $v(t) = B/2$?

8.3 Let $v_1(t) = 0$ for $t < 0$ and $v_1(t) = 2$ V for $t > 0$. What is $v_1(t = 0-)$? What is $v_1(t = 0+)$?

8.4 Let $v_1(t) = 3 \sin \omega t$. What is $v_1(t = 0+)$? What is $v_1(t = 0-)$?

8.5 In the circuit of Fig. 8.24, let $V_0 = 10$ V, $R_1 = 4,700$ Ω, $R_2 = 2,200$ Ω, and $R_3 = 3,300$ Ω. Find the voltage at A in the dc steady state.

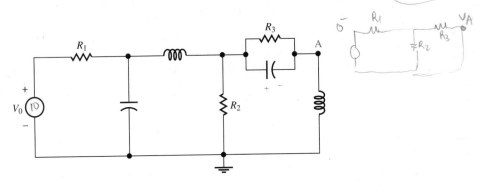

FIGURE 8.24

8.6 In the circuit of Fig. 8.25, let $v_1(t) = 1$ V for $t < 0$ and $v_1(t) = -1$ V for $t > 0$. Let $R_1 = 1,000$ Ω, $R_2 = 2,000$ Ω, and $C = 10$ μF.
 a. Find $v_A(t = 0-)$.
 b. Find $v_A(t = 0+)$.
 c. Find the current i_1 at $t = 0-$.
 d. Find $i_1(t = 0+)$.

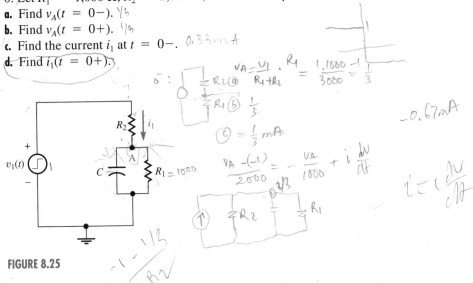

FIGURE 8.25

**8.7 For the case of Problem 8.6, find $(d/dt)(v_A)$ at $t = 0+$.[†]

[†]When analyzing first-order circuits, it is unnecessary to find the values of time derivatives at $t = 0+$. However, when second-order circuits are analyzed, these derivatives are needed to supply the second initial condition. Use the method of Example 8.2(b).

8.8 In the circuit of Fig. 8.26, let $i_1(t) = 4$ mA for $t < 0$, $i_1(t) = 0$ for $t > 0$, $R_1 = 500\ \Omega$, $R_2 = 1{,}500\ \Omega$, and $C = 1\ \mu$F.

    a. Find $v_A(t = 0-)$. 0

    b. Find $v_A(t = 0+)$.

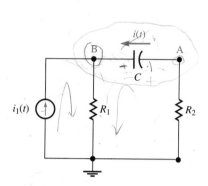

$$v_L(0) = V_B = i_1 R_1 = 2000V = V_A(0)$$

$$i_1 \text{ opens} \Rightarrow V_A(0+) = 2V.$$

$$V_A = \frac{2V}{R_1 + R_2} \cdot R_L = \frac{2 \cdot 1500}{2000} = 1.5$$

FIGURE 8.26

**8.9 For the case of Problem 8.8:

    a. Find $i(t = 0-)$.

    b. Find $i(t = 0+)$.

    c. Find $(d/dt)(v_A - v_B)$ at $t = 0+$.

    d. Find $(d/dt)(v_A)$ at $t = 0+$.

8.10 In the circuit of Fig. 8.27, let $R_1 = 2{,}000\ \Omega$, $R_2 = 3{,}000\ \Omega$, and $L = 20$ mH, and let $i_0(t) = -1$ mA for $t < 0$ and $i_0 = +2$ mA for $t > 0$.

    a. Find $i_1(0-)$.

    b. Find $i_1(0+)$.

    c. Find $i_2(0-)$.

    d. Find $i_2(0+)$.

    e. Find $v_A(0-)$.

    f. Find $v_A(0+)$.

$$a)\ \frac{-1}{3+2} \cdot 3 = -3/5 = -0.6$$

$$b)\ -0.6$$

$$c)\ -0.4$$

$$d)\ 2.6$$

$$e)\ 3 \cdot (-0.4) = -1.2$$

$$f)\ 3 \times 2.6 = 7.8$$

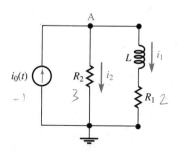

FIGURE 8.27

8.11 For the case of Problem 8.10, find $(d/dt)(v_A)$ at $t = 0+$.

8.12 In the circuit of Fig. 8.28, the switch is moved from position E to position D at time $t = 0$. (Assume the switching action is instantaneous so current through L is not interrupted.)

a. Find $v_B(0-)$. $v_0 = v_C$
b. Find $v_A(0-)$. 0
c. Find $v_B - v_A$ at $t = 0+$. $v_0 = v_C$
d. Find $v_A(t = 0+)$.

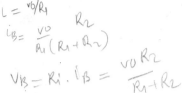

$i_L = v_0/R_1$

$i_B = \dfrac{v_0}{R_1(R_1 + R_2)} R_2$

$v_B = R_i \cdot i_B = \dfrac{v_0 R_2}{R_1 + R_2}$

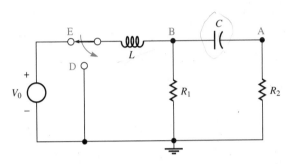

FIGURE 8.28

8.13 Verify Eq. (8.4) by substituting Eq. (8.1) into Eq. (8.3).

*8.14 For the circuit of Fig. 8.25, let $v_1(t < 0) = 1$ V and $v_1(t > 0) = -1$ V, and let $R_1 = 1{,}000\ \Omega$, $R_2 = 2{,}000\ \Omega$, and $C = 10\ \mu\text{F}$.
a. Find $v_A(t)$ for $t > 0$.
b. What is the approximate duration of the transient?
c. Sketch v_A as a function of time.

*8.15 For the circuit of Fig. 8.26, let $i_1(t < 0) = 4$ mA, $i_1(t > 0) = 0$, $R_1 = 500\ \Omega$, $R_2 = 1{,}500\ \Omega$, and $C = 1\ \mu\text{F}$.
a. Find $v_A(t)$ for $t > 0$.
b. What is the approximate duration of the transient?
c. Sketch v_A as a function of time.

*8.16 For the circuit of Fig. 8.27, let $i_0(t < 0) = -1$ mA, $i_0(t > 0) = 2$ mA, $R_1 = 2{,}000\ \Omega$, $R_2 = 3{,}000\ \Omega$, and $L = 20$ mH.
a. Find v_A for $t > 0$.
b. What is the approximate duration of the transient?
c. Sketch $v_A(t)$.

*8.17 For the case of the previous problem, find $i_1(t)$ for $t > 0$. Sketch the result.

8.18 Suppose that the coiled filament of a certain light bulb is equivalent, approximately, to a 30 Ω resistance and a 15 mH inductance connected in series. Suppose further that a certain battery is represented approximately by a Thévenin equivalent with $V_T = 1.5$ V and $R_T = 20\ \Omega$. A switch is closed suddenly, connecting the battery across the bulb. Estimate the time required for the current to increase to a sizable fraction of its steady-state value.

*8.19 Apply Thévenin's theorem to the circuit of Fig. 8.25. Find the time constant by comparison with Fig. 8.5.

*8.20 Repeat Problem 8.14 after first simplifying the circuit (Fig. 8.25) by means of Thévenin's theorem. (Replace v_1, R_1, and R_2 by a Thévenin equivalent.)

*8.21 In the circuit of Fig. 8.29, let $V_0 = 2$ V, $C = 1$ µF, $R_1 = 1,000$ Ω, and $R_2 = 3,000$ Ω. The switch is initially open. At time $t = 0$ it is closed. After a long time the switch is reopened. Make a sketch of v_A as a function of time. Indicate on your sketch the approximate durations of the two transients.

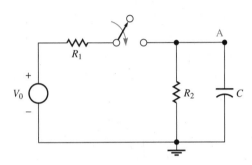

FIGURE 8.29

8.22 In Eq. (8.12), verify that the sum voltage shown for $T < t$ is equal to the sum of the responses to v_1 and v_2.

*8.23 For the circuit of Fig. 8.15(a), the duration of the rectangular input pulse [as shown in Fig. 8.14(b)] is 1 µsec and $R = 1,000$ Ω. It is required that $v_{OUT}(t = 2$ µsec) must be less than $0.1V_0$.
  a. Find two ranges in which the value of C can lie, either of which satisfy the requirement.
  b. Which of the two ranges gives less pulse distortion?
  c. Assuming C is in the range determined in part (b), what is the largest value that C can have?

*8.24 Consider the circuit of Fig. 8.15(a) with the rectangular input pulse shown in Fig. 8.14(b). Suppose that instead of being uncharged at $t = 0$, the voltage across the capacitor at $t = 0$ is xV_0, where x is a constant.
  a. Find v_{OUT} for $0 < t < T$.
  b. Find v_{OUT} for $t > T$.

**8.25 Use the results of the preceding problem to analyze the case when the input consists of an infinite sequence of pulses separated by T (in other words, a rectangular wave). Note that in this case $v_{OUT}(t = 2T)$ must be equal to xV_0. Find the minimum and maximum values of $v_{OUT}(t)$. Sketch $v_{OUT}(t)$ as a function of time.

*8.26 Consider the circuit of Fig. 8.15(a) with the rectangular input pulse shown in Fig. 8.14(b). Let us define a distortion parameter D according to

$$D \equiv \frac{1}{V_0^2 T} \int_0^\infty [v_{OUT}(t) - v_{IN}(t)]^2 \, dt$$

Assuming that T is held constant, find D as a function of RC. Find and explain the limits of D as $RC \to \infty$ and as $RC \to 0$.

Section 8.2

8.27 Find a particular solution of the equation

$$A_2 \frac{d^2v}{dt^2} + A_1 \frac{dv}{dt} + A_0 v = F$$

where A_2, A_1, A_0, and F are all constants.

8.28 Show that if $p(t)$ is a particular solution of Eq. (8.13) and $h_1(t)$, $h_2(t)$, . . . ,$h_N(t)$ are solutions of the homogeneous equation, then

$$f(t) = C_1 h_1(t) + C_2 h_2(t) + \cdots + C_N h_N(t) + p(t)$$

(where $C_1 \cdots C_N$ are any constants) is also a solution of Eq. (8.13).

8.29 Find the homogeneous solution of the equation in Problem 8.27, assuming that the roots of the characteristic equation are all different.

8.30 Find the characteristic equation for the circuit of Fig. 8.5, and solve for the time constant.

8.31 Find the characteristic equation for the circuit of Example 8.5, and solve for the time constant.

8.32 Find the time constant of the circuit of Fig. 8.5 by the alternative method using impedances.

8.33 Find the time constant of the circuit of Example 8.5 by the alternative method using impedances.

*8.34 For the circuit of Fig. 8.30, let $v_1(t < 0) = 0$ and $v_1(t > 0) = 1$ V.
a. Obtain a second-order differential equation for $v_{OUT}(t)$.

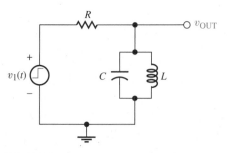

FIGURE 8.30

b. Find a particular solution for $t > 0$.

c. Find the homogeneous solution, assuming that the roots of the characteristic equation are real and unequal.

d. Find v_{OUT} and $(d/dt)(v_{OUT})$ at $t = 0+$.

e. Find $v_{OUT}(t)$ for $t > 0$.

*8.35 For the circuit of Fig. 8.30, find the roots of the characteristic equation by the alternative method using impedance.

*8.36 For the circuit of Fig. 8.31, let $v_1(t < 0) = 1$ V and $v_1(t > 0) = -1$ V.

a. Obtain a second-order differential equation for $v_{OUT}(t)$.

b. Find a particular solution for $t > 0$.

c. Find the homogeneous solution, assuming the roots of the characteristic equation are real and unequal.

d. Find v_{OUT} and $(d/dt)(v_{OUT})$ at $t = 0+$.

e. Find $v_{OUT}(t)$ for $t > 0$.

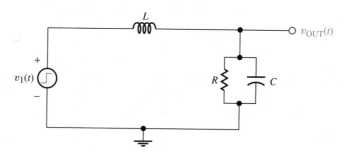

FIGURE 8.31

*8.37 Obtain the characteristic equation for Fig. 8.31 by the alternative method using impedance.

**8.38 For the circuit of Fig. 8.32, $v_1(t < 0) = -1$ and $v_1(t > 0) = 1$ V. The two resistors have equal resistance R.

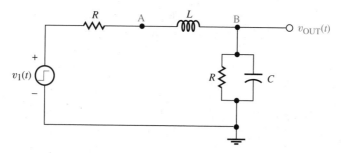

FIGURE 8.32

a. Obtain a second-order differential equation for $v_{OUT}(t)$. (*Suggestions:* Write two node equations for nodes A and B and eliminate v_A by substitution. Be careful to retain the constants of integration—initial values of current through L and voltage across C.)

b. Find a particular solution for $t > 0$.

c. Obtain the homogeneous solution, assuming that the roots of the characteristic equation are real and unequal.

d. Find v_B and $(d/dt)v_B$ as $t = 0+$.

e. Find $v_{OUT}(t)$ for $t > 0$.

*8.39 For the preceding problem, find the roots of the characteristic equation by the alternative method using impedance.

*8.40 Repeat Problem 8.34 under the assumption that the roots of the characteristic equation are complex numbers.

*8.41 Fill in the missing steps between Eqs. (8.31) and (8.32).

*8.42 For the circuit of Fig. 8.17, let us define a quality parameter $Q \equiv \omega_R L/R$ [compare Eq. (7.26)]. Assume that $\omega_R = (LC)^{-1/2}$ is a fixed number.

a. Let Q_c be the value of Q when the circuit is critically damped. Find Q_c.

b. Sketch the form of the homogeneous solution when $Q > Q_c$ and when $Q < Q_c$.

**8.43 For the case of Problem 8.42:

a. Find the roots of the characteristic equation in the limit as $Q \to \infty$.

b. Find the approximate roots of the characteristic equation, as a function of Q, in the limit as $Q \to 0$. [*Suggestion:* Use the result from the binomial theorem $(1 - x)^{1/2} \cong 1 - \frac{1}{2}x$, valid when $x \ll 1$.]

**8.44 In Eq. (8.32), the factor $e^{s't}$ is sometimes called the "envelope" of the function $i(t)$ because it encloses the sinusoid [see Fig. 8.21(b)]. The envelope indicates how the amplitude of the sinusoid varies with time. Let us define a quality factor $Q \equiv \omega_R L/R$, and let $\omega_R = (LC)^{-1/2}$ be a fixed given quantity. Let T_D be the decay time needed for the envelope to decay by a factor of $1/e$. Find the number of complete cycles of the sinusoid that take place in a time T_D, as a function of Q only. (Your answer should not contain L, C, R, or ω_R.) Find a simpler approximate form of your answer, valid when $Q \gg 1$.

8.45 For the circuit of Fig. 8.17, let LC be a fixed constant. Let $R/2L = 2(LC)^{-1/2}$. On *semilog paper* (logarithmic vertical scale, linear horizontal scale) plot the function $f(t) = 10e^{s_1 t} + e^{s_2 t}$, where s_1 is the root of the characteristic equation having the larger absolute value, over the range $0 < t < 2(LC)^{-1/2}$. A plot of this kind is said to have "two slopes." Discuss the meaning of this expression.

Section 8.3

8.46 Verify Eqs. (8.35) and (8.36). Verify that ϕ is in the third quadrant.

*8.47 For the circuit of Fig. 8.22, find the complete response $v_{OUT}(t)$ for $t > 0$, when $v_1(t < 0) = 0$ and $v_1(t > 0) = V_0 \cos \omega t$.

*8.48 For the circuit of Fig. 8.22, find the complete response $v_{OUT}(t)$ for $t > 0$, when $v_1(t < 0) = V_1$ and $v_1(t > 0) = V_0 \cos \omega t$. Find the value of V_1 for which the transient vanishes.

*8.49 Find the complete response $v_{OUT}(t)$ for the circuit of Fig. 8.22 when $v_1(t < 0) = V_0 \cos \omega t$ and $v_1(t > 0) = V_0 \sin \omega t$.

*8.50 For the circuit of Fig. 8.30, let $v_1(t < 0) = 0$ and $v_1(t > 0) = V_0 \cos \omega t$. Find the complete response $v_{OUT}(t)$ for $t > 0$. Assume the circuit is underdamped.

ANALOG SIGNALS AND TECHNIQUES

CHAPTER 9

Principles of Analog Systems

To a very large degree, electrical engineering deals with the transmission, storage, and processing of *information*. This may at first seem a surprising statement. What about electric lights and motors? But although electric power is important, it is an exception to the general rule: Most electrical engineers work on systems for handling information.

In order to handle information electrically, the information must first be represented in an electrical form. For example, suppose Mr. A and Mr. B are located far from one another, but a pair of wires runs between them. Mr. A has a voltage source (or more realistically, a battery), which he can connect to his end of the wires, as shown in Fig. 9.1(a). Mr. B, at his end, has a voltmeter. He can measure the voltage between the wires at his end, and thus discover that A's battery is connected. The question is, is information being transmitted from A to B?

Well yes, some information is being transmitted. By measuring the voltage at his end, Mr. B learns that Mr. A's battery is connected, that the battery has not run down, and that the wires are not broken. But this is not very much information, compared, for example, with transmitting the contents of an encyclopedia. The problem is that the voltage between the wires is constant. If it were time-varying, much more information could be sent, for instance as in Fig. 9.1(b). Here we have provided Mr. A with a switch, so he can turn the voltage on and off; this gives the old-fashioned communication system called the telegraph. Now Mr. B receives a time-varying voltage, by which

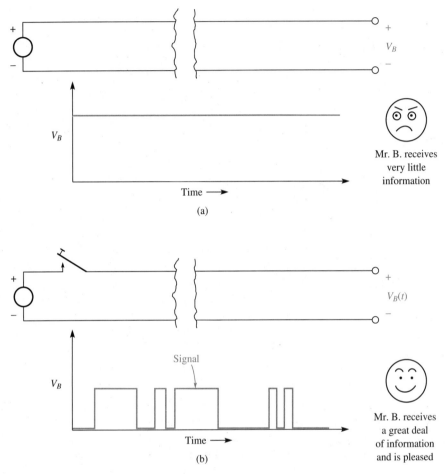

FIGURE 9.1

the information is represented. (Morse code, for example, could be used.) The time-varying voltage representing information is called a *signal*.

The telegraph and its Morse code have by now been replaced by faster systems that can transmit millions of times more information per second. However, the need for signals—time-varying voltages or currents representing information—still remains. In modern work, two basic types of signals are used: *analog* and *digital*. Associated with these two kinds of signals are two different families of electronic circuits, known as analog circuits and digital circuits. These circuits turn out to be quite different, so the subject of electronics tends to divide into two branches: analog technology and digital technology. In this chapter we shall introduce the basic principles of analog technology, and we shall go on to some of its practical aspects in Chapter 10. Digital technology will then be introduced in Chapters 11 and 12.

9.1 Analog Signals and Their Uses

From the dawn of electronics until fairly recently (1970 or so), analog signals were by far the most important type. Familiar technologies such as radio, television, and phonograph records, which have been with us a long time, are based on analog signals. However, when we ask just what an analog signal *is*, we encounter some difficulty: The word "analog" is not at all descriptive, and thus is rather hard to define. In many cases, "analog" means *proportional*. An analog signal is then a voltage (or current) that is proportional to the value of some quantity of interest. For example, imagine an electric thermometer that gives out a voltage proportional to the temperature. As time passes, the temperature might change as shown in Fig. 9.2(a). The thermometer's output voltage then changes proportionally, as shown in Fig. 9.2(b). This time-varying voltage is an analog signal; it contains information concerning temperature. Anyone with a voltmeter can measure the voltage and (assuming the constant of proportionality is known) determine the temperature at any time.

As a second example, consider the analog signal produced by a microphone. We think of this device as one that converts sound to electrical form. What it actually does, however, is to produce a voltage proportional to acoustic pressure. These variations in pressure are quite rapid; audible sounds contain frequencies up to 20,000 Hz, which means that changes in pressure occur in times on the order of 10 microseconds. The microphone produces a voltage that is proportional to this rapidly changing pressure. All the information of the original sound is contained in this analog signal, and by applying the signal to suitable devices (such as an audio amplifier and loudspeaker), we can create an accurate replica of the original sound.

Analog signals are used for a variety of purposes, some of which will now be described. However, there is nothing about the *applications* that is specifically analog;

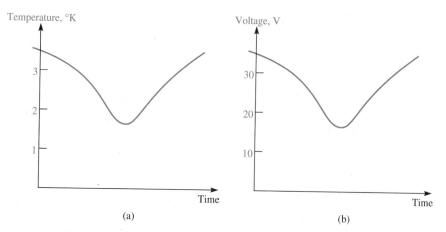

(a)

(b)

FIGURE 9.2   Representation of information by an analog signal. The time-varying temperature in (a) is described by the proportional voltage shown in (b). The time-varying voltage is an analog signal.

using analog signals is just one possible method. In many cases, one can also use digital technique to achieve the same purpose.

INSTRUMENTATION AND DATA ACQUISITION

Often one needs information about some process, while the process is taking place. One then *instruments* the process, or *acquires data* about it, in order to record what is happening or to control the process. For instance, the electric thermometer of Fig. 9.2 is an example of instrumentation. It might be used by a meteorologist. He or she could take the voltage from the thermometer and apply it to a *chart recorder*. In one form, this consists of a moving paper tape, with a pen that constantly writes upon the paper as the paper moves. The position of the pen in the direction perpendicular to the paper's motion is proportional to the voltage applied to the recorder, as shown in Fig. 9.3. In this way we can produce a record of the temperature as a function of time over days or weeks.

Temperature, of course, is a quantity that changes fairly slowly. If we wish to observe quantities that vary more quickly, we may use an oscilloscope (see Section 2.6). Fundamentally, an oscilloscope does the same thing as a chart recorder: It creates a graph of voltage versus time. However, instead of a slow-moving pen, an electron beam is used to write on a phosphor screen (just as in a television set). By this means, very rapid changes can be observed. For example, geotechnical engineers use pressure transducers (essentially the same as microphones) to study acoustic waves that pass through the earth; this technique is used in oil exploration.

A great many kinds of analog transducers are available to make measurements of different kinds. They all have the property of producing a voltage V proportional to some quantity being measured. In order to be general, let us call this quantity Q. (For instance, for the case of a temperature transducer, Q would be the temperature.) Then we define $\Re$, the *responsivity* of the transducer, by

$$\Re \equiv \frac{dV}{dQ} \tag{9.1}$$

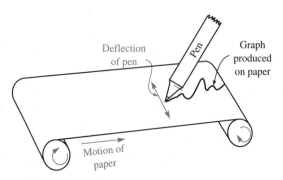

FIGURE 9.3 Diagram of a chart recorder.

The responsivity is important because it indicates whether or not, in a given measurement, enough voltage will be produced to be useful.

One might at first think that the responsivity would not be so important, because one could always use an amplifier to increase the magnitude of the signal. However, it is important here to remember the influence of *noise* on measurements. All transducers will produce some random voltages in addition to their signals. If these random voltages are comparable in size to the expected signals, measurements become more difficult, and perhaps impossible. Additional noise may be added by an amplifier or any other device to which the transducer is connected.

A peculiarity of noise voltages is that they normally depend on the bandwidth being used. A common situation is that the noise voltage V_N is given by

$$V_N = V_{NO} \sqrt{B} \tag{9.2}$$

Here the bandwidth B is determined not just by the transducer but by the system as a whole. For instance, if the transducer is followed by a narrow-band amplifier that amplifies only in the range 10 to 400 Hz, then B is 390 Hz. The quantity V_{NO} is a constant that depends on the phenomena that produce the noise. Its dimensions are conventionally said to be "volts per root Hertz." It is possible to reduce noise by reducing the bandwidth. For instance, if a filter that transmits only the range 10 to 30 Hz is introduced after the transducer, bandwidth is reduced from 390 Hz to 20 Hz, and noise voltage is reduced by a factor of 4.4. However, there is a cost to doing this: The narrow-band filter will not transmit rapid changes, which involve higher frequencies. In practice one must consider how fast the measured quantity changes, and use a bandwidth that is sufficiently wide. Note, therefore, that it is more difficult to measure rapidly changing phenomena than slow ones. The former measurements require greater bandwidth and thus are afflicted with greater noise.

The number that actually determines whether or not a measurement is feasible is the signal-to-noise ratio, usually abbreviated S/N. This is an approximate number obtained by dividing the expected signal voltage by the expected noise voltage. If S/N is considerably larger than one (say, 10), the measurement is feasible. If S/N is only one, or less, the quantity under observation will be obscured by the noise. (In some cases it may still be possible to make a measurement, even if S/N is less than one. However, advanced signal-averaging techniques would then have to be used.)

EXAMPLE 9.1

A vibration transducer (in essence a low-frequency microphone) is being used for measurements on earthquakes. It is connected to a filter, an amplifier, and a chart recorder, as shown in Fig. 9.4. The responsivity of the transducer is 0.4 V/cm, and the filter transmits signals between 1 Hz and 20 Hz. The noise of the amplifier, referred to the amplifier input, is 16 nV/$\sqrt{Hz}$. (This means that the noise coming out of the amplifier is the same as it would be if the amplifier were noiseless and a noise source of 16 nV/$\sqrt{\text{Hz}}$ were connected to its input.) Since in this case the noise of the transducer is much less than that of the amplifier, the former noise can be neglected.

Determine the amplitude of ground motions that can be observed with a signal-to-noise ratio of 10.

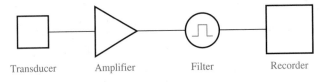

FIGURE 9.4   Typical data acquisition system.

SOLUTION

Let x be the amplitude of the ground motion. The amplitude of the signal reaching the recorder is $\Re xA$, where A is the amplification of the amplifier. The noise reaching the recorder is $V_{NO}\sqrt{B}\,A$. We require that

$$\frac{\Re xA}{V_{NO}\sqrt{BA}} = 10$$

Thus the measurable displacement x is given by

$$x = \frac{10V_{NO}\sqrt{B}}{\Re} = \frac{(10)(16 \cdot 10^{-9})\sqrt{19}}{0.4}$$

$$= 1.7 \times 10^{-6}\,\text{cm}$$

(The numbers chosen for this example were obtained as follows: The responsivity is that of a typical piezoelectric crystal pickup for phonographs—typical cost, $3. The amplifier is a typical op-amp—cost, $1. Note that the resulting system is remarkably sensitive. A distance of 10^{-6} cm is only 100 times the size of a single atom!)

INFORMATION STORAGE

We have already mentioned the chart recorder, which is a device for remembering signals, that is, for "information storage." The chart recorder is a slow device, the main advantage of which is that it records information in permanent, graphical form. For applications involving higher frequencies, other storage techniques can be used.

A familiar example is that of music recording. Phonograph records are analog information-storage devices. The spiral groove in which the phonograph needle rides is made to wiggle from side-to-side. The sideways displacement can be made proportional to the voltage from the microphone. Tape recorders work on a similar principle. A recording head is used to magnetize the tape as it moves by. The magnetism left on the tape is proportional to the voltage being recorded.

The phonograph and tape recorder are fairly old devices. Newer technologies for information storage, such as compact discs and computer memories, tend to be digital rather than analog. We shall discuss them in a later chapter.

COMMUNICATIONS

The old-fashioned telephone is an analog communication device. The analog signal from a microphone is simply carried by a pair of wires (with or without amplification) to an earphone at the receiving end. The earphone is the reverse of a microphone: a transducer that creates acoustic pressure in proportion to the analog signal voltage. Modern telephone technology, however, increasingly relies on digital techniques.

Radio communications technology is a large and interesting field, in which both analog and digital techniques are used. Broadcast AM (amplitude modulation) radio, for instance, makes use of analog signals. The radio-frequency wave being transmitted, in the absence of signal, is $A \cos 2\pi ft$, where A is an arbitrary magnitude constant and f is a frequency on the order of 1 MHz. This is known as the *carrier wave*. The analog signal $v(t)$ is then imposed on the carrier (in a process called *modulation*) in such a way that when $v(t)$ is positive the wave is made larger, and when $v(t)$ is negative the wave is made smaller. The formula for the transmitted wave is

$$A[1 + mv(t)] \cos 2\pi ft$$

where m is a constant (less than 1) called the *modulation index*. The process of modulation causes the *amplitude* of the radio wave to vary in response to the signal v, hence the term "amplitude modulation." Other sorts of modulation also exist; for example, in *frequency modulation* the frequency of the radio wave is caused to vary by the signal.

9.2 Feedback and Its Effects

One of the most characteristic features of analog technology is the technique known as feedback. The term is so common that it has made its way into non-technical speech. For example, one speaks of "giving a person some feedback," meaning, perhaps, that we tell someone how we respond to what he or she says. This usage is borrowed from the original meaning in electronics, in which a system is said to possess feedback when some of its output is "fed back" so it can affect the system's input. Presumably, when we "give a person some feedback" on his or her ideas, our advice will be used to restructure the ideas, so that they are made more perfect. As we shall now see, electronic feedback can be used in the same way, for correcting errors.

A familiar example serves to illustrate the principle. Let us consider a heating system with and without thermostatic control. A schematic illustration is shown in Fig. 9.5. In the system without feedback, Fig. 9.5(a), once the heat control is set, the furnace delivers heat at a constant rate to the room. If the outside temperature changes or the quality of the fuel changes, the room temperature will change. The addition of a simple thermostat and a feedback connection, as shown in Fig. 9.5(b) improves the temperature stability of the room considerably. If the wind blows and the temperature of the room starts to drop, the thermostat signals the furnace to increase the heat flow.

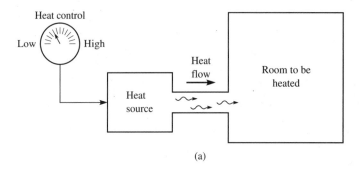

(a)

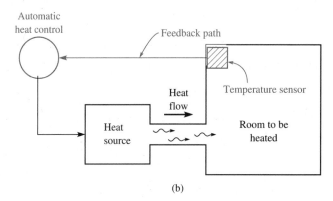

(b)

FIGURE 9.5   An example of the use of feedback. In (a) no feedback is used. In (b) the action of the heat source is regulated by information "fed back" from the temperature sensor.

Similarly, when the sun comes out and the temperature of the room begins to rise, the furnace is instructed, via the feedback path, to reduce the heat flow. This example illustrates several interesting features of a feedback system. First, the output (here, the temperature of the room) is determined by only a few simple elements in the system (in this example, the thermostatic controller). Second, the output is less sensitive to other details of the heating system than it would be if feedback were not used. For instance, note that in Fig. 9.5(a) such factors as the size of the heat pipe are important, but with feedback they become far less important.

In the example just presented, the feedback is applied in such a way as to decrease the output when the output is too high or increase the output when the output is too low. This kind of feedback, which exerts a restraining or stabilizing effect on variations of the output, is known as *negative feedback*.

Feedback is used in electronic circuits in order to obtain similar benefits. One particularly important application is in operational amplifier circuits (op-amps). Op-amps are usually made in integrated-circuit (IC) form, and thus they have the typical advantages of ICs: compactness, ruggedness, and, above all, low cost. The cost of ICs is low because of the fabrication process, by means of which identical circuits can be produced in large quantities. Interestingly, the complexity of an integrated circuit has

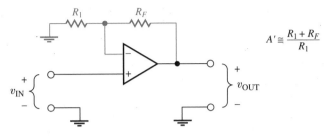

$$A' \cong \frac{R_1 + R_F}{R_1}$$

FIGURE 9.6

rather little influence on its cost. (In much the same way, the cost of printing one page of a book remains the same regardless of whether there is one word on the page or a thousand.) The availability of powerful circuit blocks in cheap IC form has had a major effect on electronics. However, there is a side effect: The exact properties of these mass-produced circuits are hard to control. Thus, although we can have cheap IC amplifiers, it would be hard to get them with a specific value of gain.

This is the point at which feedback is helpful. As we have just seen, feedback can be used to make the output of a system largely independent of the system's internal details, and this is just what is done in op-amp circuits. For example, consider the non-inverting amplifier circuit (Fig. 4.27), repeated for convenience in Fig. 9.6. The feedback path is evident: Signal is returned from the output through R_F to one of the input terminals. The fact that the feedback is returned to the *inverting* input implies that we have negative feedback, which is normally what is wanted. The function of R_1 is to siphon off some of the feedback current to ground, so that by varying R_1 and R_F we can adjust the amount of feedback that is present.

The parameters that describe the op-amp are its input resistance R_i, output resistance R_o, and open-circuit voltage amplification A. These parameters appear in the op-amp model, repeated for convenience in Fig. 9.7. In production op-amps, their values are poorly controlled; all we know is that A and R_i are large and R_o is small. Thus,

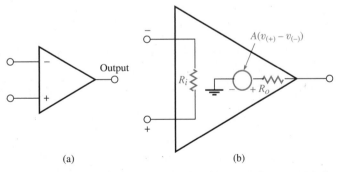

FIGURE 9.7 The operational amplifier. (a) Circuit symbol. (b) A suitable model. The symbols $v_{(+)}$ and $v_{(-)}$ stand for the voltages at the "+" (noninverting) and "−" (inverting) input terminals, respectively, measured with respect to ground.

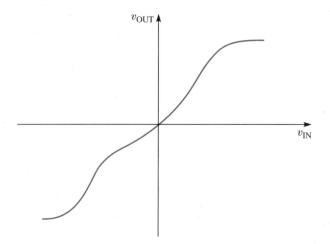

FIGURE 9.8   Example of nonlinear response.

if the amplification of the op-amp circuit depended on their values, we would never know what amplification we would get. But this is not the case. By means of the ideal op-amp technique (Eq. 4.16) we have seen that the amplification of the non-inverting amplifier circuit is given by

$$A' \cong \frac{R_1 + R_F}{R_1} \qquad (9.3)$$

The point to notice is that Eq. (9.3) does *not* depend on the op-amp parameters R_i, R_o, and A. This is the effect of negative feedback. In the furnace example (Fig. 9.5), the temperature of the room depended only on one feedback component of the system, the thermostat. Here the amplification depends only on the two feedback resistors R_1 and R_F.

Another application of feedback arises when an amplifier is *nonlinear*. For an ideal linear amplifier, output voltage would be exactly proportional to input voltage. That is, we would have $v_{OUT} = Av_{IN}$, where A is truly a constant. However, as with all ideal circuit elements, the reality is somewhat different. In practice, the number A will not be constant, and the graph of v_{OUT} vs. v_{IN} will be something other than a straight line; for instance, as in Fig. 9.8. Clearly such an amplifier will produce *distortion*: The output will not be simply an enlarged replica, but will have a different shape. In the case of sinusoidal signals, new frequencies will appear in the output that were not present in the input. In the case of a music amplifier, even small amounts of distortion are very noticeable, making a piano sound like a harpsichord or a guitar like a banjo. One answer to this problem is negative feedback. In a feedback amplifier, we recall that the output is nearly independent of A, the amplifier gain. Thus, if A varies slightly, it doesn't matter.* For a non-inverting amplifier, for example, we would still have, with

---

*There are, of course, limits to how much nonlinearity can be corrected. The slope $\frac{dv_{OUT}}{dv_{IN}}$ has to remain large.

high accuracy, $v_{OUT} = A'v_{IN}$, where A' is given by Eq. (9.3). Negative feedback is almost always used in hi-fi amplifiers to obtain good linearity, and in a great many other circuits as well.

A third application of feedback is in *control systems*. This large and very practical subject will be introduced in Section 9.4.

9.3 Frequency Response of Feedback Amplifiers

In Chapter 4 we saw that it is easy to analyze op-amp circuits with the ideal op-amp technique. Identical methods are useful in sinusoidal analysis, with impedance taking the place of resistance. This makes it easy to design filter circuits, that is, circuits that transmit certain desired frequencies while suppressing others.

As an example, we can use the circuit of Fig. 9.9. This is in essence the same circuit as Fig. 4.29, the inverting amplifier. The feedback resistances are now replaced by the impedances $\mathbf{Z}_F$ and $\mathbf{Z}_1$. Assuming that the op-amp is ideal, we apply assumption 1 (that the potential difference between the input terminals is zero) and conclude that the potential is zero at node A. Writing a phasor node equation for this node, we have

$$\frac{\mathbf{v}_{OUT}}{\mathbf{Z}_F} + \frac{\mathbf{v}_{IN}}{\mathbf{Z}_1} = 0 \tag{9.4}$$

from which we have

$$\mathbf{A}' \equiv \frac{\mathbf{v}_{OUT}}{\mathbf{v}_{IN}} = -\frac{\mathbf{Z}_F}{\mathbf{Z}_1} \tag{9.5}$$

However, the boxes marked $\mathbf{Z}_F$ and $\mathbf{Z}_1$ can contain capacitors or inductors, as well as resistances, and thus the values of $\mathbf{Z}_F$ and $\mathbf{Z}_1$ can vary with frequency. By choosing appropriate subcircuits for $\mathbf{Z}_F$ and $\mathbf{Z}_1$, we can design filter circuits with useful characteristics.

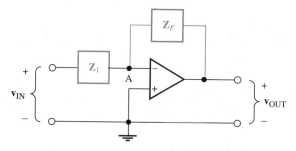

FIGURE 9.9 A generalized circuit, of which the integrator and the inverting amplifier are two special cases.

EXAMPLE 9.2

Let the input to the op-amp circuit in Fig. 9.9A(a) be sinusoidal, with amplitude 1 mV. Find the amplitude of the output sinusoid as a function of frequency. Assume that the op-amp is ideal.

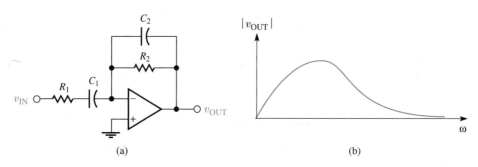

(a)   (b)

FIGURE 9.9A

SOLUTION

This circuit is of the general type shown in Fig. 9.9. Since the op-amp is ideal, we can use Eq. (9.5). In this case $\mathbf{Z}_F = R_2/(1 + j\omega R_2 C_2)$ and $\mathbf{Z}_1 = (1 + j\omega R_1 C_1)/j\omega C_1$. Thus the complex voltage gain is given by

$$\mathbf{A}' = \frac{j\omega R_2 C_1}{1 - \omega^2(R_1 C_1 R_2 C_2) + j\omega(R_1 C_1 + R_2 C_2)}$$

Accordingly,

$$|\mathbf{v}_{OUT}| = |\mathbf{A}'\mathbf{v}_{IN}| = |\mathbf{A}'| \cdot V_0$$

$$= \sqrt{\mathbf{A}' \mathbf{A}'^*} \cdot V_0$$

$$= V_0 \cdot \frac{(R_2 C_1)\omega}{\sqrt{1 + \omega^2(R_1^2 C_1^2 + R_2^2 C_2^2) + \omega^4(R_1 C_1 R_2 C_2)^2}}$$

A graph of $|\mathbf{v}_{OUT}|$ as a function of ω is given in Fig. 9.9A(b).

   An amplifier circuit that has maximum amplification in a certain range of frequencies and little or none elsewhere is known as a *bandpass amplifier*. Bandpass circuits are sometimes constructed using inductors. Inductors, however, have size and cost disadvantages. The present circuit shows how similar behavior may be obtained using only resistors, capacitors, and an op-amp. This circuit also has an advantage in that the shape of its transmission curve (the curve of $|\mathbf{v}_{OUT}|$ versus ω) is nearly unaffected by the load to which the output is connected. This would be very difficult to accomplish in a passive circuit.

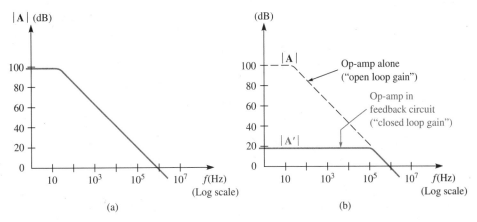

FIGURE 9.10 (a) A plot of open-loop voltage amplification versus frequency for the type 741 op-amp. (b) Plot of voltage amplification versus frequency for the amplifier of part (a), used in a feedback circuit with gain of 10 (solid curve). The open-loop response is repeated for comparison (dashed curve). Note that all scales are logarithmic.

In practice, however, the value of **A** for any amplifier block itself depends on frequency. An amplifier may be nearly ideal at low frequencies; nonetheless, at high frequencies its amplification will decrease and ultimately become less than 1. Such unwillingness to work at high frequencies is a general property of all electronic circuits, which in general places limits on their capabilities. Over the years, progress has largely depended on the invention of new devices and circuits that would work at higher frequencies. This trend is still in progress. Recently the best transistor amplifiers have been pushed to limits near 100 GHz (10^{11} Hz).

In designing feedback amplifiers, this variation of **A** with frequency has to be taken into account. For example, the frequency response of the type 741 op-amp is shown in Fig. 9.10(a). This figure shows $|\mathbf{A}|$, the gain of the op-amp all by itself, with no feedback. If we define bandwidth as the range of frequencies over which $|\mathbf{A}|$ is within 3 dB of its maximum value, we see that the bandwidth of the type 741 op-amp is about 10 Hz. The characteristic shown in Fig. 9.10(a) is typical of general-purpose op-amps. Obviously, 10 Hz is a very small bandwidth, much less than what technology permits. The op-amp is designed to have this characteristic for particular reasons. (See the discussion under "Stability" in the following section.)

It might at first be supposed that such an op-amp would be useless for signals above 10 Hz, but this is not in fact the case. As we have seen, negative feedback makes circuits insensitive to the value of $|\mathbf{A}|$. Thus, so long as $|\mathbf{A}|$ remains larger than 1, the circuit operates much as it would when $|\mathbf{A}|$ is very large. For example, let a 741 op-amp be used in Fig. 9.9, with $\mathbf{Z}_F = 5{,}000$ ohms, $\mathbf{Z}_1 = 500$ ohms, both purely resistive. The circuit amplification A' is then 10, from Eq. (9.5). This result does not cease to hold when the frequency reaches 10 Hz, because $|\mathbf{A}|$ is still very large at that frequency. However, we do not expect that the gain of the circuit can exceed the gain of the op-amp itself. When the frequency reaches a value that makes the op-amp gain $|\mathbf{A}|$

less than the low-frequency circuit gain $|\mathbf{A}'|$ (10, in this example) the circuit gain must finally begin to drop off. This is shown in Fig. 9.10(b). The solid curve in Fig. 9.10(b) is the amplification with the feedback circuit in place, or "closed." Thus this curve is known as the *closed-loop* gain. If we should break the feedback loop, we would see only the characteristic of the op-amp by itself, as shown in Fig. 9.10(a). This is known as the *open-loop* gain.

The bandwidth of the feedback circuit extends approximately to that frequency at which $|\mathbf{A}|$ is equal to $|\mathbf{A}'|$. If the closed-loop circuit gain were increased by a factor of 10, we see from Fig. 9.10(b) that the bandwidth would be reduced by a factor of 10. Thus there is a trade-off between gain and bandwidth, which may be expressed by stating that the product of gain times bandwidth, known as the *gain-bandwidth product*, is a constant. This is a common situation in amplifier design generally; that is, it is often found that increasing the gain of a given amplifier results in a proportional decrease in bandwidth. Figure 9.10(b) shows the gain-bandwidth product of the 741 amplifier to be about 10^6 Hz.

If the product of gain times bandwidth is limited, what can one do to obtain larger gain over the same bandwidth? In that case one can add additional amplifier stages to obtain the desired amplification. For example, four stages, each with the frequency response shown in Fig. 9.10(b), could be used in cascade to provide a gain of 10^4. The bandwidth will still be approximately 10^5 Hz.*

EXAMPLE 9.3

Using type 741 op-amps, design a circuit that has a gain of 100 over the frequency range 0 to 10 kHz. The circuit should have a very large input resistance ($>10^7$ Ω).

SOLUTION

The required circuit has a gain-bandwidth product of 10^6 Hz. This is just about equal to the gain-bandwidth product of a single 741 amplifier. However, use of a single amplifier stage would leave no safety factor, and it is possible that the highest frequencies would be under-amplified. This danger can be eliminated by using two amplifier stages, each with a gain of 10. From Fig. 9.10(b) we see that the bandwidth of each stage will then be nearly 10^5 Hz. The bandwidth of the two stages in cascade will be 64 kHz according to the footnote.*

---

*The bandwidth of N identical stages in cascade is actually slightly less than the bandwidth of the individual stages. If the bandwidth of a single stage is B, the bandwidth of N identical stages in cascade can be shown to be $B\sqrt{2^{1/N} - 1}$.

Since it is required that the input resistance of the circuit be high, noninverting amplifiers are suitable. The required circuit could then appear as shown in Fig. 9.11.

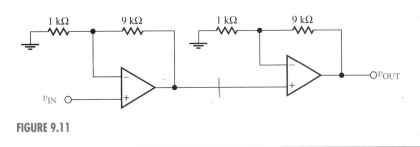

FIGURE 9.11

STABILITY

When an amplifier performs its function reliably under all normal operating conditions, it is said to be *stable*. The opposite condition, instability, can come about in various ways. Most often, an instability manifests itself by the appearance of an output signal when no input signal is applied. Usually this spurious output signal is sinusoidal in form; in that case *oscillation* is said to occur. Instability in feedback amplifiers is a very common difficulty; it usually must be prevented, since it interferes with normal operation of the circuit.

In general, instabilities arise from feedback of an improper kind. For example, usually a time delay, or phase shift, is experienced by a signal as it passes through an op-amp. This phase shift typically is nearly zero at the lowest frequencies but increases at higher frequencies. Since shifting the phase of a sinusoid by 180° is equivalent to multiplying the sinusoid by -1, we see that the negative feedback occurring at low frequencies can reverse in sign as a result of phase shift and turn into positive feedback at higher frequencies. Positive feedback is very likely to lead to instability. Speaking intuitively, this is because the signal fed back from output to input then tends to reinforce the signal already present at the input (rather than subtracting from it); the reinforced input signal makes the output still larger, which makes the input larger, and so forth; loosely speaking, the circuit "runs away."

In the following discussion we shall develop a simple criterion for the determination of stability in feedback circuits. This criterion is a variant of a general test for stability known as the *Nyquist criterion*.

First, it is necessary to express the closed-loop gain of the feedback circuit $\mathbf{A'}$ in a general form. Throughout this discussion we shall assume that the op-amp output re-sistance R_o is zero and that its input resistance R_i is infinitely large. The open-loop voltage amplification of the op-amp $\mathbf{A}$ is a complex number; its magnitude and argument are both functions of frequency. $\mathbf{A}$ is not to be regarded as infinitely large in this discussion; in general, its magnitude approaches zero as the frequency increases.

Under these assumptions, it can be shown that the gain $\mathbf{A}'$ of a negative feedback circuit can be written in the form

$$\mathbf{A}' = \frac{\mathbf{c}(\omega)\,\mathbf{A}(\omega)}{1 + \mathbf{A}(\omega)\mathbf{f}(\omega)} \tag{9.6}$$

Here, $\mathbf{c}(\omega)$ is a parameter having to do with the form of the circuit, and $\mathbf{f}(\omega)$ is another parameter called the *feedback coefficient*. For example, for the voltage-follower circuit we see from Eq. (4.6) (with R_i set to ∞ and R_o to 0) that $\mathbf{c}(\omega) = 1$ and $\mathbf{f}(\omega) = 1$. For the inverting amplifier we can proceed similarly, obtaining

$$\mathbf{A}' = -\frac{\mathbf{A}\,\mathbf{Z}_F}{(\mathbf{A} + 1)\mathbf{Z}_1 + \mathbf{Z}_F} \tag{9.7}$$

from which we have $\mathbf{c}(\omega) = -\mathbf{Z}_F/(\mathbf{Z}_1 + \mathbf{Z}_F)$, $\mathbf{f}(\omega) = \mathbf{Z}_1/(\mathbf{Z}_1 + \mathbf{Z}_F)$.

● EXERCISE 9.1

Derive the result of Eq. (9.7).

● EXERCISE 9.2

Beginning with result (9.7), verify the above expressions for $\mathbf{c}(\omega)$ and $\mathbf{f}(\omega)$ for the inverting amplifier.

From Eq. (9.6) it is evident that the circuit will be unstable if there is a frequency for which $\mathbf{A}(\omega)\,\mathbf{f}(\omega) = -1$. At such a frequency the circuit gain would be infinite, and an output signal would be produced even in the absence of any input signal. Usually at low frequencies both $\mathbf{A}$ and $\mathbf{f}$ are positive real numbers. However, as the frequency increases, phase shifts occur in both $\mathbf{A}(\omega)$ and $\mathbf{f}(\omega)$. If a phase shift of $180°$ occurs, the feedback has been changed in sign, from negative to positive. The situation $\mathbf{Af} = -1$ can then occur.

Actually, although the condition $\mathbf{Af} = -1$ implies oscillation, it is possible for oscillation to occur under more general circumstances. In most cases θ, the argument of the product $\mathbf{A}(\omega)\mathbf{f}(\omega)$, defined by $\theta = \tan^{-1}[\mathrm{Im}\,(\mathbf{Af})/\mathrm{Re}\,(\mathbf{Af})]$, has the value $-180°$ for only a single frequency. For this situation, we can state the Nyquist criterion as follows: *If, at the frequency at which $\theta = -180°$, $|\mathbf{Af}| > 1$, then the system is unstable.* We have already seen that if $|\mathbf{Af}| = 1$ when $\theta = -180°$, instability occurs. The Nyquist criterion states that if $|\mathbf{Af}|$ is greater than unity when $\theta = -180°$, instability also occurs.*

*We refer to phase shifts of $-180°$ rather than $+180°$ because in practice θ is usually negative.

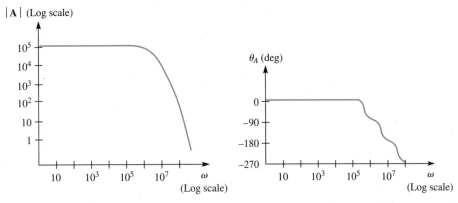

FIGURE 9.12 Graphs of |A| and θ_A versus frequency for a typical hypothetical amplifier. A voltage follower constructed using this amplifier will be unstable.

As a first example, let us consider an op-amp in which **A** has the dependence on ω shown in Fig. 9.12. Here $\mathbf{A}(\omega)$ is shown as a complete Bode plot, with both $\log|\mathbf{A}|$ and θ_A (the argument of **A**) graphed versus $\log \omega$. Suppose this op-amp is used in a voltage-follower circuit. In this circuit, as we have already mentioned, $\mathbf{f} = 1$. By the laws of multiplication of complex numbers, $\theta = \arg(\mathbf{Af}) = \theta_A + \theta_f$, and in this case $\theta_f = 0$. Hence in this case θ is identical to θ_A, shown in Fig. 9.12. We see that $\theta = -180°$ at $\omega = 3 \times 10^7$, and at this frequency $|\mathbf{Af}|$ is about 1,000, much greater than unity. Hence the voltage-follower circuit will be thoroughly unstable and will oscillate.

Evidently an op-amp with characteristics such as those of Fig. 9.12 would not be very useful. To make feedback circuits stable, op-amps usually are designed to have $\mathbf{A}(\omega)$ approximately as shown in Fig. 9.13. We see that an op-amp with this characteristic can be used to make a stable voltage-follower circuit; by the time the phase shift reaches $-180°$, $|\mathbf{A}|$ has decreased to about 10^{-1}.

An op-amp characteristic like that shown in Fig. 9.13 can be obtained by including somewhere in the op-amp an *RC* filter such as that shown in Fig. 7.11, or the equivalent of such a filter. The simple *RC* filter gives a rolloff of $|\mathbf{A}|$, above the break frequency, of 20 dB per decade and a phase shift of a safe $90°$ at frequencies above the

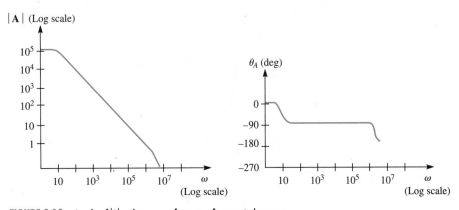

FIGURE 9.13 Graphs of |A| and θ_A versus frequency for a typical op-amp.

break frequency. In general, additional phase shifts will occur at higher frequencies, caused by small stray capacitances elsewhere in the op-amp. However, the *RC* filter should be designed so that at the higher frequencies, where these secondary phase shifts occur, the *RC* filter has already reduced $|A|$ to less than unity.

The type 741 op-amp has a characteristic much like that of Fig. 9.13, as can be seen from Fig. 9.10(a). (Actually, the latter figure shows only $|A|$ versus ω for the 741, and not θ_A. However, it can be proved that if two amplifiers have the same graph of $|A|$ versus ω, their graphs of θ_A versus ω must also be the same.) In the case of the 741, a capacitor is built right into the IC in order to give it a break frequency at 10 Hz. Some op-amps do not have this capacitor built in, since it is rather large for inclusion in an IC. Instead, such op-amps have terminals, usually designated "external frequency compensation," for external connection of the capacitor.

From the preceding discussion it can be seen that the stability of a circuit depends on the Bode plot of the op-amp and also on the form of the circuit, which determines the feedback coefficient **f**. Graphs of $|Af|$ versus ω and of θ ($= \theta_A + \theta_f$) versus ω should be used to check the stability of a given circuit.

EXAMPLE 9.4

The amplifier of Fig. 9.12 is used in the circuit of Fig. 9.9, the amplification of which was given in Eq. (9.7). Let us consider the case in which $\mathbf{Z}_1 = R_1$ and $\mathbf{Z}_f = R_F$ (that is, the inverting amplifier). Let $R_1 = 1,000 \ \Omega$. What is the smallest value R_F can have so that the system will be stable?

SOLUTION

Compare Eq. (9.7) with Eq. (9.6). We see that $\mathbf{f}(\omega) = \mathbf{Z}_1/(\mathbf{Z}_1 + \mathbf{Z}_F) = R_1/(R_1 + R_F)$. In this case θ_f, the argument of f, is zero. We must choose R_F so that $|Af| < 1$ at the frequency that makes θ ($= \theta_A + \theta_f = \theta_A$) equal to $-180°$. From Fig. 9.12 we see that this frequency is about 3×10^7 rad/sec. The value of $|A|$ at this frequency is about 1,000. In order to make $|Af| < 1$, we must have

$$|\mathbf{f}| = \frac{R_1}{R_1 + R_F} < 10^{-3}$$

Thus the smallest value of R_F that will make the circuit stable is $R_F \equiv 10^6 \ \Omega = 1 \ M\Omega$.

It is interesting to observe that for voltage amplifications of 1,000 or *more*, the circuit is stable. However, it is not possible to make a stable inverting amplifier circuit with gain *less* than 1,000, using the amplifier of Fig. 9.12.

9.4 Application: Control Systems

One very important application of electronics is in the control of mechanical systems. The resulting apparatus is then called "electromechanical" by electrical engineers; mechanical engineers call the same system "mechatronic." The most usual case is that in which some device or machine is controlled by automatic, error-correcting electronics. The ensemble is called a *control system*. Many familiar examples of electromechanical control systems come to mind. The control surfaces (rudder, ailerons, elevators) of large airplanes are moved by motors, under electronic control. Space vehicles navigate by observing the stars and using rockets—again under electronic control—to obtain the right trajectory. Machines in modern factories are becoming "robotic," their human operators replaced by electronic control systems.

From the engineering point of view, control systems are a large, active, and important field, in which electrical and mechanical engineers must work together. Both analog and digital techniques are used. In either case, a basic principle of automatic control is the use of feedback of some sort. Imagine, for example, that you are launching a missile intended to strike the moon. Without feedback, you would be unlikely to get a hit. No matter how carefully you aimed, the missile, being imperfect, would fail to travel in the predicted path; it also would be unpredictably blown off course by winds as it passed through the atmosphere. For these and other reasons, it would miss. What is needed is a feedback system to correct the errors. This could be accomplished by adding an optical device that would observe whether or not the missile is pointed at the moon, and, if not, apply a small thrust that would put it back on course. (Note the similarity to the thermostat-controlled furnace described in Section 9.2.) With such a control system, chances of success would be greatly improved. As usual in feedback systems, the output (in this case, the direction of travel) becomes nearly independent of the details of the missile, and is determined almost entirely by the feedback elements of the control system.

The missile in question is shown in Fig. 9.14. The angle by which it misses being pointed at the moon is called θ. The steering mechanism consists of a tail rocket that

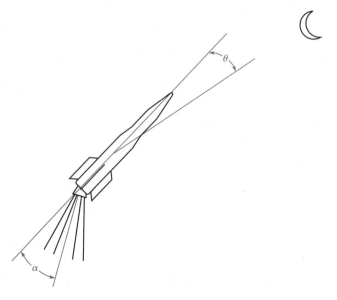

FIGURE 9.14 Missile with control system.

can be swiveled from side to side; the angle it makes with the missile axis is called α. By varying α, the control system can apply various torques to the missile, which cause it to rotate around an axis perpendicular to the page, thus, ideally, reducing the error angle θ to zero. In this case we apply the mechanical equation

$$\text{(Moment of inertia)} \times \frac{d^2\theta}{dt^2} = \text{Torque} \tag{9.7}$$

Let us assume that the torque is equal to $-C\alpha$, where C is some constant, and let the missile's moment of inertia be another constant, I. Then the mechanical system obeys

$$I\frac{d^2\theta}{dt^2} = -C\alpha \tag{9.8}$$

Now let us suppose that in the nose of the missile there is an optical sensor that measures θ. We then need to design a control system that adjusts the control force α in response to the error signal θ. We must choose a *control strategy*. Obviously a great many choices are possible. For instance, we could decide that as soon as any error at all is measured, the maximum possible α is applied to correct it. Such a system, in which the control force is always applied full force with one sign or the other, is known as a "bang-bang" control system. We shall not consider that possibility any further here.

Instead, since we have been concentrating on linear circuits, let us make the arbitrary choice of a linear control strategy, in which α is simply made proportional to θ:

$$\alpha = K_1\theta \tag{9.9}$$

where K_1 is another constant that indicates how strongly the control force is applied in response to θ. Inserting Eq. (9.9) into Eq. (9.8), we have

$$\frac{d^2\theta}{dt^2} + \frac{CK_1}{I}\theta = 0 \tag{9.10}$$

This we recognize as the equation of an undamped sinusoid. The solution is

$$\theta = M_1 \cos\left[\left(\frac{CK_1}{I}\right)^{1/2} t + \Phi_1\right] \tag{9.11}$$

where M_1 and Φ_1 are some constants that are determined by the initial conditions. In other words, the missile will oscillate back and forth around the proper direction, but will never settle down onto the correct heading. It seems our choice of a control strategy wasn't very good.

A more sophisticated control strategy, which is still linear, can be obtained if the detector in the nose detects not only the error, θ, but also its rate of change, $d\theta/dt$.* Now let the control force depend on both θ and $d\theta/dt$ according to

$$\alpha = K_1\theta + K_2\frac{d\theta}{dt} \tag{9.12}$$

---

*Alternatively, a separate device, called a "rate gyro," can be used to measure $d\theta/dt$.

where K_2 is another constant. Combining Eqs. (9.12) and (9.8), we have

$$\frac{d^2\theta}{dt^2} + \frac{CK_2}{I}\frac{d\theta}{dt} + \frac{CK_1}{I}\theta = 0 \tag{9.13}$$

This is a second-order linear equation just like those considered in Section 8.2; in fact, it is identical with Eq. (8.2), except for the names of the constants. The possible solutions (depending on the relative sizes of the constants) are shown in Fig. 8.21. If K_2 is zero we get the undamped sinusoids of Fig. 8.21(a). With positive K_2 we get the damped sinusoids shown in Fig. 8.21(b), or, as K_2 is increased, the critically damped or overdamped decays of Fig. 8.21(c) and (d). Critical damping would probably be a good choice, since it gets the motion over with quickly. One thing we should not choose is a negative value of K_2. This would result in negative damping and a *growing* sinusoid, with disastrous consequences for our mission. Evidently positive, rather than negative, feedback occurs in that case.

Control engineers are fond of diagrams like Fig. 9.15, which illustrates the system we have just designed. In addition to the control force α, there may also be other, random forces (such as winds) acting on the mechanical system, and these are also indicated on the diagram. A full analysis would have to study the response of the system when disturbed by events of various kinds. The presence of feedback in this system is apparent in the figure.

We should notice the rich array of possibilities suggested by our example. The control strategy we chose was *linear*. (That is, α was chosen to be a linear function of θ and $d\theta/dt$.) But this was a purely arbitrary choice. We might have chosen $\alpha = K_3\theta^3 + K_4(d\theta/dt)^3$. Would this have given better results, or worse? Actually, there is an infinite choice of possible control strategies. What would all these other choices do, and how

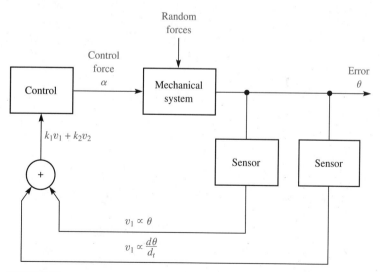

FIGURE 9.15 Block diagram of missile control system.

do you find the best one? Also, we have assumed a very simple model for our system; for example, we assumed that the rocket nozzle could turn instantaneously when the control signal is applied. In actuality it would take a certain time to move into position. What effect will this have? The subject of control systems is thus very interesting and very large, really a science in itself.

POINTS TO REMEMBER

- An analog signal is usually a voltage or current proportional to some quantity of interest.
- Feedback systems are those in which a signal derived from the output is returned to become part of the input signal.
- Feedback is useful because it makes performance independent of the system's internal details and imperfections.
- Op-amp circuits usually involve feedback. Low-cost, imperfect op-amps can be used because feedback corrects the errors.
- Filter circuits using op-amps can be designed by placing frequency-dependent impedances (capacitors or inductors) in the feedback path.
- The gain of a feedback circuit containing an amplifier (known as the *closed-loop gain*) is different from the gain of the amplifier without feedback (the *open-loop gain*). The bandwidths of the two are also different. However, the product of gain and bandwidth is nearly independent of the amount of feedback used. This number is called the *gain-bandwidth product* of the amplifier.
- An amplifier is said to be *unstable* if it can produce an output in the absence of an input. Instability arises when feedback becomes positive at some frequency. Designers of feedback systems must avoid this possibility.
- Control systems are designed to direct the operation of mechanical devices. Negative feedback enables the system to correct its own errors, so it can run by itself without external control.

PROBLEMS

Section 9.1

9.1   A photodetector has responsivity of 200 V/W. How many watts of light are required to produce a 100 mV signal?

*9.2   The *noise-equivalent power* (*NEP*) of a photodetector is the light power required to achieve a signal-to-noise ratio of unity. Find the NEP of the photodetector in the previous problem, if the constant V_{NO} in Eq. (9.2) is 10^{-8} V/$\sqrt{\text{Hz}}$ and the bandwidth is 100 Hz.

Section 9.2

*9.3 The voltage gain $\mathbf{A}'$ of the inverting amplifier circuit (with R_i assumed infinite and R_o zero) is given by Eq. (9.7). Suppose $\mathbf{Z}_F = \mathbf{Z}_1 = 1,000 \ \Omega$ (real), and $\mathbf{A} = 10^5$.
   a. What is $\mathbf{A}'$?
   b. Now suppose $\mathbf{A}$ increases by 10%. Now what is $\mathbf{A}'$?
   c. Explain what is being illustrated by this problem.

*9.4 *Desensitivity* is a measure of the ability of a circuit's performance to ignore changes in some parameter. Find the desensitivity of the gain of a voltage follower with respect to A, given by

$$D = \frac{\left(\dfrac{dA}{A}\right)}{\left(\dfrac{dA'}{A'}\right)}$$

(This is the ratio of the percentage change in A to the percentage change in A'. The larger D is, the less sensitive the circuit.) Assume $A = 10^5$, $R_i = \infty$, $R_0 = 0$ and refer to Eq. (4.6).

*9.5 An op-amp has the nonlinear input-output characteristic shown in Fig. 9.16. It is used in a voltage follower circuit. Make a graph of the output of the voltage follower vs. input voltage, over the range 0 to 8 volts. Use Eq. (4.6) and assume $R_1 = \infty$, $R_0 = 0$. Is the voltage follower more linear than the op-amp by itself?

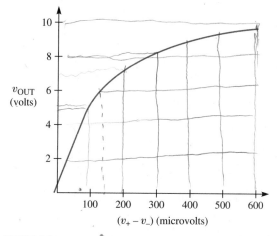

FIGURE 9.16

Section 9.3

9.6 An op-amp has the open-loop frequency response shown in Fig. 9.10(a).
  a. What is the approximate bandwidth of a circuit using this op-amp, with a closed-loop voltage gain of 100?
  b. With a voltage gain of 1,000?
  c. What is the gain-bandwidth product for this op-amp?

9.7 Use phasor analysis to find $\mathbf{v}_{OUT}$ as a function of ω for the circuit of Fig. 9.17 with sinusoidal input of constant amplitude V_0. Assume the op-amp is "ideal."

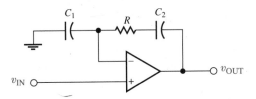

FIGURE 9.17

9.8 Repeat Example 9.4, using the op-amp of Fig. 9.13 instead of that of Fig. 9.12. What is the smallest gain that can be obtained without loss of stability?

**9.9 Refer to the integrator circuit of Problem 5.6 and Fig. 5.15. Show that the integrator circuit is stable when the op-amp of Fig. 9.13 is used.

**9.10 A differentiator circuit can be made by interchanging the resistance and capacitance in Fig. 5.15.
  a. Verify that the resulting circuit is a differentiator.
  b. Comment on the stability of the differentiator when the op-amp of Fig. 9.13 is used.

Section 9.4

9.11 If we change our control strategy, replacing Eq. (9.9) by $\alpha = K_1\theta^2$, the result is sure to be unsuccessful. Explain why.

*9.12 Assuming that all the other constants are given, find the value of K_2 needed to achieve critical damping. [See Eqs. (8.20) and (8.33).]

**9.13 A crude control system is constructed in the following way. When θ is positive, Eq. (9.9) is replaced by $\alpha = \alpha_{MAX}$, where α_{MAX} is a constant. When θ is negative, $\alpha = -\alpha_{MAX}$.
  a. At $t = 0$, $\theta = \theta_0$ and $d\theta/dt = 0$. Find t_1, the time at which θ becomes zero. What is $d\theta/dt$ at time t_1?
  b. Find t_2, the time at which $d\theta/dt$ again reaches zero. What is θ at time t_2?
  c. Sketch θ as a function of time for several cycles of the motion. Is this a good control strategy? Explain why or why not.

CHAPTER 10

PRACTICAL ANALOG TECHNOLOGY

Most readers of this book will from time to time be users of analog technology. When choosing an amplifier or other analog component, one must identify the particular characteristics that are needed for the application at hand. This is a very typical problem for engineers. For example, one cannot just go to a manufacturer and ask for "a motor." One must specify the power, the type (dc or ac? induction or synchronous?), the voltage, the speed, and any other characteristics that are relevant. The situation with analog components is similar, involving such considerations as power, frequency response, and noise performance. In Sections 10.1 through 10.3 we shall introduce the main considerations that users need to know. Since the operational amplifier is the analog component *par excellance*, we shall then go on to treat its special characteristics in greater detail. In the Application for this chapter we shall talk about high-fidelity audio, a demanding analog technology found in many readers' homes.

10.1 Power Limitations

Almost all amplifiers can be represented by the basic model shown in Fig. 4.9. In practice, however, amplifiers vary greatly in size, type of construction, and degree of sophistication. Evidently there are many amplifier characteristics beyond the values of A, R_o, and R_i. One such characteristic is the amplifier's ability to deliver output power. This matter is not always important; a great many amplifiers are used in low-power applications, where the required output power is so small that its exact value is unim-

357

portant. On the other hand, though, one might need a public-address system for speaking to crowds at a football stadium. In that case, hundreds of watts of signal power would be needed, and power would be a key consideration.

An amplifier intended to deliver a larger-than-usual amount of signal power is known as a *power amplifier*. The easy way to determine an amplifier's power capability is to simply look at the manufacturer's specifications, which might say, for example, "rated output power, 100 watts." Most amplifiers can be driven to output powers higher than the rated value by increasing the input level, but this is a dangerous practice. Eventually the amplifier may blow a fuse or simply burn up. (Therefore, when working with high-power equipment, caution is in order!) More often, however, what happens is that the amplifier is driven out of its linear range, and distortion rapidly increases. On music amplifiers, one often sees a specification such as "output power 100 watts with 0.5% distortion." This suggests that you might be able to obtain 110 watts, but it won't sound very good!

The output power of an amplifier, however, often is not specified by the manufacturer, especially for small integrated-circuit (IC) devices like op-amps. In such cases the *maximum power dissipation* of the device is usually given. This term refers to the maximum power that can be converted to heat in the amplifier before it overheats. The user then needs to calculate the power dissipation expected in the particular application, and make sure it does not exceed the limit. The allowable power dissipation can depend on the way an amplifier is mounted. For example, an IC amplifier may be rated to dissipate 50 mW when it is cooled by the surrounding air, but the allowable dissipation may increase to 750 mW when the IC is mounted, for cooling purposes, on a large piece of thermally conductive metal called a "heat sink."

We note that the power dissipation in the amplifier is *not* the same thing as the power delivered to the load. In order to calculate power dissipation, we must take the power-supply connections into consideration. Such connections must always be present. This is evident because the amplifier's output signal power exceeds the input signal power; the difference is supplied by power drawn from an external source. Usually this power is supplied by dc voltage sources of specified values, which must be connected to the appropriate terminals of the amplifier. (Free-standing units, such as commercial electronic equipment, have their own built-in power-supply circuits, which convert house-current ac power to the required dc.) Because power-supply connections to the amplifier can always be assumed to be present, they are usually not shown on circuit diagrams. Nonetheless, one has to be aware that they are necessary, and refer to the manufacturer's data to learn what power-supply voltages must be supplied, and where to connect them. It is also necessary to think about the power supply when calculating heat dissipation. A fraction of the power delivered from the power supply to the amplifier leaves the amplifier again, in the form of output power, but the rest of it is converted to waste heat. The amount of heat produced is equal to the net power (the power entering from the power supply, minus the power leaving as output; the input power is assumed to be too small to be of interest) being delivered to the amplifier. It is this waste heat on which the specification places a limit. We have already seen a convenient way to calculate the power entering a circuit element. If we know the voltages at, and the currents through, each wire entering the amplifier, we can find the power entering the amplifier (and being converted to heat) from Eq. (3.16).

EXAMPLE 10.1

Find the power dissipated in the amplifier in the circuit shown in Fig. 10.1. Assume $R_i \cong \infty$, $R_o \cong 0$, $A = 10$, $R_L = 50$, and $v_{IN} = 0.5$ V. The voltage at the power-supply terminal, V_{PS}, is 15 V. Assume that $I_{PS} = I_L$.

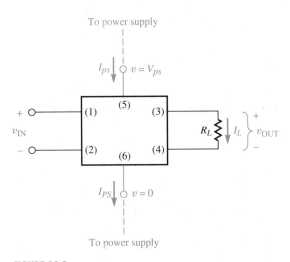

FIGURE 10.1

SOLUTION

We use Eq. (3.16),

$$P = \sum_{n=1}^{6} V_n I_n$$

where V_n is the voltage at terminal n and I_n is the current flowing *into* terminal n. Currents I_1 and I_2 are zero because $R_i = \infty$; since $R_o = 0$, $v_{OUT} = Av_{IN} = 5$ V. The power supply maintains terminal (5) at $V_5 = V_{PS} = 15$ V and terminal (6) at $V_6 = 0$. Note also that $I_3 = -I_4$ and $I_5 = -I_6$ (since whatever current leaves the power supply has to return to it.) Thus

$$
\begin{aligned}
P &= V_3 I_3 + V_4 I_4 + V_5 I_5 + V_6 I_6 \\
&= V_3(-I_L) + V_4 I_L + V_5 I_{PS} - V_6 I_{PS} \\
&= (V_4 - V_3)I_L + (V_{PS} - 0)I_{PS} \\
&= -v_{OUT}I_L + V_{PS}I_{PS} \\
&= (V_{PS} - v_{OUT})I_L
\end{aligned}
$$

Note that $I_L = v_{OUT}/R_L$; thus

$$P = 10(5/50) = 1.0 \text{ W}$$

The approximation $R_i \cong \infty$ is often justified because many amplifiers are designed with large R_i in order to keep the input power low. The assumption $I_{PS} = I_L$ is probably not precisely true, since a certain amount of additional current will pass directly through the amplifier from terminal (5) to (6) without going through the load, making I_{PS} slightly larger than I_L. This extra current may be needed to operate the amplifier, but it wastes power and creates excess heat. In a well-designed power amplifier this waste will be minimized, and I_{PS} will not be very much larger than I_L.

• EXERCISE 10.1

In the circuit of Example 10.1 let $v_{IN} = (1 - 0.5 \cos \omega t)$ V. Find the time-averaged power dissipated in the amplifier. **Answer:** 0.75 W.

10.2 Frequency Response of Amplifiers

In analog work one usually thinks of signals that either are sinusoids or are more complex waveforms that can be regarded as sums of sinusoids. The behavior of an amplifier always depends on the frequency of the sinusoidal signal in question. In general, there is always an upper frequency limit above which the block ceases to function. The upper frequency limits of various blocks are determined by the uses for which they are intended. The useful frequency range of an amplifier is called its *bandwidth*. An audio amplifier (that is, one meant for audible frequencies) typically amplifies the frequencies from near zero to the range 2 kHz to 100 kHz. Another type of amplifier, known for historical reasons as a "video amplifier," usually functions between a minimum frequency near zero and a maximum frequency of a few MHz. Microwave amplifiers can have upper frequencies ranging to tens of GHz (1 GHz $= 10^9$ Hz), but these are usually *bandpass* amplifiers—that is, they function only within a band of frequencies, limited by both a minimum and a maximum. The amplifier model can describe these frequency effects if we allow the open-circuit voltage amplification A to be a function of frequency. When frequency effects are significant, we should also regard A as being a complex number $\mathbf{A}(\omega)$. Since $\mathbf{A}(\omega) = \mathbf{v}_{OUT}/\mathbf{v}_{IN}$, the ratio of output amplitude to input amplitude is given by $|\mathbf{A}(\omega)|$. The phase angle of $\mathbf{A}(\omega)$ specifies the phase relationship between output and input sinusoids. If $\mathbf{A} = |\mathbf{A}|e^{j\theta_A}$, it is seen that the phase difference between output and input sinusoids is θ_A. Both $|\mathbf{A}|$ and θ_A are functions of frequency. However, at low frequencies θ_A is often zero (meaning that output and input are in phase with each other); thus at low frequencies $\mathbf{A}$ can often be regarded as real, as we have done until now. Figure 10.2 shows variations of $|\mathbf{A}|$ and θ_A typical of audio, video, bandpass, and operational amplifiers.

When frequency effects are important, input and output resistances should be generalized to impedances. They are also functions of frequency; thus we write $\mathbf{Z}_i(\omega)$ and

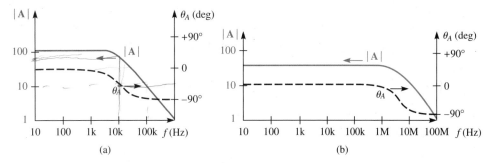

FIGURE 10.2 (a) and (b)   Typical frequency response characteristics for (a) an audio amplifier and (b) a video amplifier.

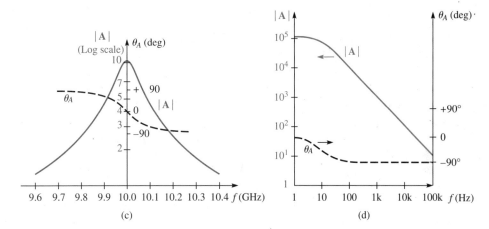

FIGURE 10.2 (c) and (d)   Typical frequency response characteristics for (c) a bandpass amplifier (microwave FET) and (d) an operational amplifier. Note the separate scales for the absolute values and phase angles of *A*.

$\mathbf{Z}_o(\omega)$. Most commercial amplifiers, however, are designed to make $\mathbf{Z}_i$ and $\mathbf{Z}_o$ real and constant over the useful frequency range of the block.

The variation of **A** (and perhaps of θ_A) with frequency for an amplifier is referred to as its *frequency response*. Just as A' for a circuit containing an amplifier can be quite different from the A of the amplifier itself, so can the frequency response of a circuit containing an amplifier be quite different from the frequency response of the amplifier by itself. This has been seen in Section 9.3.

EXAMPLE 10.2

In the circuit shown in Fig. 10.3(a), $R_s = 1 \text{ k}\Omega$, $R_L = 16 \text{ }\Omega$, and $C = 0.2 \text{ }\mu\text{F}$. The amplifier has $R_i = 10 \text{ k}\Omega$, $R_o = 0$, and $\mathbf{A}(\omega)$ as shown in Fig. 10.2(a). Let us define $G_V(f) = |\mathbf{v}_L|/|\mathbf{v}_s|$. Find and sketch $G_V(f)$ versus frequency. Let us define the circuit's "maximum usable frequency" f_{MAX} as the frequency at which G_V has dropped 6 dB below its maximum value. Find f_{MAX} for this circuit.

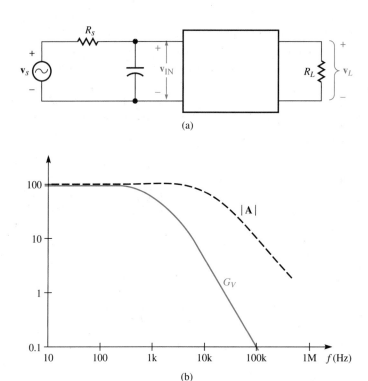

(a)

(b)

FIGURE 10.3

SOLUTION

The ratio $|\mathbf{v}_{IN}|/|\mathbf{v}_s|$ can be found using phasor analysis and the generalized voltage divider formula (7.12). In doing this analysis we note that R_i is in parallel with C. The result is

$$\frac{\mathbf{v}_{IN}}{\mathbf{v}_s} = \frac{R_i \| (j\omega C)^{-1}}{R_i \| (j\omega C)^{-1} + R_s}$$

$$= \frac{R_i}{R_i + R_s + j\omega C R_i R_s} = \frac{R_i}{R_i + R_s} \frac{1}{1 + j\omega C R_{\|}}$$

where we have defined $R_{\|} = R_i \| R_s = R_i R_s / (R_i + R_s)$. Taking the absolute value, we have

$$\left| \frac{\mathbf{v}_{IN}}{\mathbf{v}_s} \right| = \frac{R_i}{R_i + R_s} \frac{1}{\sqrt{1 + (\omega R_{\|} C)^2}} \tag{A}$$

The function we wish to plot is (since $R_o = 0$)

$$G_V(f) = \left|\frac{\mathbf{v_{IN}}}{\mathbf{v_s}}\right| \cdot |\mathbf{A}(f)|$$

It can be found by multiplying the above expression for $|\mathbf{v_{IN}}|/|\mathbf{v_s}|$ by the curve given in Fig. 10.2(a). (Note that $\omega = 2\pi f$.) The resulting frequency response is shown in Fig. 10.3(b). $\mathbf{A}(\omega)$ is shown for comparison. Clearly, the amplifier's gain is considerably decreased by the "bypassing" effect of C. We note that response is "flat" (that is, constant) up to frequencies on the order of $(2\pi R_{||}C)^{-1} = 875$ Hz. Then it goes down at approximately -20 dB per decade, until the vicinity of 10 kHz, due to the rolloff of $|\mathbf{v_{IN}}|/|\mathbf{v_s}|$. In this frequency range $|\mathbf{A}|$ has not yet begun to decrease. Above 10 kHz, G_V decreases faster because $\mathbf{v_{IN}}/\mathbf{v_s}$ and $\mathbf{A}$ are both decreasing. Since each of the two factors decreases at -20 dB per decade, G_V decreases at -40 dB per decade in the highest-frequency region.

Since $|\mathbf{v_{IN}}|/|\mathbf{v_s}|$ is decreasing in the range 875 Hz $< f <$ 10 kHz while $|\mathbf{A}|$ is nearly constant in this range, the 6 dB point is determined primarily by the former. Since a change of 6 dB corresponds to a decrease of 50%, the 6 dB rolloff frequency is the frequency $\omega_{6\,dB}$ at which, from Eq. (A),

$$\frac{1}{\sqrt{1 + (\omega_{6\,dB}R_{||}C)^2}} = \frac{1}{2}$$

Thus

$$2\pi f_{6\,dB}R_{||}C = \sqrt{3}$$

$$f_{6\,dB} = \frac{\sqrt{3}}{2\pi R_{||}C} = 1516 \text{ Hz}$$

● **EXERCISE 10.2**

Find the 6 dB rolloff frequency if the amplifier in Example 10.2 is the operational amplifier of Fig. 10.2(d). **Answer:** $f_{6\,dB} = 20$ Hz.

10.3 Noise Performance

The matter of *noise* has been mentioned earlier. In general, electronic circuits invariably produce undesired random signals. These eventually appear in their output, added on to the "true" output signal one wants to have. When the "true" signal is large this is not much of a problem, as the much smaller noise can then usually be neglected. However, when the "true" signal is small it may become lost in the noise, and thus be unusable. Thus it is noise that determines how small signals can be, while remaining useful.

In making sensitive measurements, noise must always be considered. In communications, noise is a great enemy, introducing errors and placing limits on how far information can be sent.

In the case of an analog block, such as an amplifier, noise is actually contributed by all the resistors and transistors that are inside the block. However, most users are not interested in the details of what goes on inside the block, making do, instead, with a knowledge of its model. Thus it is convenient to refine the model to include noise effects, as shown in Fig. 10.4. Here the original block (which is still represented by its usual model) is combined with two *noise sources* connected at its input. These two sources, one a current source and one a voltage source, produce random voltages and currents, which add to the intentional inputs that are applied to the block. Often the two sources are *uncorrelated*, which means that the noise power that appears at the output is the sum of the noise powers that result from each of the two sources taken separately.

Noise sources have some characteristics that are peculiar, compared with other sources. For example, how do we specify their values? We cannot specify the output voltage of a noise voltage source, because its voltage is a randomly varying number. We cannot even specify its *average* voltage, because its random voltage is positive as often as it is negative, and the average voltage is zero! However, the noise voltage source is capable of producing power; if we connect it across an ideal resistor, its random voltage will produce currents in the resistor, and the resistor will be heated. The instantaneous power dissipated in the resistor, of value R, is v_N^2/R, where v_N is the time-varying noise voltage. The time-averaged power is

$$P_{AV} = <v_N^2/R> = <v_N^2>/R \qquad (10.1)$$

where the brackets indicate time average. In order to be able to use this formula and predict the noise power, the thing we need to specify about the noise voltage source is the time average of the square of its voltage, $<v_N^2>$. Similarly, for the noise current source, $<i_N^2>$ is specified.

Another peculiarity of noise sources is that their values depend on the bandwidth of the circuit. (See Section 9.1.) We can think of the noise source as producing random sinusoids of all frequencies; the total noise power is the sum of the noise powers at all the frequencies. Thus the wider the bandwidth of the circuit, the more frequencies come through and the more noise power is produced. It is often assumed that noise

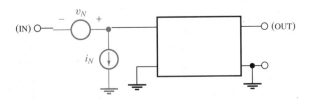

FIGURE 10.4

sources are "white"; white noise, like white light, contains equal amounts of all frequencies. Put another way, white noise contains noise sinusoids at all frequencies, with
each frequency interval of a certain size, Δf, contributing the same amount of noise
power as every other frequency interval of the same size. Hence we write

$$\langle v_N^2 \rangle = e_n^2 B \qquad (10.2)$$

where B is the bandwidth of the circuit. The constant of proportionality, e_n, and the
corresponding constant describing the current source, i_n, are characteristics of the
block, and are often specified by the manufacturer.

EXAMPLE 10.3

A bandpass amplifier is used to amplify a signal source v_S, R_S, as shown in Fig. 10.5(a). The
amplifier is represented by the noise model of Fig. 10.4 and has the following properties: $A =$
100, $R_i = 50\ \Omega$, $R_o = 50\ \Omega$, $e_n = 16\ \text{nV/Hz}^{1/2}$, $i_n = 1.0\ \text{nA/Hz}^{1/2}$; its bandpass extends between
0.7 GHz and 1.0 GHz. The circuit values are $R_S = R_L = 50\ \Omega$. Find the noise power delivered to
the load R_L.

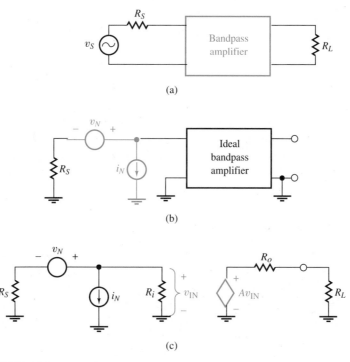

(a)

(b)

(c)

FIGURE 10.5

SOLUTION

When the noise model is substituted, we have Fig. 10.5(b), and when the amplifier model itself (Fig. 4.9) is inserted, we obtain Fig. 10.5(c). To find the noise power at the output, we set the independent signal source to zero, which means we replace it by a short circuit. (This is logical, since a voltage source with value zero would have zero potential difference across it, just as a short circuit does. If the signal source contained an independent current source, that would be replaced by an *open* circuit.) We then find the noise contribution of the noise voltage source, with the noise current source set to zero (i.e., replaced by an open circuit.) The dependent voltage source inside the amplifier now becomes a noise source with mean-square amplitude

$$<(Av_{IN})^2> = A^2 \left[\frac{R_i}{R_i + R_S}\right]^2 <v_N^2>$$

Thus the power contributed to the load by the noise voltage source is

$$P_V = 10^4 \cdot \frac{1}{4} \cdot e_n^2 B \left[\frac{R_L}{R_L + R_o}\right]^2 \cdot \frac{1}{R_L}$$

$$= 10^4 \cdot \frac{1}{4}(16 \times 10^{-9})^2 \cdot (3 \times 10^8) \cdot \frac{1}{4} \cdot \frac{1}{50} = 0.96 \ \mu W$$

Similarly, we obtain the contribution of the noise current source by replacing the voltage source with a short circuit. We find

$$P_I = 10^4 \cdot i_n^2 B \cdot (R_i \| R_S)^2 \cdot \left[\frac{R_L}{R_L + R_o}\right]^2 \cdot \frac{1}{R_L}$$

$$= 9.375 \ \mu W$$

Assuming that the two noise sources are uncorrelated, the total noise power is just the sum of P_V and P_I, or 10.335 μW. An interesting point to note from this example is that the noise of an amplifier does not depend just on the amplifier itself. It is also influenced by the form of the input circuit.

In the preceding example the signal source itself, being external to the amplifier, was treated as though it were noiseless. In practice, however, any signal source can be expected to contribute some noise along with its signal. A principal source of noise is *thermal noise*; this term refers to random signals produced by the thermal motion of particles at any frequency above absolute zero. It is difficult to escape such noise. For example, consider a radio astronomy antenna pointed upward, toward outer space. One might at first think that no noise would be received. However, the warm atmosphere radiates some noise signal into the antenna, and besides this there is the background noise of outer space, the famous 3° black-body radiation. In fact, the background temperature of the universe was first discovered by radio astronomers when they found unexplained antenna noise. (The excess noise was first thought to arise from a bird's

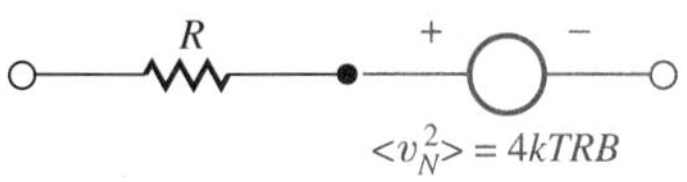

FIGURE 10.6

nest in the antenna, but this proved not to be the case.) The observed noise radiation arriving from outer space is a crucial piece of evidence supporting the theory of the Big Bang.

In electronic circuits thermal noise is produced by resistances, in which case it is also known as *Johnson noise*. A real resistor R can be modeled by a combination of an ideal noiseless resistor and a random voltage source with value

$$<v_N^2> \ = \ 4kTRB \qquad\qquad (10.3)$$

as shown in Fig. 10.6. Here k is Boltzmann's constant (1.38×10^{-23} joule/°K) and T is the absolute temperature in degrees Kelvin. Noise is also produced by semiconductor devices such as transistors, but here the noise contains other contributions besides the thermal noise, and modeling it becomes more complicated.

The main thing to watch out for in connection with noise is the signal-to-noise ratio, introduced in Section 9.1. When designing, one must bear in mind the presence of noise and if possible estimate its magnitude. The noise must not exceed the expected signal magnitude. If it does, the system will, in all likelihood, be useless!

*10.4 The Practical Op-Amp

When an analog block of some sort is needed, the usual procedure is to buy one from a suitable manufacturer and hook it up. Along with the block, the manufacturer can be expected to supply fairly detailed information on its characteristics. The user probably will not need to know much about how the interior of the block is constructed but he or she *will* need to know about all the characteristics that affect the performance of the block as seen from outside. In other words, one needs to know about its *terminal characteristics*.

The analog block most often used is the operational amplifier. This is partly because op-amps are so versatile; as we have seen, they can be used to perform many different operations. Because they are usually in integrated-circuit form, they are also rugged, compact, and, best of all, cheap. The basics of op-amp circuit design have already been discussed in Chapter 4. However, in practice one must consider some important details of op-amp performance, which will be discussed in this section. For these details it is useful to refer to the manufacturer's specification sheet, an example of which appears in Fig. 10.7.

Much of the manufacturer's information is readily understood. Information on the allowable heat dissipation (500 mW) is given under "Absolute Maximum Ratings." A plot of the open-loop frequency response is given in one of the small graphs; from this

 **National Semiconductor**

Operational Amplifiers/Buffers

BI-FET II™ Technology

LF351 Wide Bandwidth JFET Input Operational Amplifier

General Description

The LF351 is a low cost high speed JFET input operational amplifier with an internally trimmed input offset voltage (BI-FET II™ technology). The device requires a low supply current and yet maintains a large gain bandwidth product and a fast slew rate. In addition, well matched high voltage JFET input devices provide very low input bias and offset currents. The LF351 is pin compatible with the standard LM741 and uses the same offset voltage adjustment circuitry. This feature allows designers to immediately upgrade the overall performance of existing LM741 designs.

The LF351 may be used in applications such as high speed integrators, fast D/A converters, sample-and-hold circuits and many other circuits requiring low input offset voltage, low input bias current, high input impedance, high slew rate and wide bandwidth. The device has low noise and offset voltage drift, but for applica-

tions where these requirements are critical, the LF356 is recommended. If maximum supply current is important, however, the LF351 is the better choice.

Features

■ Internally trimmed offset voltage	10 mV
■ Low input bias current	50 pA
■ Low input noise voltage	$16\,nV/\sqrt{Hz}$
■ Low input noise current	$0.01\,pA/\sqrt{Hz}$
■ Wide gain bandwidth	4 MHz
■ High slew rate	$13\,V/\mu s$
■ Low supply current	1.8 mA
■ High input impedance	$10^{12}\Omega$
■ Low total harmonic distortion $A_V = 10$, $R_L = 10k$, $V_O = 20\,Vp\text{-}p$, BW = 20 Hz-20 kHz	<0.02%
■ Low 1/f noise corner	50 Hz
■ Fast settling time to 0.01%	$2\,\mu s$

Typical Connection

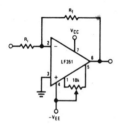

Simplified Schematic

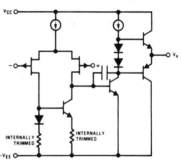

Connection Diagrams (Top Views)

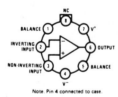

Metal Can Package

Note. Pin 4 connected to case.

Order Number LF351H
See NS Package H08C

Dual-In-Line Package

TOP VIEW

Order Number LF351N
See NS Package N08A

Absolute Maximum Ratings

Supply Voltage	±18V	Input Voltage Range (Note 2)	±15V
Power Dissipation (Note 1)	500 mW	Output Short Circuit Duration	Continuous
Operating Temperature Range	0°C to +70°C	Storage Temperature Range	−65°C to +150°C
T_j(MAX)	115°C	Lead Temperature (Soldering, 10 seconds)	300°C
Differential Input Voltage	±30V		

FIGURE 10.7   Manufacturer's specifications for a typical op-amp. (*Courtesy of National Semiconductor Corp.*)

DC Electrical Characteristics (Note 3)

SYMBOL	PARAMETER	CONDITIONS	LF351 MIN	LF351 TYP	LF351 MAX	UNITS
V_{OS}	Input Offset Voltage	$R_S = 10 k\Omega$, $T_A = 25°C$		5	10	mV
		Over Temperature			13	mV
$\Delta V_{OS}/\Delta T$	Average TC of Input Offset Voltage	$R_S = 10 k\Omega$		10		$\mu V/°C$
I_{OS}	Input Offset Current	$T_j = 25°C$, (Notes 3, 4)		25	100	pA
		$T_j \leqslant 70°C$			4	nA
I_B	Input Bias Current	$T_j = 25°C$, (Notes 3, 4)		50	200	pA
		$T_j \leqslant 70°C$			8	nA
R_{IN}	Input Resistance	$T_j = 25°C$		10^{12}		Ω
A_{VOL}	Large Signal Voltage Gain	$V_S = \pm 15V$, $T_A = 25°C$ $V_O = \pm 10V$, $R_L = 2 k\Omega$	25	100		V/mV
		Over Temperature	15			V/mV
V_O	Output Voltage Swing	$V_S = \pm 15V$, $R_L = 10 k\Omega$	± 12	± 13.5		V
V_{CM}	Input Common-Mode Voltage Range	$V_S = \pm 15V$	± 11	$+15$ -12		V V
CMRR	Common-Mode Rejection Ratio	$R_S \leqslant 10 k\Omega$	70	100		dB
PSRR	Supply Voltage Rejection Ratio	(Note 5)	70	100		dB
I_S	Supply Current			1.8	3.4	mA

AC Electrical Characteristics (Note 3)

SYMBOL	PARAMETER	CONDITIONS	LF351 MIN	LF351 TYP	LF351 MAX	UNITS
SR	Slew Rate	$V_S = \pm 15V$, $T_A = 25°C$		13		V/μs
GBW	Gain Bandwidth Product	$V_S = \pm 15V$, $T_A = 25°C$		4		MHz
e_n	Equivalent Input Noise Voltage	$T_A = 25°C$, $R_S = 100\Omega$, f = 1000 Hz		16		nV/$\sqrt{Hz}$
i_n	Equivalent Input Noise Current	$T_j = 25°C$, f = 1000 Hz		0.01		pA/$\sqrt{Hz}$

Note 1: For operating at elevated temperature, the device must be derated based on a thermal resistance of 150°C/W junction to ambient or 45°C/W junction to case.
Note 2: Unless otherwise specified the absolute maximum negative input voltage is equal to the negative power supply voltage.
Note 3: These specifications apply for $V_S = \pm 15V$ and $0°C \leqslant T_A \leqslant +70°C$. V_{OS}, I_B and I_{OS} are measured at $V_{CM} = 0$.
Note 4: The input bias currents are junction leakage currents which approximately double for every 10°C increase in the junction temperature, T_j. Due to the limited production test time, the input bias currents measured are correlated to junction temperature. In normal operation the junction temperature rises above the ambient temperature as a result of internal power dissipation, P_D. $T_j = T_A + \theta_{jA} P_D$ where θ_{jA} is the thermal resistance from junction to ambient. Use of a heat sink is recommended if input bias current is to be kept to a minimum.
Note 5: Supply voltage rejection ratio is measured for both supply magnitudes increasing or decreasing simultaneously in accordance with common practice.

Typical Performance Characteristics

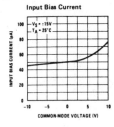

Input Bias Current

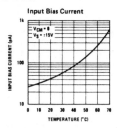

Input Bias Current

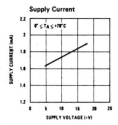

Supply Current

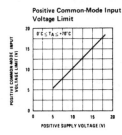

Positive Common-Mode Input Voltage Limit

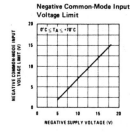

Negative Common-Mode Input Voltage Limit

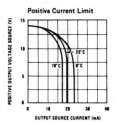

Positive Current Limit

FIGURE 10.7 (Continued)

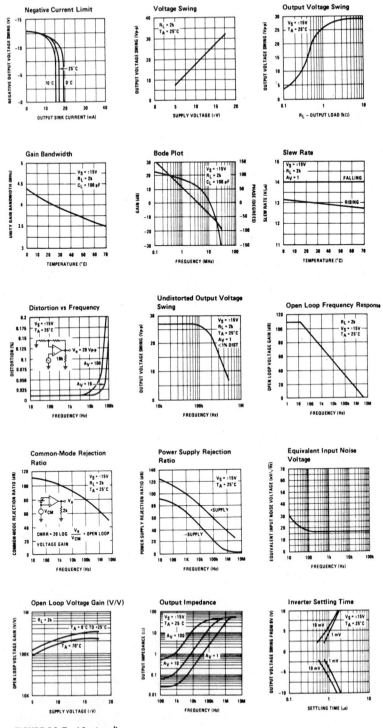

FIGURE 10.7   (Continued)

graph we see that the break frequency is about 15 Hz and the gain-bandwidth product is about 4.7×10^6. The "equivalent input noise voltage," e_n, and the "equivalent input noise current," i_n, are given under "AC Electrical Characteristics." Two other practicalities of op-amp operation will now be discussed.

OFFSET

From looking at the op-amp model, one might believe that the output voltage of the op-amp would be zero whenever $v_{(+)} - v_{(-)} = 0$. However, in practice real op-amps are always slightly "unbalanced," or asymmetrical, with respect to the two inputs. This leads to the condition known as *voltage offset*, in which the output voltage fails to be zero even when $v_{(+)} - v_{(-)} = 0$.

The amount of voltage offset existing in an op-amp can be specified by stating how large a dc voltage must be applied between the (+) and (−) input terminals in order to make the amplifier's output voltage zero. The required input voltage is called the *input offset voltage* and is typically in the range 0.5 to 10 mV.

An op-amp with nonzero voltage offset can be represented by the model shown in Fig. 10.8(a). Here the real op-amp is represented as an offset-free op-amp combined with an offset-voltage source V_{OS}. It is easily seen that when a voltage equal to the input offset voltage V_{OS} is applied between the terminals of this model, the output is reduced to zero, as required by the definition of V_{OS}.

When an op-amp containing voltage offset is used in a circuit, the offset voltage acts to produce a spurious output voltage. For example, consider the voltage-follower circuit model shown in Fig. 10.8(b), in which the op-amp model of Fig. 10.8(a) has been used. Here it is evident that the effective input voltage being applied to the op-amp is $v_S + V_{OS}$, and the output voltage will be $v_S + V_{OS}$, not v_S.

A related effect occurs because small dc currents (which, like the voltage offset, are not indicated by the op-amp model of Fig. 4.15) flow through the op-amp input terminals. These currents are required to operate the input transistors of the op-amp. The average of the two dc currents flowing through the two input terminals is known as the *input bias current* and is typically in the range 0.5 to 5 μA. The difference between the two dc bias currents, which arises from asymmetry of the amplifier, is known as the *input offset current*. It is typically 10 times less than the input bias current.

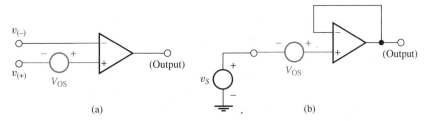

FIGURE 10.8 Voltage offset in an operational amplifier. (a) Model for an amplifier with voltage offset. The model is composed of an offset free op-amp plus an offset-voltage source. (b) Circuit model for a voltage follower when the op-amp has voltage offset.

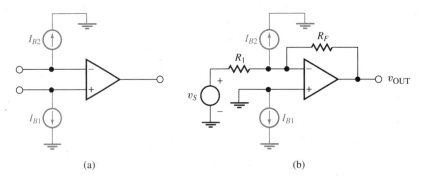

(a)   (b)

FIGURE 10.9 Current offset in an operational amplifier. (a) Model for the op-amp that gives representation to the dc bias currents that flow through the input terminals. This model consists of two current sources plus an op-amp similar to that of Fig. 8.2. (b) Use of this model in an inverting-amplifier circuit.

The dc input currents may give rise to additional offset of the output voltage, beyond that caused by V_{OS}. To observe the effects of the dc input currents, the op-amp model shown in Fig. 10.9(a) can be used. Here the op-amp is represented as an input current-free op-amp plus two current sources that represent the bias currents that flow into the real op-amp's input terminals. If the op-amp is used in an inverting-amplifier circuit, we have the circuit shown in Fig. 10.9(b). Let us calculate the output voltage of this circuit arising from I_{B1} and I_{B2}, when $v_s = 0$. Source I_{B1} has no effect, since its current flows directly to ground. However, a current equal to I_{B2} must enter the $(-)$ node through some path. No current flows through R_1; we know this because $v_{(-)} = v_{(+)} = 0$, and we have set $v_S = 0$. Thus there is no voltage across R_1 and hence no current flows through it. We may then conclude that a current equal to I_{B2} enters the $(-)$ node through R_F. Since $v_{(-)} = 0$, we must have $v_{OUT}/R_F = I_{B2}$. Thus the output voltage resulting from the bias currents in this case is

$$\Delta v_{OUT} = I_{B2}R_F \qquad (10.4)$$

If $I_{B2} = 1$ μA and $R_F = 1,000$ Ω, $\Delta v_{OUT} = 1$ mV.

We note that voltage offset and current offset are two separate effects. In the most general case the two offset current sources of Fig. 10.9 and the offset voltage source of Fig. 10.8 must all be considered to be present simultaneously. Their offsets add (algebraically; cancellation is possible).

In applications where precision is required, it may be useful to reduce offset with a special nulling circuit. Such circuits can take on various forms, depending on the rest of the op-amp circuit. A nulling circuit for an inverting amplifier is shown in Fig. 10.10(a). The adjustable current source cancels the effect of I_{B2} by removing the need for any current to flow through R_F. Additional nulling current of either sign can be added to balance out the effects of input voltage offset. A way in which the adjustable current source can be constructed is shown in Fig. 10.10(b). If R_2 and R_3 are chosen to be large, the Norton equivalent of the subcircuit shown in Fig. 10.10(b) is a current

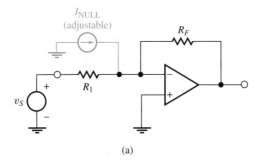

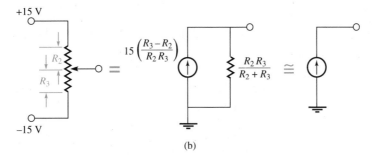

FIGURE 10.10   (a) A nulling circuit for canceling out offset in an inverting amplifier. A means of obtaining the adjustable nulling current source is shown in (b).

source in parallel with a large Norton resistance. The latter is not desired, but it has little effect on the circuit except to reduce the effective value of R_i (the op-amp's input resistance) with which it is in parallel. The arrow in Fig. 10.10(b) indicates an adjustable contact on the resistor that can be moved up or down, to adjust R_2 and R_3.*
In practice, final adjustment of R_2 and R_3 would be made while watching a voltmeter connected to the op-amp output; thus exact balance is obtained by experiment.

EXAMPLE 10.4

An op-amp used in an inverting amplifier has an input offset voltage V_{OS} of approximately 5 mV. We desire to null the output voltage with the circuit shown in Fig. 10.10. Calculate approximate values of R_2 and R_3. Assume that the input bias currents are negligible, and take $R_F = 2$ kΩ and $R_1 = 1$ kΩ.

---

*Such an arrangement is called a *potentiometer*. Potentiometers (or "pots") are widely used as adjustable voltage dividers in volume controls and so forth.

SOLUTION

To see the effect of V_{OS}, let us substitute the model of Fig. 10.8(a). The result is given in Fig. 10.11. (Since we are now concerned with the output voltage in the absence of input signal, v_S has been set to zero.) Writing a node equation for the $(-)$ node, where, according to the ideal op-amp rules, the voltage is V_{OS}, we have

$$I_{NULL} - \frac{V_{OS}}{R_1} + \frac{v_{OUT} - V_{OS}}{R_F} = 0$$

We wish to find I_{NULL} when $v_{OUT} = 0$. Setting v_{OUT} to zero in the above equation, we obtain

$$I_{NULL} = V_{OS}\left(\frac{1}{R_1} + \frac{1}{R_F}\right) = V_{OS}\frac{R_1 + R_F}{R_1 R_F}$$

Using the given values for V_{OS}, R_1, and R_F, we have $I_{NULL} \cong 7.5 \times 10^{-6}$ A.

The value of the Norton equivalent current source in Fig. 10.10(b) is to be equal to I_{NULL}. Thus

$$15\frac{R_3 - R_2}{R_2 R_3} = 7.5 \times 10^{-6} \text{ A}$$

This equation by itself does not determine the values of R_2 and R_3; there is still an element of choice possible. However, we should make both R_2 and R_3 as large as the above equation will allow, so that the Norton resistance of the combination will be as large as possible. A possible choice would be $R_3 = 10$ MΩ. Solving the above equation for R_2, we find $R_2 = 1.7$ MΩ.

In practice, one might choose for R_2 and R_3 a potentiometer whose total resistance $R_2 + R_3$ has a fixed value of, say, 11.7 MΩ. One would then obtain exact null by experiment, adjusting the movable contact on the potentiometer until zero output voltage is observed.

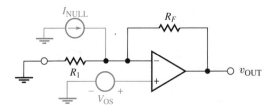

FIGURE 10.11

Although a nulling circuit can be used to set the output voltage to zero at a given temperature, changes in temperature will cause the input offset voltage and bias currents to change, thus spoiling the balance. The resulting change in the dc output voltage with temperature is known as *drift*. To aid in estimating the drift, manufacturers often specify the rate of variation of the offset voltage and bias currents per degree of change in temperature. Some typical values are as follows: for input offset voltage, 5 μV/°C; for input offset current, 4 nA/°C.

MAXIMUM SLEW RATE

As a practical matter, there is a limitation on the speed at which the output voltage of an op-amp can change. An op-amp circuit, basically, is a system in which the output takes on a value determined by an input instruction. Such a system, in general, is called a *servo system*. If the input instruction is changed suddenly, the output will require a finite time to reach its new value. This process is known as *slewing*. The maximum rate at which the output can move to its new value is called the *maximum slew rate*.

A familiar example of a servo system is a TV antenna rotator. On top of the TV set will be a box with a control that can be pointed at any compass direction. Suppose initially the control and antenna are both pointed "north." If the control is suddenly moved to "east," the antenna will begin to crank around to this new position, but since the motor has a maximum speed, the antenna will turn at some constant maximum rate. If 15 sec is required for the antenna to slew 90°, we would say the maximum slew rate is 90°/15 = 6°/sec.

Although op-amps respond much more quickly, they exhibit the same kind of behavior. Figure 10.12 shows a typical output response for a voltage-follower circuit with a step-voltage input. We see that the output requires about 5 μsec to slew from 0 to 5 V; thus the maximum slew rate is about 10^6 V/sec. Note that the curve of v_{OUT} versus time is not an exponential function of the sort seen in the transient response of RC

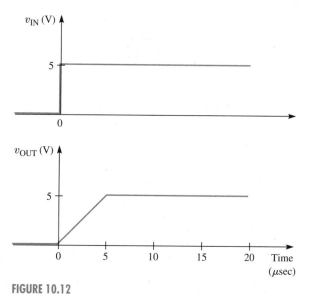

FIGURE 10.12

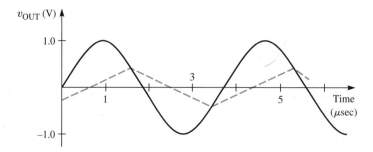

FIGURE 10.13   Solid line: undistorted output with amplitude 1 V, f = 278 kHz. Dashed line: distorted output that results if MSR is reduced to 0.5 V/μsec.

circuits. Here we are dealing with a constant maximum rate of change of v_{OUT}, and the output curve is composed of nearly straight-line segments.*

If the input voltage is a sinusoid, instead of a step function, the situation is slightly different. Let us consider a circuit with voltage gain A', and let the input voltage be $v_{IN} = V_0 \cos \omega t$ volts. If the slew rate were unlimited, the output would be $v_{OUT} = A' v_{IN}$, and dv_{OUT}/dt would be $A'(dv_{IN}/dt) = -A'\omega V_0 \sin \omega t$ volts per second. Clearly dv_{OUT}/dt is itself time-varying; its absolute value reaches maximum at times $\omega t = \pi/2$, $3\pi/2$, etc. This maximum value is $A'\omega V_0$. If the desired dv_{OUT}/dt never exceeds the maximum slew rate, the output will be a nearly perfect multiple of the input. However, suppose the product $A'\omega V_0$ is increased. The slew rate limitation begins to have effect when

$$A'\omega V_0 \geq \left(\frac{dv_{OUT}}{dt}\right)_{MAX} = \text{MSR} \qquad (10.5)$$

where MSR is the maximum slew rate for this op-amp in volts per second. Notice that MSR can be exceeded if either A', ω, or V_0 is increased. The effect of increasing the product $A'\omega V_0$ is illustrated in Fig. 10.13. In this figure the solid line represents the output for the case $A'V_0 = 1$ V and $f = 278$ kHz, assuming that MSR is so large that it has no effect. For comparison, the dashed line shows the output expected with MSR $= 0.5$ V/μsec. In this case the op-amp is slewing all the time, first trying to catch up with the solid line in the uphill direction, then chasing it downhill. The slope of each dashed line is equal to $\pm$MSR; thus by observing this output on an oscilloscope, one can measure MSR. The amplitude of the output is reduced because v_{OUT} never has time to reach the maximum before it has to reverse direction. From the figure we also observe that the output maxima are delayed in time with respect to the input maxima.

---

*Exponential behavior is characteristic of linear circuits. Although the op-amp is usually considered to be a linear device, when it is slewing it is operating in an abnormal, nonlinear regime.

It should be understood that, although there are some similarities, the slew-rate limitation on op-amp performance is a *different* limitation than that imposed by the closed-loop frequency response. The two effects arise from different causes. The closed-loop frequency response arises from the sinusoidal steady-state response of the entire op-amp. Slew rate, on the other hand, is limited by the maximum currents that can be drawn through the high-power output stage.

In both cases, the observed effect would be that the amplification of sinusoidal signals is less than expected, but in a given practical situation, performance of the op-amp might be restricted by either effect. Which one is responsible depends on the conditions of operation. A main point to notice is that, with sinusoidal signals, slew rate increases both with amplitude and with frequency, as is seen from Eq. (10.5). Thus, with signals of high frequency and low amplitude, amplification might fall off because of the closed-loop frequency response, while the slew-rate limitation might have no effect. On the other hand, at lower frequencies the frequency response limitation might have no effect, but the slew rate limitation might come into play when the amplitude became sufficiently large. If one observes the output of the circuit with an oscilloscope, the cause of reduced gain should become apparent. When frequency response is the limitation, the amplitude of the output waveform is reduced, but it still remains a perfect sinusoid. When the slew rate is the problem, the output waveform is distorted from a sinusoid into a triangular wave like that shown in Fig. 10.13.

10.5 Application: High-Fidelity Audio

Almost every reader will have some experience with high-fidelity music systems. These familiar systems, containing tape players, record players, amplifiers, and loudspeakers, are based almost entirely on analog signals. (Only the most recent additions to the technology, such as compact disc players, make use of digital techniques. For a related discussion, see Section 11.4.)

Home music systems originated in the mechanical phonographs of Thomas Edison. These first phonographs were not electrical at all; they were entirely mechanical. The record was turned by a wind-up spring. The sound vibrations were produced by a needle riding side-to-side in a wiggly groove on the record. This motion was then coupled mechanically to a vibrating diaphragm located at the small end of a cone-shaped megaphone. The machine depicted in the famous "His Master's Voice" trademark, with the dog listening attentively, is an example of this type. No doubt the dog *had* to be attentive, as the sound was weak and muddled. The listener could make out the words being spoken, or what music was being played, but would never be fooled into thinking that the actual person or orchestra was really present in the room. The reproduced sound was not very faithful to the original, so we would say this was a "low-fidelity" system.

With the advent of electronics, amplifiers began to be used. The phonograph needle (later called a stylus) was little changed, but instead of driving the sound-producing horn mechanically, it was connected to a device known as a *transducer* (the "pickup cartridge"). The function of this device was to produce a voltage proportional to the mechanical displacement of the needle—a classic example of an analog signal. The signal was then amplified (i.e., multiplied by a constant greater than one) and applied to another transducer, called a loudspeaker, which performed a function opposite to that of the pickup. That is, it created mechanical vibrations proportional to the amplified signal voltage, and these vibrations were heard in the room as sound. This kind of system was a great improvement on the all-mechanical dog-and-phonograph type, because now the reproduced sound could be made as loud as one wished. The quality of the sound was better, too.

The old mechanical phonograph, improved with an amplifier and loudspeaker, has been with us without much change from the 1920s right up to today. However, around 1950 people began to be dissatisfied. They were annoyed that although music from phonographs was perfectly understandable and recognizable, and you could even dance to it, it just didn't sound much like the real thing. The loud parts weren't very loud, the tonal qualities of musical instruments didn't sound right, and the whole sound was sunk in a kind of gooey fog that spoiled its clarity and left it blurred. The electronic technology of 1950 still used vacuum tubes, instead of transistors, but it had advanced a good deal, and engineers were ready to attack the problem of recorded music that didn't sound as it should. The result was the family of electronic techniques known as "high-fidelity audio."

The problems of the old low-fidelity audio systems were common ones in analog electronics. The more important failings were these:

POOR FREQUENCY RESPONSE

Audible sounds contain frequencies from about 15 Hz up to about 20,000 Hz. Clearly a music system should not exclude any of these frequencies. If, for example, the bandwidth of the system extended only from a few hundred Hz to 2,000 Hz, low organ notes and tinkling cymbals would be lost. (This sort of bandwidth is actually typical of telephone systems. Think of how a telephone sounds.) Actually, obtaining the necessary bandwidth is a cinch, with modern amplifiers. However, there is also the problem of making the amplification *flat*, that is, equal at all frequencies. It turns out that the ear is extremely sensitive to rough frequency response: When some frequencies are amplified more than others, the tonal quality is changed, making, for instance, a piano sound like a harpsichord. Still more subtle is the problem of phase delays, or phase shifts. If some frequencies move through the amplifier faster than others, they reach the listener's ear sooner, which also causes a deterioration of the sound. Amplifiers with nearly ideal flat frequency response are now easy to build, thanks to modern circuit design. However, the problem is still very much with us in electromechanical devices like loudspeakers, with their mechanical resonances. Furthermore, one has to live with the acoustic resonances of the listening room itself!

NONLINEARITY

An amplifier is said to be *linear* if its output is exactly linearly proportional to its input. In this case the input voltage is simply multiplied by a constant, and as a result no new frequencies are added to those present in the original signal. If the original signal happened to be

$$A \sin \omega_1 + B \sin \omega_2$$

thus containing only the two frequencies ω_1 and ω_2, it would emerge from a perfectly linear amplifier as

$$K\left(A \sin \omega_1 + B \sin \omega_2\right)$$

that is, multiplied by a constant, K, but containing exactly the same frequencies as before it was amplified. Suppose, however, that the amplification were not linear, and contained a square-law term: $v_{OUT} = K v_{IN} + K' v_{IN}^2$. It is easily shown that in that case the output will contain not only the original frequencies ω_1 and ω_2 but also the *harmonic* frequencies $2\omega_1$ and $2\omega_2$, and in addition the *intermodulation* frequencies $\omega_1 - \omega_2$ and $\omega_1 + \omega_2$. Injection of harmonic frequencies (frequencies that are multiples of those originally present) is called *harmonic distortion*, while injection of sum and difference frequencies is known as *intermodulation distortion*. Both types of distortion will occur whenever the system is anything but ideally linear.

Harmonic distortion is only mildly offensive, because the original sounds being amplified probably contained harmonics anyway. However, the new frequencies introduced by intermodulation are not harmonically related to the original sounds, and generally sound awful. To experience the horrors of intermodulation distortion, try the following experiment: Listen to a small, cheap transistor radio at low volume; it will probably sound tinny, but fairly clear. Now turn up the volume to the point at which it ceases to get any louder, and note the quality of the sound: It will probably be so muddled that it is hard to make out the words. This is because the circuit is fairly linear at low output levels, but becomes overstressed and nonlinear when made to produce large amounts of output power.

Fortunately, we have at our disposal a wonderful technique for obtaining linearity: feedback! Negative feedback is almost a "magic bullet" for analog systems, because, as we have seen, it corrects the errors in most of the system. The only errors that remain are those arising from the feedback elements themselves (like the feedback resistor, or the thermostat in Fig. 9.5). Since the feedback elements are usually very simple, one can achieve almost perfect linearity. (Flat frequency response is obtained in this way, too.) It is impossible to eliminate nonlinearity entirely, because a perfect amplifier cannot be built. However, in modern feedback amplifiers, nonlinearity at rated power is reduced to negligible levels.

POOR DYNAMIC RANGE

The term "dynamic range" refers to the range of loudness the system can produce. The loudest sounds are usually limited by the onset of distortion; the quietest sounds allowed are limited by the presence of noise and other extraneous signals such as 60 Hz power-line "hum." To avoid running into either of these limits, system designers used to introduce *compression*; that is, the quiet sounds were boosted up and the loud sounds were reduced, so all sounds emerge at nearly the same level. When compression is used, one gets an effect that is not so much unpleasant as boring; all the sounds, whether pianissimo or fortissimo, emerge sounding equally loud. Compression is not needed by modern amplifiers, which have large dynamic range. However, it remains an unsolved and unsolvable requirement of phonograph records themselves. Home music with full dynamic range has only recently been achieved, with the advent of digital recording, as in the compact disc.

The present state of high-fidelity music reproduction is highly advanced. The electronic components, in particular, have almost achieved perfection, and at remarkably low cost. However, the electromechanical components, such as the speakers, are still far from perfect. One has only to listen to a recording of piano music—the best recording available, using the best hi-fi system—close one's eyes, and ask whether this sound could possibly be an actual piano. There are still exciting major problems here, to be solved by electrical and mechanical engineers, working together!

POINTS TO REMEMBER

- An amplifier's output power is subject to a limit. Often this limit is determined by the amount of waste heat that can be dissipated by the amplifier. The power being dissipated can be calculated from a knowledge of the voltages at, and currents through, all the terminals of the amplifier. The power-supply terminals must be included!

- No amplifier can amplify signals of all frequencies. The range of frequencies over which its amplification is close to its maximum value is known as the amplifier's *bandwidth*. The absolute value of the amplification and the associated phase shift, graphed as functions of frequency, represent the amplifier's *frequency response*.

- Some types of amplifiers, distinguished by their typical frequency responses, are audio amplifiers; video amplifiers; and bandpass amplifiers. The open-loop bandwidth of an operational amplifier is usually very small, but the closed-loop bandwidth is much larger.

- The lowest signal levels at which analog blocks can be used are limited by the inevitable presence of random signals known as *noise*.

- Sources of noise are included in models by adding independent noise voltage and noise current sources. Because of the random nature of their values, the values of these sources are specified by stating the mean-square voltage or mean-square current.

- Electronic noise is usually "white," meaning that there is a certain amount of noise per unit bandwidth. Therefore the amount of noise present depends on the bandwidth of the circuit.

- Noise in amplifiers is modeled by adding equivalent noise voltage and current sources at the input. Noise is also produced by resistors; this noise is known as *Johnson noise*.

- Operational amplifiers are "dc amplifiers," meaning that their bandwidth extends downward to frequencies including zero. In general, however, an op-amp circuit will give a nonzero output even when the input voltage is zero—a condition known as "offset." Offset, which can be considered as zero-frequency noise, is modeled in a way similar to that used with noise: with dc voltage and current sources at the op-amp input. Unlike ac noise, however, offset can be corrected with a suitable compensating circuit.

- An op-amp is in actuality a kind of servo system: It is a device that constructs an output voltage proportional to the input voltage. When the input voltage is changed, the output voltage has to move to its new value; this change is known as *slewing*.

- There is a maximum rate (given in volts per second) at which the output voltage of an op-amp can slew; this is known as the amplifier's maximum slew rate. If the input is changed suddenly, the output cannot move to its correct new value instantaneously; it can only move toward the correct new value as fast as it can, that is, at the maximum slew rate. This gives rise to abnormal operation, in which the output is momentarily not related to the input by the customary $v_{OUT} = Av_{IN}$. Generally, the result is distortion of the input waveform.

- With sinusoidal signals, the slew rate required at the output is proportional to both the input amplitude and the input frequency. Increasing either of these excessively can cause the slew rate to reach its limit. The result is usually distortion of the sinusoidal output into a triangular form.

- The limitation imposed by the maximum slew rate is a different limitation than that imposed by the closed-loop frequency response. The frequency response limitation is not related to amplitude, and comes into play only when frequency is excessive. Sinusoidal signals with frequency too high for the closed-loop frequency response usually experience reduced amplification, but are not distorted out of their sinusoidal form.

- Home music systems involve primarily analog technology. Extremely accurate performance is required in these systems, if high-fidelity music reproduction is to be achieved. Faults to be avoided in such systems include (a) poor frequency response; (b) nonlinearity, which produces *harmonic distortion* and *intermodulation distortion*; and (c) poor dynamic range, which gives rise to *compression*. Electronic components of extremely high quality are now available, but electromechanical components, such as loudspeakers, are still far from perfect.

PROBLEMS

Section 10.1

10.1 Find the power being dissipated (converted to heat) in the block shown in Fig. 10.14.

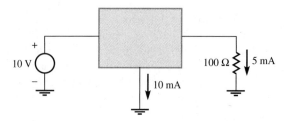

FIGURE 10.14

10.2 Find the power being dissipated in the block of Fig. 10.15. Assume $V_{OUT} = 10V_{IN}$, $V_{IN} = 1$ V, and $I_1 = 3V_{IN}$ (where V_{IN} is in volts and I_1 in amperes).

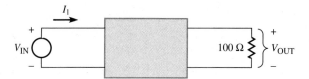

FIGURE 10.15

*10.3 Find the time-averaged power being dissipated in the case of Problem 10.2, if the source V_{IN} is changed to $2 \cos \omega t$ volts.

*10.4 The op-amp of Fig. 10.16 has power-supply connections at $+15$ V and -15 V, as shown. Estimate, approximately, currents I_1 through I_5. Explain your reasoning. Any currents less than 1 mA can be considered zero for the purposes of this problem.

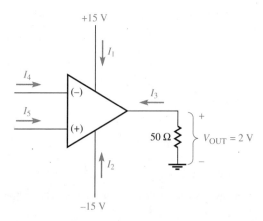

FIGURE 10.16

10.5 Suppose you plan to use the op-amp of Fig. 10.7. What power-supply voltages should be applied? To which pins? Sketch the op-amp package as seen from the top and indicate where the power-supply connections should be made.

*10.6 For the case of Example 10.1, sketch the power dissipation as a function of v_{OUT} for the range $0 < v_{OUT} < V_{ps}$. For what value of v_{OUT} is the power dissipation maximum?

*10.7 What is the smallest value of R_L that could be used in the circuit of Example 10.1, so that no matter what value v_{OUT} takes in the range $0 < v_{OUT} < 15$ V, the power dissipation in the amplifier never exceeds 50 mW? Assume $R_i = \infty$, $R_o = 0$, $A = 10$, and $V_{ps} = 15$ V.

**10.8 Suppose, in Example 10.1, that $v_{IN} = 0.75(1 - \cos \omega t)$ V. Find the time-averaged power dissipated in the amplifier.

Section 10.2

*10.9 In the circuit of Fig. 10.17, the amplifier is the bandpass amplifier shown in Fig. 10.2(c). Assume $R_i \cong \infty$, $R_1 = 50$ Ω, and $C = 4 \times 10^{-13}$ F. We consider the two-port enclosed by the dashed line to be a new amplifier block. Sketch $|\mathbf{A}'|$ and θ_A' as functions of frequency over the range 9.6 to 10.4 GHz.

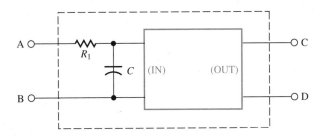

FIGURE 10.17

*10.10 The amplifier whose characteristics are shown in Fig. 10.2(a) is used in the circuit of Fig. 10.18. Assume $R_i = \infty$. Consider A,B to be the input terminals and C,D to be the output terminals of a new amplifier block and sketch $|\mathbf{A}'|$ and θ_A' as functions of frequency.

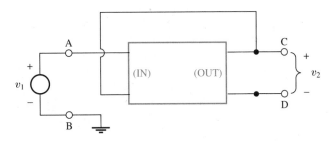

FIGURE 10.18

*10.11 Repeat Problem 10.10 using the bandpass amplifier of Fig. 10.2(c).

10.12 Let us define a *3 dB bandwidth* as the range of frequencies over which the value of $|A(f)|$ is within 3 dB of its maximum (that is, greater than 70.7% of its maximum value). Find the bandwidth of each of the amplifiers shown in Fig. 10.2.

*10.13 The amplifier of Fig. 10.2(d) is used in the circuit of Fig. 10.18. Assume $R_i \cong \infty$, $R_o \cong 0$. Find the 3 dB bandwidth of the resulting circuit. (Extrapolate the curves to higher frequencies as necessary.) Is it larger than the bandwidth of the amplifier by itself?

Section 10.3

10.14 A random noise voltage $v_N(t)$ is connected across a resistor R.
  a. What is the instantaneous current?
  b. Find the instantaneous power, expressed in terms of $v_N(t)$ and R.
  c. What is the time-averaged power, expressed in terms of $<v_N^2>$?
  d. What is the time-averaged power, expressed as the limit of an integral of $v_N^2(t)$?

10.15 When the input terminal is grounded, the output of the amplifier shown in Fig. 10.19 exhibits random noise of about 3 mV. What is the smallest signal input (approximately) that will give an output signal-to-noise ratio of 10:1? Let $R_F = 5,000\ \Omega$ and $R_1 = 800\ \Omega$.

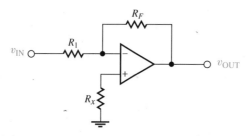

FIGURE 10.19

10.16 a. The noisy amplifier of Fig. 10.4 has finite input resistance R_i. Suppose its input terminal is grounded. Does the noise voltage source or the current source contribute more to output noise?
  b. What if the input terminal is open-circuited?

*10.17 The noisy amplifier of Fig. 10.4 has input resistance R_i and open-circuit voltage amplification A. A Thévenin signal source V_S, R_S is connected to its input. Let $<v_N^2> = R_N^2 <i_N^2>$, where R_N is a given constant. Find $<v_{OUT}^2>$, the mean-square noise voltage that appears at the amplifier's open-circuited output terminals.

**10.18 Find the signal-to-noise ratio (defined as the ratio of the mean-square signal voltage to the mean-square noise voltage) at the output for the case of problem 10.17.

*10.19 A resistor R_1, at temperature T, is represented by the noise model of Fig. 10.6. It is connected in parallel with another resistor, R_2, which is refrigerated to zero temperature. Find the power that is transferred from the hot resistor to the cold one.

**10.20 Two resistors, R_1 and R_2, both at temperature T, are connected in parallel.
   a. Explain why it is logically necessary that the noise power going from R_1 to R_2 must be equal to the noise power going from R_2 to R_1.
   b. Show using Eq. (10.3) that statement (a) is correct.

**10.21 A sinusoidal signal source is represented by its Thévenin phasor voltage V_S and Thévenin resistance R_S. It is connected to an amplifier with input resistance R_i, as shown in Fig. 10.20. The source resistor R_S produces thermal noise, but we shall assume that all parts of the amplifier are completely noiseless. The maximum power available from the signal source, $|V_S|^2/8R_S$ (see Problem 7.43) is equal to P_o, a fixed given value. Show that the output signal-to-noise ratio (ratio of signal power to noise power at the amplifier output) is equal to P_o/kTB.

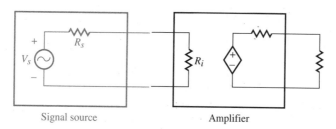

Signal source   Amplifier

FIGURE 10.20

*10.22 Suppose the signal source in Fig. 10.20 is a satellite TV antenna at an effective temperature of 300° K. The amplifier is noiseless and has bandwidth 6 MHz (sufficient for one TV channel). Use the result of problem 10.21 to find the minimum signal power that must be received from the satellite to give a signal-to-noise ratio of 10.

10.23 For the circuit of Fig. 4.27, find the error introduced in v_{OUT} if the op-amp has an input offset voltage of 2 mV. (Assume the input bias currents are zero.) Let $R_1 = R_F = 2$ kΩ. The sign of the offset voltage is as shown in Fig. 10.8.

10.24 Repeat Problem 10.23, assuming that, in addition to the voltage offset, each op-amp input has a bias current of 1 μA. The signs of the bias currents are as shown in Fig. 10.9.

10.25 For the circuit of Fig. 10.19, assume that voltage offset can be neglected and that the two input bias currents are equal. Let $R_1 = R_F = 100$ kΩ. Find the value of R_x that eliminates error in v_{OUT} arising from the bias currents.

10.26 An op-amp circuit has a voltage gain of 10. The input voltage changes suddenly from 0 to 1 volt. The output voltage increases *linearly* (not exponentially) and reaches 10 volts after a delay of 10 microseconds.
   a. Sketch the output voltage versus time.
   b. What is the maximum slew rate of this amplifier?

10.27 An amplifier with voltage gain of 10 has a sinusoidal input $v_{IN} = 0.5 \cos \omega t$ V.
   a. Sketch $v_{OUT}(t)$.
   b. Find the slew rate dv_{OUT}/dt at time $t = 0$.
   c. Find the slew rate when $\omega t = \pi/4$.
   d. What is the largest value reached by the slew rate at any time?

*10.28 For the circuit of Fig. 10.19, let $R_F = 3 \text{ k}\Omega$ and $R_1 = 900 \, \Omega$. The input voltage is $v_{IN} = 1.5 \cos \omega t$ volts, and the maximum slew rate of the op-amp is 10^6 V/sec. Find the largest input frequency (in Hz) that can be used without exceeding the maximum slew rate.

*10.29 In the circuit of Fig. 10.21, capacitor C is initially uncharged. The switch is open until time $t = 0$, when it is closed. The ideal-op-amp technique may be used. Let $R = 1,000 \, \Omega$, $C = 10^{-6}$ F, $\omega = 2,000$ rad/sec, and $V_0 = 10$ V.
   a. Make a sketch of v_{OUT} versus time.
   b. Find the smallest maximum slew rate that the op-amp can have if the slew rate limitation is not to affect v_{OUT}.

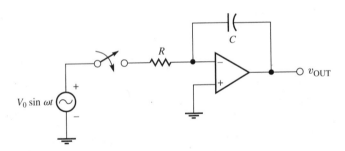

FIGURE 10.21

**10.30 In Fig. 10.22(a) the op-amp is ideal except that it has a maximum slew rate of 10^6 V/sec. The input voltage is as shown in Fig. 10.22(b). Sketch v_{OUT} as a function of time. Provide a time scale and show maximum and minimum values of v_{OUT}. Assume that $v_{IN} = 0$ for all $t < 0$.

*10.31 For the circuit and input of Fig. 10.22, assume the op-amp is ideal. (No slew-rate limitation applies.) Let the op-amp power supply voltages be ± 15 V. Since the output voltage is always positive, assume that no current flows through the $(-)$ power-supply connection. Find the time-average power converted to heat inside the op-amp.

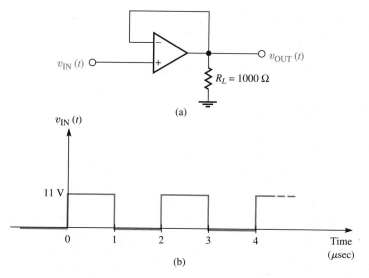

FIGURE 10.22

10.32 Suppose that for the circuit of Fig. 10.22(a) v_{IN} has the constant value of 5 V. When R_L is reduced from ∞ to 1,000 Ω, v_{OUT} is observed to decrease by 0.01 mV. What is R'_o, the output resistance of the op-amp circuit?

Section 10.5

*10.33 An imperfect amplifier has output given by

$$v_{OUT} = 10v_{IN} + \frac{v_{IN}^2}{V_o}$$

where $V_o = 10$ V. A sinusoidal signal $v_{IN} = 1.5 \cos \omega t$ volts is applied to the input. Show (by using trigonometric identities) that the output is a sum of sinusoids, with frequencies ω, 2ω, and zero. Find the amplitudes of these three sinusoids.

*10.34 The amplifier of Problem 10.33 is given an input $v_{IN} = 1.5 \cos \omega_1 t + 1.5 \cos \omega_2 t$ volts. Show that the output contains a sinusoid of frequency $\omega_2 - \omega_1$, and find its amplitude.

DIGITAL SIGNALS AND THEIR USES

CHAPTER 11

DIGITAL BUILDING BLOCKS

One of the most important developments taking place in electrical engineering is the trend toward digital technology. More and more, information is being stored and processed in digital form. Before proceeding further here, the reader may wish to briefly return to Section 0.3 to review the terms "analog signal" and "digital signal" and remember how they differ.

Because digital signals have a particular form, distinctive digital building blocks are used to process them. This chapter will introduce the major digital building blocks and the fundamentals of their use. Chapter 12 will then continue to their uses in digital systems.

Digital systems use numbers to represent information. This makes them especially suitable for applications where the information begins and ends in numerical form, as in a computer. However, one should not think of digital blocks as being useful only in computers. More and more they are being used in such traditionally "analog" applications as control systems and telephones. The primary advantage of digital technology has to do with the simplicity, cheapness, and versatility of the building blocks. Digital systems are built by repeating a very few simple blocks. The approach is powerful because the blocks are cheap, in integrated-circuit form, and can be repeated thousands of times. On the other hand, the blocks are cheap because they are mass-produced in enormous numbers. These inexpensive elements are like tiny brain cells. Like individual brain cells they each perform only simple, elementary functions, but when large numbers of them are connected in ingenious ways they form powerful systems. The trend is for more and more of these "brain cells" to be compressed into integrated-circuit chips, as the industry has progressed from SSI (small-scale integration) to MSI (medium-scale integration) to LSI (large-scale integration) and on to VLSI (very-large-scale integration). There is no telling how far this process can go. As the number of available "brain cells" increases, the problem of how they are to be connected becomes more and more complex. That problem is the basis of the subject known as computer science.

11.1 Digital Signals

Like an analog signal, a digital signal is a voltage that varies in time. However, in a digital signal the voltage must always be within one of two voltage ranges. In the usual case the possible spectrum of voltages is divided as shown in Fig. 11.1. Signal voltages are allowed to be anywhere in the "high" range or anywhere in the "low" range, but they should not be found outside the two ranges. A signal voltage that is neither in the high nor in the low range indicates a malfunction.

The actual voltages that bound the two ranges vary from one situation to another. As an example, the low range might lie between 0.0 and 1.0 V and the high range between 4.0 and 5.0 V. A voltage that lies in the low range can represent the number "zero." A voltage in the high range would then represent the number "one." We note that *any* voltage in the high range has the same significance: 4.3 V represents "one," and 4.9 V also represents "one." This is different from the analog case, where different voltages usually have different meanings. The digital signal has an advantage here; small errors or noise can disturb the signal voltage slightly, but (unless it is pushed out of the allowed range) no information is lost. The assignment just made, where "low" voltage means "zero" and high voltage means "one," is known as *positive logic*. The

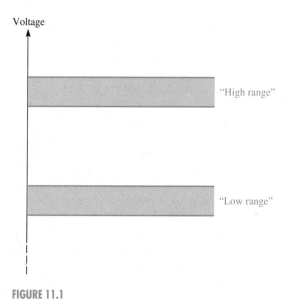

FIGURE 11.1

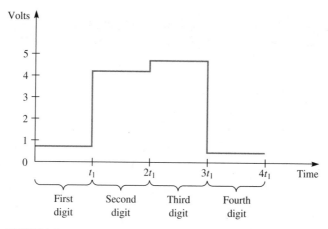

FIGURE 11.2

opposite convention, in which "low" voltage means "one," is clearly also possible; this is called *negative logic*. The "zeros" and "ones" represented by signal voltages are *not* voltages; they are special numbers known as *logical states*. To avoid confusion with other numbers, the logical states will be printed in this text in **boldface**, as **0** and **1**. Thus with the ranges 0 to 1 V and 4 to 5 V and positive logic, a voltage of 0.4 V would be said to represent **0**.

Typically a digital signal is a time-varying voltage resembling Fig. 11.2. We notice that the time axis is broken into intervals of length t_1. We can imagine that somewhere in our system is a clock that gives out pulses every t_1 seconds, so that the time intervals are well defined. The signal in the figure is "low" during the first time interval, then "high," "high," and "low." If we are using positive logic, this signal represents **0, 1, 1, 0**. Each **0** or **1** that is communicated is called a *bit* of information. ("Bit" is a contraction of "*bi*nary digi*t*.") Figure 11.2 illustrates the transmission of four bits of information. The rate at which information can be sent determines the speed of a digital system. Information rates are often stated in *bauds* (1 baud = 1 bit per second). Speeds of over 100 megabauds (10^8 bits per second) are possible.

EXAMPLE 11.1

What digits are represented by the signal of Fig. 11.2 if negative logic is used?

SOLUTION

The sequence of voltages reads low-high-high-low. In positive logic low stands for **0** and high for **1**; thus the signal stands for **0110**. In negative logic low stands for **1** and high for **0**, and hence the signal stands for **1001**.

EXAMPLE 11.2

Consider again the signal shown in Fig. 11.2. If for technical reasons the time occupied in transmitting each digit (t_1) is 2 μsec, what is the rate of information transmission? (That is, how many bits will be transferred per second?)

SOLUTION

Since the time per digit is 2×10^{-6} sec, one can transmit 5×10^5 digits per second. The information rate is 5×10^5 bits per second, or 500 kilobauds.

The binary digits **0** and **1** of course lend themselves readily to calculations in the binary number system.* For example, the four digits shown in Fig. 11.2 could represent the four-digit binary number 0110, which would be equal to the decimal number 6.

At some point A in a circuit, we can say the voltage has a certain value, say 4.3 V. We would write V_A = 4.3 V. Similarly, we can give the logic value corresponding to the voltage at A a name, for example **A**. **A** is then said to be a *switching variable*, whose value depends on the voltage at point A. Thus if V_A = 4.3 V, we would say **A** = **1**. To indicate that **A** is a switching variable it is printed in boldface type. Note that either **A** = **1** or else **A** = **0**; no other possibilities exist.

11.2 Logic Blocks

There are two basic types of digital building blocks. The first type includes the *combinational* or *logic* blocks; the second includes the *sequential* blocks known as "flip-flops." Logic blocks will be considered first, and sequential blocks in the following section.

Logic blocks normally have several inputs and one output. The voltages applied to each input must lie in one of the two allowed ranges and thus represent switching variables; for a logic block with three inputs, the three input variables might be **A**, **B**, and **C**. The output voltage will then also lie in one of the two allowed ranges and represents another variable we shall call **F**. Now the value of **F** (either **0** or **1**) will depend on the values of the three input variables. We would write **F** = **F**(**A**,**B**,**C**), meaning that **F** is a function of the three input variables **A**, **B**, and **C**. What function this is, is determined by what logic block is being used.

---

*Readers unacquainted with binary numbers may wish to look ahead to the discussion of number systems in Section 12.1.

Inputs		Output
A	**B**	**F**
0	**0**	**0**
0	**1**	**1**
1	**0**	**1**
1	**1**	**1**

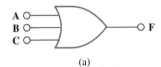

(a)   (b)

FIGURE 11.3   A two-input **OR** gate. (a) The symbol and (b) the truth table.

One important logic block is the **OR** gate, the symbol for which is shown in Fig. 11.3(a). The **OR** gate can have two or more inputs; in this example two are shown. The function of the **OR** gate is to execute the **OR** function, which is defined as follows: If and only if the signal at one or more of the inputs has the significance **1**, the output is **1**. This function may be expressed in tabular form, as shown in Fig. 11.3(b). A table of this kind is known as the table of combinations, or *truth table*, for the gate. The truth table has as many entries as there are possible combinations of inputs. We see that a truth table for a two-input **OR** gate has four entries. A three-input **OR** gate would have a truth table with eight entries, as illustrated in Fig. 11.4.*

(a)

Input	Input	Input	Output
A	**B**	**C**	**F**
0	**0**	**0**	**0**
0	**0**	**1**	**1**
0	**1**	**0**	**1**
0	**1**	**1**	**1**
1	**0**	**0**	**1**
1	**0**	**1**	**1**
1	**1**	**0**	**1**
1	**1**	**1**	**1**

(b)

FIGURE 11.4   (a) Symbol and (b) truth table for an **OR** gate with three inputs.

*Note that the eight possible combinations of inputs are written by counting from 000 to 111 in binary numbers.

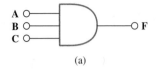

(a)

Input	Input	Input	Output
A	**B**	**C**	**F**
0	0	0	0
0	0	1	0
0	1	0	0
0	1	1	0
1	0	0	0
1	0	1	0
1	1	0	0
1	1	1	1

(b)

FIGURE 11.5   (a) Symbol and (b) truth table for an **AND** gate with three inputs.

A second important digital block is the **AND** gate. The function of this block is to execute the **AND** operation: if and only if *all* of the input voltages have the significance **1**, the output is **1**. The circuit symbol for the **AND** gate is shown in Fig. 11.5(a). Again the number of inputs is arbitrary; in this example three are shown. The truth table for the **AND** gate is shown in Fig. 11.5(b).

Another logical operation is the one known as *logical negation*, or **COMPLEMENT**. This operation simply inverts a switching variable; that is, if the input is **1**, the output is **0**, and vice versa. The **COMPLEMENT** operation is performed by a block called an *inverter*, shown with its truth table in Fig. 11.6. There is another way of indicating the **COMPLEMENT** operation in a diagram. A circle (like that on the "nose" of the inverter symbol) can be placed at any input or output terminal to indicate that the **COMPLEMENT** operation is performed on signals passing through it. For example, consider the block shown in Fig. 11.7. This is an **AND** operation followed by the

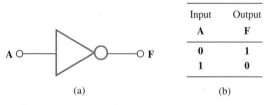

(a)

Input	Output
A	**F**
0	1
1	0

(b)

FIGURE 11.6   The **COMPLEMENT** operation. (a) Logic symbol. (b) Truth table.

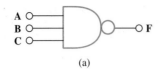

(a)

Input	Input	Input	Output
A	**B**	**C**	**F**
0	0	0	1
0	0	1	1
0	1	0	1
0	1	1	1
1	0	0	1
1	0	1	1
1	1	0	1
1	1	1	0

(b)

FIGURE 11.7   The **NAND** gate. (a) Logic symbol. (b) Truth table.

COMPLEMENT operation. The effect is that the output coming from the **AND** gate is inverted, so that the **F** column is the opposite of that shown in Fig. 11.5. The block shown in Fig. 11.7 is useful in its own right; it is known as a **NAND** gate. Similarly, the **OR** operation followed by **COMPLEMENT** is known as the **NOR** operation, shown in Fig. 11.8.

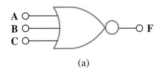

(a)

Input	Input	Input	Output
A	**B**	**C**	**F**
0	0	0	1
0	0	1	0
0	1	0	0
0	1	1	0
1	0	0	0
1	0	1	0
1	1	0	0
1	1	1	0

(b)

FIGURE 11.8   The **NOR** gate. (a) Logic symbol. (b) Truth table.

EXAMPLE 11.3

Construct the truth table for the logic block in Fig. 11.9.

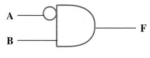

FIGURE 11.9

SOLUTION

In this case input **A** is complemented before it is applied to the **AND** gate. To construct the truth table, let us first introduce a new logical variable **A′**, which is the signal coming out of the inverter and entering the **AND** gate. Clearly **A′** is the complement of **A**; that is, when **A** = **0, A′** = **1**, and vice versa.

We begin the truth table by writing down the four possible inputs and the corresponding values of **A′**:

A	B	A′
0	0	1
0	1	1
1	0	0
1	1	0

Now **F** is found by performing the **AND** operation on **B** and **A′**. The only combination of inputs that gives an output of **1** is the combination that gives **B** = **1, A′** = **1**. Thus the completed truth table is

A	B	A′	F
0	0	1	0
0	1	1	1
1	0	0	0
1	1	0	0

Two other logic blocks that are sometimes used are the **EXCLUSIVE OR** (or **XOR** for short) and **COINCIDENCE** (also known as **EXCLUSIVE NOR**) gates. Conventional symbols for these blocks are shown in Fig. 11.10 and 11.11 together with

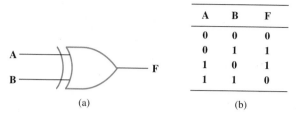

A	B	F
0	0	0
0	1	1
1	0	1
1	1	0

(a)   (b)

FIGURE 11.10   The **EXCLUSIVE OR** gate. (a) Logic symbol. (b) Truth table.

their truth tables. It is also possible to synthesize logic blocks by combining other blocks. For example, Fig. 11.12 shows a realization of an **EXCLUSIVE OR** gate. A convenient way of obtaining the truth table (which gives the functional dependence of the output **F** on the two inputs **A** and **B**) is to first calculate the intermediate variables **X** and **Y**, as shown. Once these are known, **F** is easy to find, using the truth table for the final **NOR** gate.

It is convenient to have a notation describing these logical operations. For example, if the logic variable **F** depends on two other variables **A** and **B** in the same way as the output of an **OR** gate (that is, if **F** obeys the truth table of Fig. 11.3), we write **F** = **A** + **B**. The "plus" sign here has nothing to do with ordinary addition. It is simply a conventional shorthand for representing the **OR** operation. Similarly, the **AND** operation is represented by the notation **F** = **A B**. (Sometimes this is written **F** = **A** · **B**.) The **COMPLEMENT** operation is represented by a bar over a symbol: If **F** is the complement of **A**, we write **F** = $\overline{\textbf{A}}$. The operations already introduced are summarized in Table 11.1 (see pg. 402).

If more than one operation is involved, combinations of logical expressions are used. For instance, if **F** is obtained by using an **AND** gate on inputs **A** and **B** and then complementing the result, we would write **F** = $\overline{\textbf{AB}}$. This is of course the same as the **NAND** operation, explaining the notation used for **NAND** in Table 11.1. If several operations are involved, there may be a question of the *order* in which the operations are to be performed. By convention, negation is performed first, then **AND**, and finally **OR**. A bar over an entire expression, as in $\overline{\textbf{AB}}$, means that the result of the operation

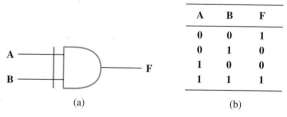

A	B	F
0	0	1
0	1	0
1	0	0
1	1	1

(a)   (b)

FIGURE 11.11   The **COINCIDENCE** (or **EXCLUSIVE NOR**) gate. (a) Logic symbol. (b) Truth table.

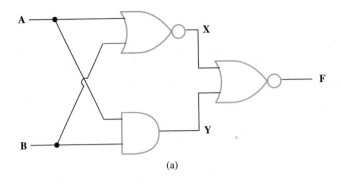

(a)

A	B	X	Y	F
0	0	1	0	0
0	1	0	0	1
1	0	0	0	1
1	1	0	1	0

(b)

FIGURE 11.12

(**AND**, in this case) is complemented. Operations in parentheses are performed before operations outside the parentheses. For example, $F = A\overline{(B + C)}$ means that **F** is obtained by first operating on **B** and **C** with an **OR** gate, then complementing the result, and then combining that result with **A** in an **AND** gate.

EXAMPLE 11.4

For the switching function $F = A(\overline{A} + B)$:

    a. Draw a corresponding set of logic blocks.

    b. Write the truth table.

SOLUTION

A suitable connection of logic blocks is shown in Fig. 11.13(a). Using the intermediate variables as indicated, we have the truth table shown in Fig. 11.13(b).

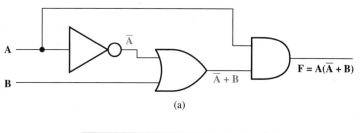

$$F = A(\overline{A} + B)$$

(a)

A	B	$\overline{A}$	$\overline{A} + B$	F
0	0	1	1	0
0	1	1	1	0
1	0	0	0	0
1	1	0	1	1

(b)

FIGURE 11.13

The reader should note that in general it is improper to draw a diagram in which the outputs of any two blocks are connected together. This is because with some inputs the output of one block might be zero and the other output one; drawing the two outputs connected together would then be equivalent to saying that the output voltage was both high and low at the same time. Two outputs should be combined by using them as two inputs to another gate.*

The logic notation just introduced is part of a mathematical system known as *Boolean algebra*. Using Boolean algebra, it can be determined when different-appearing logical expressions are in fact identical. We shall not go deeply into Boolean algebra here, but we can note that two logical expressions are equivalent if their truth tables are the same. For instance, in Example 11.4 we obtained the truth table for the expression $A(\overline{A} + B)$; we can now observe that this table is identical with the table for AB. Thus we can write $A(\overline{A} + B) = AB$. The practical importance of this result is that when we want to construct a circuit to perform the operation $A(\overline{A} + B)$ we do not need to use the three gates shown in Fig. 11.13(a); a single **AND** gate is sufficient. Many other theorems can be proved similarly, by showing that the truth tables for the expressions on either side of the equals sign are the same.

---

*In some technologies the outputs of two gates *can* be connected together and a "free" logic operation obtained—for example, the so-called "wired-**AND**." One can do this only if the internal wiring of the particular gates permits it. Since we are presently dealing with the *general* properties of logic blocks, we shall not consider this possibility.

TABLE 11.1   Logic Symbols and Notation

Symbol	Name	Notation	Truth Table		

	COMPLEMENT	$F = \overline{A}$	A	F	
			0	1	
			1	0	

A ————▷o———— F

	OR	$F = A + B$	A	B	F
			0	0	0
			0	1	1
			1	0	1
			1	1	1

	AND	$F = A B$	A	B	F
		(sometimes written $A \cdot B$)	0	0	0
			0	1	0
			1	0	0
			1	1	1

	NOR	$F = \overline{A + B}$	A	B	F
			0	0	1
			0	1	0
			1	0	0
			1	1	0

	NAND	$F = \overline{A B}$	A	B	F
			0	0	1
			0	1	1
			1	0	1
			1	1	0

	EXCLUSIVE OR	$F = A \oplus B$	A	B	F
			0	0	0
			0	1	1
			1	0	1
			1	1	0

	COINCIDENCE	$F = A \odot B$	A	B	F
			0	0	1
			0	1	0
			1	0	0
			1	1	1

EXAMPLE 11.5

Prove by comparing truth tables the theorem $A(B + C) = AB + AC$.

SOLUTION

We construct truth tables for the necessary intermediate variables and the two expressions to be compared. The result is shown in Fig. 11.14. Since the fifth and eighth columns of the table are the same, the theorem is proved. The direct representation of **AB + AC** would require three gates (two **AND**s and one **OR**), but from this theorem we see that two gates (one **OR** and one **AND**) can perform the same function.

A	B	C	B + C	A(B + C)	AB	AC	AB + AC
0	0	0	0	0	0	0	0
0	0	1	1	0	0	0	0
0	1	0	1	0	0	0	0
0	1	1	1	0	0	0	0
1	0	0	0	0	0	0	0
1	0	1	1	1	0	1	1
1	1	0	1	1	1	0	1
1	1	1	1	1	1	1	1

FIGURE 11.14

LOGIC SYNTHESIS

Thus far we have seen how to find the truth table for any combination of logic blocks. The inverse process, that of finding a connection of blocks to produce a given truth table, is known as *logic synthesis*. This is a large subject, of great interest to computer scientists; we can only touch upon it here. We can, however, give a method by which a suitable connection of gates can always be found, although it may not be the simplest one possible. Figure 11.15(a) shows a gate that gives an output of **1** only when **A** and **B** are both **1**. Figure 11.15(b) shows a gate that gives an output of **1** only if **A** is **1** and **B** is **0**. Figures 11.15(c) and (d) show the other two possible gates of this kind. Now suppose we wish to find a connection of gates that obeys the truth table shown in Fig. 11.16(a). This can be done by locating all rows in the truth table for which the output **F** is **1**. We then use the blocks from Fig. 11.15 that give outputs of **1** when the inputs of those rows are applied. In this example the second and fourth rows of the truth table give outputs of **1**; the corresponding blocks are those of Figs. 11.15(c) and (a), respec-

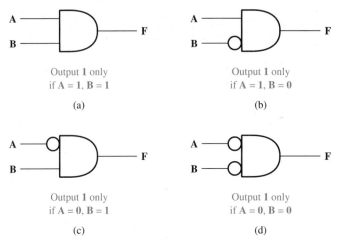

Output **1** only
if A = 1, B = 1

(a)

Output **1** only
if A = 1, B = 0

(b)

Output **1** only
if A = 0, B = 1

(c)

Output **1** only
if A = 0, B = 0

(d)

FIGURE 11.15

tively. We then combine the outputs of these blocks in an **OR** gate, as shown in Fig. 11.16(b). The result is a circuit that gives an output of **1** if inputs **0 1** *or* **1 1** are applied, which is what the given truth table requires. The circuit is said to be a *realization* of the truth table.

The method just described is known as the *sum-of-products method*. This name comes from the mathematical expression for the output. The Boolean expression for **F** in Fig. 11.16(b), for example, is $\mathbf{F} = \overline{\mathbf{A}}\mathbf{B} + \mathbf{A}\mathbf{B}$. (Of course, neither addition nor multiplication is actually implied by this logical expression.) The sum-of-products method can be used to realize any truth table, but it is inefficient; usually other realizations that require fewer blocks can be found. One way to simplify the realization is to use the "distributive law" proved in Example 11.5, plus the simple theorem $\mathbf{A} + \overline{\mathbf{A}} = \mathbf{1}$. Continuing with the example of Fig. 11.16, the expression $\mathbf{F} = \overline{\mathbf{A}}\mathbf{B} + \mathbf{A}\mathbf{B}$ can be "factored" by means of the distributive law into $\mathbf{F} = (\overline{\mathbf{A}} + \mathbf{A})\mathbf{B}$. But since

A	B	F
0	0	0
0	1	1
1	0	0
1	1	1

(a)

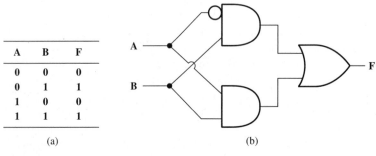

(b)

FIGURE 11.16

$\overline{\mathbf{A}} + \mathbf{A} = \mathbf{1}$, we see that $\mathbf{F} = \mathbf{B}$. Thus a sufficient realization is to simply connect input **B** to output **F**, leaving input **A** unconnected! The sum-of-products realization can usually be simplified by this "factoring" method. Finding minimal realizations is part of the large subject known as *logic design*.

EXAMPLE 11.6

Logic blocks can be used to perform binary arithmetic. For example, suppose we wish to add two binary variables, **A** and **B**. Let us call their sum (which may be a two-digit number) **CD**. Thus the summation reads

$$\begin{array}{r} \mathbf{A} \\ + \mathbf{B} \\ \hline \mathbf{CD} \end{array}$$

The necessary truth table is

Inputs		Outputs	
A	B	C	D
0	**0**	**0**	**0**
0	**1**	**0**	**1**
1	**0**	**0**	**1**
1	**1**	**1**	**0**

Using the sum-of-products method, **C** and **D** can be generated by the circuit shown in Fig. 11.17(a).

As is usually the case when the sum-of-products method is used, the resulting circuit is not minimal. The truth table can be realized with only three gates using the circuit shown in Fig. 11.17(b). The correctness of the circuit in Fig. 11.17(b) can be verified by constructing its truth table.

Addition of binary numbers is so common that the realization of the preceding truth table has become a building block in its own right. It is called a *binary half-adder* and is available in IC form, as shown in Fig. 11.17(c). It is called a "half"-adder because it is not quite capable of performing additions all by itself. The half-adder can add single-digit binary numbers, but in most systems multidigit numbers must be added. For instance, consider the addition

$$\begin{array}{r} \mathbf{WA} \\ + \mathbf{XB} \\ \hline \mathbf{YZD} \end{array}$$

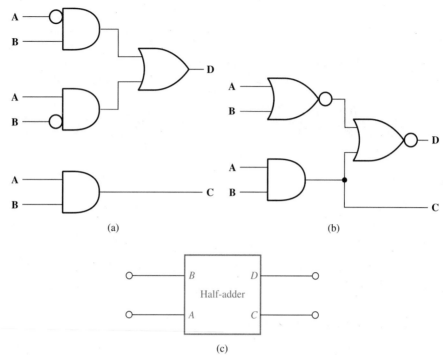

FIGURE 11.17(a-c)

To perform this addition, we must first add **A** to **B** in a half-adder, obtaining **D** as the "units" digit of the sum, and "carrying" **C**. Then we must add together **W**, **X**, and **C** to obtain **Y** and **Z**, the other two digits of the sum. Clearly the second step requires a circuit that can add *three* binary digits, according to the truth table

	Inputs			Outputs	
W	X	C		Y	Z
0	0	0		0	0
0	0	1		0	1
0	1	0		0	1
0	1	1		1	0
1	0	0		0	1
1	0	1		1	0
1	1	0		1	0
1	1	1		1	1

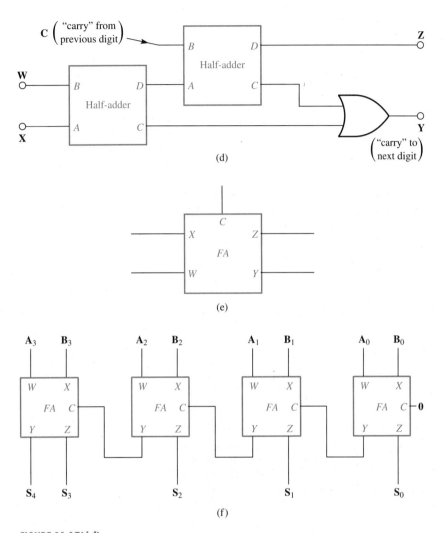

FIGURE 11.17(d–f)

Again this table could be realized from the sum-of-products method, but 10 gates would be needed. A simpler realization can be obtained using two half-adders, as shown in Fig. 11.17(d). This circuit (or any other circuit realizing the same truth table) is known as a *full adder* [Fig. 11.17(e)]. Full adders are also available in IC form. N full adders are needed to add together two N-digit binary numbers; for example, Fig. 11.17(f) illustrates the addition

$$\begin{array}{r} A_3\,A_2\,A_1\,A_0 \\ + \ B_3\,B_2\,B_1\,B_0 \\ \hline S_4\,S_3\,S_2\,S_1\,S_0 \end{array}$$

PRACTICAL LOGIC BLOCKS

Logic blocks are almost always found in IC form. They can be purchased in SSI (small-scale integrated) packages containing several gates; they also appear as parts of larger LSI (large-scale integrated) systems. Figure 11.18 shows two typical SSI packages: an IC containing four two-input **NOR** gates and one containing three three-input **AND** gates. These ICs are mounted in packages known as "DIPs" ("dual in-line packages"), each with 14 wires ("pins") meant to be plugged into a 14-pin socket. The package is about 1 inch long and $\frac{3}{8}$ inch wide. The number of gates per package is determined by the available number of connections. For example, each two-input

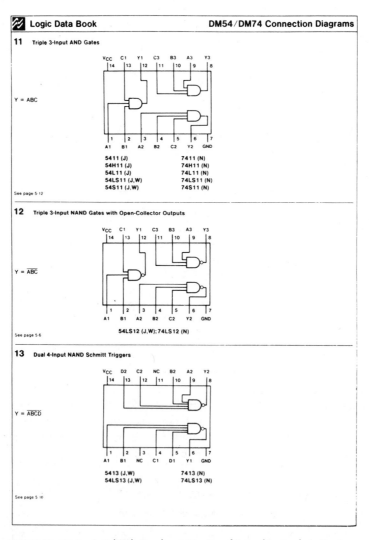

FIGURE 11.18(a)   Typical SSI logic packages. (*Courtesy of National Semiconductor Corp.*)

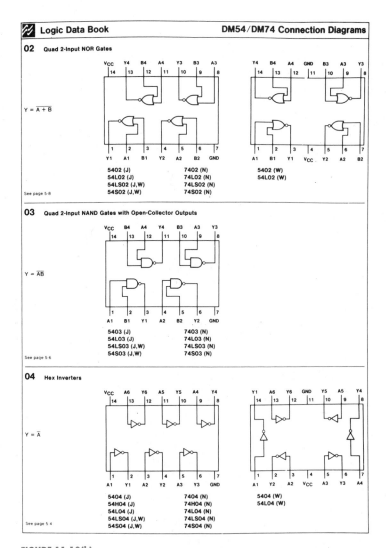

FIGURE 11.18(b)

gate requires three external connections; thus the 14-pin package can hold four of them. The remaining 2 pins are required for power-supply connections. Just as with op-amps, each IC requires connections to an external dc voltage source, although these connections are usually understood to exist and hence are not shown on diagrams. In Fig. 11.18(a) the external voltage source is to be connected between pin 14 (called "V_{CC}") and pin 7 ("GND").

Although the logic diagrams remain the same, the actual gates can be made with several different kinds of internal construction, giving rise to what are known as "logic families." Generally gates of one family are compatible with other gates of the same

SSI   **DM54/DM7400, 04, 10, 20, 30, S133 NAND Gates/Inverters**

Electrical Characteristics over recommended operating free-air temperature range (unless otherwise noted).

Parameter	Conditions	DM54/74 00, 04 10, 20, 30			DM54/74 H00, H04 H10, H20, H30			DM54/74 L00, L04 L10, L20, L30			DM54/74 LS00, LS04 LS10, LS20, LS30			DM54/74 S00, S04, S10 S20, S30, S133			Units
		Min	Typ(1)	Max	Min	Typ(1)	Max	Min	Typ(1)	Max	Min	Typ(1)	Max	Min	Typ(1)	Max	
V_{IH} High Level Input Voltage		2			2			2			2			2			V
V_{IL} Low Level Input Voltage	DM54			0.8			0.8			0.7			0.8			0.8	V
	DM74			0.8			0.8			0.7			0.8			0.8	V
V_I Input Clamp Voltage	V_{CC} = Min, I_I = -8 mA / -12 mA / -18 mA			-1.5			-1.5		N/A	N/A			-1.5			-1.2	V
I_{OH} High Level Output Current				-400			-500			-200			-400			-1000	µA
V_{OH} High Level Output Voltage	V_{CC} = Min, V_{IL} = Max, I_{OH} = Max — DM54	2.4	3.4		2.4	3.5		2.4	3.3		2.5	3.4		2.5	3.4		V
	DM74	2.4	3.4		2.4	3.5		2.4	3.2		2.7	3.4		2.7	3.4		V
I_{OL} Low Level Output Current	DM54			16			20			2			4			20	mA
	DM74			16			20			3.6			8			20	mA
V_{OL} Low Level Output Voltage	I_{OL} = Max, V_{CC} = Min, V_{IH} = 2 V — DM54		0.2	0.4		0.2	0.4		0.15	0.3		0.25	0.4			0.5	V
	DM74		0.2	0.4		0.2	0.4		0.2	0.4		0.35	0.5			0.5	V
	DM74 (I_{OL} = 4 mA)												0.4			0.4	V
I_I Input Current at Maximum Input Voltage	V_I = 5.5 V / V_I = 7 V			1			1			0.1			0.1			1	mA
I_{IH} High Level Input Current	V_I = 2.4 V / V_I = 2.7 V			40			50			10			20			50	µA
I_{IL} Low Level Input Current	V_I = 0.3 V / V_I = 0.4 V (LS30 / Others) / V_I = 0.5 V			-1.6			-2			-0.18			-0.4 / -0.36			-2	mA
I_{OS} Short Circuit Output Current	V_{CC} = Max — DM54	-20		-55	-40		-100	-3		-15	-20		-100	-40		-100	mA
	V_{CC} = Max (2) — DM74	-18		-55	-40		-100	-3		-15	-20		-100	-40		-100	mA
I_{CC} Supply Current	V_{CC} = Max									See Table							mA

Note 1: All typical values are at V_{CC} = 5 V, T_A = 25°C.

Note 2: Not more than one output should be shorted at a time, and for DM54H/DM74H, DM54LS/DM74LS and DM54S/DM74S, duration of short circuit should not exceed one second.

Figure 11.19   Manufacturer's specification sheet for TTL logic blocks. (*Courtesy of National Semiconductor Corp.*)

family, but not with those of other families. (This subject will be treated further in Chapter 15.) Figure 11.19 shows a typical set of specifications for logic blocks of the TTL ("transistor-transistor logic") family. We see, for example, that the lower limit of the "high" voltage range (here called V_{IH}) is 2 V, and the upper limit of the "low" range (V_{IL}) is 0.8 V; the manufacturer expects no signals to occur in the forbidden range between these two voltages.

Although SSI packages were common basic units at one time, the trend now is to integration on an ever larger scale. Entire digital systems, such as those to be discussed in the next chapter, are now available in IC form, and those ICs become blocks for building even larger systems. SSI packages are now mainly used for special purposes and for interconnections between larger ICs. The student should note that large ICs cost little more than small ICs or individual transistors; the production cost comes largely from the labor of assembling the packages. With these inexpensive, ever-more-powerful IC building blocks, systems of great sophistication and power can now readily be built.

11.3 Flip-Flops

As already mentioned, the basic elements of digital technology can be divided into *combinational* (or logic) blocks and *sequential* blocks. Logic blocks were discussed in the preceding section. Their distinguishing feature is the fact that the output at a given time depends only on the inputs *at that same time.** The output of a sequential block, on the other hand, depends not only on what the inputs are now, but also on what they were at earlier times. Sequential blocks have a kind of memory, and some of them are actually used as computer memories.

Most sequential blocks are of the kind known as *multivibrators*. A multivibrator is a circuit that can exist in one or the other of two states. It cannot remain for a finite length of time in any intermediate state or any other state except the basic two. An example is a common electric light switch. It can be placed in either the "up" position or the "down" position. Whichever position it is put into, it remains where it was put. It cannot be put into any intermediate position. (If it is, it will fall into one of the two "allowed" positions.) This kind of multivibrator is called *bistable* because it will remain stable in either of its two positions. Clearly a bistable multivibrator can be used as a memory device, "remembering" one bit of information. If we wish to remember that the answer to some question is **1** (as opposed to **0**), we can agree that this will be represented by placing the switch in its "up" position. Tomorrow we can return, look at the switch, and be reminded that the answer was **1**. One bistable multivibrator can store one binary digit, that is, one bit of information. If we use a bank of four switches, we can store four bits of information, and can remember any binary number between 0000 and 1111. For instance, if we wish to remember the binary number 1101 (equiva-

---

*In fact, a short time elapses after the inputs are changed before the output has time to change; this is known as a "propagation delay." Propagation delays are being neglected here.

lent to the decimal number 13), we simply set the four switches up-up-down-up. Note that the number of different numbers that can be remembered is 2^N, where N is the number of switches. (With four switches, any of 16 numbers can be stored, between zero and decimal 15.) This device would be called a simple N-bit memory.

Two other kinds of multivibrators are the *monostable* multivibrator and the *unstable* multivibrator. Suppose that mechanical switch of the last paragraph is fitted with a spring that returns it to the "down" position. Then it is "monostable" in the sense that it remains in only one of its two positions. You can put it into the "up" position, but when released it falls back to the "down" position. This is the general nature of monostable multivibrators. Sometimes there is a delay, so that when placed in the unstable state it remains for a certain length of time before returning to the stable position.

The third type of multivibrator, the unstable one, is unstable in *both* of its two states. If it is in state 1, it is unstable there and after a certain length of time changes into state 2. It is also unstable in state 2, and after a certain time it changes back to state 1; the process continues indefinitely. In our switch example, an unstable multivibrator is like a switch that continually goes up-down-up-down . . . all by itself. Such a switch is a kind of oscillator. It could be used, for example, to generate a square wave, by turning a voltage on-off-on-off.

At present we shall be concerned primarily with bistable multivibrators. The electronic versions of these devices are known as *flip-flops*. A flip-flop is a bistable electronic circuit that can be placed by the user in either of two electrical states. Once placed, it remains in that state until further instructions are received. Flip-flops are the basic sequential building blocks.

Many different kinds of flip-flop exist, such as the S-R flip-flop, the J-K flip-flop, the D flip-flop, etc. These differ primarily in the way instructions are given to the block. In general, a simple flip-flop (one that contains few transistors or other parts) is likely to require complicated instructions and thus will be clumsy for the system designer to use. More sophisticated flip-flops can be made by adding internal complexity. These are more sophisticated in the sense that they accept simpler commands and thus are easier to use. For instance, the D flip-flop is a relatively sophisticated flip-flop, which can be made by starting with an S-R flip-flop (a simple flip-flop) and adding other components. Although S-R flip-flops are somewhat awkward devices, their simplicity makes it easy to observe the fundamentals of their operation; therefore this is where we shall begin.

THE S-R FLIP-FLOP

Let us start by considering a box with a single output terminal, as shown in Fig. 11.20(a). Since this is part of a digital system, the voltage at the output terminal can only be in either the "low" range or the "high" range. It represents the logical variable **Q**, which accordingly can be **0** or **1**. This box is to represent a flip-flop. Thus we expect that if **Q** is initially **0** it will remain **0** (that is, the voltage at terminal Q will remain in the low range) until some instruction is given to the box to cause it to change. As yet, we have not said anything about how instructions are to be given, and thus Fig. 11.20(a) applies not just to the S-R but to all kinds of flip-flops.

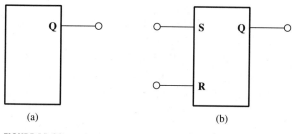

(a)   (b)

FIGURE 11.20

In Fig. 11.20(b) we have added to the basic flip-flop box two input terminals, which we shall call S and R. (These letters stand for "set" and "reset.") These terminals will be used to give instructions to the flip-flop, which is now no longer general but specifically the S-R type. The voltages at the input terminals can only be "low" or "high"; the input variables **S** and **R** can each be either **0** or **1**. We now need to state the rules by which the flip-flop will be told what to do. Since the S-R flip-flop is a simple one, its rules are slightly complicated:

RULE 1

As long as **S** = **0** and **R** = **0**, Q does not change. It keeps whichever value it initially had.

RULE 2

If **S** = **0** and **R** = **1**, then regardless of past history **Q** = **0**.

RULE 3

If **S** = **1** and **R** = **0**, then regardless of past history **Q** = **1**.

RULE 4

The input **S** = **1**, **R** = **1** is a meaningless instruction, which should not be used.

The S-R flip-flop is used as follows: Initially **Q** has some value; the block is "remembering" either **0** or **1**. As long as you wish it to remember, you keep inputs **S** and **R** both zero. Now, suppose you wish to store new information: you wish to store the number **1**. To do this you change the S input to **1**, leaving the R input equal to **0**. According to Rule 3, **Q** will now become **1**. (It may be that **Q** already was **1** before you gave your instruction. This makes no difference. Regardless of what **Q** was before you gave your instruction, after this instruction **Q** will definitely be **1**.) Next you change **S** back to **0**, with **R** still remaining **0**. Now the flip-flop is in its "remembering" mode again (see Rule 1). The output **Q** will remain **1** until new instructions are given.

If the inputs are next changed to **S** = **0**, **R** = **1**, output **Q** will become **0**, and if the inputs are then returned to **R** = **0**, **S** = **0**, the flip-flop will now "remember" **Q** = **0**. As an analogy, think of a lamp with two pull-chains hanging from it. The lamp can be on or off (corresponding to **Q** = **0** or **Q** = **1**). One pull-chain turns the lamp "on" and the other turns the lamp "off." (Pulling the "on" chain is like setting **S** to **1**. Pulling the "off" chain is like setting **R** to **1**.) If you pull the "on" chain, the light goes on (if you set **S** to **1**, **Q** becomes **1**). If you pull the "off" chain, the light goes off. (If you set **R** = **1**, **Q** becomes **0**.) If you don't pull any chains and the light was on, it remains on; if you don't pull any chains and the light was off, it remains off. If you pull the "on" chain when the light was already on, it simply remains on; there is no change. What happens if both chains are pulled at once? Then you are telling the light simultaneously to turn on and to turn off; this is a meaningless instruction, which should not be given. Similarly, the flip-flop input **S** = **1**, **R** = **1** is a meaningless instruction, as stated in Rule 4.

EXAMPLE 11.7

The inputs to an S-R flip-flop are as shown in Fig. 11.21. Find the value of **Q** at times t_1, t_2, and t_3.

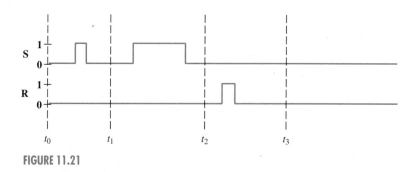

FIGURE 11.21

SOLUTION

The value of **Q** at time t_0 is not given, but it is not necessary to have this information. The first pulse of **S** "sets" the flip-flop—that is, puts it in the state **Q** = **1**; thus at t_1, **Q** = **1**. The second pulse of **S** again tries to "set" the flip-flop, but since **Q** was already **1** there is no change, and **Q** is still **1** at time t_2. The pulse of **R** then "resets" the flip-flop, and at t_3 **Q** = **0**.

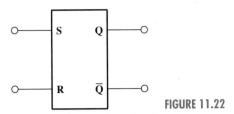

FIGURE 11.22

Most commercial flip-flops possess a second output terminal. There is nothing much more to learn about this output. The flip-flop is simply designed so that when output **Q** is **1**, this other output is **0**, and when **Q** is **0**, the other output is **1**. In the language of Boolean algebra, the other output is the complement of **Q**, or $\overline{\mathbf{Q}}$. The $\overline{\mathbf{Q}}$ output was not mentioned until now simply in order to avoid complicating the discussion. It is supplied because it makes it easier to use the flip-flop in systems. (It saves an inverter in cases when $\overline{\mathbf{Q}}$ is needed.) Furthermore, producing it adds no additional complexity to the flip-flop's internal circuit. As will be seen shortly, the flip-flop internally is a symmetrical circuit and produces $\overline{\mathbf{Q}}$ automatically. The complete symbol for the S-R flip-flop is shown in Fig. 11.22.

REALIZATION OF THE S-R FLIP-FLOP

The principal advantage of the S-R flip-flop is that it can be simply made from two **NAND** gates (plus two inverters), as shown in Fig. 11.23. Our next task is to verify that this circuit actually does operate according to the rules we have said describe an S-R flip-flop. Because of the feedback in this circuit, it is not possible to simply write its truth table, as one can do for a combinational circuit. Instead, a modified truth-table method is used, in which we "guess" an output and then go back and check it for self-consistency.

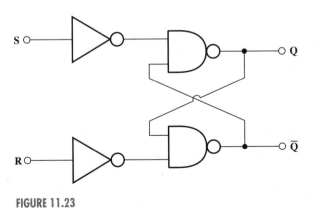

FIGURE 11.23

We begin by writing a modified truth table as shown in Fig. 11.24. Let us first find the outputs when the inputs are $S = 0$ and $R = 0$, as written in the first line of the table. We proceed by guessing an output. In this case we arbitrarily choose $Q_{GUESS} = 1$, and since we expect the terminal marked $\overline{Q}$ to give the complement of Q, we write $\overline{Q}_{GUESS} = 0$ (continuation of line 1 of the table). These are only guesses as yet; we do not know they are correct solutions. If they *are* correct, however, they must be self-consistent, which is what we now check. With our guess that $Q = 1$ the output of the lower **NAND** gate (whose inputs are **1** and **1**) must be **0**, agreeing with our guess that **Q** is **0**. The output of the upper **NAND** gate, whose inputs are **1** and **0**, is **1**, and since the output of the upper **NAND** gate is **Q**, this also agrees with our guess. Thus the last two columns of the table agree with the "guess" columns. This means that our guesses are self-consistent, and thus do represent a possible state of the system.

Suppose, however, we had made the opposite guess: $Q_{GUESS} = 0$. The result is shown in the second line of the table. Proceeding as before, we find that this guess is *also* self-consistent. Thus when both inputs are zero, the flip-flop can be in either of its two states, in agreement with Rule 1 of the previous subsection. Actually, we might have guessed this result from the symmetry of the circuit.

We can now proceed to fill out the remainder of the table by assuming the various other inputs. To improve our understanding, the third line illustrates what happens when a wrong guess of the output is made. In this case the last two columns do not agree with the middle two, informing us that a wrong guess was made. We throw this guess away (line crossed out in the truth table). The remainder of the table shows the correct outputs for the other possible inputs. The final line shows the result of the "forbidden" inputs $S = 1, R = 1$. We note that it is possible to apply this input; nothing disastrous happens electrically.

The reader should note the following rather subtle point: The circuit of Fig. 11.23 cannot accurately be said to *be* an S-R flip-flop; an S-R flip-flop is an idealization, an imaginary perfect device that obeys the S-R instruction rules. What we have in Fig. 11.23 is a *realization* of an S-R flip-flop, a physical circuit that behaves similarly to the ideal device. It is certainly possible to apply the "forbidden" input to the real circuit, but then you are not using it properly as an S-R flip-flop. Even though the instruction is improper, however, it can still be used, provided that one knows what the result will be. While the instructions $S = 1, R = 1$ are being given, both the **Q** and the $\overline{Q}$ outputs will have the value **1**. Then if we first make **R 0** while S is still **1**, **Q** will be set to **1**; if we then make **S 0**, the flip-flop will "hold" $Q = 1$. On the other hand, if we first

S	R	Q (guess)	$\overline{Q}$ (guess)	Q	$\overline{Q}$
0	0	1	0	1	0
0	0	0	1	0	1
~~1~~	~~0~~	~~0~~	~~1~~	~~1~~	~~1~~
1	0	1	0	1	0
0	1	0	1	0	1
1	1	1	1	1	1

FIGURE 11.24

go from $S = 1$, $R = 1$ to $S = 0$, $R = 1$, the flip-flop will be reset to $Q = 0$, and when we then make $S = R = 0$, the flip-flop will "hold" $Q = 0$. Thus the final result depends on whether S or R returns to 0 first. One has to be sure which it is, or else the final value of Q will be unpredictable.

EXAMPLE 11.8

The circuit of Fig. 11.23 is given the inputs shown in the first two lines of Fig. 11.25. The initial value of Q is 1. We wish to find Q as a function of time. The solution is shown in the third line. *Note:* This kind of figure is known as a *timing diagram.*

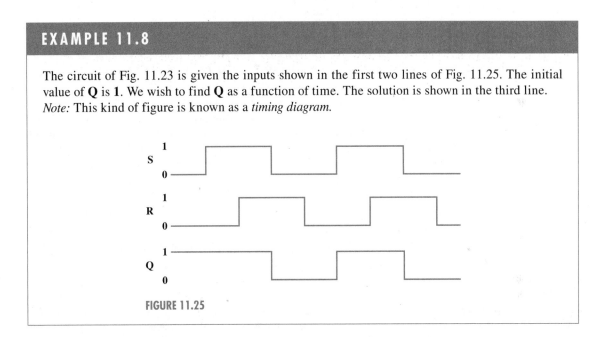

FIGURE 11.25

THE D FLIP-FLOP

There are many kinds of more advanced flip-flops, differing primarily in the rules for applying their control instructions. Rather than attempt to catalog them all, let us concentrate on a single example, the clocked D flip-flop. The symbol for this device is shown in Fig. 11.26. The two output terminals, Q and $\overline{Q}$, behave just as in the S-R flip-flop. The use of the input terminals, D and CK, will now be discussed.

The term "clocked" flip-flop means that this device cannot change its state (that is, Q cannot change) unless a specific "change" instruction is given. This instruction is

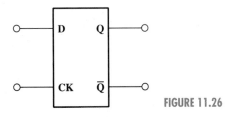

FIGURE 11.26

D	Q (before)	Q (after)
0	0	0
0	1	0
1	0	1
1	1	1

FIGURE 11.27

given through the CK ("clock") input. When a "change" instruction is given, the flip-flop *may or may not* change its state (depending on other instructions), but in the absence of a "change" instruction no change of **Q** can occur.

When a "change" instruction is received, the output **Q** takes on the value that the **D** input has at that instant of time. This is the case irrespective of what value **Q** had before the change instruction was received. Figure 11.27 shows the values taken by **Q** after the change instruction, for various inputs **D** and prior values of **Q**. Note that the value of **Q** after the change instruction is equal to the value of **D** at the time the change instruction is received. The value of **Q** before the change instruction does not matter.

We still need to specify how the change instruction is to be given. There are several variations of the device, but let us consider the rising-edge-triggered flip-flop. In this variation, a change instruction is effected whenever the CK input *makes a change from 0 to 1*. A constant CK input is not a change instruction, even if **CK = 1**; only a positive-going transition of CK is a change instruction.

EXAMPLE 11.9

The edge-triggered D flip-flop is given the inputs shown in the first two lines of Fig. 11.28. The initial value of **Q** is **0**. We wish to find **Q** as a function of time. The result is shown in the third line. The vertical dashed lines indicate the times of change instructions.

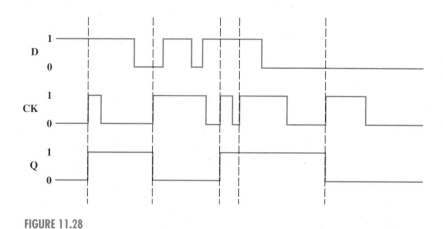

FIGURE 11.28

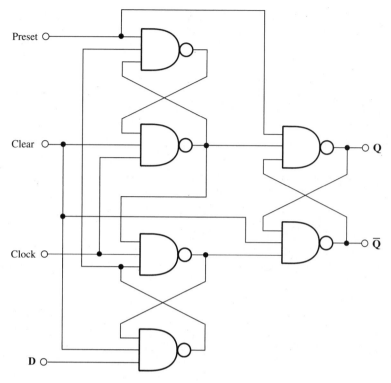

FIGURE 11.29

It is important to know that flip-flops have *propagation delay*. This means that there is a small delay between the change instruction and the time **Q** actually changes. The value of **D** that matters is its value *when the change instruction is received*, not its value at the later time when **Q** changes. The propagation delay is so short (about 20 nsec) that it does not show up in a timing diagram like Fig. 11.28. However, one must know that it exists, or else some circuits cannot be understood!

Figure 11.29 shows a realization of a D flip-flop using six **NAND** gates. The output part of the circuit is very similar to the S-R flip-flop of Fig. 11.23. The other four gates are used to translate the D-type control instructions into a form that the S-R flip-flop can use. This circuit has two additional input terminals called "preset" and "clear." These inputs are used to manually set (make **Q** = **1**) or reset (make **Q** = **0**) the flip-flop. If nothing is connected to these inputs, they default to high (which means the gates consider a disconnected input to be a "high" input, or **1**). If the preset input is grounded (made "low" or **0**), **Q** is forced to become **1**, irrespective of other inputs. Similarly, grounding "clear" forces **Q** to become **0**.

PRACTICAL FLIP-FLOPS

As with logic blocks, flip-flops appear almost exclusively in IC form. The situation regarding packaging, logic families, power supplies, and so forth is quite similar. This is not surprising, because in many cases flip-flops are actually composed, internally, of the logic blocks of the same family, as illustrated in Fig. 11.29. One difference is that flip-flops are even more likely to be found in LSI and VLSI form. A very important application is in computer memories. A typical *random-access memory* (or "RAM") is an IC containing a very large number of flip-flops: the so-called "256k RAM," for example, consists of approximately 256,000 flip-flops in a single IC! It is interesting that the number of blocks per IC is steadily increasing; the 16k RAM represented the state of the art only a few years ago. As microfabrication technology develops, there is really no telling how powerful these IC blocks may eventually become.

11.4 Application: Digital Communications

A communication system, such as a telephone, can be either analog or digital. In an analog system, one transmits a voltage proportional to the acoustic pressure of the speaker's voice. In a digital system, however, one somehow turns the time-varying pressure into a string of numbers and transmits the numbers. Then, at the receiving end, the incoming numbers are used to reconstruct a replica of the original acoustic signal.

The sequence of binary numbers sent through the communication system would look like Fig. 11.2. The numbers in this figure are *binary*, which means that the only digits that can be sent are "zero" and "one." The reader may be aware, however, that binary numbers can be converted to ordinary decimal numbers. (This will be discussed in the following chapter.) Thus, at the receiving end, the incoming data might be interpreted as a sequence of numbers like 467, 578, 704, 695, . . . The numbers must arrive quite close together; for a telephone we might use 10,000 numbers per second. This seems like a lot of numbers, but in fact 10,000 per second is quite a modest rate, as data communications go. Electronic devices are very fast!

The exact way in which information is encoded in the string of numbers is a matter to be decided by the system designer. A simple way to do this might be to divide the possible range of acoustic pressures into 1,000 equal parts. Then we could let the transmitted number 000 indicate that the pressure at the time the number was sent was 00.0% of the maximum pressure. If the transmitted number 536 arrives, it would mean that the pressure at the time of the number's transmission was 53.6% of the maximum, and so on. The machinery at the receiving end has to follow the instructions contained in the numbers and build a sequence of acoustic pressures according to the numbers that arrive. One will then have a replica of the original sound. (See Fig. 0.4.)

The reader may think that this is awfully complicated. Why not just use an old-fashioned analog telephone? In fact, however, telephone systems are increasingly becoming digital. This is because digital systems offer some important advantages:

RESISTANCE TO ERRORS

The transmission system between sender and receiver inevitably introduces noise. Random noise signals are added to the signals being sent, sometimes positive, sometimes negative; gradually the signal becomes distorted, and if the process continues, the meaning of the signal will be lost. In the case of analog signals, there is little that can be done to prevent degradation of this kind. The only defense is to make the signal very large in comparison with the noise. Digital signals, however, can be somewhat resistant to noise, as can be seen from Fig. 11.2. In this figure, the "low" range is 0 to 1 V, and the "high" range is 4 to 5 V. A small noise voltage added to a signal in middle of the "low" range will still be in the low range; it will still be interpreted as a zero, so no information has been lost! One can even "clean up" the signal by taking any voltages that have been "noised" into the "forbidden" range between 1 and 2 V and resetting them back down to 0.5 V. This approach to signal distortion cannot completely eliminate errors. Occasionally an extra large burst of noise or a series of additive small ones may move a signal all the way from the low range to the high range; but with good system design, nearly ideal performance can often be achieved. Furthermore, many additional tricks are available to the system designer, in the form of error-correcting codes. For example, in a system in which high accuracy was vital, the system might send each number *twice*, and automatically recheck if the two transmissions were different.

VERSATILITY

The idea just suggested, that of sending each number twice as a check on accuracy, is an example of something that would be hard to do in an analog system but is fairly easy to do in a digital one. Analog systems are usually "real-time"; that is, the information is sent out steadily as it comes in. In the digital system, however, you can easily store the numbers or repeat them. You can send several messages simultaneously, by interleaving the numbers: that is, you would send a number from the first phone conversation, then a number from the second conversation, then one from the third; then another number from the first conversation, and so on. Provided your system can send enough numbers per second, a lot of calls can be sent simultaneously in this way. In fact, telephone calls have become indistinguishable from other computer data, and with modern computer power available, you can manipulate the data in all sorts of ways. With analog signals, usually the only thing you can do is amplify to make them louder. Moreover, your digital telephone line can conveniently be used to transmit other kinds of numbers, such as actual computer data.

ECONOMY

Digital blocks are remarkably cheap. Thanks to IC technology, millions of digital blocks are available to the system designer at low cost. It is true that each of these blocks—gates or flip-flops—is extremely simple, but then so, probably, are brain cells. The cleverness is all in how the blocks are hooked up. Because they are plentiful, small, and cheap, the communication system can be made complex and powerful, performing any computer-like functions that may be needed. Furthermore, digital systems are typically programmable, just as computers are. Thus they can be redesigned to perform new functions just by changing the software! Analog systems, by contrast, tend to be "hard-wired"; that is, they are designed to perform one specific function, and that is all.

It is for reasons such as these that digital communication systems are taking over. Designing them for optimum performance is an active field in electrical engineering, and a great deal of interesting work is being done. The rest of us can at least admire the power and cleverness of the technology, the next time we talk to someone far away. It may sound like the voice of your Aunt Mabel, but in fact what you hear is an ingenious robot that knows how to sound just like her. It is synthesizing a replica of her voice, using, as instructions, only a stream of numbers arriving in thousands each second. Perhaps someday they will be able to send Aunt Mabel herself in this way!

POINTS TO REMEMBER

- Digital building blocks perform functions quite different from those of analog blocks. Digital blocks usually are designed to process binary signals. Binary signals must fall into one or the other of two ranges, known as the high and low ranges. In positive logic a signal in the high range represents the binary digit **1**, and a signal in the low range represents the digit **0**. In negative logic the reverse is true.

- Digital blocks divide into *combinational* (or *logic*) blocks and *sequential* blocks. The output of a logic block depends on the inputs at the same instant of time. Sequential blocks remember inputs applied at earlier times.

- Some of the more common logic blocks are the **AND, OR, NAND, NOR, EXCLUSIVE OR**, and **COINCIDENCE** gates, and the inverter, which performs the operation known as **COMPLEMENT**.

- Mathematical expressions for logical operations are written using the notation of *Boolean algebra*.

- The operation of a logic block, or combination of blocks, is specified by a table showing the outputs that result from various combinations of inputs. This table is known as a *truth table*.

- Two logical operations are identical if their truth tables are the same.

- Any truth table can be realized with an appropriate connection of inverters, **AND** gates, and one **OR** gate, using the *sum-of-products method*. However, the resulting circuit is usually not the simplest one possible.

- The most common sequential block is the flip-flop. A flip-flop is a bistable circuit that remembers a single binary digit according to instructions.

- Various types of flip-flops exist, such as the S-R flip-flop and the D flip-flop. These differ from one another in the way instructions for storing information are applied.

- Flip-flops can be constructed using combinations of logic blocks. The S-R flip-flop, which can be built from just two **NAND** gates, is relatively clumsy to use.

Other flip-flops, such as the D-type clocked flip-flop, have more internal complexity but are easier to program.

• Digital blocks are usually found in IC form, ranging from small-scale integration (SSI, 2 to 4 blocks per IC) to very-large-scale integration (VLSI, 250,000 or more blocks per IC). Large, powerful ICs cost very little more than small ICs.

• Digital blocks made with one particular fabrication technology are said to belong to a particular *logic family*. Several different logic families exist and are used for different purposes. In general, blocks of one family are compatible with other blocks of that family but not with those of other families.

PROBLEMS

Section 11.1

11.1 A time-varying voltage represents a series of binary numbers. Let the "high" range be 4 to 5 V and the "low" range 0 to 0.8 V. Assume that positive logic is used. Sketch a possible waveform representing the binary digits 10011010.

11.2 Explain the following:
  a. The binary digit 1.
  b. A switching variable.
  c. The logical state **1**.
  d. The "high" range.

Section 11.2

11.3 Obtain the truth tables for the blocks shown in Fig. 11.30(a), (b), and (c).

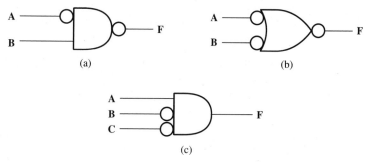

(a)

(b)

(c)

FIGURE 11.30

11.4 Obtain the truth table for the connections shown in Fig. 11.31(a), (b), and (c). For parts (b) and (c) explain why the answers come out as they do.

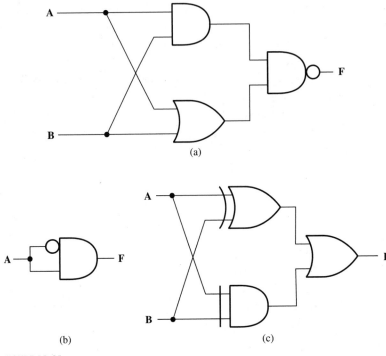

(a)

(b)   (c)

FIGURE 11.31

11.5 Write logical (Boolean algebra) expressions for the blocks shown in Fig. 11.30.

11.6 Write logical (Boolean algebra) expressions for the connections shown in Fig. 11.31.

11.7 Let the electrical output of a block be related to the electrical inputs as shown in Fig. 11.32. What logical block or combination of blocks describes this electrical operation if the negative-logic convention is adopted? If the positive-logic convention is adopted?

Input	Input	Output
A	B	F
low	low	high
low	high	low
high	low	low
high	high	high

FIGURE 11.32

11.8 Find the values (**0** or **1**) of the following:
a. $(1 + 0) 1$

b. $0 + (1\overline{1})\,1$

c. $\overline{(1 + 0)}\,(1 + 1)$

11.9 Write the truth tables for the following switching functions:

   a. $F = (X + Y)\,\overline{X}$

   b. $F = (Y \cdot Z)\,\overline{X} + X$

11.10 Any truth table defines a switching function.

   a. How many *different* switching functions of two input variables exist?

   b. Give an example of one that is not in Table 11.1.

   c. How many different switching functions of three variables exist?

11.11 Show, by constructing truth tables, the following:

   a. $(A)(1) = A$

   b. $A + \overline{A} = 1$

   Note: $(A)(1)$ means **A AND 1**.

11.12 Show, by means of truth tables, that $A(A + B) = A$.

11.13 Show, by means of truth tables, that $A + BC = (A + B)(A + C)$.

11.14 Show, by means of truth tables, that $\overline{A\ B} = \overline{A} + \overline{B}$. This important result is known as *De Morgan's theorem*.

11.15 Using De Morgan's theorem (previous problem), design an **OR** gate using one **NAND** gate and two inverters.

11.16 Show that the circuits shown in Fig. 11.33(a) and (b) are each equivalent to inverters. In Fig. 11.33(b) the input marked V_{CC} is connected to the highest power-supply voltage and is therefore a permanent **1**.

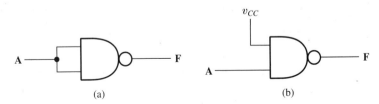

FIGURE 11.33

11.17 Using the results of Problems 11.14 and 11.16 and Fig. 11.12, show how to construct an **EXCLUSIVE OR** gate using only **NAND** gates. (Note that in a similar way *any* switching function can be realized using only **NAND** gates.)

11.18 Design a circuit equivalent to a two-input **COINCIDENCE** gate using two **AND** gates, one **OR** gate, and inverters.

11.19 Let *ABC* be a three-digit binary number. The number is said to have *even parity* if an even number of its digits (or none of them) is 1; otherwise it has *odd parity*. (Examples: 110 has even parity; 010 has odd parity.) Design a circuit whose output **F** is **1** if and only if the three inputs **A B C** have odd parity. (*Suggestion:* First construct the desired truth table, then realize it. Three-input gates may be used.) This circuit is called a *parity checker*.

*11.20 The voltages appearing on four wires are V_A, V_B, V_C, and V_D, corresponding to the four switching variables **A**, **B**, **C**, and **D**. Design a circuit that gives output **F** = **1** if and only if **A B C D** = **1010**. Use only **NAND** gates with two inputs each. (See Problems 11.14 to 11.16.)

*11.21 An important circuit for use in computers is a comparator, which has zero output unless the inputs are identical. Synthesize a circuit that compares A_1 A_2 (a two-digit binary number) with B_1 B_2 (another two-digit binary number). Use only **NAND** gates.

*11.22 Many times it is useful to have a circuit that indicates if two binary numbers are equal, or, if not, which one is greater. Design a circuit which has inputs **A** and **B** and three outputs **F**, **G**, and **H**; **F** = **1** only if **A** > **B**, **G** = **1** only if **A** = **B**, and **H** = **1** only if **A** < **B**.

*11.23 Figure 11.34 shows, in symbolic form, a device often needed in digital systems, called a *multiplexer*. The purpose is to make output **F** equal to *either* input **A** or input **B**, depending on the value of control variable **C**. When **C** = **0** we want **F** = **A**, and when **C** = **1** we want **F** = **B**. Realize this device using logic blocks.

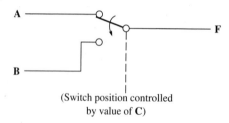

(Switch position controlled
by value of **C**)

FIGURE 11.34

11.24 For the circuit shown in Fig. 11.35(a):
a. Construct the truth table.
b. The values of the four input variables **A B C D** as a function of time are shown in the timing diagram, Fig. 11.35(b). Fill in the variation of **F** as a function of time.

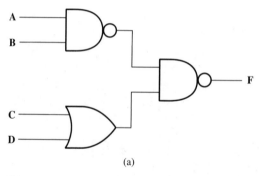

(a)

FIGURE 11.35(a)

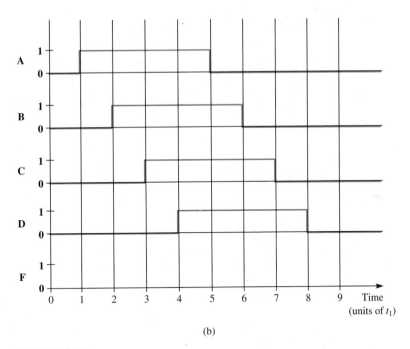

(b)

FIGURE 11.35(b)

Section 11.3

11.25 An S-R flip-flop is used to control a camera shutter. The circuit is arranged so that when output **Q** is **1**, the shutter is opened; when **Q** = **0**, the shutter is closed. We wish to open the shutter for 1 msec exposures and to make 300 exposures per second. Make a timing diagram showing suitable **S** and **R** inputs to operate the shutter.

11.26 In Fig. 11.36, V_0 is in the "high" range and corresponds to logical **1**, while 0 V corresponds to **0**. The switch is connected to input **S** by a short wire and to input

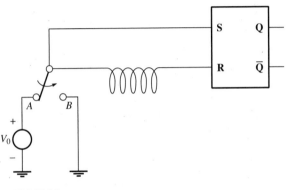

FIGURE 11.36

R by a very long wire, which produces a delay. The switch is in position A until time $t = 0$, when it is moved to B.

a. What is **Q** for $t < 0$?

b. What is the final value of **Q**?

*11.27 An alternative realization of an S-R flip-flop using **NOR** gates is shown in Fig. 11.37. By constructing a table similar to Fig. 11.24, verify that this realization is correct.

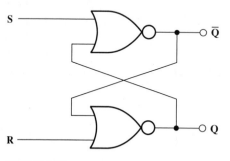

S

$\overline{Q}$

R

Q

FIGURE 11.37

11.28 The circuit shown in Fig. 11.38(a) is a *clocked S-R flip-flop*.

a. Verify that when CK $= 0$, the value of **Q** does not change.

b. Verify that when CK $= 1$, the circuit acts like an ordinary S-R flip-flop.

c. Fill in the values of **Q** in the timing diagram, Fig. 11.38(b). Initially **Q** $= 0$.

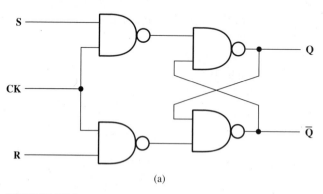

S

CK

R

Q

$\overline{Q}$

(a)

FIGURE 11.38(a)

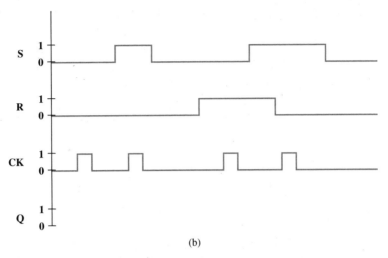

(b)

FIGURE 11.38(b)

11.29 Figure 11.39(a) shows a simple realization of a D-type flip-flop. (Note, however, that this is not an *edge-triggered* D-type flip-flop, which was the example discussed in the text.) Fill in the timing diagram of Fig. 11.39(b). Initially **Q** = **0**.

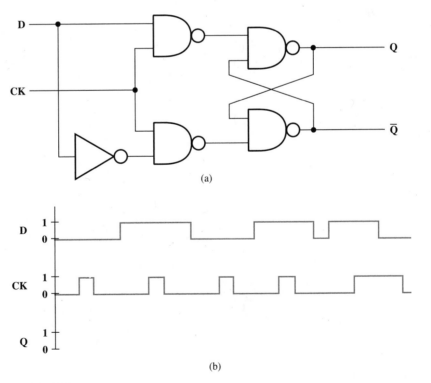

(a)

(b)

FIGURE 11.39

11.30 The inputs of Fig. 11.39(b) are applied to the positive-edge-triggered D flip-flop discussed in the text. Initially **Q** is **0**. Construct the timing diagram for **Q**. If you have worked Problem 11.29, compare the outputs of these two kinds of D flip-flops.

11.31 The purpose of this problem is to verify that Fig. 11.29 is a realization of a positive-edge-triggered D flip-flop. For simplicity we shall assume that the "preset" and "clear" inputs are always **1, so that they do not affect operation. The four left-hand gates then are as shown as in Fig. 11.40(a).

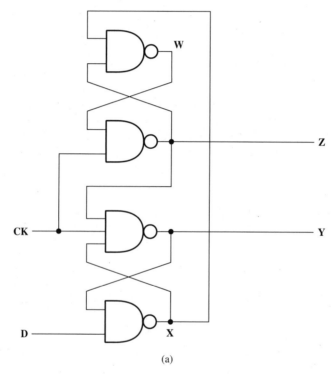

(a)

FIGURE 11.40(a)

a. Fill in the timing diagram of Fig. 11.40(b). *Note:* This problem is rather difficult. Here are some helpful observations. (1) The output of a **NAND** gate is **1** if *any* of its inputs is **0**. (2) The top two gates are an S-R flip-flop whose inputs are **X** and **CK**.

b. Now, referring to the complete circuit, Fig. 11.29, find **Q** as a function of time. (Note that the right-hand two gates also form an S-R flip-flop.) The initial value of **Q** is **0**.

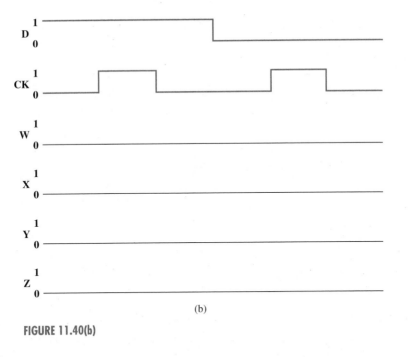

(b)

FIGURE 11.40(b)

11.32 Refer to the manufacturer's specifications for type DM54 TTL logic blocks (Fig. 11.19). State the most conservative (safest or "worst case") answers for all the blocks of the series:

a. What is the lowest input voltage that will be recognized as "high" by the blocks?

b. What is the highest input voltage that will be recognized as a "low?"

c. When the output is "high," what is the minimum output voltage?

d. When the output is "low," what is the maximum output voltage?

e. Let the answers to parts (a) to (d) be V_a, V_b, V_c, and V_d. Suppose the output of one block is connected to the input of another block of this same type. State two conditions on V_a, V_b, V_c, and V_d that must be satisfied for reliable operation. Are these conditions in fact satisfied?

CHAPTER 12

INTRODUCTION TO DIGITAL SYSTEMS

In Chapter 11 we introduced the basic combinational and sequential building blocks. These blocks seldom are used individually, however, or even in small groups. Most often they occur in medium-, large-, or very-large-size systems. In fact, the *reason* that digital systems are so inexpensive and yet so powerful is that they consist of very large numbers of just a few building blocks, repeated in simple ways. (The repetitions *have* to be simple, of course; no one could design a system containing half a million blocks if each block were different from all the others.) In this chapter we first introduce some basic systems concepts (Section 12.1). Then we shall go on to some important small systems (Section 12.2), and finally to large systems (Section 12.3). We cannot go very deeply into computer design in this introductory treatment. We shall be able, however, to develop enough of the basic ideas to explain, in a general way, how these remarkable machines can be built up from the simple blocks of Chapter 11.

12.1 Bits and Bytes

We have already seen how one binary digit—1 *bit*—can be stored in a single switch or flip-flop. Often, however, it is convenient to handle information in larger units than single bits. Suppose we consider a set of four binary elements, such as the set of four signal flags shown in Fig. 12.1. Let us agree that a raised flag means **1** and a lowered

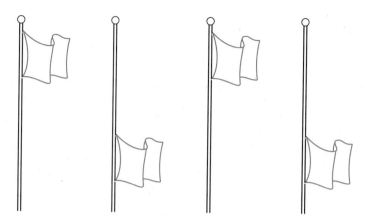

FIGURE 12.1   Four flags whose positions represent the binary number 1010.

flag means **0**. Then the situation shown in Fig. 12.1 represents the binary number 1010 (equivalent to the decimal number 10).* If all the flags were lowered they would represent 0000 (decimal 0), and if they all were raised they would represent 1111 (equivalent to decimal 15). Each individual flag represents a single binary digit. The set of four flags represents four binary digits, and hence is a *4-bit memory*. The 4-bit memory can remember any number between 0000 and 1111—16 different numbers. In general, an *N*-bit memory can remember 2^N different numbers.

In most cases information is handled in units of several bits, known as *bytes*. There is no fundamental rule as to the number of bits in each byte; however, in practice 8-bit bytes are very common. Figure 12.1 illustrates a 4-bit byte. In a digital system one would have groups of flip-flops, each group containing enough flip-flops to store 1 byte. For example, a computer based on 4-bit bytes would have its flip-flops arranged in groups of four.

EXAMPLE 12.1

Suppose we wish to store English-language writing, with 1 byte representing each letter. What is the minimum number of bits per byte that could be used?

---

*See Table 12.1. The relationship among binary, decimal, and hexadecimal numbers is discussed at the end of this section.

SOLUTION

There are 26 letters in the alphabet; thus if N is the number of bits per byte, we must have $2^N >$ 26. Since N is an integer, the smallest possible value for N is 5. In this case we could, for example, let 00000 represent the letter A, 00001 represent B, and so on up to 11001, representing Z. The remaining six combinations could be used to represent punctuation marks or spaces. (This is, in fact, the way letters are represented in teletype systems.) Of course, one might also wish to store *capital* letters, in which case there are 52 letters instead of 26. In that case 6-bit bytes would have to be used.

Suppose we are dealing with 4-bit bytes. There are two different ways 4 bits can be transmitted. We could send the bits one after another (*serially*). This, for example, is the way information is sent over a radio link or a telegraph wire (see Fig. 11.2 on p. 393). On the other hand, if we have four wires available, we can send the 4 bits *simultaneously*, as shown in Fig. 12.2. In the figure a 4-bit byte (determined by the positions

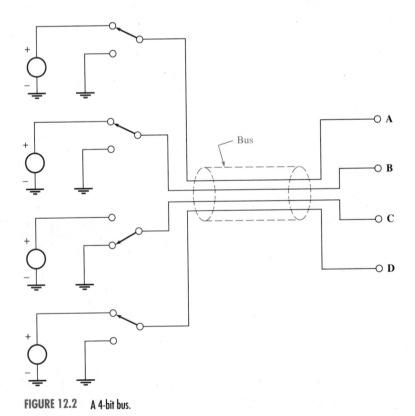

FIGURE 12.2   A 4-bit bus.

of the switches—1101 as shown) is being transmitted simultaneously over a set of four wires. (Actually there must be a ground wire also, but it is usually assumed to exist and therefore is not shown on diagrams.) The set of four wires is known as a *bus*. By means of the bus the entire byte is transmitted at the same time. After a certain time, the switches may be reset to new values and another byte sent. In this kind of operation, which is usual inside digital systems, bytes are transmitted serially, but an entire byte is sent simultaneously through the bus each time.

Normally a byte containing N bits is stored in a set of N flip-flops. This set of flip-flops is then referred to as a *register*. Often we wish to take a byte stored in one register and store it in a second register. This process is shown in Fig. 12.3, for the case $N = 4$. Initially a byte of information is stored in the left-hand register (flip-flops 1 to 4). The **Q** outputs are connected through a four-wire bus to the **D** inputs of the second register (flip-flops 5 to 8.) Now when the second register receives a "change" instruction through the "clock" wire, the **Q** output of FF5 takes on the value that was stored

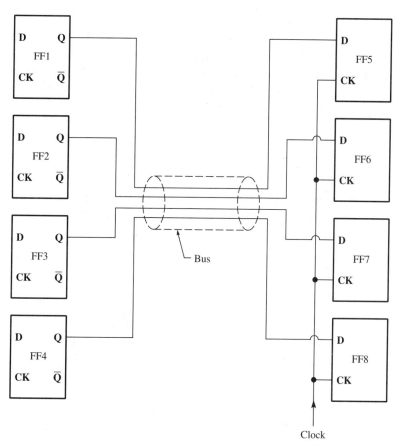

FIGURE 12.3   A 4-bit data latch.

in FF1, and so forth. After the clock pulse, the byte has been transferred to the right-hand register, where it will continue to be stored, at least until another clock pulse is received. Meanwhile, if no clock inputs have been sent to the left-hand register, the byte continues to be stored there as well. Used in this way, the register FF5–FF8 is referred to as a *storage register* or *data latch*.

BINARY NUMBERS

The two-state nature of digital technology makes the binary number system a natural tool. In the binary system there are only two digits, which are 0 and 1. When we count in binary, we begin with 0, and the next number is 1. Then, since there is no "two" in this system, we start the units column over again at 0 and carry a "1" over to the next column, which is the "two's" column. Continuing in this way, the first few binary numbers are 0, 1, 10, 11, 100, 101, 110, and so on. Binary numbers can be added, subtracted, multiplied, and divided using methods completely analogous to those of ordinary arithmetic.

HEXADECIMAL NUMBERS

It is rather clumsy to write the value of a 4-bit byte as, for example, 0110, and even clumsier to represent an 8-bit byte as 10110101. It is more convenient to have a shorthand. For instance, to represent 0000 we might write %, for 0001 we might write $, and so on, until we had 16 symbols representing the 16 possible 4-bit bytes. This is what is done, except that the 16 symbols actually used are the following: 0, 1, 2, 3, 4, 5, 6, 7, 8, 9, A, B, C, D, E, F. These 16 characters are known as the 16 hexadecimal digits. The character "0" represents the byte 0000, "9" represents 1001, and "E" represents 1110.

The hexadecimal digits are actually something more than a shorthand notation; they are the digits of the hexadecimal number system. In this system we can construct multidigit numbers just as in the decimal system, except that we count by 16's instead of 10's. The right-hand digit represents 1's, the second digit 16's, the third digit 256's, and so forth. The right-hand digit is known as the *least-significant digit*. The leftmost digit is the *most-significant digit*. It is perhaps unfortunate that the first 10 symbols of the hexadecimal system look the same as the 10 decimal symbols, leading to possible confusion. When the possibility of confusion exists, we add subscripts "10" or "16" to indicate that the number is decimal or hexadecimal. (Sixteen is said to be the *base* of the hexadecimal system.) Since the number 8 in the decimal system means the same as 8 in hex, we can write $8_{10} = 8_{16}$. However, $11_{10} = B_{16}$. Also, $23_{10} = 17_{16}$ and $198_{10} = C6_{16}$.

If one tried to use a system like this for 8-bit bytes, it would not work out well, because $2^8 = 256$ different symbols would be required! However, an 8-bit byte can be considered as two 4-bit bytes. (ABCDWXYZ can be written ABCD, WXYZ.) Hence 8-bit bytes are usually designated by double hex symbols, one for the first 4-bit byte and one for the second. Thus the 8-bit byte 11001001 would be represented in hex notation as C9.

CONVERSION BETWEEN NUMBER SYSTEMS

The reader may be interested in the way a number expressed in terms of one base can be written in terms of another. Table 12.1 shows the first 20_{10} numbers expressed in decimal, binary, and hexadecimal forms. (Note that this list of 20 numbers ends with 19_{10}, not 20_{10}. This is because the first number on the list is zero, instead of one — a universal practice.) Numbers less than or equal to 19_{10} can be converted simply by looking at the table, but for larger numbers techniques for conversion are needed.

Conversion from binary to hex and vice versa is very easy. One simply replaces each group of four binary digits with the appropriate hex digit. (Counting off of the

TABLE 12.1 The First 20 Numbers Expressed in Decimal, Binary, and Hexadecimal Form

Decimal	Binary	Hexadecimal
0	0000	0
1	0001	1
2	0010	2
3	0011	3
4	0100	4
5	0101	5
6	0110	6
7	0111	7
8	1000	8
9	1001	9
10	1010	A
11	1011	B
12	1100	C
13	1101	D
14	1110	E
15	1111	F
16	10000	10
17	10001	11
18	10010	12
19	10011	13
⋮	⋮	⋮

groups of four digits begins from the *right*. Add zeros on the left as necessary to complete groups of four.) For example,

$$01101111010_2 = (0)\, 011 \quad 0111 \quad 1010$$
$$= \quad 3 \qquad 7 \qquad A$$
$$= 37A_{16}$$

An example of conversion from hex to binary:

$$BC_{16} = 1011 \qquad 1100$$
$$= 10111100_2$$

Conversion from binary to decimal is done by recalling that the right-hand binary digit represents the number of 1's, the next digit 2's, the next 4's, then 8's, and so on. Example:

$$
\begin{aligned}
1101_2 = \quad & (1 \times 1) = \quad 1 \\
+ \, & (0 \times 2) \quad\; +0 \\
+ \, & (1 \times 4) \quad\; +4 \\
+ \, & (1 \times 8) \quad\; \underline{+8} \\
& \qquad\qquad = \; 13_{10}
\end{aligned}
$$

Conversion from hex to decimal is done similarly, except that in hex the right-hand digit is 1's, the next is 16's, the next is 256's ($16^2 = 256$), the next 4096's ($16^3 = 4096$), and so on. Example:

$$
\begin{aligned}
DE4_{16} = \quad & (4 \times 1) \qquad = \qquad 4 \\
+ \, & (14 \times 16) \qquad + \;\; 224 \\
= \quad & (13 \times 256) \qquad \underline{+\, 3328} \\
& \qquad\qquad\qquad 3556_{10}
\end{aligned}
$$

TABLE 12.2   An Example of the Conversion of a Decimal Number to its Binary Equivalent

Decimal number to be converted	Quotient		Remainder
→ 245 ÷ 2	122	+	1
122 ÷ 2	61	+	0
61 ÷ 2	30	+	1
30 ÷ 2	15	+	0
15 ÷ 2	7	+	1
7 ÷ 2	3	+	1
3 ÷ 2	1	+	1
1 ÷ 2	0	+	1
Binary result			1 1 1 1 0 1 0 1

$$245_{10} = 11110101_2$$

Conversion from decimal to binary is a less obvious procedure. Perhaps the easiest method is the one involving repeated divisions by two, as shown in Table 12.2. (Notice that when using this method the procedure does not stop until a quotient of zero is obtained.) Conversion from decimal to hex is most easily done by converting from decimal to binary using the method of Table 12.2 and then performing the simple conversion from binary to hex.

Arithmetic operations are easily performed in binary, using essentially the same operations as in decimal arithmetic. Calculations are more difficult in hex because we do not know the hexadecimal addition and multiplication tables by heart. It is easiest to convert to binary or decimal for the calculation and then convert back.

EXAMPLE 12.2

Multiply $9C_{16}$ by $A3_{16}$. Express the answer in hex.

SOLUTION

We proceed by converting to binary and carrying out the long multiplication in binary form. The result is then converted back to hex.

$$
\begin{array}{rl}
9C_{16} = & 1001\ 1100 \\
A3_{16} = & \underline{1010\ 0011} \\
& 1001\ 1100 \\
& 1\ 0011\ 100 \\
& 1\ 0011\ 100 \\
& \underline{100\ 1110\ 0} \\
& 110|0011|0101|0100 \\
& \ \ 6\quad\ 3\quad\ \ 5\quad\ \ 4 \quad = 6354_{16}
\end{array}
$$

• EXERCISE 12.1

Divide 258_{16} by A. Express the result in hex. **Answer:** 3C.

12.2 Small Digital Systems

As our first example of a small digital system, let us consider the device known as a *shift register*. The need for this arises when data are being transmitted serially (for instance, over a radio channel). In that case the arriving signal voltage may appear as

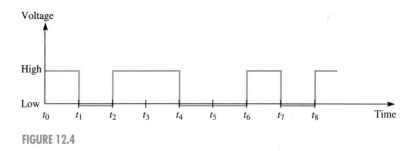

FIGURE 12.4

in Fig. 12.4. Let us assume that the system contains a "clock," which emits regular pulses available to both sender and receiver.* Thanks to the clock, the receiver knows that the first bit of information arrives between times t_0 and t_1, the next between t_1 and t_2, and so forth. If it is arranged that the first-arriving bit is least significant, Fig. 12.4 represents the arrival of the number 01001101, or 4D.

We now wish to store the incoming number 4D, in order to use it for some purpose. The problem is that it arrives 1 bit at a time. There is no way of telling that the

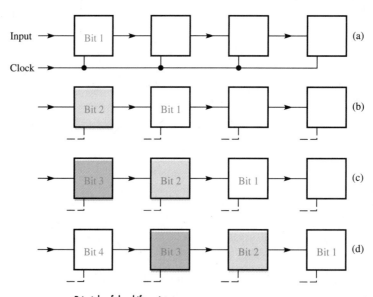

FIGURE 12.5   Principle of the shift register.

___

*There are several ways of arranging this. A system based on the regular ticking of a clock is said to be *synchronous*. Systems without clocks are *asynchronous*.

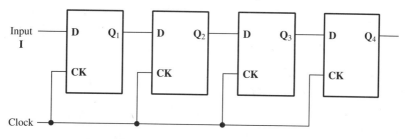

FIGURE 12.6 Realization of a 4-bit shift register with D flip-flops.

incoming data represent 4D unless we can observe all 8 bits at the same time. The purpose of the shift register is to capture the bits as they come in and hold them until the entire byte is there.

The operation of a 4-bit shift register is shown schematically in Fig. 12.5. In this case four storage cells (nature as yet unspecified) are required because there are 4 bits per byte. When time t_0 arrives, the clock informs the first storage cell to note the number present at the input (bit 1) and store it, as shown in Fig. 12.5(a). The contents of the other three cells are irrelevant at this time. When the clock ticks again at time t_1, storage cell 2 records the information that had been in storage cell 1, and storage cell 1 records the new bit present at the input (bit 2), as shown in Fig. 12.5(b). After two more ticks of the clock all 4 bits of the byte have been stored, and are simultaneously available for use, as shown in Fig. 12.5(d). The byte must be used before the next tick, or the next incoming bit will be stored in cell 1 and bit 1 of the old byte will be lost.

It is very convenient to construct a shift register using D flip-flops. The system is shown in Fig. 12.6. After four clock pulses, an incoming 4-bit byte will be simultaneously available at the **Q** outputs of the four flip-flops. The bit that arrived first will be in the flip-flop on the right.

Eight-bit shift registers are conveniently available in MSI IC form (as are the other small systems discussed in this section). They are packaged in the usual 14-pin dual-in-line packages, which are about 1 cm wide and 2 cm long. Eight-bit data latches are also available in MSI form.

EXAMPLE 12.3

Serial data are arriving on wire I, synchronized with clock pulses arriving on wire CL_1. After each 4-bit byte, an end-of-byte pulse arrives on wire CL_2. We wish to capture the incoming bytes in a shift register and then store them in a data latch so that a time equal to four "ticks" of CL_1 is available to make use of each byte. Design an appropriate system using D flip-flops.

SOLUTION

A suitable system is as shown in Fig. 12.7.

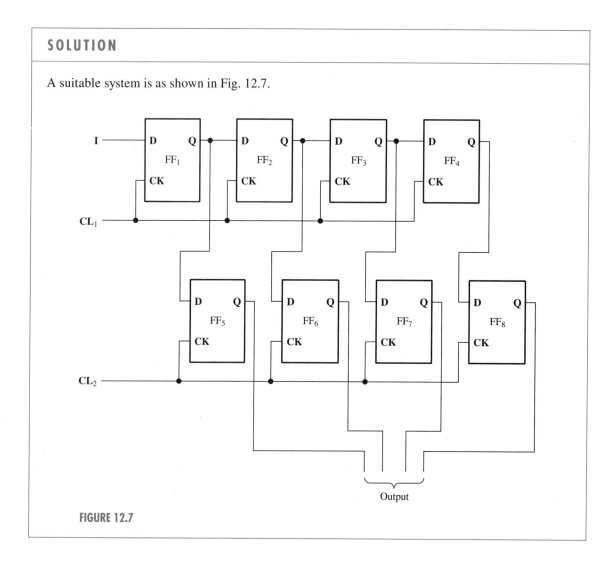

FIGURE 12.7

As our next small system, let us consider the devices known as *counters*. These are used when the input is a series of pulses and we wish to know how many input pulses have been received. A very simple counter is shown in Fig. 12.8(a). The flip-flop is the rising-edge-triggered D flip-flop discussed in Chapter 11. To understand the operation, we refer to the table shown in Fig. 12.8(b). In the left-hand column, N is the number of pulses that have been received. Q_N, in the next column, is the value of Q after N pulses, and D_N is the value of D after N pulses. At the start of operation, no pulses have been received, and $N = 0$ (first line of table). We assume that the initial value of Q, Q_0, is 0, perhaps because the flip-flop has been manually set to 0 using the "clear" input.

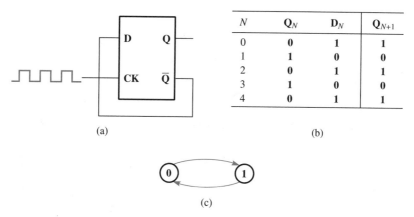

N	Q_N	D_N	Q_{N+1}
0	0	1	1
1	1	0	0
2	0	1	1
3	1	0	0
4	0	1	1

(a)   (b)

(c)

FIGURE 12.8   A 1-bit counter.

The input **D** is connected to $\overline{\mathbf{Q}}$; thus since $\mathbf{Q}_0 = \mathbf{0}$, $\mathbf{D}_0 = \mathbf{1}$. Now when the next pulse is received, activating the clock input of the flip-flop, the output **Q** will take on a new value, $\mathbf{Q}_1$. According to the rules of the D flip-flop, this new value will be the same as the value of **D** *before* the clock input; therefore $\mathbf{Q}_1 = \mathbf{1}$. This is the meaning of the last column of the table. (Since for the top line $N = 0$, $\mathbf{Q}_{N+1}$ in the top line means $\mathbf{Q}_1$.) Now we can proceed to fill in the table. Since we have found $\mathbf{Q}_1$ already, we can enter it on the second line, proceed to compute $\mathbf{Q}_2$, and so on.* This kind of table, which is called a *state table*, is very useful for analyzing sequential systems.

● EXERCISE 12.2

Figure 12.9 is a state table for the shift register of Fig. 12.6. (In this table $\mathbf{I}_N$ is the value of **I** just before the $(N + 1)$th clock pulse is received.) The values of $\mathbf{I}_N$ are to be taken from Fig. 12.4. The first clock pulse comes between t_0 and t_1, the

---

*Note that here $\mathbf{D}_N$ is the value of **D** *before* the $(N + 1)$th clock pulse is received. *After* the $(N + 1)$th clock pulse, **D** changes to a new value, but by this time the flip-flop has already changed. It is the value of $\mathbf{D}_N$ *before* the clock pulse that determines $\mathbf{Q}_{N+1}$.

The time that elapses between the change instruction and the actual change of **Q** is known as the *propagation delay* of the flip-flop. In this circuit the propagation delay is useful, even necessary, because it makes **D** have a constant value until after the change instruction is over. If the propagation delay of the flip-flop is too small, additional delay can be inserted in the wire leading from $\overline{\mathbf{Q}}$ to **D**.

second between t_1 and t_2, and so forth. The flip-flops are all initially cleared to $\mathbf{Q} = \mathbf{0}$. Fill in the state table.

N	$\mathbf{I}_N$	$\mathbf{Q}_{1N}$	$\mathbf{Q}_{2N}$	$\mathbf{Q}_{3N}$	$\mathbf{Q}_{4N}$
0		**0**	**0**	**0**	**0**
1					
2					
3					
4					
5					
6					

FIGURE 12.9

From Fig. 12.8(b) we see that Q is initially 0, becomes 1 after the first input pulse, 0 after the second, and so on. In other words, when the system is in the state Q 5 0 it moves (on receipt of a clock pulse) into the state Q 5 1, and when it is in the state Q 5 1 it moves next into the state Q 5 0. This behavior is illustrated in Fig. 12.8(c), which is called a state diagram.

The reader may be unimpressed by the ability of this circuit to act as a counter. However, it *is* a counter, in fact. After zero pulses have been received, **Q** reads **0**, and after one pulse has been received, **Q** correctly reads **1**. But since **Q** is only a single binary digit, **1** is the highest number to which it can count. Thus on the second pulse it goes back to $\mathbf{Q} = \mathbf{0}$ and begins to count to **1** over again. This simple device would be called a 1-bit counter.

In order to produce a counter that can count to higher numbers, we can use several flip-flops, as shown in Fig. 12.10(a). Here FF1 operates exactly as described for the 1-bit counter. After two pulses $\mathbf{Q}_1$ goes from **1** back to **0**, causing $\overline{\mathbf{Q}}_1$ to go from **0** to **1**. This provides a clock pulse to FF$_2$, which therefore counts once every *two* input pulses. Similarly every second transition of FF2 clocks FF3, which therefore counts once for every *four* input pulses. The result is that $\mathbf{Q}_3\mathbf{Q}_2\mathbf{Q}_1$ is a three-digit binary number (note that $\mathbf{Q}_3$ is the most-significant digit) registering the number of input pulses that have been received. The highest number to which this 3-bit counter can reach is 111_2, or 7_{10}; in general an N-bit counter can count from 0 to $2^N - 1$. This particular type of counter is known as a *ripple counter*. We note that unlike the shift register, this is an asynchronous system; that is, it does not require synchronization to a regularly ticking clock. The pulses to be counted can arrive at random times.

A state table for the ripple counter is shown in Fig. 12.10(b). This table has a somewhat different form than that for the shift register. (In general state tables all tend to be different; they have to be designed to suit the system at hand.) The horizontal arrows indicate times when clock inputs are applied to FF$_2$ and FF$_3$. These times are located by noting that every time $\mathbf{Q}_1$ makes a transition *from* **1** *to* **0**, FF$_2$ is clocked, and when $\mathbf{Q}_2$ goes from **1** to **0**, FF$_3$ is clocked. The state diagram is shown in Fig. 12.10(c).

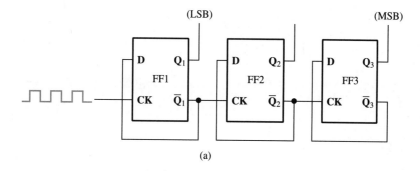

(a)

N	Q_{1N}	Q_{2N}	Q_{3N}	$(Q_3Q_2Q_1)_N$
0	0	0	0	000
1	1	0	0	001
2	0 ⟶	1	0	010
3	1	1	0	011
4	0 ⟶	0 ⟶	1	100
5	1	0	1	101
6	0 ⟶	1	1	110
7	1	1	1	111
8	0 ⟶	0 ⟶	0	000

(b)

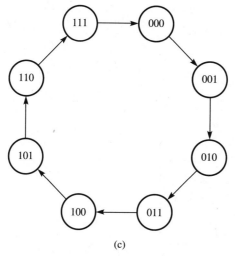

(c)

FIGURE 12.10 A 3-bit ripple counter.

Here the eight states of the system are indicated by the values of the three-digit binary number $Q_3Q_2Q_1$.

EXAMPLE 12.4

Design a system that gives an output of **0** until six input pulses have been received, at which time the output becomes **1**.

SOLUTION

We can use a 3-bit ripple counter (which can count as high as 7_{10}). Let the three outputs of the counter be $Q_3Q_2Q_1$ (Q_3 most significant). Then we must use a three-input logic circuit that gives an output of **1** if and only if $Q_3 = 1$, $Q_2 = 1$, $Q_1 = 0$ (110_2 being equal to 6_{10}). From Chapter 11 we know how to synthesize any truth table by the sum-of-products method. (See Figs. 11.15 and 11.16. In this case we do not need the **OR** gate, since only one combination of inputs needs to have an output of **1**.) Thus we have the circuit shown in Fig. 12.11.

For the sake of simplicity in this example we have not said what the circuit should do when pulses *after* the sixth one arrive. The circuit shown will give outputs of **1** after the 6th, 14th, 22nd, etc. pulses. However, it would be possible to redesign the circuit to make it reset itself to zero after the sixth pulse. In that case it would give outputs of **1** after the 6th, 12th, 18th, . . . pulses, which might be more useful for some purposes. We note, for example, that a clock could be made by using a counter to count pulses of the 60 Hz power line (which are extremely accurately spaced). The circuit would use a 6-bit ripple counter to count 60 pulses. Then an **AND** gate would be used to derive an output **1** that would cause a second hand on a clock face to move ahead one notch and reset the counter to zero. This is exactly the way digital clocks are made.

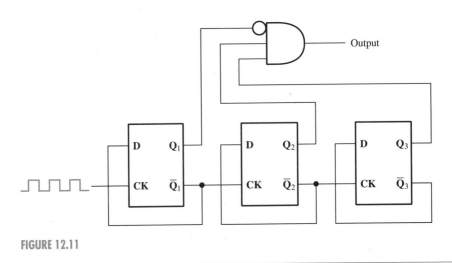

FIGURE 12.11

EXAMPLE 12.5

Another type of counter, known as a *synchronous counter*, is shown in Fig. 12.12(a). Fill in the timing diagram shown in Fig. 12.12(b), with Q_1, Q_2, Q_3 initially 0. Show that the number of counted pulses is the binary number $Q_3Q_2Q_1$.

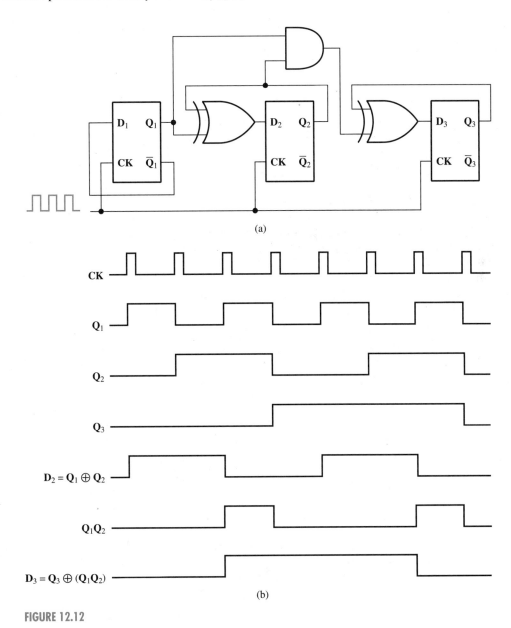

(a)

(b)

FIGURE 12.12

SOLUTION

One advantage of this type of counter, as compared with the ripple counter, is that it reduces problems with propagation delays. In the ripple counter the input triggers the first flip-flop, whose output changes and triggers the second, whose output changes and triggers the third, and so on. There is a certain small delay, the propagation delay, in each flip-flop; in a 16-bit counter there could be an undesirably long time between arrival of a pulse and stabilization of the counter. In the synchronous counter, on the other hand, all the flip-flops change at the same time; thus the total delay is the same as the delay of a single flip-flop.

• EXERCISE 12.3

Figure 12.13 shows a circuit known as a *divide-by-three* counter. Initially Q_1 and Q_2 are zero. First construct a timing diagram for Q_1 and Q_2. Then construct a state diagram for $Q_1 Q_2$. Show that the circuit returns to its original condition after every *three* pulses.

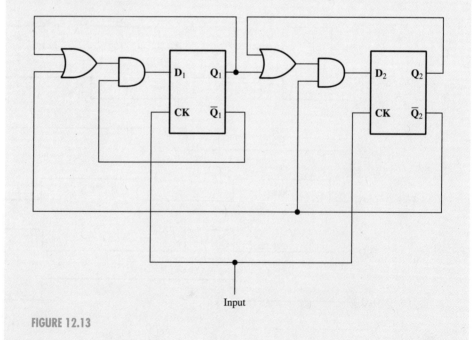

FIGURE 12.13

12.3 Digital Signal Processing

Information is increasingly being processed in digital form. However, many signals originate as analog signals and must be converted to digital signals before digital processing technology can be used. For instance, consider the digital telephone system discussed in Section 0.3. Here the microphone produces an analog signal, proportional to the acoustic pressure. This signal is *sampled* periodically. After each sampling, or measurement, a set of binary digits representing the measurement is generated. These sets of digits, generated periodically after each sampling, become the new digital signal. The device that constructs the equivalent digits is known as an *analog-to-digital* (or A/D, pronounced "A-to-D") *converter*.

SAMPLING

As we shall see, a certain time is required for A/D conversion to take place. However, the input analog signal is changing constantly. Thus at each sampling we must capture the instantaneous value of the input signal and retain it while digitization is taking place. This is the function of the *sample-and-hold circuit* shown in Fig. 12.14. A mechanical switch is not really used in the circuit, as this would be much too slow; a field-effect transistor circuit, which acts as a fast-acting electronic switch, is used instead.* Regularly spaced pulses from the clock briefly close the switch and then reopen it. While the switch is closed, C charges to the then-existing input voltage. Then, while the switch is open, the capacitor, which discharges very slowly, holds the input voltage of the voltage follower nearly constant, providing a constant voltage out-

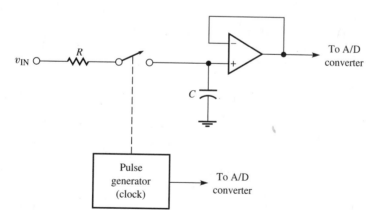

FIGURE 12.14   Sample-and-hold circuit. These are available in IC form.

---

*Field-effect transistors (FETs) are discussed in Chapters 13 and 15.

put that the A/D converter can convert. Pulses from the same clock are sent to the A/D converter so that it operates in synchronism with the sampling process.

One may ask how frequently the sampling must occur. There is a mathematical theorem, the Nyquist sampling theorem, which states that an analog signal having frequency components up to a highest frequency f_{max} can in principle be converted to digital form without any loss of information, provided that the sampling interval T_s satisfies

$$T_s < \frac{1}{2f_{max}} \tag{12.1}$$

In practice, sampling usually is performed at two to five times the minimum rate determined by Eq. (12.1).

A/D AND D/A CONVERTERS

We have already mentioned A/D conversion at the start of a digital telephone link. At the receiving end of the telephone the reverse process must take place. After each set of binary digits arrives, an output voltage must be generated that is proportional to the number that was received. Since new numbers are received every few microseconds, the output voltage varies constantly according to the incoming instructions, giving an analog output signal. In the telephone system this voltage is proportional to the original acoustic pressure and is applied to the earphone. The device used on the receiving end is a *digital-to-analog* ("D-to-A" or D/A) *converter.*

Neither the D/A nor the A/D is, strictly speaking, a digital system; both are androgynous, half digital and half analog. Moreover, the A/D converter is not really a very small system. However, these devices are important enough to require discussion, and they do resemble the other devices discussed in this section in that they are available as integrated-circuit blocks.

A D/A converter is shown in Fig. 12.15 in schematic form. This converter is based on the summing amplifier op-amp circuit. From Eq. (4.26) we have

$$v_{OUT} = -\tfrac{1}{4}[v_1 + 2v_2 + 4v_3] \tag{12.2}$$

The values of v_1, v_2, and v_3 are determined by the switches. Their operation is assumed to be controlled by the input digits Q_1, Q_2, and Q_3. When $Q_1 = 1$, the upper switch is in the left position, and $v_1 = 4$ V; when $Q_1 = 0$, $v_1 = 0$ V. The other two switches operate similarly. If the three-digit binary number applied to the input is $Q_3Q_2Q_1$ (Q_3 being the most-significant bit), the output voltage is given by

$$v_{OUT} = -[4Q_3 + 2Q_2 + Q_1] \tag{12.3}$$

Clearly the output voltage is proportional to the decimal equivalent of the binary number applied to the input.

Figure 12.16 illustrates the operation of the D/A converter. We imagine that inputs Q_1, Q_2, Q_3 are connected to the output of a three-bit counter like that of Fig. 12.10. As

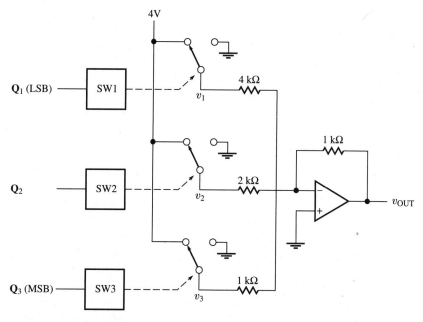

FIGURE 12.15 A 3-bit D/A converter.

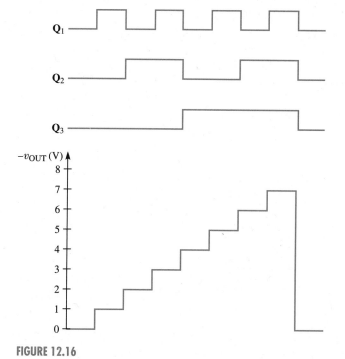

FIGURE 12.16

the counter's output increases from 000 to 111, the output of the converter is as shown. The switches are FET switches, as mentioned earlier. The entire device, including op-amp and FET switches, is available in IC form.

The inverse device, the analog-to-digital converter, is somewhat more complicated. Its operation is shown in outline form in Figure 12.17. The process of A/D conversion is normally repetitive. After each tick of the clock at the top of the diagram the process begins anew. When the clock pulse is received, the control circuitry resets the counter to zero and starts the pulse generator. The output of the counter then consists of a steadily increasing train of binary numbers, which appear on its output bus. For simplicity Fig. 12.17 shows a 3-bit system; thus the output of the counter is first 000, then 001, then 010, and so forth. The multiplexer (MUX) is a switch that on instructions from the controller sends the counter output to the D/A converter (which is like the one described in the preceding paragraph). The output of the D/A converter is the voltage v_1. The comparator block then compares v_1 with the input voltage v_{IN}. So long as $v_1 < v_{IN}$, the comparator tells the controller to let the process continue; the counter output and v_1 continue to increase. However, when v_1 becomes larger than v_{IN}, the output of the comparator changes, telling the controller that the correct binary number has been reached. At this time the binary number at the output of the counter is just larger than v_{IN}. The controller now tells the multiplexer to switch these binary digits to the output terminals, where they are available for use until the next tick of the clock. When the clock ticks again, the process repeats, converting whatever value v_{IN} now has at this later time.

The process of A/D conversion has many steps, and it might seem that it would take a long time. In fact A/D conversion *is* a comparatively slow process, but thanks to

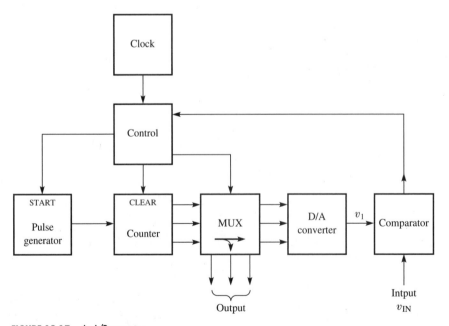

FIGURE 12.17   An A/D converter.

the high speed of its individual parts it is still quite possible to obtain 100,000 conversions per second. The A/D converter is also comparatively large and complicated. (The controller is in fact a very small computer in itself.) Nonetheless, complete A/D converters are now also available in IC form.

It is interesting to consider the accuracy of the 3-bit A/D converter. The steadily increasing v_1 is like the v_{OUT} shown in Fig. 12.16. Suppose v_{IN} is 4.1 V. The digital output of the counter would increase until $v_1 > v_{IN}$, which happens when $v_1 = 5$ V. The corresponding digital output is 101. Now if we send this digital output back through the D/A converter, the resulting v_{OUT} would be 5 V. It seems reasonable that when an analog voltage is converted to digital and back to analog, the original voltage should be recovered, but clearly a large *quantization error* has been introduced. This error arises because in this example the size of the digital steps is large, 1.0 V out of a maximum 7.0 V. All voltages within the 1 V step size are treated as identical by the A/D converter. Thus expressed as a fraction of the full-scale analog voltage, 7 V, the maximum error introduced in a complete A-to-D-to-A conversion is 1/7, or 14 percent. This is much too large for most purposes. In general, the percentage quantization error of an N-bit system is given by

$$\text{Percentage quantization error} = 100(2^N - 1)^{-1} \qquad (12.4)$$

If the input to the D/A converter is an eight-wire bus, v_{OUT} as shown in Fig. 12.15 will have $2^8 - 1 = 255$ steps, instead of seven. The error will thus be reduced from 1/7 to 1/255, or 0.39 percent.

EXAMPLE 12.6

Suppose the speed of an 8-bit A/D converter is limited by the counter, which has a maximum speed of 4×10^7 counts per second. Estimate the maximum number of A/D conversions per second that can be obtained.

SOLUTION

What we are calculating here is the rate of the clock at the top of Fig. 12.17. This rate will be constant, independent of v_{IN}, and thus must be slow enough to allow the counter to count up to the highest possible input voltage. This will require 255 counts, which will take 6.375 μsec. Thus the process can be repeated 157,000 times per second. Actually the process will be slightly slower than this because of the time required for the controller to operate at the start and end of each cycle.

As an example of the use of digital signals, let us consider digital phonograph recording for home audio use. This technique offers tremendous advantages over the old-fashioned long-playing record. The lp is purely analog, the displacement of the groove being proportional to the signal voltage. Unfortunately, errors creep into the analog process. Dirt and groove imperfections produce minute changes in signal voltage, which collectively result in sound distortion. Furthermore, very loud passages cannot be recorded because the side-to-side displacements become so large that the stylus cannot stay in the groove. Now let us consider a digital phonograph record. Information is stored as a sequence of **1**s and **0**s. No signal on the record has a larger amplitude than the others (since all signals are either "high" or "low"). Thus loud sounds are as easy to record as soft ones. Moreover, small errors will in most cases have no effect, because any voltage in the high range will be interpreted as a **1**, irrespective of its exact value. (This insensitivity to noise is a major advantage of digital systems.) One such system uses a laser pickup, which uses light to read out the depths of tiny pits in the record. (The depths are either "deep" or "shallow," corresponding to **1** and **0**.) In this system there is no physical contact with the record, which, besides being almost free of audible distortion, should last practically forever!

A diagram illustrating the design of a digital phonograph player is shown in Fig. 12.18. The shift register and data latch perform the same function as the sample-and-hold circuit did in the case of A/D conversion: they retain each byte of information while it is being converted by the D/A converter.

One advantage of digital signal processing, its comparative immunity to noise, has already been mentioned. Perhaps a greater advantage, however, is the versatility of digital systems. Once information has been converted to digital form, it can be manipulated and processed very readily. For example, in modern automobiles efficient engine operation is extremely important in order to reduce emissions and fuel consumption. In older cars the carburetor was adjusted to some compromise setting and left there.

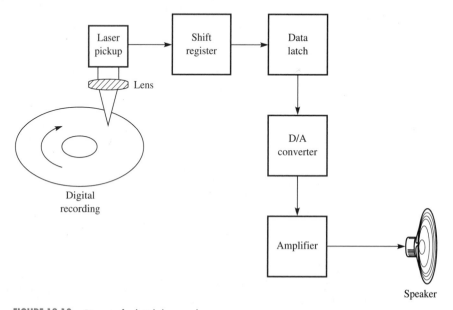

FIGURE 12.18   Diagram of a digital phonograph.

The resulting operation was simple but inefficient. Today's cars, however, are often equipped with sensors to monitor such variables as engine temperature, speed, and vacuum pressure. The resulting information is digitized and applied to a very small computer, which then calculates the correct engine settings: timing, choke setting, and so forth. What we have here is a "smart" automobile, one that constantly adjusts itself for optimum operation. The key to this technique is the digital computer module, which can be programmed to generate the correct outputs in response to the sensor data. It would be nearly impossible to perform this function with analog blocks.

Many other kinds of equipment can also be made self-optimizing by this approach, which represents an important advance in the development of technology. Of course the technique would not be so useful, if the computer module had to be large and expensive. However, modern IC technology has succeeded in producing the microprocessor, known as the "computer on a chip." These are small and inexpensive, and thus are ideal for all sorts of "smart" machines. Microprocessors will be discussed in Section 12.4.

*12.4 Introduction to Computer Architecture

VLSI fabrication techniques have made it possible to construct systems with thousands of components in IC form. Design and development costs are high for such large ICs, and thus the variety of available circuits is not very great. On the other hand, once designed, the fabrication costs are not much higher than for small ICs. Thus when a large IC becomes popular and many are sold, the development costs can be worked off and the price can come down. Eventually some of these extremely powerful building blocks become available at very reasonable cost. Physically these ICs are available in packages that resemble the 14-pin DIP, but they are somewhat larger: The semiconductor IC itself may be 7 mm square, and the package 5 cm long, with (often) 40 pins.

MEMORIES

We have already seen how eight flip-flops can form a storage register capable of remembering an 8-bit byte of information. The important system known as a *random-access memory* (or RAM) is simply a large collection of such registers, fabricated in one IC. Memory ICs are described by the number of kilobits they can store. For example, a "256k RAM" can store 32,000 bytes of 8 bits each.*

Such a large array of flip-flops presents problems of organization. Obviously it is impractical to bring out the **D** and **Q** connections of each flip-flop, since this would

---

*The RAM is not to be confused with the ROM, or read-only memory. In the latter, permanently stored information can be read out, but new information cannot be entered. Computer memories are described in terms of bytes rather than bits. A 16k computer memory would use N 16k memory ICs, where N is the number of bits per byte.

require half a million connections to the IC. The technique of *addressing* is used to bring the number of connections down to a manageable size.

Let us consider a hypothetical memory containing a large number of 8-bit registers. We assign a number, called an *address* or *memory location*, to each register. These numbers are usually given in hexadecimal. The first register is designated 0000, the second 0001, and so on up to FFFF. Using four-digit hexadecimal addresses we can number 16^4 (= $65,536_{10}$) addresses. Each register can, of course, store an eight-digit binary number, but it is more convenient to think of the stored number as a two-digit hex number. We might think of the memory as being like a set of numbered pigeonholes; into each pigeonhole can be placed a slip of paper on which a two-digit hex number is written. Thus, for instance, it makes sense to say, "the number 3C is stored at location 2C4F."

Suppose we wish to read out the number stored at a certain memory location. The memory will have to have an eight-wire data bus coming out of it, over which the number can be transmitted. However, we also must be able to tell the memory *which* of the 65,536 numbers it is to send. This can be done using an additional 16-wire bus known as an *address bus*. If we wish to interrogate memory location 2C4F we simply place the binary numbers 0010 1100 0100 1111 on the address bus. In addition, there will probably be another wire to "strobe" the memory; that is, the command to act (send the desired number on the data bus) is given by placing a **1** on the strobe wire. One more wire will be needed for the read/write instruction. The reading-out operation, for example, could be elected by placing a **1** on the read/write wire. We must also have a way of putting information *into* the memory. This would be done by selecting the address for storage with the address bus, placing the number to be stored on the data

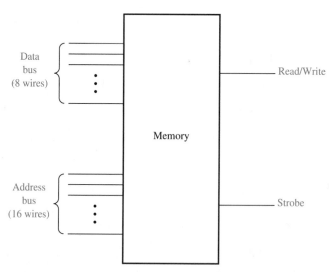

FIGURE 12.19   Diagram of a hypothetical memory storing 65,536 bytes of 8 bits each.

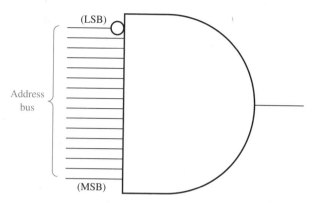

(LSB)

Address
bus

(MSB)

FIGURE 12.20   A decoder for a 16-bit address bus. This particular decoder is
for memory location FFFE$_{16}$.

bus, setting read/write to **0** to select the writing mode, and applying a **1** to the strobe.
A diagram of this hypothetical memory is shown in Fig. 12.19.

Internally, the address bus is brought to each memory location. At each register is
a *decoder*, a simple device like that shown in Fig. 12.20. The decoder shown in Figure
12.20 gives an output of **1** when the address bus is excited by the digits FFFE$_{16}$; the
output from this decoder enables that particular register. (*Enables* means "allows it to
respond.") The decoders on all the other registers have different combinations of
inverters on their inputs and are enabled only by their own particular addresses. The
rest of the switching functions (enabling the flip-flops, read/write, strobe) are accomplished with a few additional FET switches and logic blocks. Thus the basic principles
of a large memory IC are quite simple. On the other hand, the tremendous number of
internal components and interconnections make production a very formidable task.

DISPLAYS

An interesting application of the decoder involves digital displays. These displays have
become familiar as the devices that "read out" information from pocket calculators,
wristwatches, and countless other products. A typical seven-segment display is shown
in Fig. 12.21(a). In operation, different groups of segments light up to represent the
decimal numerals from 0 through 9. For example, to display the decimal number "1"
we light segments S_1 and S_2; to display "2" we light segments S_0, S_1, S_3, S_4, and S_6.

The number to be displayed arrives in the form of binary digits on a data bus. Let
the incoming binary number be *WXYZ*. Furthermore, let us define state variables $\mathbf{S_0}$,
$\mathbf{S_1}$, etc., such that when $\mathbf{S_1} = \mathbf{0}$, segment 1 is lit (and when $\mathbf{S_1} = \mathbf{1}$, segment 1 is
dark), and so forth. Then a truth table can be constructed relating inputs **W**, **X**, **Y**, and
Z to the outputs $\mathbf{S_0} \ldots \mathbf{S_6}$. This table is shown in Fig. 12.21(b). In order to realize this
table, we can use a combination of decoders and **OR** gates. For example, from the table

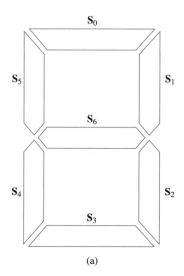

(a)

Decimal integer	W	X	Y	Z	S_0	S_1	S_2	S_3	S_4	S_5	S_6
0	0	0	0	0	0	0	0	0	0	0	1
1	0	0	0	1	1	0	0	1	1	1	1
2	0	0	1	0	0	0	1	0	0	1	0
3	0	0	1	1	0	0	0	0	1	1	0
4	0	1	0	0	1	0	0	1	1	0	0
5	0	1	0	1	0	1	0	0	1	0	0
6	0	1	1	0	1	1	0	0	0	0	0
7	0	1	1	1	0	0	0	1	1	1	1
8	1	0	0	0	0	0	0	0	0	0	0
9	1	0	0	1	0	0	0	1	1	0	0

(b)

FIGURE 12.21   The seven-segment display.

we see that S_0 is to equal **1** if *WXYZ* is 0001, 0100, or 0110. A circuit to realize this is shown in Fig. 12.22(a). (In this figure all inputs marked "**W**" are connected together, etc.) A different combination of decoders is used to generate S_1 and the other **S** outputs. Clearly, a rather large collection of gates is required just to operate a single seven-segment display, and of course *N* seven-segment displays are needed to read out an *N*-digit decimal number. Fortunately, these circuits do not have to be built from individual gates. Since they are used in great numbers, they are available prepackaged as integrated circuits, known as seven-segment decoder/drivers. The decoder/driver block is shown in Fig. 12.22(b). It is an IC with four inputs and seven outputs, each of which has sufficient current capacity to control the appropriate segment of the display.

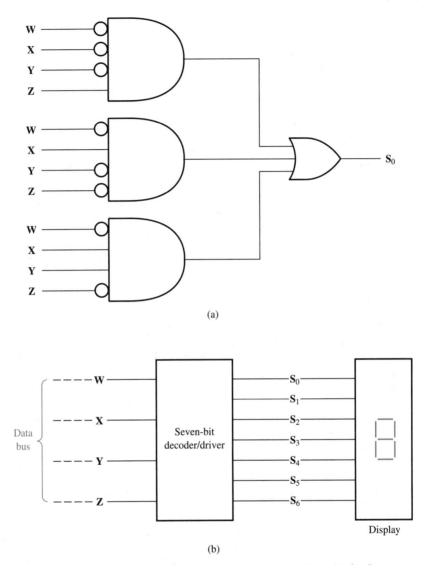

(a)

(b)

FIGURE 12.22 Seven-segment decoder/driver. (a) Circuit for segment S_0. (b) Decoders for all seven segments in IC form.

(In general, IC outputs with greater-than-usual output power capacity are known as "drivers.") Thanks to this convenient IC, it is only necessary to insert the one IC block between the data bus and the display, and all connections are made automatically. (Often the decoder/driver is built right into the display unit, which is still more convenient.) Such simplifications make a great contribution to digital technology because they allow complicated functions to be performed with just a few parts and minimal wiring.

The display itself can be of several kinds. In the past gas-discharge tubes were used as displays. These, however, were bulky and power-consuming, and they have largely given way to *light-emitting diodes* (LEDs) and *liquid-crystal displays* (LCDs). LEDs are special *pn* junctions (see Chapter 13) made in the shapes of the segments shown in Fig. 12.21(a). They cannot be made from silicon; gallium arsenide and its chemical relatives (known collectively as III–V semiconductors) must be used instead. When a large current is passed through the junction, green, red, or infrared radiation is emitted, depending on which semiconductor is used. LEDs are quite similar in construction to semiconductor lasers. They can provide considerable light and thus make excellent displays. Their principal disadvantage is their high power consumption, which makes them unsuitable for battery-operated equipment.

An alternative with low power consumption is the liquid crystal display. These devices do not emit radiation at all; like the moon, they operate by reflected light. Liquid crystals are complex organic compounds that change their optical properties when placed in an electric field. An LCD display contains liquid crystal between transparent electrically conductive electrodes. When no voltage is applied to the electrodes, the liquid crystal has no effect and light is reflected in the usual way. However, when voltage is applied to a segment of the display, the liquid crystal rotates the polarization of light passing through it. Incoming light is made to pass through a polarizer before arriving at the crystal. When light has passed through the crystal and been reflected back, it cannot pass back through the polarizer if its polarization has been changed. Thus a segment that has voltage applied to it appears *dark*. Only about 4 V is required to operate typical LCD displays. They are very common in portable equipment, but they are less easily readable than LED displays, and are also rather slow, requiring a sizable fraction of a second to change from one digit to another.

MICROPROCESSORS

The best-known large digital system is undoubtedly the computer. The subject of computers is of course a very large one. Our object here is to introduce some basic principles, and in particular to relate the computer to the smaller digital systems already considered. In keeping with our emphasis on IC building blocks, we shall be thinking primarily in terms of *microprocessors*. A microprocessor is a building block in which the central part of a computer is realized in IC form. Each microprocessor is different in its details. Our discussion will be based on a hypothetical microprocessor with "typical" properties.

Fundamentally, a computer is a kind of automaton, or robot. All a computer can do is perform a small set of very simple tasks. What, then, distinguishes a computer from an old-fashioned mechanical adding machine? The most important difference is that a computer can perform a predetermined series of steps according to a set of instructions called a *program*. A second difference, less fundamental but still decisive, is that computers perform their simple operations very fast. The power of a computer

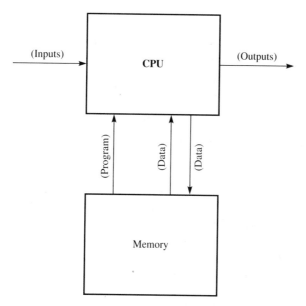

FIGURE 12.23 Typical outline of a computer.

comes from its ability to do simple, standardized operations, over and over at a prodigious rate, in the proper order.* It is interesting that a computer may be able (for example) to play chess well enough to defeat any human player, while inside, at the deepest level, everything is being done with flip-flops and gates. A computer, one might say, consists of digital blocks—flip-flops and gates—plus a lot of organization. Primarily the electrical engineer is concerned with making the building blocks faster and cheaper, and with organizing them for the greatest efficiency.

At the heart of a computer is the *central processor unit*, or CPU. This is the part of the computer that directs the flow of operations and performs them. (A microprocessor is in essence a realization of a CPU in IC form.) Usually the CPU works in cooperation with a memory, such as the RAM discussed earlier. There will also be input/output devices (keyboard, video screen, etc., and their supporting circuits). A typical block diagram of a computer is shown in Fig. 12.23.

The computer's program is usually stored in the memory. (It is recorded into the memory at some earlier time using the input/output circuitry.) However, we cannot store an *English-language* program in the memory; the only things that can be stored in

---

*Note how this relates to digital technology in general. Digital hardware consists of a few cheap and simple building blocks, cleverly interconnected and repeated thousands of times.

8-bit memory registers are two-digit hex numbers.* Thus we use a kind of code, known as *machine language,* for our program. In this code, instructions are represented by two-digit symbols. For example, $3A_{16}$ might mean "stop." A program written in machine language is then just a series of two-digit hex numbers, which can conveniently be stored in the memory. We should note, however, that not everything stored in the memory is program. There will also be numbers (*data*) arising from the calculations themselves. Data also consist of two-digit hex numbers, which look just the same as the numbers used for programs. It is impossible to tell whether, for example, "AB" is a number or an instruction in a program until we pay attention to the location where it is stored. The programmer must keep track of such things, and know whether the quantity stored in location *WXYZ* is program or data.

To carry out its operation a typical CPU makes use of several special registers.† One of these is the *accumulator.* Most operations performed on numbers are performed by first storing the number to be acted upon in the accumulator. Table 12.3 lists some typical CPU operations, most of which involve the accumulator. Three-letter "names" of the operations are given in the left-hand column of the table. (These names are usually *mnemonics*—memory aids—because they contain some initials of the words that describe the operation.) The right-hand column gives the two-digit hex symbol of the operation in an imaginary machine language. Each CPU can have its own set of symbols for operations; thus there is no universal machine language. Moreover, each CPU will have a different selection of operations. Table 12.3 shows only 5, but a typical microprocessor may have 30 or so. All of the operations are very simple, like those shown. There will not be an instruction for "take the cosine of the number stored in the accumulator." There may not even be an instruction for multiplying two numbers. The most basic set of operations does not need to include such operations because they can be synthesized out of more basic operations. For example, a number can be multiplied by three by adding it to itself twice. The cosine of a number must be generated by adding up the power series ($\cos x = 1 - x^2/2! + x^4/4! - x^6/6! + \cdots$). In fact, even large computers generate trigonometric functions in this way, and it is possible to do so with a CPU that has no arithmetic abilities beyond those of Table 12.3.†† This may seem very slow and clumsy, but it should be remembered that each of the operations listed in Table 12.3 takes very little time, perhaps 3 microseconds in a typical microprocessor and much less in a large computer. If a multiplication requires 30 operations, even a microprocessor can multiply 10,000 numbers every second.

---

*We are continuing to use, as an example, the hypothetical memory introduced earlier, which stores 8-bit bytes.

†Physically these registers may be located in the memory, but from the point of view of computer architecture they are considered part of the CPU. Different microprocessors may have different architectures than described here; we are considering a simple example.

††These five instructions would not be sufficient by themselves. However, the other necessary instructions would be ones affecting the order of the program (analogous to IF, GO TO, and DO in FORTRAN), not ones that perform arithmetic on the data. For example, x^2 can be generated (at least when x is an integer) by adding x to itself $x - 1$ times.

TABLE 12.3   Some Typical Microprocessor Operations*

Mnemonic Name	Meaning of Operation	Symbol in a Hypothetical Machine Language
LDA	Take a number stored somewhere in the memory and *load* it into the *a*ccumulator. (It remains stored in the memory.)	AD
ADC	Take a number stored somewhere in the memory and *ad*d it to the number stored in the accumulator. The sum is stored in the accumulator. (The number originally stored in the memory remains there unchanged.)	6D
STA	Take the number stored in the *a*ccumulator and *st*ore it somewhere in the memory. (It remains unchanged in the accumulator.)	8D
ASL	*S*hift number in *a*ccumulator *l*eft one bit. (The accumulator contains an 8-bit binary number. This instruction shifts each digit one place to the left, and enters a **0** for the least-significant bit. The most-significant bit disappears from the accumulator, although it may be stored elsewhere. For example, 0110 0110 becomes 1100 1100; or equivalently in hex notation 66 is replaced by CC.)	0A
SBC	Take a number stored somewhere in memory and *sub*tract it from the accumulator. (The difference is stored in the accumulator; the number in memory remains there.)	ED

*Examples in this chapter are taken from the MCS 6500 microprocessor family.

It will be noticed that some of the instructions in Table 12.3 are not complete in themselves. For instance, STA "takes a number stored in the accumulator and stores it somewhere in the memory," but in which memory location is the number to be stored? To provide this necessary information, the instruction is followed by an address. In our hypothetical memory an address consists of four hex digits; thus the entire instruction will consist of six hex digits, which themselves will occupy three memory locations. For example, suppose the instruction "store the contents of the accumulator in register *ABCD*" is itself to be stored in memory locations 0000, 0001, and 0002. We would record 8D in location 0000, CD in location 0001, and AB in location 0002. (We adopt the common convention that the least-significant byte of the address comes first, followed by the most-significant byte.) A program for transferring a number from memory location 31F2 to location AC4B would be

Program Storage Location	Operation	Machine-Language Codes
0018	LDA	AD
0019		F2
001A		31
001B	STA	8D
001C		4B
001D		AC

Note that we have arbitrarily placed the first instruction in memory location 0018. The program can be stored anywhere, provided that the locations are not already being used for something else.

To execute the program, we start the CPU at location 0018. It reads the instruction AD, recognizes that this instruction requires an address, and thus reads the next two bytes as well. Now it can perform the LDA operation, after which it goes on to the next instruction, 8D, and so on, following the instructions in the order in which they are recorded in the memory. Execution continues in this way until a STOP instruction is reached or a branching or GO TO instruction changes the sequence of operations.

EXAMPLE 12.7

Using the hypothetical machine language of this section, write a program for multiplying the two-digit hex number in location 12A6 by 2 and storing the result in location E34B.

SOLUTION

We arbitrarily choose to begin the program storage at location 1000. A suitable program is shown in the table.

Program Storage Location	Operation	Machine-Language Codes
1000	LDA	AD
1001		A6
1002		12
1003	ADC	6D
1004		A6
1005		12
1006	STA	8D
1007		4B
1008		E3

It is instructive to follow the operation of the CPU as it executes a program. In order to do this we will need to consider some of the other special registers. The *program counter* contains the address of the next instruction to be read and performed. For instance, in Example 12.7 we store the number 1000 in the program counter before calculation begins. (Clearly the program counter must actually be a double register, containing 16 flip-flops.) The CPU then increments (adds one to) the program counter as necessary to move through the program. The program counter is useful because it allows the CPU to keep track of where to look for the next instruction. This is particularly useful in the case of GO TO or branch instructions. A GO TO instruction, for example, sets the program counter to a new address, out of sequence with the addresses used until then. The mnemonic name of this operation is JMP ("jump"); let its machine-language code be 4C, followed by the address to which the jump takes place. Then a possible sequence might be as shown in the table.

Program Storage Location	Operation	Machine-Language Codes
A32B	JMP	4C
		31
		14
1431	LDA	AD
		CD
		AB

In this example the JMP operation actually stores a new number (1431_{16}) in the program counter; the CPU then reads the address in the program counter and goes there for the next instruction.

Branch operations are similar except they are *conditional* jumps (analogous to IF in FORTRAN). In a *branch* instruction the program counter is set to one new value if something is true or to a different new value if the thing is not true. For example, let us imagine an instruction BEQ ("*b*ranch if *eq*ual,") whose code is F0 followed by a two-digit hex number. The effect of this instruction is that if the number in the accumulator is not zero, the program counter increments normally to the next operation. However, if the number stored in the accumulator is zero, a "branch" occurs: the two-digit hex number is added to the program counter.* Consider the following sequence:

Program Storage Location	Operation	Machine-Language Code
0000	LDA	AD
0001		3B
0002		10
0003	SBC	ED

---

*The two-digit number is still considered to be the address associated with the BEQ instruction, but it is *not* a memory location; it is the number of program locations to be skipped. Hence it is called a *relative* address.

Program Storage Location	Operation	Machine-Language Code
0004		3C
0005		10
0006	BEQ	F0
0007		03
0008	LDA	AD
0009		00
000A		00
000B	STA	8D
000C		00
000D		10

This program will do one of two different things, depending on whether the numbers initially stored in locations 103B and 103C are equal or unequal. If they are unequal, BEQ will not affect the program counter and LDA will load the contents of register 0000 into the accumulator. This number is AD, since 0000 happens to be one of the locations in the program. Then STA transfers AD to location 1000. On the other hand, if the numbers stored in 103B and 103C are equal, BEQ will add "three" to the program counter, when the program counter would normally increment to 0008. As a result, the number that appears in the program counter is not 0008 but instead 000B. The LDA instruction is skipped, and the number in the accumulator (zero) is stored in location 1000.

The *instruction register* contains the latest instruction read out of memory. Every time the program counter advances to a new instruction, its code is stored in the instruction register. Similarly, the *address register* contains the address (if any—some instructions, such as STOP, do not need an address) associated with the last instruction that was read. The actions of the various special registers can now be followed in detail as we execute the simple program in the following example.

EXAMPLE 12.8

Consider the following program:

Program Storage Location	Operation	Machine-Language Codes
0000	LDA	AD
0001		00
0002		10
0003	STA	8D
0004		00
0005		20

0006	ADC	6D
0007		00
0008		10
0009	STA	8D
000A		01
000B		20

Let us assume that the number initially stored in location 1000 is AA. Fill in the following table, showing the contents of the special registers after each step.

After This Many Steps Have Been Performed	Present Contents of Registers				The Next Operation to Be Performed Is	Machine-Language Codes for Next Operation
	Program Counter	Instruction Register	Address Register	Accumulator		
0	0000	—	—	—	LDA	AD 00 10
1	0003	AD	1000	AA	STA	8D 00 20
2					ADC	6D 00 10
3						
4					—	—

SOLUTION

Before any steps have been performed, the program counter is set to 0000, the address of the first instruction. The other three special registers contain random numbers left over from previous calculations—their exact contents do not matter for present purposes, as indicated by the dashes. The CPU then proceeds to read the contents of memory locations 0001, 0002, and 0003 and stores the codes·it finds there in the instruction and address registers. Then it follows these instructions, with the result that AA is stored in the accumulator. Then the process continues. It is left to the reader to fill out the remainder of the table as an exercise.

It is certainly possible for a user to program a microprocessor (or even a large computer) in machine language, but programming in machine language is laborious and prone to errors. Instead, it is usual to program in assembly language or a higher-level language such as BASIC, Pascal, or FORTRAN. When a higher language is used, it usually is not necessary to keep track of memory locations; also, the program can be shorter than it would be in machine language because one higher-level-language instruction may execute many machine-language instructions. However, we note that in any case the CPU is run by a machine-language program. The assembly-language or higher-level language must first be translated into machine language by a special program, known respectively as an *assembler* or *compiler*. Microprocessors have now become sufficiently powerful to run their own compiler programs, so that even very small computers can now be programmed in higher languages.

Although rapid technological change is characteristic of electronics generally, the area of microprocessors has developed especially rapidly. In the last few years the cost of a moderate-size computer has decreased, and so has the physical size of the machine; meanwhile, software and peripherals (disk drives, light pens, and so on) have steadily gained in power and flexibility. This is certainly one of the most exciting areas of modern engineering, but the student who makes it his or her specialty must plan to keep up to date!

12.5 Application: Digital Process Control

"Hey, I'm studying to be a mechanical (civil, industrial, agricultural, . . . or whatever) engineer. Is this electrical stuff going to be of any use to *me*?"

Yes, you can bet it will be useful, in many different ways. But digital process control is more than just useful; it is a revolution taking place before our eyes. The changeover from "dumb" machines to "smart" machines is a step comparable in importance to the industrial revolution itself. Ever since machines were invented, they have been designed for compromise, with inefficiency built in. Only now has it become possible for machines to regulate themselves for optimum operation, making tremendous improvements possible.

For example, consider an automobile engine. It has a number of adjustments, traditionally set by a mechanic: spark timing, fuel-air mixture, idle setting, and so forth. The way that these adjustments are set determines how well the engine will run: how much power it produces, and with what fuel efficiency, and how much smog it creates in the process. Now, in theory, one could find the right set of adjustments and set them to make power, efficiency, and smog all as good as possible. But then what happens? A hot day comes, and the engine works differently when it is warmer. The driver buys some lower-octane gas. The barometric pressure changes. And the engine gets older and drifts out of adjustment. In this situation, the engine still may be able to run, inefficiently and with clouds of smog, but it will certainly not be giving the best performance that in theory it could give.

A clear solution is to build some sort of automatic control into the engine, so that it adjusts itself for changing circumstances. The vacuum spark advance, which changes the timing in response to the gas pedal, is a mechanical control device, which has been in use for decades. But with electronics, you can do much more. One can continually measure the vacuum pressure, engine temperature, engine speed and torque, and even the amount of pollutants in the tailpipe. This information can be automatically converted to digital form and applied to a computer, which automatically adjusts the engine for best performance. One part of the system can be a feedback loop; for example, the system can monitor the level of exhaust pollutants, and continually readjust the engine to minimize them.

The beauty of systems like this lies in their power and versatility. Using a microcomputer (in a car, this would be a computer built as a single integrated circuit), one can generate any kind of complicated input/output relationships one wants. For instance, if the pressure needs to depend on temperature according to $P = (T^3 - 1)/7$, the computer can be programmed to do that. It can also be programmed to change P to its new value slowly, or after a time delay; any control logic that one can imagine can be programmed.

The extension to other kinds of machines is evident. Suppose you are spray-painting objects as they come down an assembly line. Your system can continuously monitor the temperature and humidity and adjust the paint mixture accordingly. Or suppose you are designing clothes washers: Why design to get average performance with average loads? The machine can weigh the clothes and adjust the water level and soap. It is easy to think of other examples, because almost any machine or system will work better if it is "smart," that is, if it optimizes itself in response to conditions. In this age in which we have to conserve resources and minimize pollution, optimization and efficiency will be more and more important.

"All right," the reader may say, "I get the general idea, but I still don't have a clue how to set up such a system in practice." Fortunately, the pieces needed for setting up such systems are commercially available, and come with convenient instructions and design information. For example, the Hewlett-Packard Corporation makes a wide range of data acquisition systems. Their Model 3421A Data Acquisition/Control Unit measures up to 30 variables, including dc and ac voltages, frequency, and temperature. It can make up to 24 voltage measurements each second, and is meant to work with Hewlett Packard computers and other components, providing compatible, modular systems for all kinds of process control.

Once you see the possibilities, you will be ready to look for new ways to apply this new technology. It will not be hard to find opportunities. There are a lot of old-fashioned, inefficient machines out there for you to optimize!

POINTS TO REMEMBER

- A register is a set of N flip-flops, which can store one N-bit byte of information. Bytes can be transmitted from one register to another through an N-wire *bus*.

- Four-bit bytes are conveniently represented by *hexadecimal* digits. A number can be represented in binary, decimal, or hexadecimal form.

- A *shift register* is a small system that collects bytes of data when the bits are arriving serially.

- *Counters* are small systems that move through a cycle of states, moving one step for each input.

- Some useful tools for analyzing sequential systems are *timing diagrams*, *state tables*, and *state diagrams*.

- *Analog-to-digital* and *digital-to-analog converters* are used as interfaces between analog and digital systems.

- *Memories* are large digital systems used primarily in computers. They consist of large arrays of registers combined with *decoders* for *addressing*.

- *Displays* are visible output devices for reading out digital information. Seven-segment LED and LCD displays are common.

- *Microprocessors* are realizations of computer *central-processor units* (CPUs) in IC form.

- Computers are automatic machines that perform a small number of simple operations very quickly, in an order determined by a program.

- The CPU executes programs in *machine language*. Assembly language or higher-level languages must be translated into machine language before they can be used.

PROBLEMS

Section 12.1

12.1 Suppose 10^6 8-bit bytes per second are being transmitted through an eight-wire bus. How many bits per second are being transmitted? How many hex numbers per second?

12.2 What is the largest decimal number that can be represented by a 100-digit binary number? Based on this, approximately how many decimal digits does one obtain for each binary digit? Does one arrive at the same ratio if one begins with a 200-digit binary number? A 20-digit binary number?

12.3 Add the following pairs of binary numbers:
  a. 11 plus 11.
  b. 11011 plus 00110.

12.4 Perform the following subtractions involving binary numbers:
  a. 101 from 110.
  b. 1111 from 1100.

12.5 Multiply the following pairs of binary numbers:
  a. 10 by 11.
  b. 10110 by 101.

12.6 Find the hexadecimal equivalents of the following binary numbers:
  a. 1101.
  b. 11101.
  c. 110111101.

12.7 Find the decimal equivalents of the following binary numbers:
- **a.** 1100.
- **b.** 1011100110.

12.8 Find the decimal equivalents of the following hexadecimal numbers:
- **a.** 4B.
- **b.** F6D.
- **c.** C91B.

12.9 Find the binary equivalents of the following hexadecimal numbers:
- **a.** 1F.
- **b.** C03.

*12.10 Find the binary equivalents of the following decimal numbers:
- **a.** 33.
- **b.** 100.
- **c.** 4673.

*12.11 Find the hexadecimal equivalents of the three decimal numbers in Problem 12.10.

12.12 Add the following pairs of hexadecimal numbers:
- **a.** F plus B.
- **b.** 88 plus B7.
- **c.** C3FB plus E6D.

12.13 Subtract FE_{16} from $34A_{16}$. Express the result in hex.

12.14 Multiply $4C3_{16}$ by $B4_{16}$. Express the result in hex.

12.15 What hex number is represented by the flags in Fig. 12.1 if the most significant bit is on the left? If it is on the right?

Section 12.2

12.16 Consider the shift-register/latch system shown in Example 12.3. Assume $Q = 0$ initially for all flip-flops. Binary digits arrive on line D equivalent to the hex number B9, with the least significant bit arriving first. CL_1 pulses during each data bit, and CL_2 pulses after every four pulses of CL_1. Construct a table showing the Q values for each of the eight flip-flops, after each pulse of CL_1.

12.17 Construct a state diagram for a 4-bit shift register when the input is an infinitely repeating series of 4-bit bytes (the least-significant bit arriving first) equivalent to the following:
- **a.** 9_{16}.
- **b.** A_{16}.
- **c.** F_{16}.

*12.18 For the system of Fig. 12.24:
- **a.** Construct a state table similar to that of Exercise 12.2. Assume that initially $Q_1 = 1, Q_2 = Q_3 = Q_4 = 0$.
- **b.** Construct a state diagram for $Q_1 Q_2 Q_3 Q_4$.

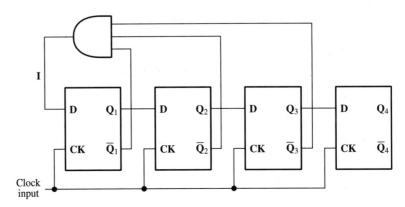

FIGURE 12.24   A 4-bit ring counter.

*12.19 For the system of Fig. 12.24, construct a complete state diagram showing all 16 states of the system. Your diagram should have 16 circles, numbered 0000 through 1111, connected with arrows showing which state each state goes to after one tick of the clock. Interestingly, this system has a *limit cycle:* that is, no matter which state it starts in, it ends up going repetitively through the same sequence of states. Find the limit cycle.

*12.20 The system of Fig. 12.24 is known as a 4-bit *ring counter* because it can be used to give an output of **1** after every fourth clock pulse. Show how it can be used in this way. Can you design a system that gives an output of **1** after every *fifth* clock pulse?

*12.21 We wish to design a system that gives an output of **1** after every 60 input pulses. (This system, which would be called a *divide-by-sixty counter*, could be used in a household clock. It would move a second hand one notch after every 60 cycles of the 60 Hz electric power line.) Design a suitable system using three ring counters. Do not use more than a total of 12 flip-flops.

*12.22 Figure 12.25 shows a 3-bit *Johnson counter*. Construct a complete state diagram showing all eight possible states. Show that the system has two different limit cycles. Show how (with the addition of one gate) it can be used as a divide-by-six counter. (See Problem 12.21.)

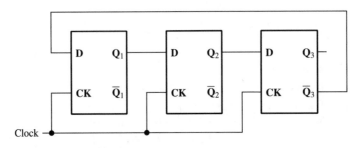

FIGURE 12.25   A 3-bit Johnson counter.

*12.23 A 5-bit Johnson counter is shown in Fig. 12.26. Show that, depending on the initial values of $Q_1 \cdots Q_5$, the device finds itself in one of four different limit cycles (repetitive sequences of states). Find the limit cycle that contains only two states.

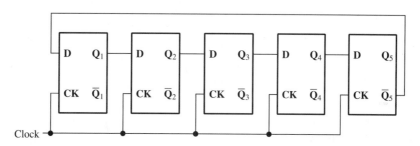

FIGURE 12.26 A 5-bit Johnson counter.

*12.24 In Fig. 12.27 the clock input is a train of regularly spaced pulses. Construct state diagrams for Q_1Q_2 as follows:

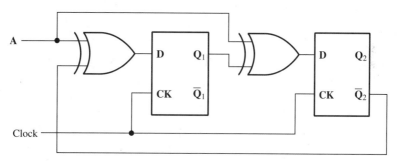

FIGURE 12.27

a. When input $A = 1$.
b. When input $A = 0$.
c. Sketch timing diagrams showing Q_1 and Q_2 as functions of time for the two cases $A = 1$ and $A = 0$.
d. Explain in words how the timing diagrams differ in the two cases.

12.25 Design an 8-bit digital-to-analog converter for which the full-scale voltage (output corresponding to an input FF_{16}) is 10 V.

12.26 Suppose that the speed of an 8-bit D/A converter is limited by the switching time of its FET switches. Let this time be 3 μsec. How many 8-bit bytes per second can be converted?

*12.27 A problem arises with the D/A converter of Fig. 12.15 if the time required for a switch to move from right to left is different than the time required to move from left to right. Let us suppose these times are 3 μsec and 6 μsec, respectively. Suppose we convert the byte $Q_3Q_2Q_1 = 100$ and then at $t = 0$ we change the input to $Q_3Q_2Q_1 = 011$. Make a sketch of the analog output versus time. Do you see the problem?

12.28 Consider the supply-voltage connection marked "4 V" at the top of Fig. 12.15. Suppose that instead of applying a constant 4 V at this point we apply $4 \cos \omega t$ volts. Find v_{OUT} for the following cases:

a. $Q_3Q_2Q_1 = 000$.
b. $Q_3Q_2Q_1 = 001$.
c. $Q_3Q_2Q_1 = 010$.
d. $Q_3Q_2Q_1 = 100$.

Section 12.3

*12.29 Consider the digital phonograph of Fig. 12.18, and assume that 16-bit bytes are being used. The highest audio frequency of interest is 20 kHz, and the sampling interval will be 10 μsec.

a. How many bits per second must be read from the record?
b. If the record turns at 4 revolutions per second and the diameter of the smallest track on the record is 10 cm, what length of track is available for each bit?
c. Let the length found in part (b) be called x. If we assume that one bit occupies an area $4x^2$ on the record, what must the outside diameter of the record be so that each side can play for 1 hour?

**12.30 Suppose a cosine wave with 1 V amplitude and frequency f is digitized with an ideal 8-bit A/D converter and then reconverted to analog form with an 8-bit D/A converter. Let us suppose also that a graph of the output of the D/A converter consists of straight line segments connecting the points that are produced at each sampling.

a. Sketch the output of the D/A if the sampling interval is $1/2f$ and if it is $1/8f$. Assume one sampling point is always located at $t = 0$.
b. From part (a) we see that the reconstructed waveform differs from the original and that the error decreases when the sampling interval decreases. Find the approximate maximum sampling interval that can be used if the error introduced in this way is not to exceed the 0.0039 volt error caused by 8-bit quantization.

Mathematical suggestions: $\cos(A + \epsilon) = \cos A \cos \epsilon - \sin A \sin \epsilon$. When $\epsilon \ll 1$, $\sin \epsilon \cong \epsilon$, $\cos \epsilon \cong 1 - \epsilon^2/2$.)

12.31 A certain microcomputer has memory addresses numbered 0000 to FFFF. However, because a small memory chip is used, only locations 0000 to 04FF are actually represented physically by registers. Moreover, of the existing registers, 266_{10} are used by the microprocessor for its own organization purposes and are thus unavailable to the programmer. How many programmable locations are there? Give your answer as a decimal number.

*12.32 Let the number of bits (not bytes) stored in a memory be 2^N. We wish to store a 30 minute musical selection in digital form. Let us use 8-bit bytes, and let the sampling interval be 20 μsec.

  a. Find the minimum integer value of N.

  b. Suppose that the largest available ICs have $N = 21$. How many such ICs would be required?

Section 12.4

For the following questions involving machine-language programming, use the codes given in Table 12.3 and in the text.

12.33 Write a machine-language program that takes a number originally stored in location 043C and stores it in location 03AA.

12.34 Write a machine-language program that multiplies a one-digit hex number stored in location 02A7 by 3 and stores the result in location 0000.

12.35 Write a machine-language program that does the following: The number in location 0034 is subtracted from the number in 0266. If the result is zero, the number 01 is stored in location 0002. If the result is not zero, the number 00 is stored in location 0002.

12.36 Write a program that multiplies a one-digit number stored in location 03CC by 8. Use the ASL instruction.

12.37 Write a program that multiplies N, a one-digit number stored in location 03CD, by 6. Do this by using ASL to obtain $2N$ and again to obtain $4N$, and adding the results.

12.38 Complete the table in Example 12.10.

12.39 Construct a table like that of Example 12.10 for the following program:

LDA	AD
	00
	02
BEQ	F0
	03
ADC	6D
	01
	02
STA	8D
	02
	02

Assume that the number stored in location 0200 is zero and that the number in 0201 is 0A. Store the program in consecutive memory locations beginning with 0000.

Advanced Exercises in Machine-Language Programming

Most CPUs include one or more special registers known as *index registers*. These can be used to keep track of the number of times a repetitive operation has been performed. For example, suppose we wish to add a number to itself three times. We store the number in the accumulator and set the number in the index register (which we shall call X) equal to zero. Then we perform one addition using ADC, add one to X, and perform a branch instruction, which stops the process if X now equals three. If $X \neq 3$, a JMP instruction takes the program back to ADC, and the process repeats until three additions have been performed. Some useful instructions related to the index register: DEX (subtract one from X), CA; INX (add one to X), E8; LDX plus four-digit address (load number stored at address into X), AE; STX plus four-digit address (store X at address), 8E; TAX (transfer accumulator to index X), AA; TXA (transfer X to accumulator), 8A; BX0 plus relative address (branch if $X = 0$), 00.*

12.40 Write a program that stores the contents of 02AA in the index register and then reduces X by one.

12.41 Write a program that subtracts X, the number in the index register, from the number stored in the accumulator and stores the result in the index register.

12.42 Write a program that adds the number stored in location 0000 to itself three times, using the index-register method outlined in the introductory comments above Problem 12.40.

12.43 Write a program to calculate the square of a one-digit hex number stored in location 2000. Store the result in location 2001.

12.44 Assume that the one-digit hex number stored at 0200 divides evenly into the two-digit hex number stored at 0201. Write a program that finds the quotient and stores it in 0202.

12.45 At the start of the following program, the number F6 is stored in location 0200, 13 is stored in 0201, and 0B is stored in 0202. (All numbers are hex.) Find the (hex) number stored in location 0203 after the program is run.

Program Storage Location	Operation	Codes
0000	LDA	AD
0001		00
0002		02

(continued)

---

*All these operations are taken from the 2500 microprocessor language, except BXO, which is contrived for simplicity. The reason for this is that we have chosen not to complicate matters by mentioning the zero flag. In 2500-language the same effect could be produced by using first CPX plus a four-digit address (compare X with the contents of address), EC, followed by BEQ plus relative address, which will branch if the two numbers just compared are equal.

(Problem 12.45 continued)

Program Storage Location	Operation	Codes
0003	LDX	AE
0004		01
0005		02
0006	SBC	ED
0007		02
0008		02
0009	DEX	CA
000A	BXO	00
000B		03
000C	JMP	4C
000D		06
000E		00
000F	STA	8D
0010		03
0011		02

ACTIVE DEVICES AND CIRCUITS

CHAPTER 13

SEMICONDUCTOR DEVICES

W e now turn our attention to the internal structure of IC building blocks. Here we encounter a new family of circuit elements known as *semiconductor devices*. This group includes diodes and also transistors of various kinds. The main reason that we have not already introduced these circuit elements is that they are nonlinear in their *I-V* characteristics; as a result, circuit analysis is less straightforward when such elements are present. On the other hand, they are extremely important.

In general, the subject of electric circuits can be divided into the subfields of *passive circuits* and *active circuits*. Passive circuits are those that contain only *passive circuit elements*, such as resistors, capacitors, or inductors. Active circuits contain not only passive circuit elements but also *active elements*, which in most cases are transistors. Circuits containing transistors are also known as *electronic circuits*.* All of the analog and digital blocks discussed in Chapters 9 to 12 are active circuits, and transistors are essential to their internal construction. In this chapter we shall introduce the most important semiconductor devices. Chapters 14 and 15 will deal with their applications in analog and digital circuits.

---

*Electronic circuits can in principle contain vacuum tubes or some other active elements in place of transistors, but the most common active elements today are transistors, by far.

13.1 Semiconductors

Semiconductors are crystalline solid materials with electrical properties midway between those of insulators and metals. By far the most important semiconductor material at present is silicon.* The prevalence of silicon technology follows not only from its good electrical characteristics, but also from a large number of other favorable properties that affect fabrication of devices. Although we shall not emphasize the subject here, fabrication technology is a very complex and subtle art. (For example, the material used in semiconductor devices must consist of perfect single crystals. Impurity atoms must be removed to obtain a purity of one part in 10^{10}!) Sometimes one observes that a device is designed in a certain way when other ways appear to be just as good. The difference may lie in details of fabrication that are by no means obvious.

Electrical conduction in semiconductors can take place when unbound electrons (which are negatively charged) move through the crystal, as shown in Fig. 13.1(a). We note that a flow of negative charges from left to right gives rise to a current from right to left. In semiconductors, however, there can also be mobile positive charges known as *holes*. Physically a hole is a vacancy that arises when a bound electron that should be present is missing. For our purposes, however, we can imagine that holes are positively charged particles that act very much like electrons, with charge equal in magnitude to the electronic charge but opposite in sign. A current can also arise from a motion of holes, as shown in Fig. 13.1(b). Both holes and unbound electrons are referred to as *charge carriers*, or simply *carriers*.

Very pure semiconductor (known as *intrinsic* semiconductor) has very few charge carriers of either kind and hence is such a poor electrical conductor that it is almost an insulator. However, by adding tiny controlled amounts of impurities, a semiconductor can be made to contain desired numbers of either holes or free electrons. Material con-

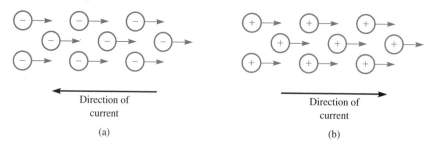

FIGURE 13.1   (a) Electrons moving from left to right give rise to a current directed from right to left. (b) Holes moving from left to right give rise to a current directed from left to right.

---

*Germanium is used for a few special purposes. Gallium arsenide has properties that could in principle make it superior to silicon, but gallium arsenide technology is beset with practical difficulties.

taining primarily holes is known as *p-type semiconductor* and that containing primarily free electrons is known as *n-type semiconductor*. It is not possible for the material to be both *n-type* and *p-type* because holes, we remember, are absences of electrons; thus when a hole and electron meet, they annihilate each other and both disappear. *n*-type material normally contains very few holes, and *p*-type material normally has very few free electrons. In *p*-type material, holes are thus said to be the *majority carriers*. However, we shall soon see that it is possible to artificially inject electrons into *p*-type material, in which case they become excess *minority carriers*. The life of an excess minority carrier cannot be very pleasant, since it is in great danger of meeting a majority carrier and being annihilated. Nonetheless, minority carriers play a vital role in certain devices.

13.2 Diodes

When *p*-type and *n*-type materials are placed in contact,* the resulting structure is called a *pn junction*. With the addition of two wires for external connection, this simple structure becomes a useful circuit element known as a *pn-junction diode*, as shown in Fig. 13.2(a). The circuit symbol for the diode is shown in Fig. 13.2(b).

PN JUNCTIONS UNDER BIAS

The *pn* junction is not only a useful circuit element in itself, but also appears as a component of other devices. Thus it is worthwhile to investigate its internal behavior when external voltages are applied. In particular, we shall be interested in the following three phenomena: (1) one-way conduction; (2) injection of minority carriers; and (3) collection of minority carriers.

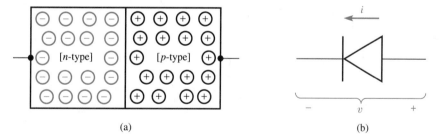

(a)   (b)

FIGURE 13.2 The *pn* junction. (a) Physical structure. (b) Circuit symbol. Note that the triangle in the symbol points from *p* to *n*. Sign conventions for current and voltage are as shown.

---

*"Placed in contact" is really a figure of speech. Actually the material must be a single continuous crystal. The junction is made by doping the two sides of the crystal with different impurities.

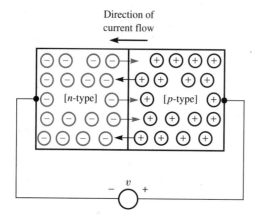

FIGURE 13.3   *pn* junction with forward bias.

A voltage applied across a semiconductor device is often called a *bias*. Figure 13.3 shows a *pn* junction with bias applied in such a way that the *p*-type material is made positive with respect to the *n*-type. The effect of the bias is to cause both holes and electrons to move across the junction, holes from right to left and electrons from left to right. The reason for this motion can be thought of as electrostatic attraction: Negative-ly charged particles are attracted to the side made positive by the external voltage source, and positively charged holes are attracted to the negative potential applied on the left. Both kinds of particles contribute to a total current from right to left, as shown. A voltage applied in this way is called a *forward bias* because when the voltage has this sign, current flows easily. Note that the direction of current flow (right to left) is the same as the direction the triangle points in the symbol for the device [Fig. 13.2(b)].

If the sign of the applied voltage is reversed, we have the situation of Fig. 13.4. Now the applied potentials tend to pull the particles away from the junction. Very few particles cross the junction, and hence very little current flows. A voltage applied with this sign is known as a *reverse bias*. We see that the junction has the property of one-way conduction: Voltages applied with one sign produce current flow, but those with the opposite sign do not. (This is quite different from a resistor, which conducts equal-ly well in either direction.) The direction in which current flows easily is indicated by the triangle in the circuit symbol.

When the junction is forward biased, as in Fig. 13.3, carriers cross it in both direc-tions. This results in the appearance of minority carriers on both sides of the junction. The effect, known as *injection of minority carriers*, will be seen below to be important in bipolar transistors.

When the junction is reverse biased, as in Fig. 13.4, no minority carriers are injected. On the contrary, if a minority carrier should somehow be placed near the junction, the applied voltage has the proper sign to pull it back across the junction, thus converting it back into a majority carrier. This effect, called *collection of minority car-*

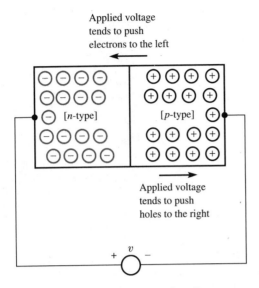

Applied voltage
tends to push
electrons to the left

[n-type]   [p-type]

Applied voltage
tends to push
holes to the right

v

FIGURE 13.4 *pn* junction with reverse bias. Charge carriers are pulled away from the junction, and very little current flows.

riers, usually takes place at the collectors of bipolar transistors, as we shall see in Section 13.3.

I-V CHARACTERISTIC OF THE PN JUNCTION

The theory of semiconductor devices shows that the *I-V* characteristic of an ideal *pn* junction obeys the equation*

$$i = I_s(e^{qv/kT} - 1) \tag{13.1}$$

Here the sign conventions for *i* and *v* are as shown in Fig. 13.2(b), *q* is the electronic charge, *k* is Boltzmann's constant, and *T* is the temperature in degrees Kelvin. The quantity *kT/q* has the dimensions of voltage, and we may define a quantity V_{T0} (the *thermal voltage*) according to $V_{T0} = kT/q$. It is useful to remember that at room temperature (300 K) $V_{T0} = 0.026$ V. The factor I_s is a constant (the *saturation current*) that is independent of applied voltage; its value depends on such things as the size of the junction, the impurity concentrations, and the temperature. Typically its value is in the range 10^{-8} to 10^{-14} A.

---

*In our notation, capital-letter symbols represent constant voltages and currents, and lowercase-letter symbols represent time-varying voltages and currents. Where no distinction is being made, the lowercase symbol is used.

$$i = 10^{-13}\left(e^{v/0.026} - 1\right)$$

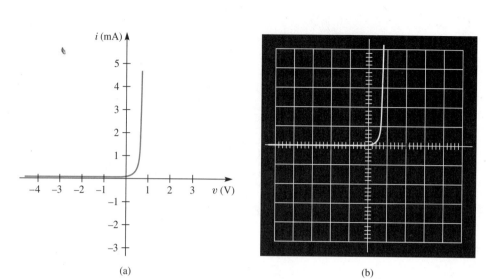

(a)   (b)

FIGURE 13.5   The current-voltage characteristic of a *pn* junction. (a) According to Eq. (13.1), with $I_s = 10^{-13}$ A. (b) The actual characteristics of a silicon diode (curve-tracer photograph). Vertical scale: 1 mA per large division; horizontal scale: 1 V per large division.

The *I-V* characteristic of Eq. (13.1) is graphed in Fig. 13.5(a). In this figure a value $I_s = 10^{-13}$ A has been assumed. Figure 13.5(b) shows an experimental *I-V* characteristic for comparison.* The shape of these curves is easily understood. When v is a positive number, we have forward bias. In this case the power to which e is raised in Eq. (13.1) is a positive number. The exponent is $qv/kT = v/V_{T0} = v/0.026 = 38.5v$; this becomes a fairly large exponent for modest forward bias voltages. For instance, if the forward bias voltage is 0.7 V, $38.5v = 26.95$, and $e^{v/V_{T0}} = 5.06 \times 10^{11}$. Thus with increasing forward bias the "one" in Eq. (13.1) quickly becomes negligible, and i is an exponentially growing function of v. This is what is seen on the right-hand side of Fig. 13.5. On the other hand, when v is negative, we have reverse bias. In this case the exponent is negative and $e^{v/V_{T0}}$ becomes negligible compared with the one. Thus for negative values of v, $i \cong -I_s$. Since $I_s \cong 10^{-13}$ A, the value of i in reverse bias is nearly zero. This is of course consistent with the idea of one-way conduction.

It should be mentioned that the *I-V* characteristics just described apply to an ideal diode. Real-world diodes are described rather well by the theory, except at high forward or reverse voltages. At high forward voltages i increases more slowly than Eq. (13.1) predicts, and at sufficiently high reverse voltages there is an onset of reverse current, an effect known as *reverse breakdown*.

---

*The experimental *I-V* characteristic is measured with a special oscilloscope known as a *curve tracer*. This convenient instrument automatically displays on its screen the *I-V* graph of any device plugged into a socket on its front panel.

EXAMPLE 13.1

A diode is very definitely not a resistor. A resistor has the property that v/i is always the same, while for a diode the value of the ratio v/i depends on what value of v is applied. Calculate the ratio v/i for an ideal diode [i.e., a diode assumed to obey Eq. (13.1)] with $I_s = 10^{-13}$ A for the following applied voltages: $-2, -0.5, +0.3, +0.5, +0.7, +1.0, +1.5$ V.

SOLUTION

Using Eq. (13.1), the results are as follows:

v (V)	i (A)	v/i
-2	-10^{-13}	2×10^{13}
-0.5	-10^{-13}	5×10^{12}
$+0.3$	1.0×10^{-8}	3×10^{7}
$+0.5$	2.2×10^{-5}	2.3×10^{4}
$+0.7$	0.050	14
$+1.0$	5.1×10^{3}	2.0×10^{-4}
$+1.5$	1.1×10^{12}	1.4×10^{-12}

We observe that the ratio v/i is a very pronounced function of the applied voltage. This illustrates the useful one-way conduction property of the diode. Reverse voltages give rise to very small reverse currents; thus the ratio v/i is very large. On the other hand, moderate forward biases give rise to very large currents, making the ratio v/i small when v is a forward voltage greater than 0.7 V or so. The last two lines of the table are unrealistic, as one might expect from the sizes of the indicated currents. Real diodes deviate from Eq. (13.1) before such unreasonably large currents are reached.

Since the *pn* junction is a nonlinear circuit element, its presence complicates circuit analysis. Node and loop equations become transcendental. To solve them one can use various approximation methods; this is usually best when one is trying to gain a general idea of how a circuit works. When precise answers are required, one can use convenient computer programs set up for this purpose. The graphical method of Chapter 3 works well on circuits where only one nonlinear element is present. This, it should be noted, is *not* an approximation method; in principle it is exact. Its results are not very precise, however, because of random errors in drawing and reading the graphs. Whether one uses computation, graphing, or approximation methods will depend on what one is doing and how accurate the answers need to be.

EXAMPLE 13.2

We wish to find the current i_D through the diode in the circuit of Fig. 13.6(a). Obtain a numerical answer by means of the graphical load-line method of Chapter 3. Let $R_1 = 1{,}000\ \Omega$ and $V_1 = 5$ V. The diode's saturation current I_s is 10^{-12} A.

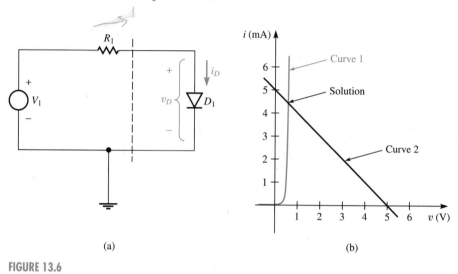

(a)   (b)

FIGURE 13.6

SOLUTION

First let us see what happens if we attempt to find i_D by means of a conventional node equation. A correct node equation for V_D is

$$\frac{v_D - V_1}{R_1} + i_D = 0$$

Expressing v_D in terms of i_D using Eq. (13.1) results in

$$\frac{kT}{q}\ln\left(\frac{i_D}{I_s} + 1\right) + R_1 i_D = V_1$$

This is a transcendental equation of a sort that in general can only be solved numerically.

The load-line technique is based on the concept of identifying two parts of the circuit whose characteristics may be separately graphed. An imaginary break is indicated by the dotted line in the circuit in Fig. 13.6(a). We sketch, on one set of axes, i_D versus v_D for the subcircuits to the left and right of the dotted line; the solution is found at the intersection of the two curves. For the diode,

$$i_D = I_s[\exp(qv_D/kT) - 1]$$

This characteristic is graphed as Curve 1 in Fig. 13.6(b). For the resistor and voltage source (as seen from the diode terminals),

$$v_D = V_1 - i_D R_1$$

This characteristic is graphed as Curve 2 in the same figure. (Note that the positive sense of i_D must be the same for both curves—clockwise in the circuit diagram in this example.) It is seen that the diode current is about 4.5×10^{-3} A.

MODELS FOR THE DIODE

Various degrees of idealization of the diode may be used to simplify diode circuits. The more the actual behavior of the diode is simplified, the easier it is to visualize the operation of the diode in the circuit. However, the greater the idealization, the less accurate and the less widely applicable it becomes. For example, if one were to proceed to the extreme of saying that a diode, since it conducts current, acts like a resistor, then the most essential property—that of conducting in only one direction—would be lost.

The term *rectifier* refers to a device that passes current in only one direction; hence the *pn* junction is a kind of rectifier. Let us define a *perfect rectifier* as having the following properties:

1. When $i > 0, v = 0.$
2. When $v < 0, i = 0.$

Property 1 states that conduction in the forward direction can occur with no voltage drop across the diode; property 2 states that if voltage is applied with the sign that tends to produce reverse current, no current flows. It is easily seen that the *I-V* characteristic shown in Fig. 13.7(a) agrees with properties 1 and 2. To indicate that we mean the idealized perfect rectifier rather than a real diode, we can use the special symbol shown in Fig. 13.7(b).

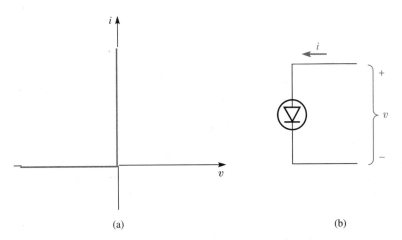

(a)   (b)

FIGURE 13.7 The perfect rectifier. (a) *I-V* characteristic. The vertical portion, corresponding to forward bias, is the same characteristic as that of a short circuit. The horizontal portion, corresponding to reverse bias, is the same characteristic as that of an open circuit. (b) Circuit symbol for the perfect rectifier.

A comparison of the *I-V* characteristic of the perfect rectifier with the characteristic of a typical *pn* junction diode, such as that of Fig. 13.5, shows that the perfect rectifier is an idealization of the diode. In many cases in the analysis of diode circuits it is useful to replace the diodes by perfect rectifiers. The operation of the circuit can then be quickly visualized, and an approximate analysis can be made. It should be kept in mind that in using an idealization, some features of the real device have been lost. One must be sure that in the problem under consideration the omitted features do not make a vital difference. Since the essential difference between the perfect rectifier and an actual diode is the small but nonzero forward voltage drop of the actual diode, the effect of this voltage drop should be considered.

EXAMPLE 13.3

The input voltage v_1 in the circuit in Fig. 13.8(a) is a pure sinusoid. Find the output voltage v_2 (that is, graph v_2 versus time). Treat the problem approximately by replacing the diode by a perfect rectifier.

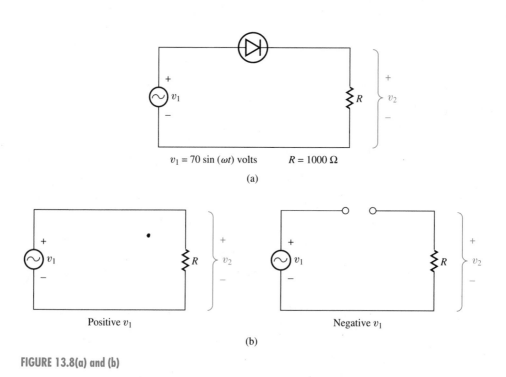

$v_1 = 70 \sin(\omega t)$ volts $R = 1000 \ \Omega$

(a)

Positive v_1 Negative v_1

(b)

FIGURE 13.8(a) and (b)

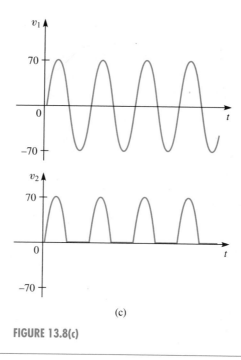

(c)

FIGURE 13.8(c)

SOLUTION

Whenever the voltage v_1 is positive, the junction is forward biased; whenever v_1 is negative, it is reverse biased. Replacing the diode by a perfect rectifier is equivalent to replacing the diode by a short circuit for positive v_1 and an open circuit for negative v_1 [Fig. 13.8(b)].

Accordingly, whenever v_1 is positive, $v_2 = v_1$. When v_1 is negative the model requires that the diode current, and consequently the current through the resistor, be zero. In this case v_2 must be zero. The graphs of the input voltage and output voltage are given in Fig. 13.8(c).

Since the voltages present in the circuit during most of the cycle are large compared to the actual diode forward drop of about 0.7 V, the answer as obtained with the perfect-rectifier model is quite reasonable. For example, at the instant the input voltage is +70 V, the output voltage in this approximation is +70 V. The actual output voltage at this instant must be somewhat less— say, 69.3 V—because of the voltage drop that occurs across a forward-biased diode. But while this 0.7 V error is very small, it is relatively larger for smaller input voltages. For example, at the instant that the input voltage equals +2 V, we indicate an output voltage of +2 V. The actual output voltage would be closer to 1.3 V, again because of the small but nonzero voltage drop of a real diode. At this particular instant, the approximation is responsible for a large error, about 35 percent.

THE LARGE-SIGNAL DIODE MODEL

The essential approximation that is made in treating a *pn* junction diode as a perfect rectifier amounts to neglecting the forward voltage drop. It is often the case that the error thus introduced is unacceptably large; yet we hesitate to use the actual diode characteristic because of the analytical difficulties introduced by its nonlinearity. In such instances a better approximation to the actual diode characteristic can be made. In a real diode the current increases so rapidly with voltage that the forward voltage drop in silicon diodes is almost always in the range of 0.5 to 0.8 V for practical currents. For example, we see that for the diode of Fig. 13.5 the forward voltage drop is between 0.6 and 0.7 V for currents in the range of 0.5 to 5 mA. This observation prompts an improved idealization, which we shall call the *large-signal diode model*, as follows:

RULE 1

When significant current flows through the diode in the forward direction ($i > 0$), the voltage drop across the diode, v, is 0.7 V.

RULE 2

When $v < 0.7$ V, $i = 0$.

It is easily seen that these rules agree with the *I-V* characteristic shown in Fig. 13.9(a). This *I-V* characteristic can be modeled by a perfect rectifier in series with a 0.7 V source, as shown in Fig. 13.9(b).

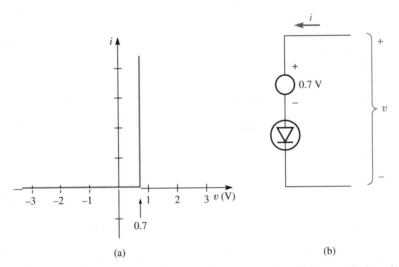

(a)   (b)

FIGURE 13.9   The large-signal diode model. (a) *I-V* characteristic. For reverse voltages and for forward voltages less than 0.7 V, the current is zero. When the current is not zero, the forward voltage is 0.7 V. (b) Circuit model. The combination of a 0.7 V voltage source and a perfect rectifier produces the *I-V* characteristics shown in (a).

Solving diode circuit problems with the large-signal diode model is almost as straightforward as using the perfect rectifier. The constant voltage of 0.7 V merely adds to the signal appearing at the terminals of the perfect rectifier.*

EXAMPLE 13.4

Find the current i_D in the circuit in Fig. 13.10(a) using the large-signal diode model. This is the same problem solved in Example 13.2 using the graphical method. Compare the answers obtained by the two methods.

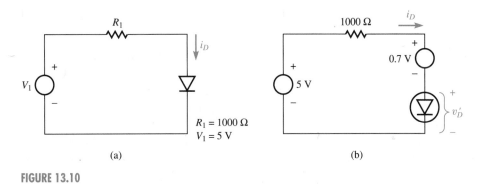

FIGURE 13.10

SOLUTION

The circuit is first redrawn, replacing the diode by the large-signal diode model, Fig. 13.10(b). The voltage drops around the single loop circuit must sum to zero; hence

$$-5 + 1000i_D + 0.7 + v'_D = 0$$

It may be seen, either from this equation or directly from the circuit, that 4.3 V must be dropped across the combination of resistor and the perfect rectifier. Furthermore, the polarity is such that the rectifier can only be forward biased. Since in forward bias there is no voltage drop across a perfect rectifier, the 4.3 V must be dropped across the resistor. From Ohm's law

$$i_D = 4.3/1000 = 4.3 \text{ mA}$$

This approximate answer is seen to be in good agreement with the result 4.5 mA obtained graphically in Example 13.2.

---

*The 0.7 V forward drop is appropriate for silicon diodes. In germanium diodes the forward drop is more typically 0.3 V and in gallium arsenide diodes 1 V.

The principal use of the large-signal diode model stems from the rapid circuit analysis made possible by its simplicity. However, its limitations should be kept in mind. The idealized characteristics given in Fig. 13.9 are not exactly like any real diode characteristics. For example, the model incorrectly implies that the current is zero whenever the forward bias is less than 0.7 V. Actually, the change in conduction is more gradual. Furthermore, according to the model the voltage drop in forward bias never exceeds 0.7 V, whereas in real diodes the voltage drop increases gradually with current and may exceed this value. The model is used, then, as a tool for quickly finding approximate values of desired circuit unknowns. In the analysis of some diode circuits, one may substitute the perfect rectifier alone, ignoring completely the voltage drop across the forward-biased diode. Such a substitution may be made if the voltages in the circuit are so large that a correction of about 0.7 V would make no difference. This is often the case in rectifier circuits, to be discussed below. On the other hand, the 0.7 V drop across the forward-biased diode does become important in diode logic circuits. Finally, there are infrequently some cases in which neither approximation is warranted, and the exact diode characteristics, along with numerical or graphical techniques, must be used.

VOLTAGE ACROSS A FORWARD-BIASED DIODE

In most electronic circuits, currents are of moderate size, seldom exceeding, say, 50 mA. The current through a forward-biased diode is an exponential function of the diode voltage. If the diode voltage were to become as large as even 1 V, the current would grow excessively large. Since this does not happen in well-designed circuits, the forward voltage must have an upper limit. This limit is on the order of 0.7 V. On the other hand, if the forward voltage is much less than 0.7 V, the current is very small, almost zero. These features can be seen in Fig. 13.5(a).

On the basis of this reasoning, we state the following approximate rule: *If a silicon diode conducts significant forward current, the forward voltage is approximately 0.7 V.* This rule is really just a verbal formation of the large-signal diode model. It is extremely useful in estimating the operation of diode and transistor circuits.

EXAMPLE 13.5

Find the approximate values of i_1 and i_2 in the circuit shown in Fig. 13.11. Use the rule stating that the voltage across a forward-biased diode is approximately 0.7 V.

SOLUTION

The sign of the applied voltage is such as to forward bias the diode. (We can see this by noting that if the diode were a resistor, current would flow through it in the direction toward which the diode symbol points.) Thus we expect that $v_A \cong 0.7$ V:

$$i_1 \cong \frac{10 - 0.7}{300} = 31 \text{ mA}$$

$$i_2 \cong \frac{0.7}{500} = 1.4 \text{ mA}$$

Clearly a current of $31 - 1.4 = 29.6$ mA is flowing through the diode.

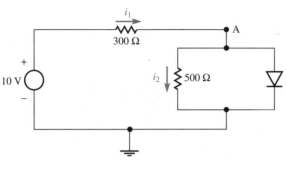

FIGURE 13.11

$\dfrac{V_A - 10}{300} + \dfrac{V_A - 0.7}{200} + \dfrac{V_A - 0.7}{100}$

$2V_A - 20 + 3V_A - 2.1$

$+ 6 V_A - 4.2$

$11 V_A - 26.3$

$V_A = \dfrac{26.3}{11} = 2.39$

● EXERCISE 13.1

Find the voltage at A in Fig. 13.12. Let $R_1 = 300 \ \Omega$, $R_2 = 200 \ \Omega$, and $R_3 = 100 \ \Omega$. **Answer:** $v_A \cong 2.39$ V.

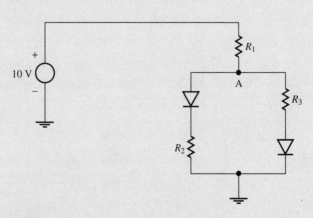

FIGURE 13.12

RECTIFIER CIRCUITS

One of the oldest and most familiar applications of the diode is as a rectifier. This name is applied to a device for converting ac power to dc. Inasmuch as almost all commercial electric power is ac while almost all electronic circuits require dc, there is a rectifier circuit in nearly every piece of electronic equipment.

A simple rectifier circuit is shown in Fig. 13.13(a). This circuit is familiar from the examples already given. The ac voltage source (which in this case might represent the ac power line) supplies the voltage v_1 shown in Fig. 13.13(b). The effect of v_1 is to alternately apply forward and reverse bias to the diode. When the diode is forward biased, a current flows through the resistance, which is the load. Since the voltage drop across a forward-biased diode is usually small compared with v_1, the current that flows on this part of the cycle is essentially the input voltage divided by the value of the resistance. On the other half of the cycle, when the diode is reverse biased, negligible current flows. Thus the current flowing through the resistance has the form of the top half of the sine wave, as shown in Fig. 13.13(c). This resulting current is called pulsating dc. It is considered dc rather than ac because the current always flows in the same direction and thus never changes sign. A rectifier circuit of this type is known as a half-wave rectifier because current flows to the load, at most, on only one-half of the operating cycle.

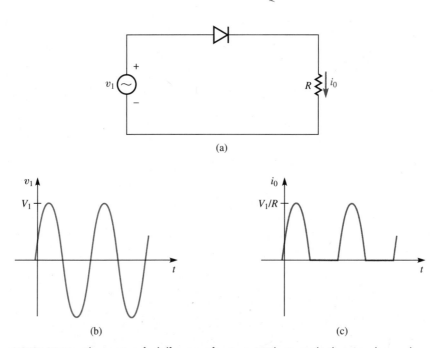

FIGURE 13.13   The operation of a half-wave rectifier circuit. (a) The circuit. (b) The input voltage, such as might come from an ac power line. The amplitude of the sinusoid is V_1. (c) The rectifier current (pulsing dc) flowing in resistance R. Its amplitude is V_1/R, approximately.

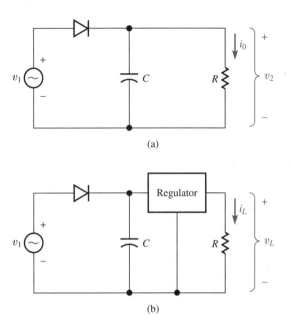

(a)

(b)

FIGURE 13.14 (a) A rectifier circuit with filter capacitor. The function of the capacitor is to supply current to the load during the period when the diode is reverse biased. (b) Addition of a three-terminal IC voltage regulator to stabilize output voltage.

In practical circuits it is usually necessary to obtain smooth dc, that is, nearly constant current, through R; thus the pulsations of the rectified current must be smoothed out. This may be done by adding a capacitor, called a filter capacitor, as shown in Fig. 13.14(a). To understand the effect of C on i_0, let us imagine that v_1 is turned on at time $t = 0$, as shown in Fig. 13.13(b), and that the diode is a perfect rectifier. Until $v_1 = V_1 \sin \omega t$ reaches its first maximum, the diode is forward biased and acts as a short circuit, so that v_2, the voltage across the capacitor, equals v_1 during this period (Fig. 13.15). If a sufficiently large capacitor is provided, then after the first maximum of v_1 the capacitor remains charged nearly to the value V_1, while v_1 decreases. This makes the diode reverse biased, and it conducts no current. The load resistance across the capacitor causes the voltage v_2 to decay exponentially during this period with an exponential time constant equal to RC. When v_1 returns to its next positive maximum, the capacitor is recharged to the value V_1, and the process repeats.

If the time constant RC is chosen to be long compared with the period of the sinusoid, the current i_0 is nearly constant. However, to achieve a large value for RC, either we must make C large (which has cost and size disadvantages) or else we must make R large, in which case little power (v_2^2/R) is delivered to the load. In practice this dilemma is often solved through the addition of an active block called a *voltage regulator*, as shown in Fig. 13.14(b). This block, which is available in IC form, has the property that its output voltage is nearly constant irrespective of the input voltage and output cur-

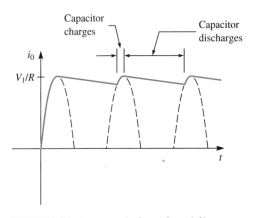

FIGURE 13.15 Output current of a rectifier with filter capacitor, Fig. 13.14(a). The solid line indicates the current through the load resistance *R*. For comparison the dotted line shows the unsmoothed pulsing dc of Fig. 13.13(c).

rent. Negative feedback is used inside this block, just as it is in op-amp circuits, to make the output voltage independent of disturbing variables. (See Problem 13.14.)

EXAMPLE 13.6

Sketch the load current for the rectifier circuit of Fig. 13.14(a) for the case $C = 50 \times 10^{-6}$ F, $R = 1{,}000$ Ω, and $v_1 = 165 \sin 377t$ V. (The last is the standard U.S. line voltage.) Assume for simplicity that the diode is a perfect rectifier.

SOLUTION

We first sketch the input voltage [see Fig. 13.16(a)]. According to the discussion already given, the capacitor is charged to approximately 165 V at time t_1. In the time interval starting just after t_1 and ending just before $5t_1$, the input and the diode are effectively removed from the circuit. During this time the capacitor discharges into the load, supplying the load current, and the circuit reduces to a resistance in parallel with a capacitance [see Fig. 13.16(b)]. We know from the properties of *RC* circuits that the voltage v_C decays exponentially with time constant *RC*. The voltage is thus of the form $v_C = 165 \exp(-t'/RC)$, where $t' = 0$ is the instant when the input was disconnected, that is, when $t \cong t_1$. In other words, $v_C \cong 165 \exp[-(t - t_1)/RC]$. This equation is plotted in Fig. 13.16(c).

   This equation describes the output only up to time t_c in the figure because the charging of the capacitor starts again at this instant and the cycle is repeated. Thus the actual waveform is as given in Fig. 13.16(d). The load current has the same form as the output voltage because $i_0 = v_c/R$.

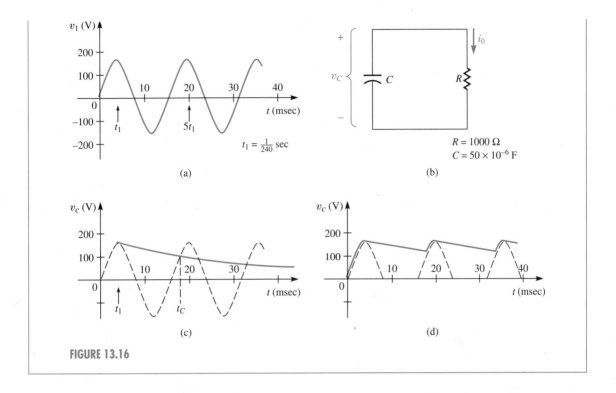

FIGURE 13.16

13.3 Bipolar Junction Transistors

The active devices in integrated circuits (as well as most other circuits) are members of the family known as *transistors*. This family can be divided into two categories, *bipolar junction transistors* (BJTs) and *field-effect transistors* (FETs), which will be considered in Section 13.4. Bipolar junction transistors came into use earlier than FETs, and they are still used extensively in ICs and in general applications.* As the name suggests, the essential components of a BJT are a pair of *pn* junctions.

A typical BJT consists of a single crystal of semiconductor doped alternately *p*- and *n*-type, as shown in Fig. 13.17. (This drawing gives only the general structure; actual shapes and proportions are different.) Operation is based on the ability of a *pn* junction to inject or collect minority carriers, as described in Section 13.2. When the junction on the left of the figure is forward biased, electrons are injected into the *p*-

---

*BJTs are in common use in both analog and digital circuits. Although there is no predicting the future, the impact of FETs to date has been primarily on digital circuits. Thus in Chapter 14 (Analog Circuits) we have emphasized BJTs, while Chapter 15 (Digital Circuits) deals with both.

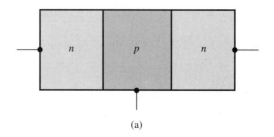

(a)

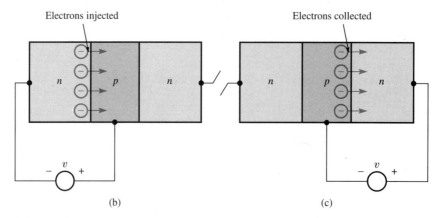

(b)   (c)

FIGURE 13.17   An *npn* transistor and the basic processes involved in its operation. (a) The transistor consists of a crystal containing two junctions in close proximity. (b) When one of the junctions is forward biased, it can inject electrons into the *p* region. (c) When one of the junctions is reverse biased, it can collect the electrons.

type region, as shown in Fig. 13.17(b). If the junction on the right is reverse biased, any electrons that approach it from the *p* side are collected, as shown in Fig. 13.17(c).

A transistor is made by placing the two junctions close enough together that carriers injected by one may be collected by the other. Refer to Fig. 13.18. Electrons injected into the *p* region by the emitting junction diffuse to the collecting junction and are collected (swept into the *n* region). The source of minority carriers (the *n* region on the left in this example) is called the *emitter*. The other *n* region is called the *collector*. The region through which the minority carriers must diffuse to pass from emitter to collector (the *p* region in this example) is called the *base*. The base is essentially a "hostile territory" through which the electrons must journey before reaching the collector. It is desirable to collect as large a fraction as possible of the emitted carriers; therefore the base is made as narrow as possible. Its width is typically in the range 10^{-5} to 10^{-3} cm, which is small enough that minority carriers can diffuse across it before they die by recombination with holes.

Most of the current through the emitter-base junction consists of electrons injected into the base, and most of these electrons reach the collector-base junction and flow out through the collector. Consequently, when the currents are measured in the external wires connected to the three regions (Fig. 13.19), it appears that most of the cur-

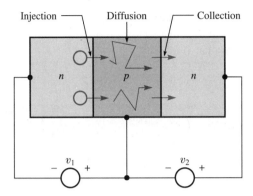

FIGURE 13.18 The operation of a transistor. An n^+p junction is forward biased and therefore injects electrons into the p region. The electrons move through the p region. Most of them reach the opposite junction (collector junction) before recombining with holes, and therefore are swept on into the n region.

rent in the transistor flows directly from collector to emitter (in the direction opposite to that of the electron motion). Large currents can flow in the circuitry connected to the collector and emitter, while only a relatively small current flows in the base circuit. A base current that is only 1 percent of the collector current is typical. The very useful control, or amplification, action of the transistor arises out of the fact that a small base current may be used to control much larger currents in the emitter or collector circuits.

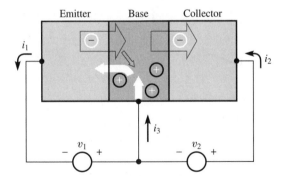

FIGURE 13.19 Current flow in the transistor of Fig. 13.18. The electrons flowing from emitter to collector produce a current i_2 into the collector and i_1 out of the emitter. Because most of the emitter current consists of electrons that transit the base and flow out the collector, i_2 is only slightly smaller than i_1. Consequently the current i_3 flowing into the base is quite small. It consists of a small flow of holes, some of which are injected into the emitter and some of which recombine with electrons in the base.

Moreover, the voltage required to produce the emitter current (the emitter-base voltage) can be much smaller than the voltage produced by the collector current as it flows through an external load resistance. Thus the bipolar transistor can produce both current gain and voltage gain.

EXAMPLE 13.7

A bipolar transistor is connected as shown in Fig. 13.20. Let us assume that the collector current i_C is related to the base current i_B according to $i_C = 100 i_B$. I_{BB} is constant, and i_S is a small time-varying input signal current described by $i_S = I_0 \cos \omega t$. Find $v_{OUT}(t)$ if $I_{BB} = 10 \ \mu A$, $I_0 = 4 \ \mu A$, $V_{CC} = 10 \ V$.

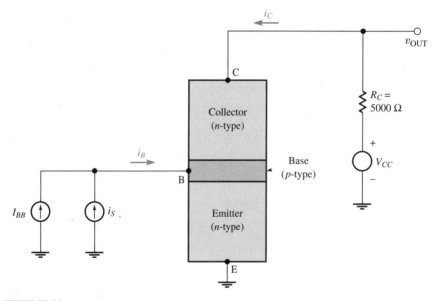

FIGURE 13.20

SOLUTION

From Ohm's law we have

$$v_{OUT} = V_{CC} - R_C i_C$$
$$= V_{CC} - 100 i_B R_C$$
$$= 10 - (100)(10^{-5} + 4 \times 10^{-6} \cos \omega t)(5000)$$
$$= 5 - 2 \cos \omega t \ V$$

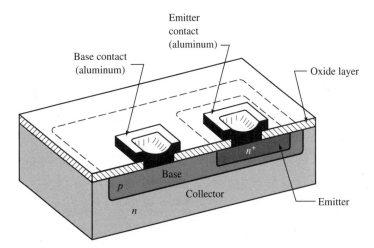

FIGURE 13.21 A cross-section of a planar *npn* silicon transistor. The base region is incorporated into the collector region, and the emitter is incorporated into the base. A protective oxide covers the surfaces, except where contacts are made to the emitter and base. The collector contact is applied to the bottom.

The structure that has just been described is called an *npn transistor*, meaning that the emitter and collector regions are *n*-type and the base *p*-type. Normally, the mode of construction is not a simple arrangement of parallel planes, as suggested by Fig. 13.17, although this is a useful approximation. A more typical configuration of an *npn* transistor is shown in Fig. 13.21. A *p*-type base region and an *n*-type emitter region are incorporated into an *n*-type crystal, and suitable contacts are applied. An oxide layer covers and protects the top surface, except in the region of the contacts. Although the structure of Fig. 13.21 is more realistic, we shall continue to use the simpler diagram of Fig. 13.17 where we are concerned only with the principles of operation.

The complementary device, the *pnp transistor*, is also used. The principles of operation are the same, the main differences being the algebraic signs of the voltages and currents. (See Table 13.1.) For reasons having to do with fabrication, the *npn* device is the more common of the two.

TRANSISTOR CONVENTIONS AND SYMBOLS

The standard circuit symbols for bipolar transistors are given in Fig. 13.22. The emitter has an arrow indicating the current direction in normal operation, in which the emitter injects minority carriers into the base. For example, when the emitter-base junction of an *npn* transistor [Fig. 13.22(a)] is forward biased, electrons flow from the emitter toward the collector, requiring a current *out* of the emitter terminal in the external emitter wire. Hence the symbol for an *npn* transistor has an arrow pointed outward from the emitter. Similarly, in normal operation of a *pnp* transistor, there is a flow of holes from emitter to collector. Since holes carry a positive charge, this flow requires

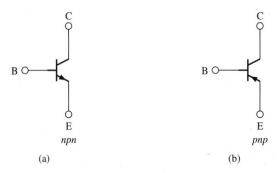

FIGURE 13.22 The standard circuit symbols for bipolar transistors. The transistor type, (a) *npn* or (b) *pnp*, is identified by an arrow on the emitter that indicates the direction of current flow in normal operation.

an external current flowing *into* the emitter terminal, and the symbol of Fig. 13.22(b) is adopted.

To specify completely the circuit currents and voltages of a three-terminal device, three terminal currents and three interterminal voltages must be specified. For the transistor, then, the emitter, base, and collector currents, as well as the emitter-to-base, emitter-to-collector, and base-to-collector voltages, must be specified. Independent of the actual direction of charge motion, the reference directions of all the terminal currents are conventionally *directed inward*, as illustrated in Fig. 13.23(a). Voltages between the terminals are named according to the formula $v_{ij} = v_i - v_j$. (For example, v_{BE} stands for the potential of the base minus that of the emitter.) The interterminal voltages are labeled according to this notation in Fig. 13.23(b). We see that one must specify six variables (three currents and three voltages) to describe the precise

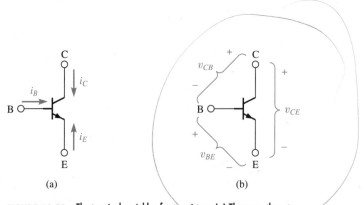

FIGURE 13.23 The terminal variables for transistors. (a) There are three terminal currents. Their reference directions are directed inward for both *npn* and *pnp* transistors. (b) There are three interterminal voltages whose signs are defined by the order of the subscripts.

operating point of a transistor. However, two relationships between the variables can be obtained directly from the basic circuit laws discussed in Chapter 1. From Kirchhoff's current law,

$$i_E + i_B + i_C = 0 \tag{13.2}$$

and from Kirchhoff's voltage law,

$$v_{EB} + v_{BC} + v_{CE} = 0 \tag{13.3}$$

These equations are valid for both *npn* and *pnp* transistors. The problem of finding the operating point of a transistor is thus reduced to finding four additional equations relating the six terminal variables.

The analysis to determine the transistor operating point is easiest when certain restrictions are placed on the range of the terminal variables. These restrictions are not arbitrarily invented but correspond to particular modes of operation that occur in real circuits. For example, in amplifier circuits (as well as many other kinds of circuits) transistors are generally operated in the so-called active mode.

TRANSISTOR OPERATION IN THE ACTIVE MODE

The active mode of transistor operation is the mode in which only the emitter injects carriers into the base. In other words, in the active mode only the emitter-base junction is forward biased. Let us consider specifically the *npn* transistor. For the active mode of operation v_{BE} must be positive so that electrons are emitted into the *p*-type base region. (Remember, the *p* side of a junction is positive with respect to the *n* side in forward bias.) The biasing polarity and the corresponding particle flow are indicated in Fig. 13.24(a). The current directions are indicated by the signs of i_E, i_B, and i_C in Fig. 13.24(b). The voltage source V_1 must be sufficiently positive to cause a current of usable magnitude to flow through the emitter-base junction. The emitter current consists almost entirely of electrons injected into the base, which subsequently diffuse across to the collector-base junction [Fig. 13.24(a)]. The collector-base junction is either zero biased or reverse biased ($v_{CB} \geq 0$), and therefore in the absence of an emitter current the collector current would be essentially zero. However, in the presence of an emitter current, the electrons injected into the base by the emitter diffuse across the base and are collected by the collector. Hence as illustrated in Fig. 13.24(a) negative charges flow from the base into the collector, after which they must flow out through the external collector wire. Therefore a collector current flows. The collector current is positive in the active mode of operation.

Not all of the electrons injected by the emitter are collected by the collector, but a large fraction of them are. This fraction is given the symbol α. Thus for the active mode of operation we have the relationship

$$i_C = -\alpha i_E \tag{13.4}$$

The minus sign in Eq. (13.4) appears because of the sign convention under which currents entering the transistor are taken to be positive. Typical values of α are in the range 0.98 to 0.999.

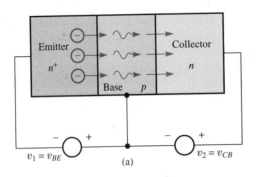

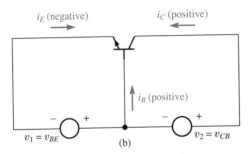

FIGURE 13.24 An *npn* transistor biased in the active mode. (a) The flow of electrons from emitter to collector. (b) The circuit shown with appropriate polarity for active mode operation. The currents' reference directions are inward; however, i_E is negative.

Another important relationship can be obtained by using Eq. (13.2) to eliminate i_E from Eq. (13.4). This results in

$$i_C = \beta i_B \tag{13.5}$$

where by definition

$$\beta = \frac{\alpha}{1 - \alpha} \tag{13.6}$$

β, like α, is a characteristic parameter of a given transistor.* As Eq. (13.5) suggests, β can be measured by observing the ratio of i_C to i_B when the transistor is biased in the active mode. Since α is nearly unity, the denominator of Eq. (13.6) is small, and β has a large value. Values in the range 50 to 1,000 are typical. The reader should care-

---

*Manufacturers' data sheets usually use the symbol h_{fe} to stand for $\Delta i_c/\Delta i_b$ and h_{FE} to stand for i_C/i_B. In this book we have chosen to neglect the rather unimportant difference between the two and refer to both quantities as β.

fully note that *Eqs. (13.4) and (13.5) apply only when the transistor is operating in the active mode*, that is, when the emitter junction is forward biased and the collector junction is either unbiased or reverse biased. We note that in the active mode the existence of forward bias on the base-emitter junction means that $v_{BE} \cong 0.7$ V.

EXAMPLE 13.8

Show that the transistor in Fig. 13.25(a) is operating in the active mode, and find the base current.

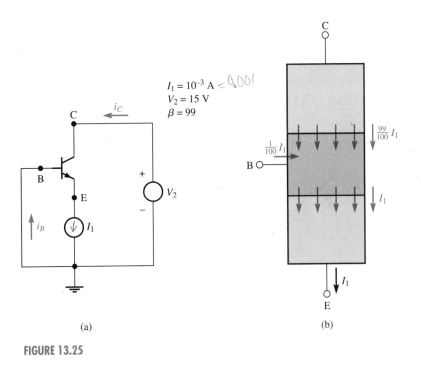

$I_1 = 10^{-3}$ A $= 0.001$
$V_2 = 15$ V
$\beta = 99$

(a)   (b)

FIGURE 13.25

SOLUTION

We must verify (1) that the emitter junction is forward biased and (2) that the collector junction is either unbiased or reverse biased. Beginning with the emitter, we see that the current i_1 is flowing through it in the direction of the emitter arrow. This means that current is flowing through the base-emitter junction in the easy, or forward, direction, which means it is forward-biased.

We see that the voltage across the collector-base junction is $v_C - v_B = 15 - (0) = 15$ V. The sign of this voltage is such as to make the *n*-type collector positive with respect to the *p*-type base, corresponding to reverse bias (see Fig. 13.4). Thus the transistor is biased in the active mode.

$i_B = i_C/\beta$

$i_C = -\alpha i_E$

$\beta = \dfrac{\alpha}{1-\alpha} = 99$

To find i_B we use Eq. (13.5):

$$i_B = i_C/\beta$$

We then express i_C in terms of i_E by Eq. (13.4) and solve for α in terms of β by Eq. (13.6), which leads to

$$\alpha = \frac{\beta}{1 + \beta}$$

With these substitutions, the base current is found to be

$$i_B = -\frac{i_E}{1 + \beta} \cong 0.01 \text{ mA} \qquad = -\frac{0.001}{100}$$

We can now find the collector current, which is

$$i_C = \beta i_B = 0.99 \text{ mA}$$

Figure 13.25(b) shows the *currents* flowing in the transistor. The manner in which i_E divides into i_B and i_C should be particularly noted. (Since the carriers are negatively charged electrons, the motion of the carriers is opposite to the direction of the arrows.)

● EXERCISE 13.2

What is the voltage at the emitter terminal E in Example 13.8? **Answer:** -0.7 V, approximately.

EXAMPLE 13.9

Find the collector voltage v_C and collector current i_C in the circuit of Fig. 13.26. Let $\beta = 150$, and assume the transistor operates in the active mode.

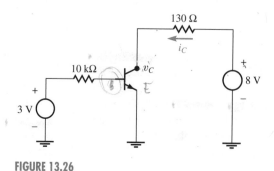

FIGURE 13.26

SOLUTION

Since operation is in the active mode, $v_{BE} = 0.7$. In this case $v_E = 0$; thus $v_B = 0.7$. The base current, from Ohm's law, is

$$i_B = \frac{3 - 0.7}{10^4} = 2.3 \times 10^{-4} \text{ A}$$

Since operation is in the active mode, $i_C = \beta i_B = (150)(2.3 \times 10^{-4}) = 0.0345$ A. The collector voltage can now also be obtained from Ohm's law:

$$\frac{8 - v_C}{130} = i_C = 0.0345 \text{ A}$$

from which we have $v_C = 3.515$ V. We can now verify the assumption that operation is in the active mode. The collector-base junction is reverse biased, as required, by $3.515 - 0.7 = 2.815$ V.

In Example 13.7 it was demonstrated that in the active mode of operation, the emitter current gives rise to a base current many times smaller (in the example, 100 times smaller). Therefore the reverse statement is also true: In normal active operation, a small base current implies an emitter current that is many times larger. Similarly, the collector current is much larger than the base current. Thus we see that, in the active mode of operation, the transistor can be used as a current amplifier by means of which a small base current can be used to control a larger current in the emitter or collector circuits. When used this way the transistor is said to have current gain.

In the active mode, the emitter-base junction acts very much like any other forward-biased junction. By analogy with Eq. (13.1), for an *npn* transistor we have

$$i_E \cong -I_{ES} e^{q v_{BE}/kT} \tag{13.7}$$

Here I_{ES} is a constant determined by the design of the transistor. The (-1) of Eq. (13.1) has been dropped because in forward bias the exponential is much larger than unity, and the minus sign in Eq. (13.7) results from the convention that the reference direction of i_E is directed inward. Since the forward-biased emitter-base junction acts much like a forward-biased diode, the value of v_{BE} in the active mode is approximately 0.7 V. Since q/kT is rather large — about 40 V^{-1} — small changes in v_{BE} result in fairly large changes in i_E. In fact, if v_{BE} increases by 0.018 V, i_E doubles. The large changes in i_E can in turn cause large changes in v_C. The upshot is that small changes of v_B can be used to produce large changes of v_C, thus providing voltage gain. This is illustrated by the following example.

EXAMPLE 13.10

The transistor in the circuit in Fig. 13.27 is operating in the active mode. Find the base current, collector current, and collector-to-emitter voltage, assuming that $I_{ES} = 10^{-14}$ A. Find the change in collector-to-emitter voltage assuming that the input voltage increases by 0.018 V. Let $v_{IN} = 0.7$ V, $R_L = 1,000$ Ω, $V_{CC} = 15$ V, $\beta = 200$, and $kT/q = 0.026$ V.

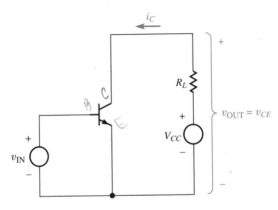

FIGURE 13.27

SOLUTION

The emitter-base junction is forward biased by 0.7 V. According to Eq. (13.7) this would give

$$i_E = -10^{-14} \exp \frac{0.7}{0.026} = -5 \times 10^{-3} \text{ A}$$

Since $\beta = 200$, α is close to unity; in fact, from Example 13.7 $\alpha = \beta/(1 + \beta) = 0.995$. Thus i_C is nearly equal in magnitude to i_E, that is, $i_C = 5 \times 10^{-3}$ A. We can solve for v_{CE} by a single loop equation:

$$v_{CE} - V_{CC} + i_C R_L = 0$$

Thus

$$v_{CE} = 15 - 5 \times 10^{-3} \cdot 1,000 = 10 \text{ V}$$

This voltage is sufficient to assure that the transistor is indeed biased in the active region, since $v_{CB} = v_{CE} - v_{BE} = 9.3$ V, and the collector-base junction is therefore reverse biased. The base current is given by

$$i_B = \frac{i_C}{\beta} = 2.5 \times 10^{-5} \text{ A}$$

We now recompute the output voltage with an 18 mV increase in the input voltage. Such an increase approximately doubles the emitter current:

$$i_E = -10^{-14} \exp \frac{0.718}{0.026} = -10^{-2} \text{ A}$$

Therefore the base current and collector current double. From the loop equation given earlier in this discussion, $v_{CE} = 15 - 10^{-2} \times 10^3 = 5$ V. (Note that the transistor remains in the active mode.)

Note: For a change in v_{BE} of 0.018 V, there is a change in v_{CE} of -5 V. That is, input voltage changes are multiplied by a factor of $-5/0.018 = -280$. We have a rudimentary voltage amplifier.

For reference, in Table 13.1 we summarize the polarity of the various transistor terminal parameters. The voltage polarities are obtained from the requirement that the emitter-base junction be forward biased and the collector-base junction be reverse biased. The emitter current direction is obtained from the emitter-base forward bias requirement, and the base and collector current directions are obtained from Eqs. (13.4) and (13.5).

TABLE 13.1 Bias Polarities for Transistors Operating in the Active Mode

	Polarity	
Parameter	*npn*	*pnp*
v_{CE}	Positive	Negative
v_{BE}	Positive	Negative
v_{CB}	Positive	Negative
i_E	Negative	Positive
i_B	Positive	Negative
i_C	Positive	Negative

GRAPHICAL CHARACTERISTICS

We are now ready to consider more general cases in which operation is not necessarily in the active mode and the relationship $i_C = \beta i_B$ does not necessarily apply. Instead, we shall make use of graphs that show the relationships of transistor variables to one another.

To begin, we recall that analysis of a transistor circuit requires calculation of four different voltages and currents. (There are three voltages and three currents associated with the transistor, but one current can be expressed in terms of the other two by means of Kirchhoff's current law and one voltage in terms of the others by Kirchhoff's

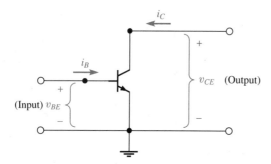

FIGURE 13.28

voltage law.) Presumably each of the variables could depend on all three of the others. This would make the subject complicated and fortunately is not the case. The usual variables chosen to describe operation are i_B, v_{BE}, i_C, and v_{CE}; as we shall see, these depend on each other in fairly simple ways.

The currents and voltages in question are indicated in Fig. 13.28. Most often the high-power signal being controlled comes out of the right-hand port; thus we shall refer to it as the output port. The low-power control signal normally enters at the left, which we shall call the input port. We note that the emitter terminal is common to (meaning it is connected to) both the input and the output ports; hence this is referred to as the *common-emitter connection*. We are now ready to specify how i_B and i_C depend on the voltages.

We have already noted that in the active mode i_E is approximately an exponential function of v_{BE}, as given in Eq. (13.7). A typical relationship between i_B and v_{BE} for an

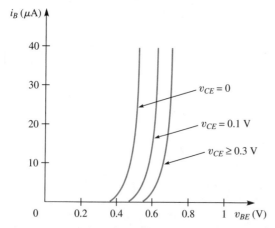

FIGURE 13.29 The input characteristics of a typical *npn* silicon transistor in the common emitter configuration. If $V_{CE} \geq 0.3$ V, the curve on the right applies.

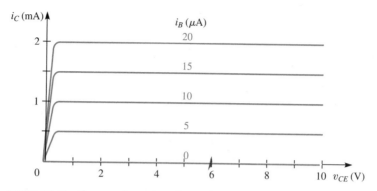

FIGURE 13.30 The output characteristics of a typical *npn* transistor in the common-emitter configuration.

npn transistor is shown somewhat more generally in Fig. 13.29. The value of i_B might be expected to depend not only on v_{BE} but also on v_{CE}, and from the figure we see that this is the case; however, the dependence is not very strong. In fact, whenever $v_{CE} \geq$ 0.3 V, the graph of i_B is nearly identical with the curve marked $v_{CE} \geq$ 0.3 V in the figure. Note that this is the case in the active mode, since in that mode $v_{CE} \geq$ 0.7 V. The right-hand curve in Fig. 13.29 is the exponential function (13.7). As $\overline{v_{CE}}$ becomes less than 0.3 V we see that the $i_B - v_{BE}$ curve does shift slightly, but this shift is not very important for circuits treated in this book. In fact, for our purposes Eq. (13.7) and the right-hand curve of Fig. 13.29 are always good approximations.

We now turn our attention to graphing the collector current i_C. A typical dependence of i_C on v_{CE} for an *npn* transistor is shown in Fig. 13.30. Several curves are shown in this graph, corresponding to different values of the base current. It is necessary to use this complete set of curves because i_C depends strongly on i_B as well as on v_{CE}. For example, if i_B is known to be 20 μA and v_{CE} is 6 V, we use the top curve in the

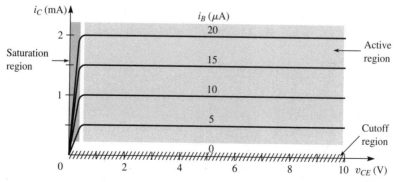

FIGURE 13.31 The output characteristics of a typical *npn* transistor. The regions of operation are indicated.

figure and read off that $i_C = 2$ mA. If i_B is reduced to 15 μA and v_{CE} is 6 V, we use the second curve from the top and find $i_B = 1.5$ mA. If i_B has some value between 15 and 20 μA, one can interpolate a suitable curve between the 15 μA and 20 μA curves.

We note that, for $v_{CE} > 0.7$ V, the curves are nearly horizontal. This means that when $v_{CE} > 0.7$, i_C is nearly independent of v_{CE} and depends only on i_B. This part of the graph corresponds to the active mode of operation. (See Fig. 13.31.) We observe that in this part of the graph $i_C \cong 100i_B$, independent of v_{CE}, in agreement with Eq. (13.5). Evidently Fig. 13.30 has been drawn for a transistor for which $\beta = 100$.

EXAMPLE 13.11

The transistor in Fig. 13.32(a) has the output characteristic shown in Fig. 13.30. Let $V_{CC} = 6$ V. Sketch a graph of i_C versus i_1 and thus find β for this transistor.

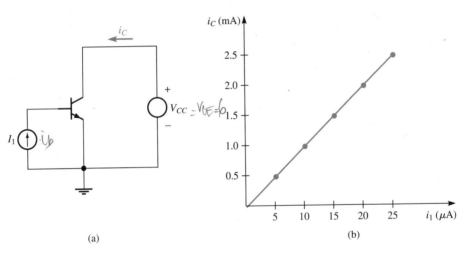

(a)   (b)

FIGURE 13.32

SOLUTION

The value of i_C corresponding to each i_B is located by finding the intersection between a vertical line at $v_{CE} = 6$ V and the curve corresponding to each value of i_B. (Note that $i_1 = i_B$ and $v_{CE} = V_{CC}$.) Points corresponding to $i_B = 0, 5, 10, 15,$ and 20 μA are plotted, and a line is drawn through them. The slope of this line, which is the ratio i_C/i_B, is equal to β. In this case we see that $\beta = 100$.

● EXERCISE 13.3

Find the value of i_C in the circuit shown in Fig. 13.33. Assume that the transistor's output characteristics are as shown in Fig. 13.30, and use the rule that the voltage across a forward-biased junction is approximately 0.7 V. Let $V_{CC} = 9$ V, $V_{BB} = 5$ V, and $R_1 = 3 \times 10^5$ Ω. **Answer:** Approximately 1.4 mA.

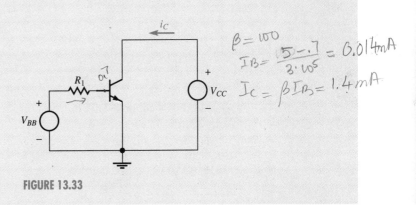

$$\beta = 100$$
$$I_B = \frac{(5 - .7)}{3 \cdot 10^5} = 0.014\text{mA}$$
$$I_C = \beta I_B = 1.4\text{mA}$$

FIGURE 13.33

Besides the active region, two other regions are noted in Fig. 13.31. The *cutoff* region is simply the curve corresponding to $i_B = 0$. This curve, which lies very close to the horizontal axis, shows that when $i_B = 0$, i_C is also very small. This condition, known as cutoff, is used in digital circuits, as we shall see in a later chapter.

At the left of Fig. 13.31 is the *saturation* region. This region corresponds to low values of v_{CE}, so low, in fact, that the collector-base junction has lost its reverse bias. In this case the collector loses its ability to collect, and i_C becomes less than βi_B. Any time v_{CE} becomes less than 0.7 V, a transistor is likely to leave the active mode and become saturated. The effect is undesirable in amplifier circuits but figures usefully in digital circuits.

How saturation comes about is illustrated in Fig. 13.34. Suppose initially the transistor is operating in the active mode. Then from Ohm's law

$$v_{CE} = V_{CC} - i_C R_C \tag{13.8}$$

Since in the active mode $i_C = \beta i_B$,

$$v_{CE} = V_{CC} - \beta i_B R_C \tag{13.9}$$

Now, as we increase i_B, v_{CE} decreases, eventually approaching 0.7 V. At this point the collector junction has lost its reverse bias and the transistor enters saturation. Further increases of i_B now do not result in proportional increases in i_C because the relationship

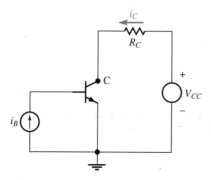

FIGURE 13.34

$i_C = \beta i_B$ no longer applies. This prevents v_{CE} from decreasing indefinitely, and in fact v_{CE} usually approaches a minimum value in the neighborhood of 0.2 V. The minimum value reached by v_{CE} in saturation is usually referred to as V_{CESAT}.

To review, we note that in saturation (1) i_B is *greater* than i_C/β and (2) $v_{CE} = V_{CESAT} \cong 0.2$ V.

EXAMPLE 13.12

Find the collector current of the transistor in Fig. 13.35(a) for the following three values of i_1: 0, 5, and 20 μA. Assume the transistor is the "typical" transistor whose characteristics are illustrated in Figs. 13.29 and 13.30.

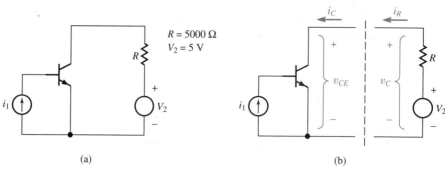

(a)   (b)

FIGURE 13.35(a) and (b)

SOLUTION

To solve this problem we will use the graphical load-line technique. An imaginary break is made in the circuit, as shown in Fig. 13.35(b). The break in the circuit is purely imaginary, and in reality i_C equals i_R and v_C equals v_{CE}. We plot the curve of i_C versus v_{CE} and the curve of i_R versus v_C on the same graph; the point of intersection of the curves is the desired solution. A set of curves of i_C versus v_{CE} for this transistor is already given in Fig. 13.30; we need only choose the appropriate curve from the known base current ($i_B = i_1$). The characteristic i_R versus v_C is obtained from Kirchhoff's voltage law: $v_C = V_2 - i_R R$. We now plot i_C versus v_{CE} and i_R versus v_C on the same graph, as in Fig. 13.35(c), for the specified values of R, V_2, and i_1. For the case $i_1 = 5$ μA, the point at which the conditions $i_C = i_R$ and $v_{CE} = v_C$ are satisfied is the point $i_C = 0.5$ mA, $v_{CE} = 2.5$ V. The transistor is operating in the active region. For the case $i_1 = 20$ μA, the solution is $i_C = 0.95$ mA, $v_{CE} = 0.25$ V, and the transistor is in saturation. When $i_1 = 0$, $i_C = 0$ and $v_{CE} = 5$ V, and the transistor is cut off.

If the entire set of curves of Fig. 13.30 is superimposed on the load line, as in Fig. 13.35(d), then i_C may be quickly determined for any value of i_1. It is noteworthy that the transistor behavior is quite different in the different regions of operation. In the active region (that is, for $v_{CE} > 0.7$ V), the collector current and collector-emitter voltage depend strongly on i_1. In the saturation region the collector current becomes almost independent of i_1, and the collector-emitter voltage is about 0.2 V. In the cutoff region, when i_1 becomes very small, the collector current goes to zero and the collector-emitter voltage equals V_2.

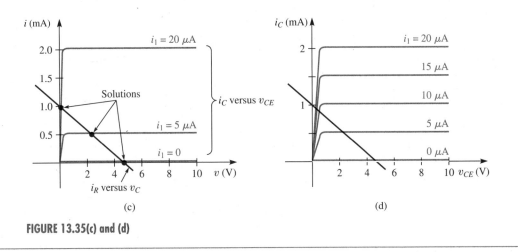

(c)　　　　　　　　　　　　　　　(d)

FIGURE 13.35(c) and (d)

It should be mentioned that the input and output characteristics shown in Figs. 13.29 and 13.30 are somewhat idealized. For comparison, experimental characteristics obtained with a curve tracer for a typical transistor are shown in Fig. 13.36. The most

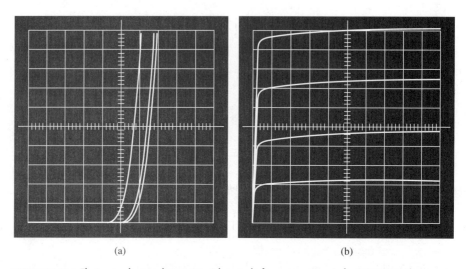

(a)   (b)

FIGURE 13.36   The input and output characteristics (photographs from a curve tracer) of a 2N3568 *npn* silicon transistor. (a) The input characteristics i_B versus v_{BE}. Vertical scale: 10 μA per large division. Horizontal scale: 0.1 V per large division. The three curves, from left to right: $v_{CE} = 0$ V, $v_{CE} = 0.1$ V, $v_{CE} = 0.3$ V. (b) The output characteristics i_C versus v_{CE}. Vertical scale: 0.2 mA per large division. Horizontal scale: 1 V per large division. The five curves from top to bottom: $i_B = 20, 15, 10, 5,$ and 0 μA.

noticeable difference is seen in the output characteristics, which are not quite horizontal but trend slightly upward. This slight dependence of i_C on v_{CE} is known as the "Early effect."

13.4 Field-Effect Transistors

Those transistors that are not bipolar belong to the family called *field-effect transistors*, or *FETs*. There are several subspecies, and the terminology is a veritable alphabet soup: one hears of JFETs, IGFETs, and MOSFETs, the last with subspecies such as NMOS and CMOS; all of these are used for various special purposes. Since we are mainly interested in principles, we shall not attempt to discuss all the many types. As our main example we shall choose the *n*-channel IGFET and its close relatives. These devices are widely used in digital integrated circuits.

All FETs make use of the principle that currents near the surface of a semiconductor can be affected by an applied electric field, as shown in Fig. 13.37. The effect occurs because the current is a motion of charge carriers (holes or electrons). The applied field can pull additional carriers into the conduction path, making it more conductive, or the field can be used to push carriers out of the conduction path, making it less conductive. It is even possible for the field to attract electrons into material that is otherwise *p*-type; when electrons thus outnumber the holes, the material begins to act as though it were *n*-type. (This is known as *inverting* the material.) The field does not penetrate very far into the semiconductor; thus only current conduction near the sur-

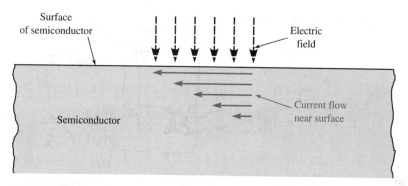

FIGURE 13.37 General principle of field-effect transistors. Current flowing parallel to the surface is influenced by an electric field applied perpendicular to the surface, as shown. Current flow is confined to a region near the surface of the semiconductor, where the electric field can reach it.

face is affected; compare the BJT of Fig. 13.18, which shows conduction throughout the entire volume of the device. For this reason an input voltage is able to control less current in an FET than in a BJT with comparable characteristics. On the other hand, very little input current is required to establish the field; thus the power gain of FETs is higher than in BJTs.

As a specific example, let us consider the device shown in Fig. 13.38. Here the electric field is supplied by an electrode, called the *gate*, that is insulated from the rest of the device. (Hence we call the device an *insulated-gate field-effect transistor*, or IGFET.) The external voltage source v_G charges the gate and produces a field between the gate and the semiconductor, as shown. Meanwhile, the external voltage source v_{DS} tends to produce a current through the device, from right to left. This current must go from the *n*-type region on the right to the one on the left, passing through the *p*-type region in the middle on the way. We note that such a current would have to flow through two *pn* junctions, and for the junction on the right the flow would be in the reverse direction. Since reverse currents are very small, very little current can flow in this case. However, if a region near the surface can be converted from *p*-type to *n*-type, the junctions will disappear, and there will be an unobstructed path for current flow through *n*-type material. This conversion of the surface layer from *p*-type to *n*-type is accomplished by the applied electric field. In the device of Fig. 13.38, electrons move from the *n* region on the left, called the *source*, to the *n* region on the right, which is called the *drain*. (Since electrons are negatively charged, they move in the opposite direction from current flow.) The path through which they travel is called the *channel*. Since conduction occurs when the channel becomes *n*-type, this is called an *n-channel device*. Amplification is obtained because small changes in v_G produce substantial changes in the current that flows from drain to source. We note that, since the gate is insulated, almost no current flows from the external source v_G; thus the external input power needed for current control is very small.

In practice, this device is sometimes made by growing an insulating layer of silicon dioxide over the channel and then depositing a metal layer on top of the oxide for

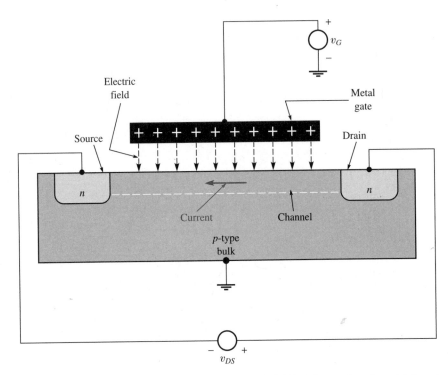

FIGURE 13.38   *n*-channel insulated-gate field-effect transistor. The field from the gate electrode repels holes and attracts electrons, thus inverting the channel and creating a conduction path from source to drain.

the gate. From the initials of "metal-oxide semiconductor," the device is then correctly called a MOSFET. It is usual at present to refer to all IGFETs as MOSFETs, and we shall follow this practice. It should be noted, however, that in many IGFETs the gate is not metal but some other material, such as polycrystalline silicon.

EXAMPLE 13.13

The *n*-channel IGFET of Fig. 13.38 is connected as shown in Fig. 13.39. When $v_{IN} = 3$ V, there is almost no conduction between source and drain. However, when v_{IN} is increased to 5 V, an *n*-type channel is induced, and 0.18 mA of current flows from drain to source through the device. What is the voltage gain? The current gain?

SOLUTION

When v_{IN} increases from 3 to 5 V, v_{OUT} decreases from 10 to $10 - (1.8 \times 10^{-4})(50 \times 10^3) = 1$ V. Thus

$$G_V = \frac{\Delta v_{\text{OUT}}}{\Delta v_{\text{IN}}} = \frac{-9}{2} = -4.5$$

The current gain is

$$G_I = \frac{\Delta i_D}{\Delta i_G}$$

where i_D and i_G are the drain and gate currents, respectively. The change in i_D is 0.18 mA. However, the input current i_G is almost exactly zero under all circumstances because the gate is insulated from the rest of the structure. Thus the current gain is nearly infinite! Since the input power is almost zero, the power gain is also very large.

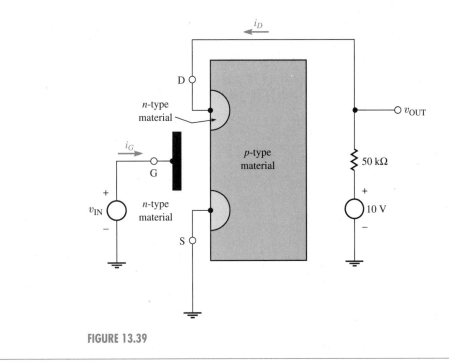

FIGURE 13.39

The type of MOSFET just discussed operates in what is known as the *enhancement mode*. This term is used because when no v_G is applied, very little drain current can flow, but when positive v_G is applied, conduction between source and drain is enhanced. There is another version of the device in which the surface layer is naturally inverted when no v_G is applied. Source-to-drain conduction then can take place when v_G is zero, but application of a *negative* v_G can now push electrons out of the channel, *reducing* its ability to conduct. Since in this version of the device the effect of v_G is to deplete the carriers in the channel, it is said to operate in the *depletion mode*.

It is possible to interchange the *p*-type and *n*-type materials, thus obtaining the complementary devices known as *p*-channel MOSFETs. The principles of the *p*-channel devices are the same, the main differences arising in signs of voltages and currents (just as with *pnp* and *npn* BJTs). The circuit symbols for four types of MOSFETs are shown in Fig. 13.40. The *B* connections are the connections to the bulk semiconductor (shown grounded at the bottom of the device in Fig. 13.38). The arrow indicates the easy-current-flow direction of the *pn* junction between the bulk and the channel; for instance, in an *n*-channel device the bulk is *p*-type, and the arrow points from *p* to *n* and therefore from the bulk toward the channel. Enhancement mode devices are distinguished by a broken line that represents the channel. This symbolizes the fact that in enhancement devices the current path through the channel is broken unless gate voltage is applied. The source is always the region from which carriers enter the channel. Thus with the usual sign conventions (Fig. 13.40) i_s is negative for an *n*-channel device and positive for a *p*-channel device.

I-V CHARACTERISTICS

Like junction transistors, FETs are nonlinear circuit elements. Thus their *I-V* characteristics are most easily specified in graphical form. The MOSFET is actually a four-terminal element because the *B* connection to the bulk is also available. At present,

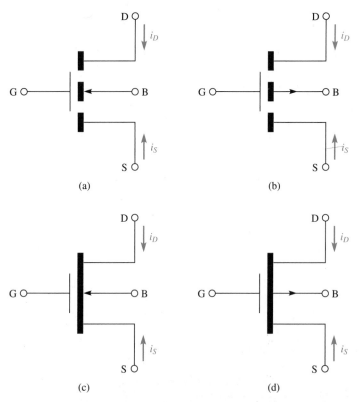

FIGURE 13.40 Four types of MOSFET: (a) *n*-channel enhancement mode; (b) *p*-channel enhancement mode; (c) *n*-channel depletion mode; (d) *p*-channel depletion mode.

however, we shall assume that the B terminal is connected to the source, as it often is in practice. The *I-V* curves are then obtained by measurement. Typical curves for an *n*-channel enhancement-mode MOSFET are shown in Fig. 13.41. Only one graph, showing the output characteristics, is needed for the MOSFET, as compared with the two (input and output) required for the BJT. This is because in dc operation of the MOSFET no current flows through the G lead. (The gate, it will be recalled, is insulated from the rest of the device.) In operation, the external circuit supplies some input voltage v_G to the gate terminal. The value of v_{GS} determines which curve of Fig. 13.41 is to be used. Then the value of v_{DS} is related to i_D by the appropriate curve. We notice

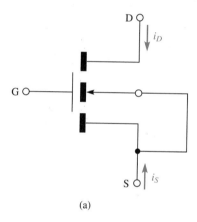

(a)

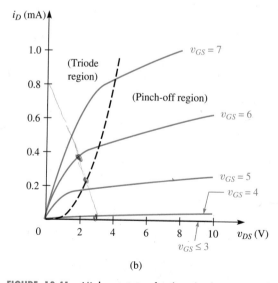

(b)

FIGURE 13.41 *I-V* characteristics of *n*-channel enhancement-mode MOSFET. (a) Connection for measurements; (b) experimental characteristics. In (b) the *I-V* characteristics are the solid lines; the dashed line is the boundary between the triode region and the pinch-off region.

that for $v_{GS} \leq 3$ V, i_D is always zero. This is because for this particular transistor a minimum of 3 V is needed to produce the *n*-type channel needed for conduction between source and drain. This minimum voltage, known as the threshold voltage V_T, can take on different values for different devices, depending on fabrication technology. As v_{GS} increases beyond V_T, current conduction becomes progressively easier. For each value of v_{GS} the drain current is the nonlinear function of v_{DS} given by the appropriate curve.

EXAMPLE 13.14

The MOSFET of Fig. 13.41 is connected as shown in Fig. 13.42(a). Find i_D and v_D.

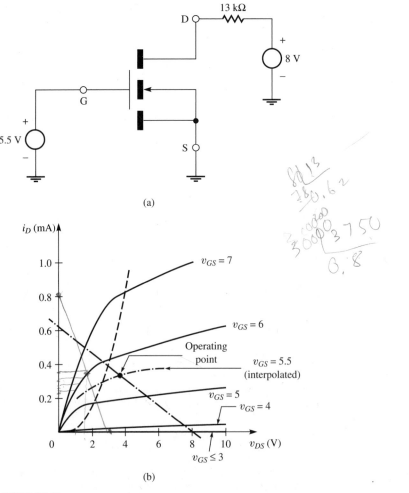

(a)

(b)

FIGURE 13.42

SOLUTION

The load line is drawn over the output characteristics as shown in Fig. 13.42(b). A curve for $v_{GS} = 5.5$ must be interpolated as shown. The operating point is approximately at $v_D = 3.8$ V, $i_D = 0.32$ mA.

It is possible to predict the *I-V* characteristics of MOSFETs from semiconductor-device physics. Highly accurate theories, however, lead to rather complicated results. In design work it is often sufficient to use approximate equations, which are easier to use. (The same is true of BJTs. For instance, the equation $i_C = \beta i_B$ is only an approximation; i_C actually is affected by v_{CE}, as can be seen in Fig. 13.36.) Different equations are used, depending on the relative sizes of the voltages being applied. The following applies to an *n*-channel enhancement MOSFET: First, if $v_{GS} < V_T$, we simply have $i_D = 0$. Second, if $v_{GS} \geq V_T$ and also $v_{DS} \leq v_{GS} - V_T$, we have the approximate equation

$$i_D \cong K[(v_{GS} - V_T)v_{DS} - \frac{1}{2}v_{DS}^2] \quad \text{(triode region)} \quad (13.10)$$

Here K is a constant that depends on the size, shape, and structure of the device. This range of operation is called the *triode* or *ohmic region*. It is the region above and to the left of the dashed line in Fig. 13.41. However, if $v_{DS} > v_{GS} - V_T$, operation is in what is known as the *pinch-off region*.* In this case the approximate expression for i_D is

$$i_D = \frac{1}{2}K(v_{GS} - V_T)^2 \quad \text{(pinch-off region)} \quad (13.11)$$

The pinch-off region is to the right of the dashed line in Fig. 13.41.

Inspecting Eq. (13.10) (for the triode region), we see that if v_{GS} is held constant, i_D is a parabolic function of v_{DS}. The parabola has a maximum at $v_{DS} = v_{GS} - V_T$; for values of v_{DS} larger than this, Eq. (13.10) indicates that i_D would start to *decrease*. This would seem unphysical, and in fact it does not happen because at the maximum of the parabola we leave the triode region and Eq. (13.11) becomes the relevant equation instead of Eq. (13.10). Equation (13.11) predicts that for $v_{DS} > v_{GS} - V_T$, i_D ceases to depend on v_{DS} at all. A glance at Fig. 13.41 shows that this is not really true. However, the predicted quadratic dependence of i_D on $v_{GS} - V_T$ in the pinch-off region is not too inaccurate and gives a useful rule for calculations.

---

*The term "saturated region" is often used for this regime. We prefer not to use this term because of possible confusion with the saturation region of BJTs.

EXAMPLE 13.15

Obtain an approximate value of K for the transistor of Fig. 13.41.

SOLUTION

In principle one can find K by choosing any point on Fig. 13.41, noting the corresponding values of i_D, v_{GS}, and v_{DS} and solving for K from Eq. (13.10) or (13.11), as appropriate. This would present no problems if Eqs. (13.10) and (13.11) really described the experimental curves, but they do not do this exactly. As a result, the foregoing procedure will give slightly different values of K, depending on what point (i_D, v_{GS}, v_{DS}) is chosen. The less the value of K varies, the more confident we can feel in the accuracy of the theory.

Arbitrarily let us choose, for our calculation, several points along the dashed line that divides the triode and pinch-off regions. On the dashed line, *both* Eq. (13.10) and Eq. (13.11) apply, and $v_{DS} = v_{GS} - V_T$. Substituting $v_{DS} = v_{GS} - V_T$ into either Eq. (13.10) or Eq. (13.11) gives

$$i_D = \frac{1}{2} K v_{DS}^2$$

or

$$K = \frac{2i_D}{v_{DS}^2}$$

Let us choose the points $v_{DS} = 2, 3$, and 4. The results are

V_{DS}(V)	i_D(mA)	K
2	0.17	0.085
3	0.44	0.098
4	0.82	0.1025

As an approximate result we can choose the average value, $K \cong 0.095$ mA/V². Since the whole procedure is approximate, however, it might be better to use an accuracy of only one decimal place and say $K \cong 0.1$ mA/V².

As we have seen, the gate lead of the FET draws very little current. This is a useful characteristic because it prevents loading of whatever circuit is connected to its input. Other advantages have to do with fabrication. MOS devices are smaller than the corresponding BJTs; thus more of them can be gotten into an integrated circuit. Furthermore, fewer fabrication steps are needed to make them, so they have fewer production defects. Hence MOS technology is widely used in the largest digital ICs. One variation, known as CMOS (to be discussed in Chapter 15), is noteworthy for its low power consumption; thus it is used in battery-powered devices such as wristwatches and pocket calculators. The traditional drawback of MOS technology has been low speed, but this disadvantage has tended to disappear as technology has improved. At present MOS digital blocks operate at rates around 20 megabits per second, within a factor of 2 of the speeds obtainable with BJT technology. MOSFETs also tend to have high noise at low frequencies; this has interfered with their use in low-level analog circuits. However, as technological progress continues, it seems likely that MOS will gradually come into wider use.

*13.5 Integrated Circuits

It is possible to build more than one transistor on the same piece of silicon. By combining several transistors, resistors, wires, and perhaps other components, we obtain an *integrated circuit*. Integrated circuits are ideal building blocks for electronic systems. They are manufactured in large quantities; one can almost think of them as being stamped out like phonograph records. Interestingly, the cost of production is not much affected by the complexity of the circuit; it costs about the same to stamp a complicated pattern as a simple one. Often the cost of the actual semiconductor device is a small part of the sales price, the largest cost being that of packaging the device and attaching the wires! But although these building blocks are available at low cost, they bestow great power on the systems designer, who has prewired blocks containing thousands of transistors at his or her disposal.

A typical integrated circuit is a silicon square a few millimeters on a side and about one-half millimeter thick. In order to handle and use such small devices, they are mounted in *packages*, protective containers with wires or pins for external connection. Three typical packages are shown in Fig. 13.43. Identical ICs can be sold in different-shaped packages for the user's convenience. The round package is the same one used for individual transistors, except that it has more wires. Figure 13.44 shows the inside of the transistor-type package when the top has been removed. The tiny silicon chip at the center is connected to the binding posts with thin bonding wires. (It is easy to see why these delicate wires contribute to the cost.) At the left a single transistor in a similar package is shown for comparison. The costs of the IC and single transistor are probably about the same.

The low cost of IC production is a result of *planar processing*. Fabrication begins with a very flat disc of silicon, called a *wafer*, 5 to 10 cm in diameter and only half a millimeter thick. The small electronic structures to be built on it are then produced photographically. One process uses a projector something like a photographic enlarger working backward. It takes large-scale patterns and projects them, in much reduced scale, on the surface of the wafer. The resulting patterns of light and dark are then converted into the desired structures by a technique known as *photolithography*.

FIGURE 13.43 Three typical IC packages. Clockwise, the round transistor-type package (diameter about 1 cm), the flat-pack, and the dual-in-line package. (*Courtesy of Texas Instruments Incorporated.*)

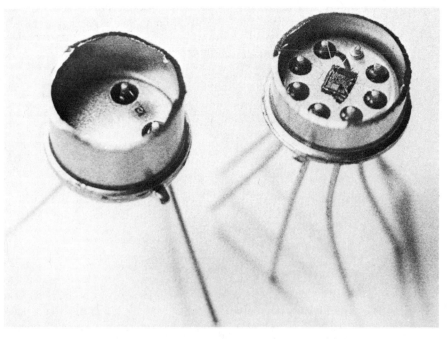

FIGURE 13.44 On the right is an IC package (diameter about 1 cm) with its top removed, showing the silicon chip connected with bonding wires. At the left is a single transistor in a similar package.

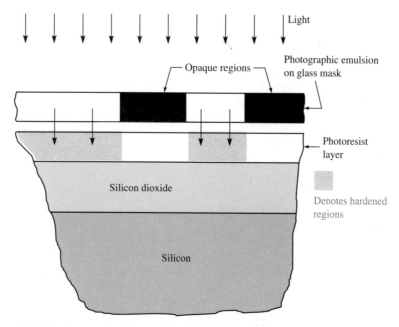

Light

Opaque regions

Photographic emulsion
on glass mask

Photoresist
layer

Silicon dioxide

Denotes hardened
regions

Silicon

FIGURE 13.45 Exposure of photoresist through a photographic mask. Wherever the mask is transparent, light causes the photoresist to harden, making it insoluble in a developer.

A key element in the process is a photosensitive lacquer known as *photoresist*. This material has the property of hardening when struck by light.* The unhardened portions can then be dissolved, leaving a pattern of hardened photoresist on the surface. Now some parts of the underlying material can be treated with chemicals while the patterned area is protected from the chemicals by the hardened photoresist.

Suppose, for example, we wish to make a small *n*-type region on a *p*-type silicon wafer. (Two such regions could be the source and drain of the FET shown in Fig. 13.38. Although for simplicity we shall talk about only one *n*-type region, in an IC the sources and drains of many transistors would be made simultaneously.) The first step would be to grow an oxide layer on the silicon by heating it in the presence of oxygen or water vapor. A layer of photoresist is then applied over the SiO_2. The desired device pattern then illuminates the photoresist, which is hardened wherever it is hit by light. In our example we will harden it everywhere *except* where we want the *n*-region to be created. This step is illustrated in Fig. 13.45. Here instead of projecting the pattern, an actual-size photographic negative is placed over the wafer; this process is known as

---

*We are speaking of what is called "negative" photoresist. "Positive" photoresist, which hardens everywhere it is *not* struck by light, also exists.

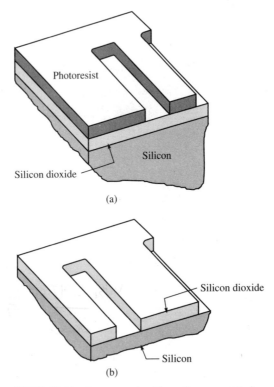

Photoresist

Silicon

Silicon dioxide

(a)

Silicon dioxide

Silicon

(b)

FIGURE 13.46 Stages in a photolithographic process: (a) after development of the photoresist; (b) after etching of the oxide and removal of the remaining resist.

"contact printing." After exposure, the wafer is placed in a solvent known as "developer." The developer removes the photoresist where it has not been hardened, leaving the structure of Fig. 13.46(a). This is followed by use of another solvent, which removes the SiO_2 wherever it is exposed by a hole in the photoresist. Still another solvent then dissolves the remaining photoresist. The result is a pattern etched into the SiO_2, as shown in Fig. 13.46(b).

Now we are ready to create the desired n region. This is done by exposing the windows in the SiO_2 to n-type impurity atoms, which diffuse into the silicon and convert it from p-type to n-type. A rather high temperature—800 to 1200°C—is required to enable the impurity atoms to move through the silicon crystal. This temperature is not high enough to melt either Si or SiO_2, but it would decompose photoresist if any were still present; that is why the SiO_2 is used as a mask, instead of simply diffusing through holes in the photoresist. The diffusion process is illustrated in Fig. 13.47. The resulting structure has diffused-in n-type regions wherever the original photographic mask (Fig. 13.45) was opaque. If a layer of metal were now placed on top of the SiO_2 between the n regions in Fig. 13.47, it could serve as a gate, the two diffused-in regions would be the source and drain, and the device would be a MOSFET (Fig. 13.38).

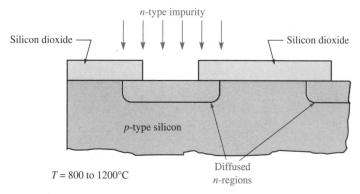

FIGURE 13.47 Diffusion of *n*-type impurities using a silicon-dioxide mask.

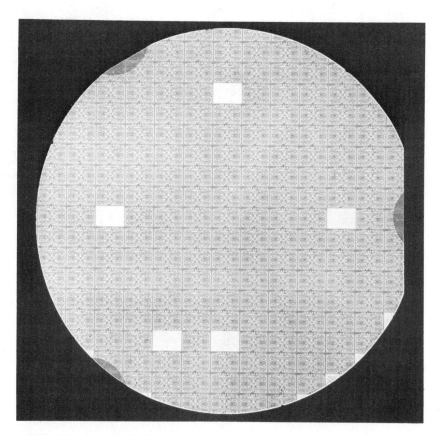

FIGURE 13.48 A completed wafer ready to be sawed apart into individual IC chips. (*Courtesy of Siemens Aktiengesellschaft, Munich.*)

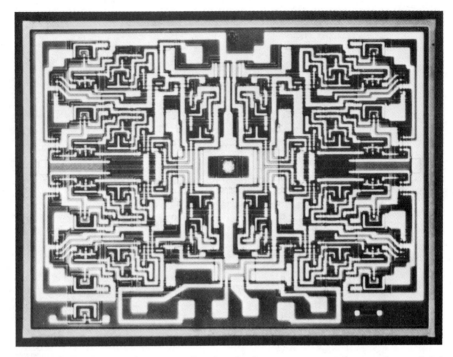

FIGURE 13.49   Magnified view of a single IC from the wafer of Fig. 13.48. (*Courtesy of Siemens Aktiengesellschaft, Munich.*)

Various sequences of fabrication steps are of course required, depending on what kinds of ICs are being made. This subject is referred to as *microfabrication*, a highly experimental art in which rapid progress is always being made. Microfabrication and circuit design are the two parts of integrated circuit technology. The subject of circuit design, analog and digital, will be introduced in the next two chapters.

Figure 13.48 shows the appearance of the wafer when processing is complete. Numerous identical circuits have been simultaneously fabricated in a single series of processing steps. All that now remains is to saw the wafer apart into individual circuits and encapsulate them, as in Fig. 13.44. A magnified view of a single IC chip is shown in Fig. 13.49.

13.6 Application: Microwave Transistors

Different types of transistors are used for different purposes. For VLSI, MOSFETs are most useful, because of their low power consumption and ease of fabrication. Bipolar transistors are used in high-power applications, and in general applications when high gain is required and power consumption is not a key issue. Other kinds of FETs, such as junction-FETs (JFETs), in which the insulated gate is replaced by a reverse-biased junction, are used in a few special applications, especially when low noise is required.

Meanwhile, new kinds of transistors are being invented. Some of the most interesting work is being done with devices intended to work at extremely high frequencies. Present-day silicon bipolar technology reaches its high-frequency limit somewhere around 5 GHz. (1 GHz = 10^9 Hz.) Above that there have been MESFETs (special JFETs in which the gate is a reverse-biased *ME*tal-*S*emiconductor junction) up to 20 or 30 GHz, and above that, no transistors at all. The part of the frequency spectrum above 1 GHz, known as the *microwave* region, is valuable for communication at very high data rates. All else being equal, the number of bits that can be sent through a communications channel is proportional to the bandwidth of the channel. This makes the microwave region very attractive; there is a lot of bandwidth up there! In fact, in the frequency band between 1 and 10 GHz, there is nine times as much bandwidth as in all the frequencies below the microwave region. (And in the range 10 to 100 GHz, there is 10 times more than that!) However, there is no way to exploit all this potential bandwidth, if transistors cannot be made to work at those frequencies. Thus development of ultra-high-frequency transistors is a problem at the cutting edge of electronic technology.

At present it appears that silicon transistors are just too slow to be promising at microwave frequencies. Fortunately, however, a new family of semiconductor materials, known as the III-V compounds, have become available. The name "III-V" refers to the fact that these are compounds in which one element comes from group III of the periodic table and the other from group V. Gallium arsenide is presently the most important of these materials; gallium phosphide and indium arsenide are other examples. Compounds containing more than two elements, such as gallium aluminum arsenide, are also important. In one sense, *n*-type gallium arsenide is inherently more suitable for microwave devices than silicon because, for a given doping level, its resistance is lower. But more important is that with III-Vs, one is not confined to just a single semiconductor (as one is with silicon); one can build structures containing several different III-V semiconducting compounds. This gives a whole range of possibilities for new devices, most of which cannot be made in silicon at all.

One of these novel devices is the High Electron Mobility Transistor, or HEMT, a special MESFET in which the GaAs channel is kept free of donors. This is desirable because collisions between electrons and donors reduce the all-important channel conductivity. How does the channel *get* any electrons if it is free of donors? The electrons are cleverly "imported" from surrounding GaAlAs regions. HEMTs are very promising ultra-high-frequency devices; they have been used up to around 100 GHz, which is as high as electronic technology presently goes. Interestingly, they also turn out to be useful at lower microwave frequencies, because they make extremely good low-noise amplifiers. Thus, for example, they are useful in satellite TV receivers, where the problem is to capture enough signal to overcome receiver noise. A few years ago, enormous backyard "dish" antennas were needed, but with low-noise HEMT receivers it is possible to use a much smaller antenna, about the size of a dinner plate. That is a great commercial advantage!

Another interesting new III-V device is the Heterojunction Bipolar Transistor, or HBT. This is similar to an ordinary bipolar transistor, except that the base is made out of a different III-V compound from that used for the emitter. This gives what is known as a "heterojunction"—a *pn* junction between two different semiconductors—in place of the ordinary junction usually found between base and emitter. A useful property of the heterojunction emitter is that, unlike an ordinary forward-biased junction, it can inject minority carriers from the emitter into the base, *without simultaneously injecting minority carriers the other way, from the base into the emitter*. This eliminates a wasted part of the emitter current, and thus increases α and β. In ordinary BJTs this waste is prevented by using weakly doped material for the base, but that has the disadvantage that the internal resistance of the base becomes high. In the HBT one can have it both ways: no wasted emitter current and a nice, low base resistance. These characteristics make the HBT very promising, especially as a high-power device between 10 and 100 GHz.

The history of electronics has always been determined by the invention of new and better active devices. First there was the vacuum triode tube: large, breakable, and power-wasting, but nonetheless the only available device from its invention in 1906 clear into the 1960s. Then came transistors, first germanium bipolar transistors, then silicon bipolar transistors, and eventually silicon and GaAs FETs. With better and faster devices has come the use of higher and higher frequencies (thus more band-width, thus more information per second). The new III-V devices are the latest step in this progression, but who knows what tomorrow will bring?

POINTS TO REMEMBER

- Semiconductor devices are nonlinear, active circuit elements found in electronic circuits. The most important semiconductor material is silicon.

- Currents in semiconductors arise from motion of negatively charged electrons and/or motion of positively charged "holes." Materials with a preponderance of electrons are said to be *n*-type; those with a preponderance of holes are *p*-type. Semiconductors can be made *n*-type or *p*-type by "doping" them with impurity atoms.

- A *pn* junction is obtained by placing *p*- and *n*-type materials in intimate contact. Three important properties of *pn* junctions are (1) one-way conduction; (2) injection of minority carriers when forward biased; (3) collection of minority carriers when zero or reverse biased.

- The voltage across a forward-biased silicon *pn* junction is about 0.7 V.

- A bipolar junction transistor consists of two *pn* junctions separated by a thin region called the base.

- The input characteristic of a bipolar transistor in the common-emitter connection resembles the *I-V* characteristic of a diode.

- The transistor operates in the active mode when the emitter junction is forward biased and the collector junction is unbiased or reverse biased. If (and only if) operation is in the active mode, $I_C = \beta I_B$.

- When both the emitter junction and the collector junction are forward biased, the transistor operates in the saturated mode. In this case $I_C < \beta I_B$ and $V_{CE} = V_{CESAT} \cong 0.2$ V.

- The operating point of the transistor is found by drawing a load line over the output characteristics.

- Field-effect transistors have the useful property that very little current flows through their input (gate) terminals.

- Integrated circuits contain semiconductor devices, resistors, wires, and sometimes other components, all fabricated in a single chip of semiconductor. They are made by *microfabrication* techniques, the most important of which is *photolithography*.

PROBLEMS

Section 13.2

13.1 A certain diode has a saturation current I_s equal to 10^{-13} A at room temperature. Calculate the current for applied forward voltages of -10, -1, -0.1, 0, $+0.1$, $+0.5$, $+0.7$, and $+1.0$ V. Plot i versus v on linear graph paper.

*13.2 Repeat Example 13.2 and find the current i_D if
  a. The diode is modeled as a perfect rectifier (Fig. 13.7).
  b. The diode is represented by the large-signal diode model (Fig. 13.9).
  c. As compared with the "exact" answer obtained in Example 13.2, what percentage error is introduced by the approximations in parts (a) and (b)?

*13.3 In the circuit of Fig. 13.50(a), the time-varying voltage source $v_0(t)$ produces the square wave shown in Fig. 13.50(b). Let $V_1 = 15$ V and $R = 10$ kΩ. Make a sketch of the diode current $i_D(t)$ versus time if
  a. The diode has the "exact" *I-V* characteristic of Fig. 13.5.
  b. The diode is modeled as a perfect rectifier (Fig. 13.7).
  c. The diode is represented by the large-signal diode model (Fig. 13.9).
  d. Discuss the relative accuracy of the three methods.

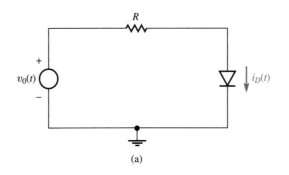

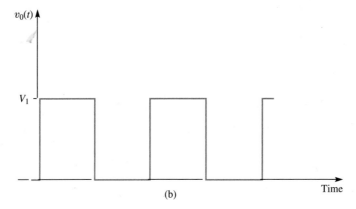

FIGURE 13.50

*13.4 Repeat Problem 13.3 with $V = 1.0$ V and $R = 1,000$ Ω. Discuss the relative accuracy of the three methods in this case.

13.5 The purpose of this problem is to demonstrate the rule that the voltage across a forward-biased silicon diode is approximately 0.7 V. A diode whose I-V characteristic is shown in Fig. 13.5 is connected as shown in Fig. 13.51. By drawing load lines on Fig. 13.5, find the diode voltage v_D if:

   a. $V_0 = 2$ V, $R = 400$ Ω.
   b. $V_0 = 10$ V, $R = 10$ kΩ.
   c. $V_0 = 1$ V, $R = 200$ Ω.

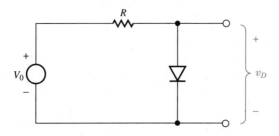

FIGURE 13.51

13.6 In the circuit of Fig. 13.52, let $I_0 = 3$ mA and $R = 1,000$ Ω. Find v_D if the diode's *I-V* characteristic is:

a. Figure 13.5.

b. That of a perfect rectifier.

c. That of the large-signal diode model.

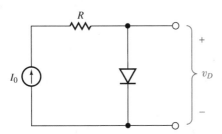

FIGURE 13.52

*13.7 In the circuit of Fig. 13.53, let $V_0 = 5$ V, $R_1 = 1,000$ Ω, and $R_2 = 1,500$ Ω, and let the diode *I-V* characteristic be that of Fig. 13.5. Find the diode current i_D. Use the following method: Replace the part of the circuit to the left of *AA'* by its Thévenin equivalent, thus reducing the problem to that of Fig. 13.51. Then solve graphically.

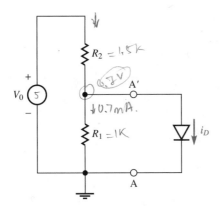

FIGURE 13.53

*13.8 Repeat Problem 13.7 using the large-signal diode model to represent the diode. Find i_D by means of straightforward linear-circuit analysis (node or loop methods).

13.9 Find the current i in Fig. 13.54 by means of the assumption that the voltage across a forward-biased diode is 0.7 V.

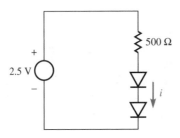

FIGURE 13.54

13.10 Find the current i in Fig. 13.55 by means of the assumption that the voltage across a forward-biased diode is 0.7 V.

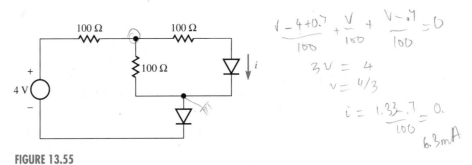

$$\frac{V-4+0.7}{100} + \frac{V}{100} + \frac{V-.7}{100} = 0$$

$$3V = 4$$
$$V = 4/3$$

$$i = \frac{1.33-.7}{100} = 0.$$
$$6.3 mA$$

FIGURE 13.55

13.11 In the half-wave rectifier circuit of Fig. 13.56, assume the diode is a perfect rectifier (Fig. 13.7).
  a. Sketch $v(t)$ as a function of time.
  b. Sketch $i_d(t)$ as a function of time.
  c. Write a mathematical expression for $i_d(t)$, valid for times between $t = 0$ and the first time after $t = 0$ when the diode ceases to conduct.
  d. Find the first time after $t = 0$ at which the diode stops conducting. (*Hint:* This is the time at which i_d becomes 0.)

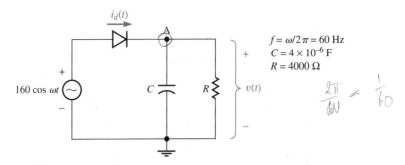

$f = \omega/2\pi = 60$ Hz
$C = 4 \times 10^{-6}$ F
$R = 4000 \ \Omega$

$$\frac{2\pi}{\omega} = \frac{1}{60}$$

FIGURE 13.56

**13.12 In the circuit of Fig. 13.56, let the *time average* of $v(t)$ be V_{AV}. Find the time-averaged current i_{AV} through the diode. *Hint:* The sum of the time-averaged currents entering node A is zero. What is the time-averaged current through a capacitor?

*13.13 Suppose in Fig. 13.56 the time-averaged current through the diode is known to be 0.5 A. Find the time-averaged power dissipated in the diode. Use the large-signal diode model.

*13.14 The three-terminal block in Fig. 13.57 is an IC voltage regulator.

    a. When $i_{OUT} = 0$, $v_{OUT} = 10.0$ V. When $i_{OUT} = 0.1$ A, v_{OUT} is found to be 9.9 V. What is the output resistance (same as the Thévenin resistance) of the regulator as seen from its output terminals?

    b. The 12 V unregulated input voltage has a peak-to-peak ripple $\Delta v_{IN} = 1$ V. Let us define the *ripple rejection ratio*, expressed in decibels, as

$$RRR = 20 \log_{10} \frac{\Delta v_{IN}/v_{IN}}{\Delta v_{OUT}/v_{OUT}}$$

    Find the output ripple Δv_{OUT} if RRR = 50 dB.

    c. Suppose the maximum heat that can be dissipated in the regulator is 1 W. If the input voltage is a constant 12 V, what is the maximum allowable constant value of i_{OUT}?

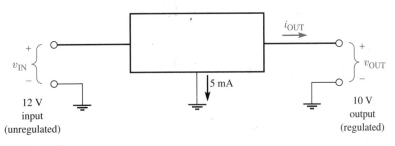

FIGURE 13.57

*13.15 In the circuit of Fig. 13.58, v_{IN} is negative with respect to ground, and the diode is characterized by Eq. (13.1), where I_S and $V_{T0} \equiv kT/q$ are given constants.

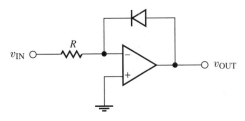

FIGURE 13.58

a. Using the ideal-op-amp technique, find a mathematical expression for v_{OUT} in terms of v_{IN}.

b. Assuming that $|v_{IN}/RI_s| \gg 1$, this circuit performs a useful mathematical operation. What is it?

*13.16 It is sometimes useful to define the "small-signal resistance" of a *pn* junction by the relationship $r \equiv dv/di$.

a. Find an exact expression for r in terms of the diode current i (*not* the voltage), V_{T0}, and I_S.

b. Simplify the expression in the case $i \gg I_S$.

c. Compute r in a diode at a forward bias of 1 mA. [If the expression derived in part (b) is correct, r does not depend on I_S.] Assume $T = 300$ K.

*13.17 The circuit of Fig. 13.59(a), known as a "clipper," is intended to protect the circuit that follows it from excess voltages. Assume the diodes are perfect rectifiers (Fig. 13.7), and let the input voltage $v_1(t)$ be as shown in Fig. 13.59(b). Sketch the output voltage $v_{OUT}(t)$.

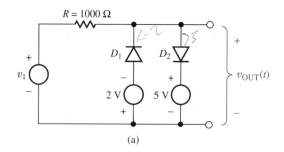

(a)

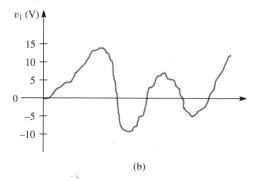

(b)

FIGURE 13.59

Section 13.3

13.18 In Fig. 13.60, determine whether or not the transistor is biased in the active mode (let $R = 10$ kΩ in all cases), if:

$V_{BE} > 0$

a. V_{CC} = +10 V, V_{BB} = +5 V. $\Rightarrow V_{CE} = 10$
b. V_{CC} = +10 V, V_{BB} = +0.3 V.
c. V_{CC} = 0, V_{BB} = +4 V.
d. V_{CC} = +0.7 V, V_{BB} = +4 V.

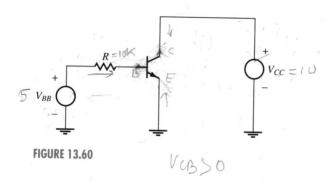

$R = 10K$

$V_{CC} = 10$

$5\ V_{BB}$

FIGURE 13.60

$V_{CB} > 0$

13.19 Determine whether or not the transistors in Fig. 13.61(a) and (b) are biased in the active mode. (Note that in Fig. 13.61(b) it is a *pnp* transistor.)

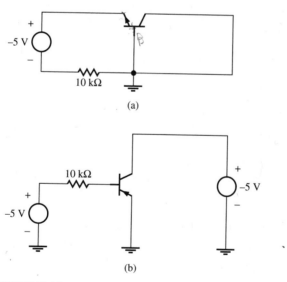

−5 V

10 kΩ

(a)

10 kΩ

−5 V

−5 V

(b)

FIGURE 13.61

13.20 In Fig. 13.62, assume that the transistor operates in the active mode. Find the collector voltage v_C in terms of i_1. Let β = 100.

13.21 What is the largest value (approximately) that i_1 can have in Fig. 13.62 consistent with operation in the active mode? Let β = 100.

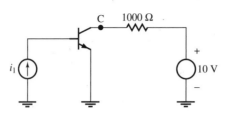

FIGURE 13.62

13.22 Assume that the transistor in Fig. 13.63 operates in the active mode. Let $\beta =$ 100 and $R_B = 100 \text{ k}\Omega$.

    a. Using the 0.7 V approximation for the forward-biased emitter junction, find i_B.

    b. Find i_C.

    c. Find v_C.

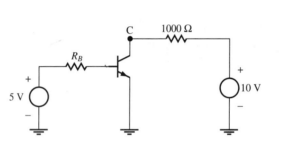
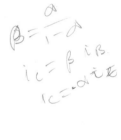

FIGURE 13.63

13.23 Suppose the base voltage source in Fig. 13.63 is increased from 5 V to higher values. What is the approximate highest value it can have consistent with operation in the active mode? Take $\beta = 100$, $R_B = 100 \text{ k}\Omega$.

13.24 A certain *npn* transistor has a reverse-bias voltage of 5 V applied to its collector-base junction. It is found that the emitter current is -0.8 mA, the base current is 5 μA, and $v_{BE} = 0.65$ V. Find β and I_{ES} for this transistor.

*13.25 Suppose we repeat Example 13.9 in order to calculate v_C with the base resistance changed from 10 kΩ to 1,000 Ω. Does your answer make sense? Explain.

13.26 Which of the three curves in Fig. 13.29 is most appropriate for a transistor operating in the active mode?

13.27 By drawing a load line on Fig. 13.29, find v_B and i_B for the circuit of Fig. 13.60. Let $V_{BB} = 1$ V, $R = 33 \text{ k}\Omega$, and $V_{CC} = 9$ V.

13.28 By drawing a load line on Fig. 13.29, find v_B, i_B, v_C, and i_C for the circuit of Fig. 13.63. Let $\beta = 200$ and $R_B = 167 \text{ k}\Omega$.

13.29 Using the output characteristics of Fig. 13.30, find:

    a. i_C if $v_{CE} = 7$ V, $i_B = 7 \mu$A.

b. i_B if $v_{CE} = 4$ V, $i_C = 1.3$ mA.

c. v_{CE} if $i_C = 1$ mA, $i_B = 20$ μA.

13.30 Suppose the transistor in Fig. 13.63 has the output characteristics of Fig. 13.30. Find i_C. Let $R_B = 287$ kΩ.

13.31 Suppose the transistor of Fig. 13.34 has the output characteristics shown in Fig. 13.30. Let $V_{CC} = 10$ V, $R_C = 14$ kΩ. Find v_C if i_B equals:

   a. 0.

   b. 5 μA.

   c. 10 μA.

   d. 20 μA.

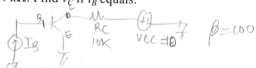

*13.32 Suppose the transistor of Fig. 13.34 has the output characteristics of Fig. 13.30. Let $V_{CC} = 8$ V and $R_C = 10$ kΩ.

   a. Make a sketch of v_C as a function of i_B over the range $0 \leq i_B \leq 100$ μA.

   b. Suppose the current source i_B is replaced by an adjustable voltage source V_{BB} and a 50 kΩ resistor connected in series. Make a sketch of v_C versus V_{BB} over the range $0 \leq V_{BB} \leq 5$ V.

*13.33 Find the power that is converted to heat in the transistor of Example 13.9. How much power is converted into heat in the entire circuit?

*13.34 Find the power dissipated as heat in the transistor of Fig. 13.63. Let $\beta = 100$ and $R_B = 86$ kΩ.

*13.35 The circuit of Fig. 13.64 is a voltage regulator. Its function is to supply an output voltage that is nearly constant, independent of output current. A *reference voltage source* is used. This is a voltage source that has a constant value but that is unable to provide much current. (It could be a small battery.) Find v_{OUT} with

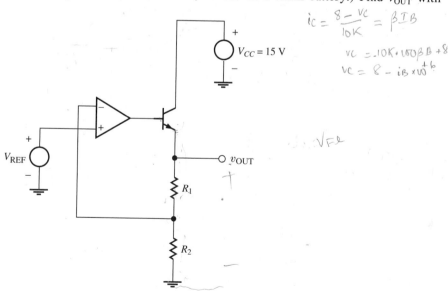

FIGURE 13.64

$v_{REF} = 6$ V, $R_1 = 300$ Ω, and $R_2 = 500$ Ω. Assume that the transistor operates in the active mode, and use the ideal-op-amp technique.

*13.36 Determine the approximate value of β for the transistor whose experimental characteristics are shown in Fig. 13.36:

a. With $v_{CE} = 2$ V, $i_B \cong 7.5$ μA.

b. With $v_{CE} = 9$ V, $i_B \cong 17.5$ μA.

Is β really a "constant"?

Section 13.4

13.37 The FET in Fig. 13.65 has the characteristics shown in Fig. 13.41. Let $V_{DD} = 6$ V and $V_{GG} = 6.3$ V. Find i_D.

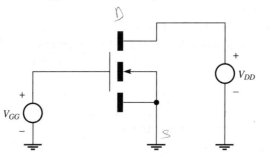

FIGURE 13.65

13.38 For the circuit of the previous problem, let V_{GG} be replaced by a sinusoidal voltage source with amplitude 6 V. Make a sketch of drain current versus time.

*13.39 Find the drain current and drain voltage for the circuit of Fig. 13.66. The FET has the characteristics shown in Fig. 13.41. Let $I_0 = 800$ μA, $R_D = 3,750$ Ω, and $V_{GG} = 6$ V. (*Suggestion:* Replace R_D and I_0 by a Thévenin equivalent.)

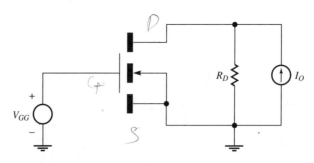

FIGURE 13.66

*13.40 Sketch the output characteristics of a *p*-channel enhancement-mode MOSFET.

Assume the device is like that of Fig. 13.41 except for appropriate changes of signs.

*13.41 Using Eqs. (13.10) and (13.11), graph the theoretical output *I-V* characteristics of an *n*-channel enhancement MOSFET. Let $V_T = 3$ V and $K = 0.095$ mA/V²; construct curves for $v_{GS} = 3, 4, 5, 6,$ and 7 V. Compare your graph with the experimental curves of Fig. 13.41.

*13.42 Assuming that the necessary alterations are only in algebraic signs, rewrite Eqs. (13.10) and (13.11) to obtain equations for a *p*-channel enhancement MOSFET. Let K and V_T be positive numbers.

13.43 Repeat Example 13.15 using as the data points $v_{GS} = 4, 5, 6,$ and 7 V and:
a. $v_{DS} = 8$ V.
b. $v_{DS} = 1$ V.

*13.44 Using the data of Fig. 13.41, graph i_D versus $(v_{GS} - V_T)^2$ for $v_{DS} = 4$ V and for $v_{DS} = 8$ V. According to Eq. (13.11) these graphs should be straight lines with slope ½K. Are they in fact nearly straight? To what values of K do they correspond?

*13.45 Repeat Example 13.14 without referring to Fig. 13.41 by instead using Eq. (13.11) to describe the transistor. Assume $V_T = 3$ V and $K = 0.1$ mA/V². How well do your answers agree with the graphical results in the example?

*13.46 Find the operating point of the circuit of Fig. 13.66 analytically using either Eq. (13.10) or Eq. (13.11), whichever is appropriate. Let $I_0 = 800$ µA, $R_D = 3,750$ Ω, $V_{GG} = 6$ V, $V_T = 3$ V, and $K = 0.1$ mA/V². Verify that your choice of Eq. (13.10) or (13.11) was the correct one.

*13.47 Figure 13.67 shows a biasing circuit for an *n*-channel enhancement MOSFET. Let $V_{DD} = 10$ V, $R_D = 10$ kΩ, $K = 0.1$ mA/V², and $V_T = 3$ V.
a. Does the transistor operate in the triode region or the pinch-off region?
b. Using either Eq. (13.10) or (13.11), as determined by part (a), find the operating point. (You may end up with a quadratic equation that has two solutions. Which one is the correct one?)

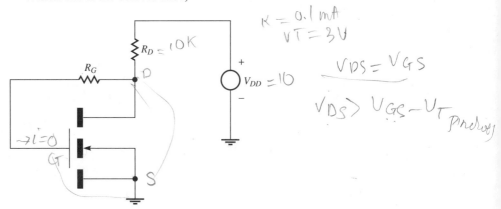

FIGURE 13.67

*13.48 Figure 13.68 shows another FET biasing circuit. Let V_{DD} = 18 V, V_{GG} = 9 V, R_S = 3.5 kΩ, R_D = 3.5 kΩ, K = 0.6 mA/V², and V_T = 2.2 V. Using Eq. (13.11) find v_{GS}, v_{DS}, and i_D. Verify that Eq. (13.11) [and not Eq. (13.10)] is the correct equation to use.

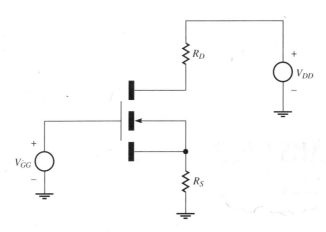

FIGURE 13.68

TRANSISTOR AMPLIFIERS

W e are now ready to apply our knowledge of semiconductor devices to circuit design. Our concern now is with the internal construction of the building blocks of Chapters 9 to 12. The subject naturally divides into analog circuits and digital circuits. Analog circuits are the subject of this chapter; digital circuits will be treated in Chapter 15.

A major problem in the analysis of transistor circuits arises from transistors' non-linearity. To avoid clumsy graphical solutions, an approximate method called *small-signal analysis* is often used. This method simplifies problems and allows them to be solved with ordinary circuit analysis, with which the reader is already familiar.

14.1 Small-Signal Analysis

Let us first begin with the simple circuit shown in Fig. 14.1. In this circuit a voltage $V_0 + v_S$ is applied to the base of the transistor. The time-varying voltage v_S is the input signal, and V_0 is a constant voltage* whose function will be explained shortly.

---

*Note our use of capital-letter symbols for constant voltages and currents and lowercase-letter symbols for time-varying voltages and currents.

547

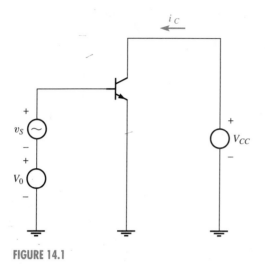

FIGURE 14.1

The collector voltage source V_{CC} causes the collector-base junction to be reverse biased.

Let us first consider the case of $v_S = 0$. The forward-bias voltage on the emitter junction is then just equal to V_0. Refer to Fig. 14.2(a), which shows the transistor's input characteristic. We find that the base current has the value I_0. Since we are operating the transistor in the active mode, the collector current now has the value $i_C = \beta i_B = \beta I_0$.

We now turn on the signal voltage v_S. If $v_S(t)$ is a sinusoid with amplitude less than V_0, its effect is alternately to increase and decrease the base-to-emitter voltage v_{BE} ($\equiv v_B - v_E$), as shown in Fig. 14.2(b). This causes the base current to vary as shown. The variations in base current give rise to variations of the collector current that are larger by a factor of β. The larger time-varying collector current is sent to a load and constitutes the output of the amplifier.

From Fig. 14.2(b) it now can be seen why the constant voltage V_0 is needed. If V_0 were made zero, v_{BE} would equal v_S. Since the slope of the input characteristic is very nearly zero at $v_{BE} = 0$, the base voltage variations v_S would have almost no effect; the base current would remain zero. If the amplitude of the sinusoidal signal v_S were increased to the order of 0.7 V or so, some base current would indeed appear at the time v_S was most positive. However, when v_S became negative, the emitter junction would be reverse biased. Thus no base current would ever flow during the negative part of the cycle of v_S. Any time v_S was negative, its value would be unable to influence the collector current; hence those parts of the input signal would not be amplified. The use of V_0 allows us to avoid this problem. From Fig. 14.2(b) we see that through the use of V_0, v_S exerts a controlling effect on i_B (and hence on i_C) at all times.

From the preceding discussion we see that it is desirable in an amplifier for the time-varying signals to be accompanied by constant voltages and currents, known as

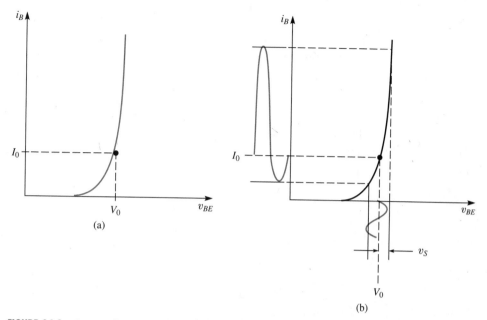

FIGURE 14.2 Input *I-V* characteristic of the transistor circuit of Figure 12.1. If $v_s = 0$, the base current determined by V_o has the value I_o, as shown in (a). If the small-signal voltage v_s is now applied, the base current fluctuates in response to it, as shown in (b).

bias voltages and currents. For instance, the dc base current (I_0 in the example just discussed), is called the *base bias current*, and the dc voltage V_0 is the *base bias voltage*. The point (V_0, I_0) in Fig. 14.2, where operation rests in the absence of signals, is called the *operating point* (or "*Q*-point") of the circuit.

LINEARIZATION

Figure 14.2(b) shows that the changes in the base current are not linearly proportional to the signal voltage v_S. Because the transistor *I-V* characteristic is nonlinear, distortion of the signal waveform occurs. However, we can show that the device approaches linearity more closely as v_S is made smaller.

Let us expand the base current i_B as a Taylor series about the operating point. In general, if $y(x)$ is any nonsingular function of x and $y(x_0) = y_0$, a power series expression of y is

$$y(x) = y_0 + \frac{dy}{dx}\bigg|_{x = x_0} \cdot (x - x_0) + \frac{1}{2!}\frac{d^2y}{dx^2}\bigg|_{x = x_0} \cdot (x - x_0)^2 + \cdots \quad (14.1)$$

Here the notation

$$\frac{dy}{dx}\bigg|_{x = x_0}$$

stands for the derivative, evaluated at $x = x_0$. Expression (14.1) for $y(x)$ is in the form of an infinite series. However, if $|x - x_0|$ is a small quantity, its higher powers rapidly become insignificantly small, and one obtains an approximate expression for $y(x)$

$$y(x) \cong y_0 + \left.\frac{dy}{dx}\right|_{x = x_0} \cdot (x - x_0)$$

This approximation becomes more accurate as $|x - x_0|$ becomes smaller, because the neglected higher terms of the series become less significant.

In the case of our transistor circuitry, the role of the function $y(x)$ is taken by the function $i_B(v_{BE})$, shown in Fig. 14.2. We know that when $v_{BE} = V_0$, $i_B = I_0$. Expanding in Taylor series about this point, we have the following expression for i_B:

$$i_B = I_0 + \left.\frac{di_B}{dv_{BE}}\right|_{v_{BE} = V_0} \cdot (v_{BE} - V_0) \tag{14.2}$$

$$+ \frac{1}{2!} \left.\frac{d^2 i_B}{dv_{BE}^2}\right|_{v_{BE} = V_0} \cdot (v_{BE} - V_0)^2 + \cdots$$

In our present case we are interested in finding the value of i_B when $v_{BE} = V_0 + v_S$. Making this substitution, we have

$$i_B = I_0 + \left.\frac{di_B}{dv_{BE}}\right|_{v_{BE} = V_0} \cdot (v_S) + \frac{1}{2!} \left.\frac{d^2 i_B}{dv_{BE}^2}\right|_{v_{BE} = V_0} \cdot (v_S)^2 + \cdots \tag{14.3}$$

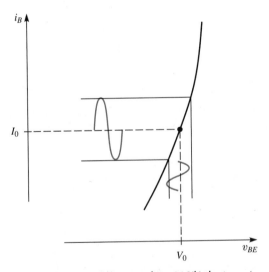

FIGURE 14.3 Magnified portion of Fig. 14.2(b) showing variations in base current due to small v_s. The I-V curve is nearly straight over a small portion of its length. Consequently $I_B - I_0$ is more nearly proportional to v_s than in Fig. 14.2(b), where v_s is larger.

Now if v_S is a small number, an approximate expression for i_B can be found by dropping the quadratic and all higher terms of Eq. (14.3):

$$i_B \cong I_0 + \left.\frac{di_B}{dv_{BE}}\right|_{v_{BE} = V_0} \cdot v_S \tag{14.4}$$

Equation (14.4) states that if v_S is made sufficiently small, the variations of the base current around the operating point are, to a good approximation, linearly proportional to the signal voltage v_S. The significance of Eq. (14.4) is shown graphically in Fig. 14.3. This figure is an expanded version of Fig. 14.2(b), showing *small* variations around the operating point. The result, Eq. (14.4), basically comes from the fact that even a curved line looks nearly straight if one considers only a small region along its length.

EXAMPLE 14.1

The transistor in the circuit of Fig. 14.4 is operating in the active mode, with $\beta = 100$. Assume that $di_B/dv_{BE} = 3.85 \times 10^{-4}\ \Omega^{-1}$. The signal voltage v_S increases from 0 to 10 mV. What is the change in v_{OUT}? What is the voltage gain? Take $V_{CC} = 10$ V and $R_C = 5,000\ \Omega$.

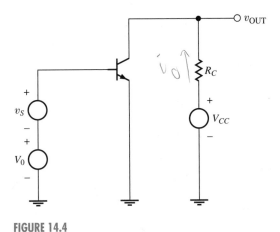

FIGURE 14.4

SOLUTION

From Eq. (14.4),

$$\Delta i_B = \frac{di_B}{dv_{BE}} v_S$$

$$= (3.85 \times 10^{-4})(10^{-2}) = 3.85\ \mu A$$

The change in the output voltage can be found as follows:

$$v_{OUT} = V_{CC} - i_C R_C$$

$$\Delta v_{OUT} = -\Delta i_C R_C = -\beta \Delta i_B R_C$$

$$= -(100)(3.85 \times 10^{-6})(5,000) = -1.93 \text{ V}$$

The voltage gain is

$$G_V = \frac{\Delta v_{OUT}}{\Delta v_{IN}} = -\frac{1.93}{0.010} = -193$$

The minus sign indicates that the amplification is accompanied by a sign inversion; that is, positive-going changes of the input voltage cause negative-going changes of v_{OUT} that are 193 times larger.

From Eq. (14.4) we might conclude that, to construct highly linear circuits, we should restrict their operation to small signals. However, even when v_S is not extremely small, it is useful to pretend that Eq. (14.4) is true. The *I-V* characteristic of the transistor, after all, is nonlinear; analysis of circuits containing nonlinear elements is a slow numerical procedure. But Eq. (14.4) offers a great simplification. It is a linear equation, and solutions for linear equations are easy to find. Thus Eq. (14.4) is made the starting point for the technique known as *small-signal analysis*, which is widely used in connection with analog circuits. With small-signal analysis we shall be able to treat transistors as though they were linear circuit elements, constructing linear models for them, just as we did for amplifier blocks in Chapter 4. Of course, Eq. (14.4) is not strictly accurate if v_S is not small. Nonetheless, small-signal analysis is such a powerful method that it is most often used anyway, at least for preliminary analysis. Often the approximation is good enough, and further calculations are not necessary.

In a typical amplifier circuit, provision must be made to apply the desired dc voltages and currents to the transistors, so that with signal absent they are biased at the chosen operating point. This part of the circuit will be considered first, in Section 14.2. In Section 14.3 we shall proceed with the subject of small-signal analysis by developing a linear model for the transistor. As in Chapter 4, the model for the transistor will be a collection of linear circuit elements designed to mimic its operation. The model can be substituted for the transistor in the circuit, so that the techniques of linear circuit analysis can then be used. In the following sections the modeling technique of the previous section is applied to analysis of simple amplifier circuits.

14.2 Biasing Circuits

In each linear amplifier circuit, provision must be made for adjusting the values of the dc currents and voltages so that under conditions of no signal the circuit rests at the desired operating point. The choice of what point is to be the operating point is a matter of discretion for the circuit designer. Several considerations enter. For one thing, the operating point must be chosen so that the circuit can accommodate the signals of interest. Even though the analysis is based on small deviations from the operating point, the circuit is nonetheless intended for use with signals of finite size. If, for example, we anticipate a base current signal of $10^{-4} \sin \omega t$ amperes, we must first be sure that the operating point base current is *greater* than 10^{-4} A. The base current can never be negative; therefore if it is to decrease by 10^{-4} A from the operating-point value, the current at the operating point must be greater than 10^{-4} A. A second consideration in the choice of operating point has to do with power consumption. Large dc bias currents may cause the circuit to consume excessive power. A third consideration has to do with the desired operating characteristics of the transistor in the circuit. When we come to the subject of transistor models, it will be seen that the parameters of the model depend on the choice of operating point. These and perhaps other considerations must be weighed by the circuit designer. There is no unique "right answer"; the choice of operating point is an "engineering decision."

Since both time-varying signals and non-time-varying bias quantities are usually present in an amplifier, it is helpful, as well as customary, to use notation that emphasizes the difference between the two. In the following discussion, bias voltages and currents, which are constant in time, will be represented by capital-letter symbols. Time-varying voltages and currents will be represented by lowercase-letter symbols.

To illustrate in a simple way how a circuit may be adjusted to its operating point, let us consider the elementary circuit of Fig. 14.5. The notation V_{CC} by the terminal at the top indicates, according to convention, that a voltage source is connected between this point and ground, so that the potential at this terminal has the value V_{CC}. To avoid cluttering the diagram that source is not shown.

The values of R_B and R_C are selected so as to place the transistor at the desired operating point. Let us assume it has been decided that the dc collector current at the operating point I_C is to have the value 1 mA. The transistor is assumed to have $\beta =$

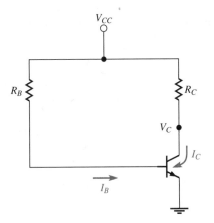

FIGURE 14.5 A simple biasing circuit for an *npn* transistor. In a discussion of biasing, only dc voltages and currents are considered.

100 and $V_{CC} = 10$ V. The circuit designer must then choose R_B and R_C so that $I_C = 1$ mA. Because the transistor is to operate in the active mode, the base current is given by

$$I_B = \frac{I_C}{\beta} \tag{14.5}$$

Therefore $I_B = 10^{-5}$ A.

For this amount of base current to flow, the emitter-base junction must be forward biased. As we have seen, this means that there must be a forward-voltage drop across it of approximately 0.7 V. Thus $V_B \cong 0.7$ V. The base current is the same current as that flowing through R_B. But according to Ohm's law, the current through R_B is

$$I_B = \frac{V_{CC} - V_B}{R_B} \tag{14.6}$$

We have already decided that I_B is to have the value 10^{-5} A, and we have adopted 0.7 V as the approximate value of V_B. Thus we have from Eq. (14.6) the result

$$R_B = \frac{10 - 0.7}{10^{-5}} = 9.3 \times 10^5 \ \Omega$$

Note that the somewhat arbitrary assumption that $V_{BE} = 0.7$ V has little effect on the result just obtained. Inspection of Eq. (14.6) reveals that if the assumed value of V_B were changed to 0.75 V, the value obtained for R_B would change by only about 0.5 percent. This is a desirable characteristic for the circuit to have. The value of V_{BE} is dependent, for instance, on temperature; it is found to decrease about 2 mV for each temperature increase of 1°C. To make circuit operation as independent of temperature as possible, it is desirable that the operating point not be a strong function of V_{BE}.

Next, the value of R_C will be found. Before doing this, it is necessary to decide what the value of the dc collector voltage V_C should be when the transistor is biased to its operating point. As with I_B, several considerations may bear on this decision. For the present, let us use the following simple line of reasoning. When signals are eventually applied to the circuit, we shall want the collector voltage to vary in response to them. It is desirable that the collector voltage v_C be able to vary as far in the positive direction, in response to signals of one sign, as it can in the negative direction, in response to signals of the opposite sign. However, v_C cannot be allowed to become less than zero, or reverse bias on the collector junction will be lost. It will not be possible for v_C to exceed 10 V either, since this is the largest voltage supplied to the circuit. Therefore the permissible range of v_C is approximately 0 to 10 V. It is reasonable to place the dc operating-point voltage V_C at the middle of this range, so that the excursions around the operating point due to signal may go equally far in either the positive or negative direction. Thus we choose $V_C = V_{CC}/2 = 5$ V.

Now we may proceed to calculate the value of R_C by Ohm's law. It has already been decided that $I_C = 1$ mA. The current I_C flows through R_C; therefore

$$I_C = \frac{V_{CC} - V_C}{R_C} \tag{14.7}$$

Solving, we have $R_C = (10 - 5)/10^{-3} = 5,000 \ \Omega$.

EXAMPLE 14.2

Find the dc operating point values of the base current, collector current, and collector voltage for the transistor in the circuit in Fig. 14.6.

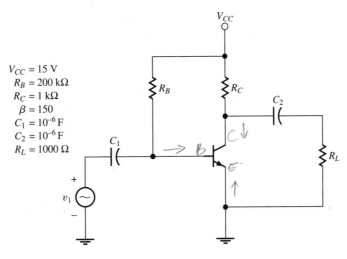

$V_{CC} = 15$ V
$R_B = 200$ kΩ
$R_C = 1$ kΩ
$\beta = 150$
$C_1 = 10^{-6}$ F
$C_2 = 10^{-6}$ F
$R_L = 1000$ Ω

FIGURE 14.6

SOLUTION

Although the circuit is fairly complex, at present we need be concerned only with the central part of the circuit. The capacitors C_1 and C_2 are open circuits for dc, and therefore v_1 and R_L do not affect the dc operating point.

To find the base current, we write a node equation for the base terminal of the transistor:

$$I_B + \frac{V_B - V_{CC}}{R_B} = 0$$

or

$$I_B = \frac{V_{CC} - V_B}{R_B}$$

If we again adopt, as our standard approximate value, $V_{BE} = 0.7$ V, we have

$$I_B = \frac{15 - 0.7}{2 \times 10^5} = 7.2 \times 10^{-5} \text{ A} = 72 \text{ μA}$$

Since the transistor operates in the active mode, $I_C = \beta I_B$ and thus $I_C = 150(72 \text{ μA}) = 10.8$ mA. To find the dc collector voltage V_C, we write a node equation for the collector terminal:

$$I_C + \frac{V_C - V_{CC}}{R_C} = 0$$

or

$$V_C = V_{CC} - I_C R_C$$
$$= 15 - (10.8 \times 10^{-3})(10^3) = 4.2 \text{ V}$$

14.3 The Small-Signal Transistor Model

The previous section was concerned with the dc operation, or "biasing," of an amplifier circuit. Now we are ready to deal with its ac behavior. To do this, we shall construct a device model for the transistor. This model is based on a linearized description of the transistor, like that used in obtaining Eq. (14.4). Consequently, the model is a collection of ideal linear circuit elements. It is intended to be substituted for the transistor in the circuit diagram to facilitate analysis. However, the model is intended to model *only the ac, or signal operation of the transistor.* Our point of view is that analyses of the circuit for dc (biasing) and ac (signals) are separate operations. The ac device model to be considered in this section deals only with the response of the circuit to small signals. *The small-signal model cannot be used to obtain information about biasing.*

Let us consider an *npn* transistor biased in the active region. To model the small-signal behavior of the device, we may first draw upon the input characteristics discussed in Chapter 13. From Eqs. (13.7) and (13.4) to (13.6) we have

$$i_B = \frac{i_E}{\beta + 1} \cong \frac{I_{ES}}{\beta + 1} \exp \frac{q v_{BE}}{kT} \tag{14.8}$$

This equation states the mathematical relationship between i_B and v_{BE}; it is graphed in Fig. 14.2(a). As explained earlier, Eq. (14.8) can be linearized by expanding in Taylor series around the operating point:

$$i_B \cong I_B + \frac{di_B}{dv_{BE}} \bigg|_0 \cdot (v_{BE} - V_{BE}) \tag{14.9}$$

Here the notation $(di_B/dv_{BE})|_0$ means that the derivative is to be evaluated at the operating point; I_B and V_{BE} are the base current and base-to-emitter voltage at the operating point. Equation (14.9) is, as previously explained, an approximation, most valid when $v_{BE} - V_{BE}$ is small.

Let us now introduce an important new notation. We shall define

$$i_b \equiv i_B - I_B \tag{14.10}$$

$$v_{be} \equiv v_{BE} - V_{BE}$$

The new quantities defined here have lowercase-letter subscripts to indicate that they represent small deviations of current or voltage away from the operating point. We shall refer to them as *small-signal variables*. [Later we shall introduce other small-signal variables; these too will be represented by symbols with lowercase-letter subscripts and will be defined analogously with Eq. (14.10).] In terms of the small-signal variables, Eq. (14.9) becomes

$$I_B + i_b = I_B + \frac{di_B}{dv_{BE}}\bigg|_0 \cdot v_{be} \tag{14.11}$$

Subtracting I_B from both sides, we have

$$i_b = \frac{di_B}{dv_{BE}}\bigg| \cdot v_{be} \tag{14.12}$$

This equation expresses a linear proportionality between the small-signal base voltage and the small-signal base current.

The quantity di_B/dv_{BE} may next be evaluated by differentiation of Eq. (14.8):

$$\frac{di_B}{dv_{BE}} = \frac{I_{ES}}{\beta + 1}\frac{q}{kT}\exp\frac{qv_{BE}}{kT} \tag{14.13}$$

We evaluate the derivative at the operating point by setting $v_{BE} = V_{BE}$ in Eq. (14.13). Substituting Eq. (14.8) into Eq. (14.13) to eliminate the exponential function, we find

$$\frac{di_B}{dv_{BE}}\bigg|_0 = \frac{qI_B}{kT} \tag{14.14}$$

The value of the derivative, Eq. (14.14), can now be substituted into Eq. (14.12). This leads to

$$i_b = \frac{qI_B}{kT}v_{be} \tag{14.15}$$

We observe that the quantity qI_B/kT must have the units of reciprocal resistance. Accordingly, let us define a resistance r_π by

$$r_\pi \equiv \left|\frac{kT}{qI_B}\right| \tag{14.16}$$

[There is of course no present need for the absolute-value sign in the definition of r_π, since in *npn* circuits the value of I_B is positive. However, with the absolute-value sign Eq. (14.16) also holds for *pnp* circuits.] In terms of r_π we have, from Eq. (14.15),

$$v_{be} = i_b r_\pi \tag{14.17}$$

Thus we see that as far as the small-signal variables v_{be} and i_b are concerned, the input of the transistor obeys Ohm's law, where the proportionality constant is r_π, given by Eq. (14.16). For the device model to exhibit this behavior, the input terminals (base and emitter) must be connected by the resistance r_π. This first stage in the development of the model is illustrated in Fig. 14.7.

We now consider the output portion of the model. Since we are dealing with the active mode of operation, $i_C = \beta i_B$. We define the small-signal collector current i_c

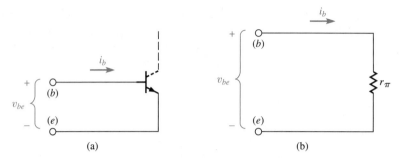

FIGURE 14.7   Input part of small-signal model of the transistor. The relationship between the small-signal voltage and current (a) is modeled by the connection shown in (b).

to be equal to the deviation of the collector current from its value at the operating point:

$$i_c \equiv i_C - I_C \tag{14.18}$$

Then in terms of the small-signal quantities, the equation $i_C = \beta i_B$ becomes

$$I_C + i_c = \beta(I_B + i_b) \tag{14.19}$$

Subtracting from Eq. (14.19) the equality $I_C = \beta I_B$, we have

$$i_c = \beta i_b \tag{14.20}$$

a linear relationship between the small-signal base and collector currents.

The relationship (14.20) can be modeled by a dependent current source connected to the collector terminal. Thus we can complete the small-signal device model as shown in Fig. 14.8. The dependent source imitates the action of the transistor [Eq. (14.20)] by causing a small-signal current βi_b to flow through the collector terminal.

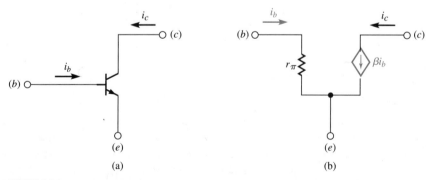

FIGURE 14.8   Completion of the small-signal transistor model. (a) Transistor symbol, showing small-signal base and collector currents. (b) The simplified-π model for the transistor, based on Eqs. (14.17) and (14.20).

The model just obtained is somewhat idealized, but it is sufficiently accurate for almost all calculations except for those involving questions of frequency response. We shall refer to it as the *simplified-π model.** It has been derived for an *npn* transistor; however, it is applicable both to *npn* and *pnp* transistors with no modifications.

USE OF THE DEVICE MODEL

We have already remarked that the function of the device model is to act as a "stand-in" for the transistor. That is, the transistor is replaced by the model in the circuit diagram, and then the resulting circuit model is analyzed. It should be remembered that the small-signal model we have developed only models the behavior of the small-signal variables. On the other hand, the actual circuit contains not only variable quantities but also dc biasing quantities. Thus the subject of how the model is to be used requires some further discussion.

The function of the small-signal model is to represent the relationships of small-signal quantities to each other. It says nothing about dc quantities; thus no reference to dc quantities need be made when the model is used. If a certain wire carries a current $i = I_B + i_b$, for small-signal calculations we simply ignore the dc part and say that $i = i_b$. Similarly, a constant voltage is the same as zero voltage insofar as small-signal calculations are concerned.

The latter statement has an important corollary: *When proceeding from the original circuit to the small-signal circuit model, dc voltage sources become short circuits.* The reasoning behind this statement is illustrated by Fig. 14.9. Suppose the value of the dc voltage source in the figure is V_0, and let v_X, the voltage at terminal X, consist of a dc part V_X plus a small-signal part v_x. The voltages at terminal Y are similarly defined. The basic property of the voltage source is that $v_X + V_0 = v_Y$, or, equivalently, that

$$V_X + v_x + V_0 = V_Y + v_y \tag{14.21}$$

Now when all signals are turned off, Eq. (14.21) reduces to $V_X + V_0 = V_Y$; since V_X, V_0, and V_Y are non-time-varying quantities, this equality must hold whether the signal is off or on. Subtracting this equality from Eq. (14.21), we have $v_x = v_y$. The way to

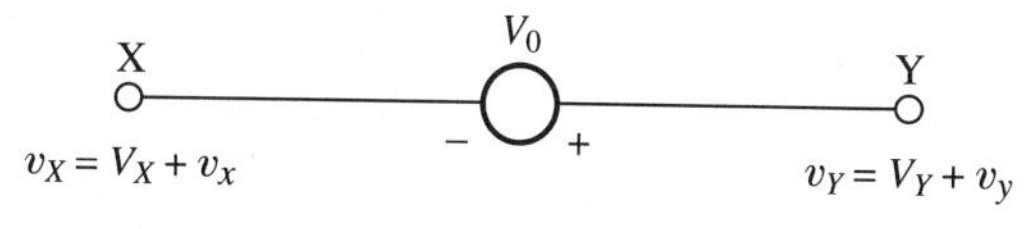

$$v_X = V_X + v_x \qquad v_Y = V_Y + v_y$$

FIGURE 14.9

---

*This name derives from a more general form of this model, known as hybrid-π model, which has a π-shaped arrangement of components.

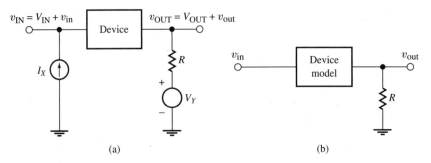

FIGURE 14.10 Comparison of a circuit (a) and its small-signal model (b). The dc current source I_X appears as an open circuit in the circuit model. The dc voltage source V_Y appears as a short circuit.

incorporate into a model the statement that two voltages are equal is to connect them by a wire. Thus it has been shown that in the small-signal circuit model a dc voltage source is replaced by a short circuit.

A similar rule applies to dc current sources. If a branch of the original circuit contains a dc current source, the current through this branch must be constant. But if the current in the branch is constant, the time-varying small-signal current in the branch must be zero. In the model, a branch through which no small-signal current can flow is correctly represented by an open circuit. Therefore, *when proceeding from the original circuit to the small-signal circuit model, dc current sources become open circuits.*

The principles of this section are illustrated by Fig. 14.10. Figure 14.10(a) shows a circuit containing a device, with dc and ac voltages and currents present. Figure 14.10(b) shows the small-signal circuit model for the circuit.

For reference, the meanings of the various types of letter symbols are reviewed in Table 14.1.

TABLE 14.1 Symbols Used in Small-Signal Circuit Analysis

Type Style	Meaning	Example
Capital-letter V or I with any subscript	Non-time-varying quantity	Collector current at operating point I_C
Lowercase-letter v or i with capital subscript	Time-varying quantity	Total collector current i_C
Lowercase-letter v or i with lowercase subscript	Small-signal quantity	Small-signal collector current i_c

Note: The total current (or voltage) equals the dc bias current (or voltage) plus the small-signal current (or voltage). For example, $i_C = I_C + i_c$.

EXAMPLE 14.3

Obtain the small-signal model for the circuit in Fig. 14.11(a) using the simplified-π model of the transistor. V_0 is a constant-bias voltage, and v_S is a small-signal voltage.

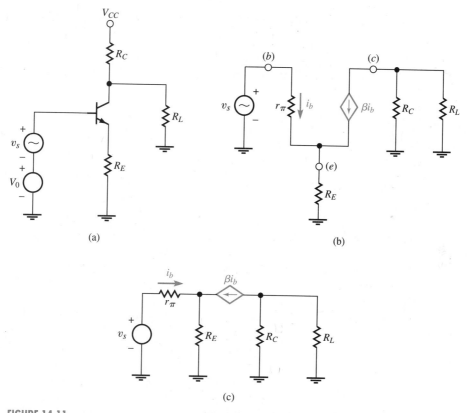

FIGURE 14.11

SOLUTION

In constructing the circuit model, we replace the dc source V_0 by a short circuit. There is also a second hidden dc voltage source in the circuit: A source of magnitude V_{CC} is understood to be connected between the terminal marked V_{CC} and ground. Thus when the small-signal circuit model is constructed, the V_{CC} terminal is connected to ground.

The model is substituted for the transistor, with care that the (b) terminal of the model is connected where the base lead of the transistor was, and so forth. After the model has been substituted, the circuit model appears as in Fig. 14.11(b). Here the locations of the (e), (b), and (c) terminals of the model have been designated to aid in visualizing the process of substitution. Now the circuit may be rearranged, if desired, to tidy it up. One possible arrangement is shown in Fig. 14.11(c). It is now no longer necessary to keep track of the locations of the (e), (b), and (c) terminals. It is, however, still necessary to show the location of the current i_b, since this current controls the dependent current source.

EXAMPLE 14.4

Suppose that in the circuit of Example 14.3 the dc base current at the operating point is 10 μA. What is r_π?

SOLUTION

We recall that the value of r_π depends on the base bias current through Eq. (14.16):

$$r_\pi = \frac{kT}{q} \frac{1}{I_B}$$

The question does not specify the temperature; in that case it is reasonable to assume the question means room temperature, 300 K. The value of kT/q at room temperature is 0.026 V. Thus

$$r_\pi = (0.026)(10^5) = 2.6 \times 10^3 \ \Omega = 2.6 \ k\Omega$$

14.4 Common-Emitter Amplifier Circuits

We are now ready to use the principles of the two preceding sections to analyze amplifier circuits. Let us consider the simple circuit of Fig. 14.12. The parameters assumed for the transistor are given on the right. Resistors R_B and R_C are present in the circuit because of the necessity for biasing; however, they do affect the ac operation of the circuit, as will be seen. The input signal is applied to the base through the input capacitor C_i. Since the capacitor is an open circuit for dc, the presence of C_i does not disturb the biasing of the transistor. However, we shall assume that the frequency of the signal is high enough that the capacitor can be considered a short circuit as far as signals are concerned. The same arguments apply to the output capacitor C_o. Capacitors used to transmit signals in and out of circuits while leaving the dc bias unaffected are known as *coupling capacitors*.

As a first step in the analysis, let us find the dc operating point of the transistor. Since no dc current flows through C_i, the current through R_B is identical to I_B. This current can be found by Ohm's law, if we make our usual assumption that $V_{BE} = 0.7$ V:

$$I_B = \frac{(V_{CC} - V_{BE})}{R_B} = \frac{15 - 0.7}{3 \times 10^6} = 4.8 \times 10^6 \ A = 4.8 \ \mu A$$

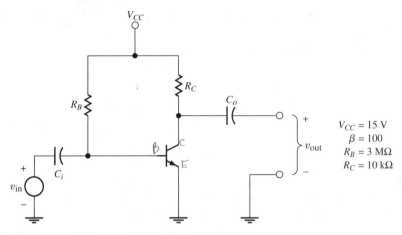

FIGURE 14.12 A simple transistor amplifier circuit.

Assuming for the moment that the transistor is operating in the active mode, we may write $I_C = \beta I_B$. The collector voltage V_C may be found by a node equation. Setting the sum of the currents leaving the collector node to zero, we have $I_C + (V_C - V_{CC})/R_C = 0$. Thus $V_C = V_{CC} - I_C R_C = V_{CC} - \beta I_B R_C = 10.2$ V. We now note that $V_{CB} = V_C - V_B = 10.2 - 0.7 = 9.5$ V. Since the collector junction is therefore strongly reverse biased, our result is consistent with our earlier assumption that operation is in the active mode.

Next we shall calculate the circuit's small-signal output voltage. Substituting the simplified-π model into the circuit, replacing dc voltage sources by short circuits, and—under our assumption that the signal frequency is sufficiently high—replacing the capacitors by short circuits, we obtain the small-signal circuit model of Fig. 14.13. The resistance r_π is evaluated by Eq. (14.16): $r_\pi = kT/qI_B = 0.026/4.8 \times 10^{-6} = 5,400\ \Omega$.

We observe that the voltage v_{in} appears across r_π. Therefore $i_b = v_{\text{in}}/r_\pi$, and

$$v_{\text{out}} = -\beta i_b R_C = \frac{-\beta R_C}{r_\pi} v_{\text{in}} \tag{14.22}$$

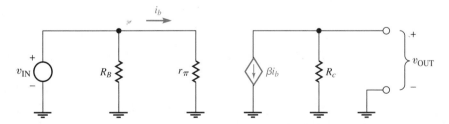

FIGURE 14.13 Small-signal circuit model for the circuit of Fig. 14.12. The simplified-π model has been used.

The open-circuit voltage amplification A of the circuit may be defined as v_{out}/v_{in}. Therefore

$$A \equiv \frac{v_{out}}{v_{in}} = \frac{-\beta R_C}{r_\pi} \tag{14.23}$$

Using the values $\beta = 100$, $R_C = 10,000 \ \Omega$, and $r_\pi = 5,400 \ \Omega$, we find that $A = 185$. The negative sign of A simply means that an increase in v_{in} gives rise to a decrease in v_{out}.

THE AMPLIFIER AS A BUILDING BLOCK

The amplifier we have been discussing can be used as a building block in a larger system. Its function as an analog building block can be more easily understood if the amplifier as a whole is represented by a simplified circuit model. It is true that the circuit of Fig. 14.13 is already a circuit model, but a simpler model can be constructed just as a simpler Thévenin equivalent can be found for a complicated two-terminal subcircuit. A model similar to that of Fig. 4.9(b) is what we have in mind. In Chapter 4 the discussion concerned the form and uses of the amplifier block, with its parameters A, R_i, and R_o taken as given. Now we are ready to evaluate these three parameters of the model by analyzing the transistor circuit.

One minor difference exists between Chapter 4 and the discussion here. The model to be developed here is a small-signal model. Strictly speaking, a small-signal model only describes the input and output small-signal variables and not, for example, large changes of dc voltage. The form of the small-signal amplifier model we shall now develop is shown in Fig. 14.14. Analogously with the discussion of Chapter 4, we define the input resistance R_i by the formula

$$R_i = v_{in}/i_{in} \tag{14.24}$$

where the sign conventions for v and i are as shown in Fig. 14.14. The output resistance R_o is defined by

$$R_o = -\frac{(v_{out})_{\text{output terminals open-circuited}}}{(i_{out})_{\text{output terminals short-circuited}}} \tag{14.25}$$

The open-circuit voltage amplification A is defined as

$$A = \left(\frac{v_{out}}{v_{in}}\right)_{\text{output terminals open-circuited}} \tag{14.26}$$

We observe that the value arrived at for A in Eq. (14.23) is the same A as that defined here. Thus one of the three parameters for the model of the amplifier in Fig. 14.12 has already been found.

By inspection of Fig. 14.13, we see that R_i is equal to the parallel resistance of the 3 MΩ biasing resistor and r_π. But $(3 \ \text{M}\Omega \| r_\pi)$ is very nearly equal to r_π, or 5,400 Ω. Thus for the circuit of Fig. 14.12

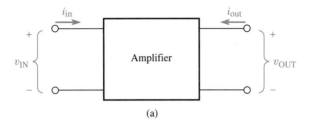

(a)

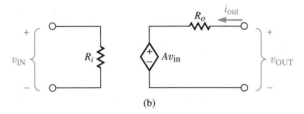

(b)

FIGURE 14.14 The amplifier regarded as a building block. (a) The form of the block, showing sign conventions for voltages and currents. (b) The small-signal model for the block.

$$R_i \cong r_\pi \qquad (14.27)$$

To find the output resistance of this circuit we use Eq. (14.25). Referring again to Fig. 14.13, the short-circuit output current is seen to be βi_b. The open-circuit output voltage can be found by writing a node equation for the node at the output terminal. This equation is $\beta i_b + v_{out}/R_C = 0$, from which we have $v_{out} = -\beta i_b R_C$. Then from Eq. (14.25), $R_o = -(-\beta i_b R_C)/(\beta i_b) = R_C$. Thus

$$R_o = R_C \qquad (14.28)$$

For this circuit $R_o = R_C = 10$ kΩ. The three parameters for the model of Fig. 14.14(b) have now been found. In further work with the amplifier as a part of a larger system, the amplifier can be represented by the model, with its parameters taking on the values we have found.

EXAMPLE 14.5

Calculate the operating point of the circuit in Fig. 14.15(a). Then find its small-signal open-circuit voltage amplification, input resistance, and output resistance. Use the simplified-π transistor model of Fig. 14.8. The capacitors may be regarded as short circuits at the signal frequency.

$$V_{CC} - \frac{V_C}{R_{\mathcal{U}}} - \frac{V_e}{R_{C2}}$$

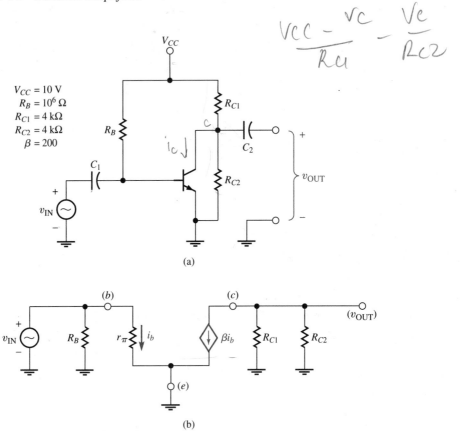

(a)

(b)

FIGURE 14.15

SOLUTION

First let us find the operating point. The capacitors C_1 and C_2 are regarded as open circuits for this (dc) part of the calculation. Writing a node equation for the node at the base terminal of the transistor, we have

$$I_B + (V_B - V_{CC})/R_B = 0$$

Making our usual assumption that $V_{BE} = 0.7$ V and noting that in this circuit (although not in general!) $V_B = V_{BE}$, we find

$$I_B = \frac{V_{CC} - 0.7}{R_B} = 9.3 \ \mu A$$

Since the transistor operates in the active mode, $I_C = \beta I_B$; thus $I_C = (9.3 \times 10^{-6}) \cdot (200) = 1.86$ mA. The collector voltage may be found by a node equation for the collector node: $I_C + (V_C - V_{CC})/R_{C1} + V_C/R_{C2} = 0$. The solution is

$$V_C = \frac{R_{C2}}{R_{C1} + R_{C2}} V_{CC} - \frac{R_{C1}R_{C2}}{R_{C1} + R_{C2}} I_C$$

$$= 5 \text{ V} - 2{,}000 I_C = 1.28 \text{ V}$$

Now that I_B has been found, r_π is found by $r_\pi = kT/qI_B = 2{,}800 \ \Omega$.

After substitution of the transistor model, we obtain the small-signal circuit shown in Fig. 14.15(b). This circuit is seen to be identical to that of Fig. 14.13, except that instead of R_C in Fig. 14.13 we now have $(R_{C1}\|R_{C2})$. Thus the calculations proceed exactly as for that circuit, and we find

$$A = -\beta(R_{C1}\|R_{C2})/r_\pi = -140$$

$$R_i = R_B\|r_\pi \cong r_\pi = 2{,}800 \ \Omega$$

$$R_o = R_{C1}\|R_{C2} = 2{,}000 \ \Omega$$

AN IMPROVED BIASING SCHEME FOR THE COMMON-EMITTER AMPLIFIER

Upon examination of the equations governing the biasing of the circuit just considered, we shall see that there is a strong dependence of I_C and V_C on the value of β. We found that $I_B = (V_{CC} - V_{BE})/R_B$; therefore

$$I_c = I_B \beta$$

$$I_C = \frac{\beta(V_{CC} - V_{BE})}{R_B} \tag{14.29}$$

In general, for a given transistor type, β is only specified by the manufacturer within about a factor of 2. (For example, β might be specified to lie in the range 100 to 200.) Thus unless one resorts to preselection of transistors, the circuit designer cannot be sure of I_C for this circuit to better than about a factor of 2. One generally likes the operation of a circuit to be more predictable than this. Thus it is desirable to design a circuit that is less sensitive to variations of β than the simple circuit just analyzed.

Let us consider the circuit of Fig. 14.16(a). The capacitors C_i and C_o serve to isolate the biasing circuit, which consists of R_{B1}, R_{B2}, R_E, and R_C. The capacitor C_E is used to "bypass" the resistor R_E at the signal frequency, as will be shown subsequently, but can be regarded as an open circuit now, when biasing is being considered. The biasing circuit is redrawn in Fig. 14.16(b). To make the analysis of the biasing circuit somewhat simpler, we can first replace the subcircuit inside the dotted line in Fig. 14.16(b) by its Thévenin equivalent. The biasing circuit is redrawn in Fig. 14.16(c) with the Thévenin equivalent inserted. The Thévenin resistance R_B' is found to be equal to the parallel combination of R_{B1} and R_{B2}, and the Thévenin voltage V_{BB} is found to be $V_{CC} \cdot R_{B2}/(R_{B1} + R_{B2})$.

We now find the bias point for the circuit of Fig. 14.16(c). Starting at V_{BB} and going toward ground, the voltage drops must sum to V_{BB}. Thus

$$I_B R_B' + V_{BE} - I_E R_E = V_{BB} \tag{14.30}$$

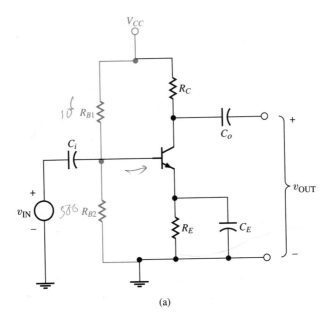

(a)

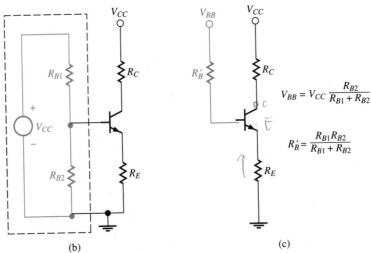

(b)   (c)

FIGURE 14.16   An improved biasing scheme for the common-emitter amplifier. (a) The complete circuit; (b) the biasing circuit; (c) the simplified biasing circuit.

We may substitute I_C/β for I_B and $-I_C/\alpha$ for I_E; α may also be written as $\beta/(\beta + 1)$. Making these substitutions and solving for I_C, we find

$$I_C = \frac{(V_{BB} - V_{BE})\beta}{R_E(\beta + 1) + R'_B} \tag{14.31}$$

We see that if $(\beta + 1)R_E \gg R_B'$, then Eq. (14.31) gives a much reduced β dependence of I_C. In fact, if $(\beta + 1)R_E \gg R_B'$, and $\beta \gg 1$, we have

$$I_C \cong \frac{V_{BB} - V_{BE}}{R_E} \tag{14.32}$$

Thus, in this limit, the value of I_C is not a function of β at all!

EXAMPLE 14.6

Find the possible range of values for I_C and V_C in the circuit of Fig. 14.16(a), if β is in the range 100 to 200. Let $V_{CC} = 15$ V, $R_{B1} = 1$ MΩ, $R_{B2} = 500$ kΩ, $R_C = 10$ kΩ, and $R_E = 10$ kΩ.

SOLUTION

The values of the Thévenin source V_{BB} and R_B' are determined first:

$$R_B' = 10^6 \| (0.5 \times 10^6) = 3.3 \times 10^5 \ \Omega$$

$$V_{BB} = 15 \times 0.5/1.5 = 5 \ \text{V}$$

From Eq. (14.31), we have

$$I_C = \frac{(5 - 0.7)\beta}{(\beta + 1 \times 10^4 + 0.33 \times 10^6)} \qquad \begin{array}{l} \text{If } \beta = 100, I_C = 0.32 \ \text{mA} \\ \text{If } \beta = 200, I_C \cong 0.37 \ \text{mA} \end{array}$$

The collector voltage is given by $V_C = V_{CC} - I_C R_C$. Therefore it is between 11.8 and 11.3 V.

The small-signal circuit model for the circuit of Fig. 14.16(a) is given in Fig. 14.17(a). Combining the parallel resistors and assuming that R_E is short-circuited by C_E as far as signals are concerned, the circuit of Fig. 14.17(b) is obtained. Comparing Figs. 14.17(b) and 14.13, we see that the two small-signal circuit models have the same form, in spite of the improvements made in the biasing. Thus the calculations previously made for A, R_i, and R_o apply to this improved circuit as well.

THE MAXIMUM VOLTAGE GAIN OF A COMMON-EMITTER AMPLIFIER

It is useful to estimate the maximum voltage gain obtainable from a common-emitter amplifier such as we have been considering—for example, that of Fig. 14.12. The open-circuit voltage gain was found in Eq. (14.23) to be equal to $-\beta R_C/r_\pi$. We can eliminate r_π from Eq. (14.23) by means of Eq. (14.16) and replace I_B by I_C/β. The absolute value of the open-circuit voltage gain is then found to be

$$|A| = \frac{I_C R_C}{kT/q} \tag{14.33}$$

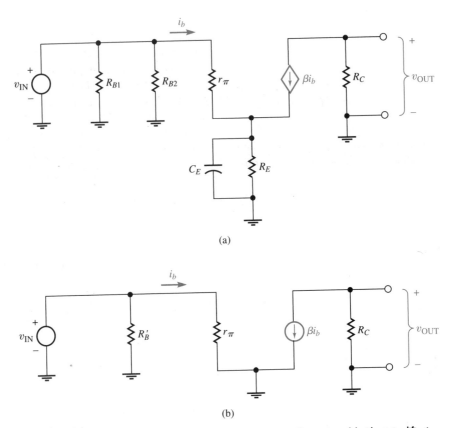

(a)

(b)

FIGURE 14.17 The small-signal circuit for the circuit of Fig. 12.16. (a) The circuit model without simplification. (b) The simplified model obtained by combining R_{B1} and R_{B2}.

At first glance one might think it possible to make A as large as desired by increasing I_C and/or R_C. However, it is necessary to keep the transistor operating in the active mode. We have already seen that $V_C = V_{CC} - I_C R_C$. Certainly, V_C must be greater than zero, or reverse bias on the collector junction will be lost. Thus we must require that $I_C R_C < V_{CC}$. The voltage gain of a single common-emitter amplifier stage therefore is limited in accordance with the expression

$$|A_{max}| < \frac{V_{CC}}{kT/q} \qquad (14.34)$$

If V_{CC} is 10 V and kT/q has its room-temperature value of 0.026 V, the maximum voltage gain is found to be ~400. Interestingly, this result is independent of β.

The value calculated in Eq. (14.34) is an upper limit for the amplification and not necessarily a typical value. Note that in Example 14.4, where V_{CC} was 10 V, we found $|A| = 140$, while the maximum possible with that value of V_{CC} is 400.

14.5 Multistage Common-Emitter Amplifiers

As an example of a simple analog system, let us consider a *multistage* amplifier. A multistage amplifier consists of several single-transistor amplifiers, called *stages*, connected one after another. Multistage amplifiers are used when more amplification is needed than can be obtained with a single stage. By using the simple model for the amplifier block developed in Fig. 14.14 it is possible to quickly determine the overall gain for any number of stages. It will be shown that because of interaction of stages with those that precede and follow, the overall gain of several stages is in general less than the product of the open-circuit amplifications of the individual stages.

Let us first compute the voltage gain of a single stage when its output is connected to a load resistance R_L and it is driven by a source with finite Thévenin resistance R_S, as shown in Fig. 14.18. We shall compute v_o/v_i, rather than v_o/v_s, because this quantity will be needed later when we consider the multistage amplifier. The output v_o from the voltage-divider formula equals

$$v_o = Av_i \frac{R_L}{R_L + R_O} \tag{14.35}$$

Therefore the single-stage gain, when load R_L is present, is given by

$$\frac{v_o}{v_i} = \frac{AR_L}{R_L + R_O} \tag{14.36}$$

Now let us consider several amplifier stages connected head-to-tail, as shown in Fig. 14.19(a). Although this circuit may at first appear formidable, its operation can be easily understood by thinking of it as an assemblage of amplifier blocks. Each amplifier may be replaced by its small-signal circuit model, given in Fig. 14.14. The simpler circuit of Fig. 14.19(b) then results.

Now we can proceed to compute the overall gain of the circuit. Just what is meant by calculating the "overall gain" is not in itself clear; what one calculates depends on what one needs to know. In this case we shall calculate the quantity v_l/v_s. In Fig.

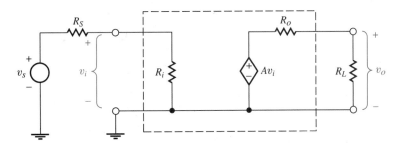

FIGURE 14.18 A very simple system consisting of an amplifier block, signal source, and load. The source is represented by its Thévenin equivalent (v_s, R_s), and the load is represented by the resistance R_L.

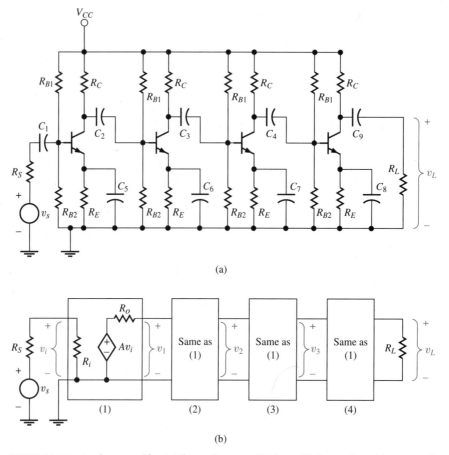

FIGURE 14.19   A multistage amplifier. (a) The complete circuit. (b) The simplified circuit obtained by using small-signal models to represent the amplifier blocks.

14.19(b) we can identify three kinds of stages: input, intermediate, and output. We compute the gain of each kind of stage separately.

Any intermediate stage has as its load R_i, the input resistance of the next stage. Therefore we can use Eq. (14.36) with $R_L = R_i$. The intermediate stage gain A_{it} is

$$A_{it} = A \frac{R_i}{R_i + R_o} \tag{14.37}$$

We have already analyzed the individual amplifier stage (Fig. 14.14) and have found that $A = -\beta R_C/r_\pi$, $R_i \cong r_\pi$, and $R_o \cong R_C$. Thus

$$A_{it} \cong - \frac{\beta R_c}{r_\pi + R_C} \tag{14.38}$$

Clearly, the maximum possible gain of an intermediate stage is β. In practice A_{it} is usually a good deal less than β, since r_π is usually of the same order of magnitude as R_C.

The output stage has as its load resistance R_L. Therefore the gain of the output stage A_{op} is given by $A_{op} = AR_L/(R_L + R_o)$, or

$$A_{op} = \frac{-\beta R_C}{r_\pi} \frac{R_L}{R_L + R_C} \tag{14.39}$$

The input stage has as its load the input resistance of the next stage. A complication occurs in this stage because we have decided to calculate v_l/v_s. The source resistance R_S and the input resistance R_i form a voltage divider that reduces the input voltage to the input stage by the fraction $R_i/(R_i + R_S)$. Therefore the gain A_{ip} of the input stage is the product of this factor and the stage gain, Eq. (14.36), with $R_L = R_i$:

$$A_{ip} = \frac{R_i}{R_i + R_S} \cdot A \cdot \frac{R_i}{R_i + R_o} \tag{14.40}$$

$$= \frac{r_\pi}{r_\pi + R_S} \cdot \left(\frac{-\beta R_C}{r_\pi} \right) \cdot \frac{r_\pi}{r_\pi + R_C}$$

$$= - \frac{\beta R_C r_\pi}{(r_\pi + R_S)(r_\pi + R_C)}$$

The overall gain of the amplifier of Fig. 14.19(b) can now be found. In terms of the intermediate voltages v_1, v_2, and v_3 indicated on the diagram,

$$\frac{v_l}{v_s} = \frac{v_l}{v_3} \cdot \frac{v_3}{v_2} \cdot \frac{v_2}{v_1} \cdot \frac{v_1}{v_s} \tag{14.41}$$

$$= A_{op} \cdot A_{it} \cdot A_{it} \cdot A_{ip}$$

$$= (\beta R_C)^4 \frac{R_L}{(r_\pi + R_C)^3 (r_\pi + R_S)(R_L + R_C)}$$

As an example, we may take $R_C = R_L = r_\pi = R_S$. Then the result would be $v_l/v_s = \beta^4/32$. If $\beta = 100$, the overall gain for the four stages is about 3×10^6, or 130 dB.

The reader may, if he or she wishes, feel pleased at having progressed to an understanding of such a complex circuit as that of Fig. 14.19(a). This illustrates the power of the building-block approach.

14.6 Frequency Response of Amplifier Circuits

All amplifiers exhibit variations of performance as the signal frequency is changed. Invariably there is a maximum frequency above which amplification does not occur; depending on the design of the circuit, there may also be a lower frequency limit below which amplification disappears. In general, one is concerned with calculating the *frequency response* of an amplifier, which may be defined as the functional dependence of output amplitude and phase upon frequency, for all frequencies. However, it is often necessary only to know the *passband* of the amplifier, which may be defined as the

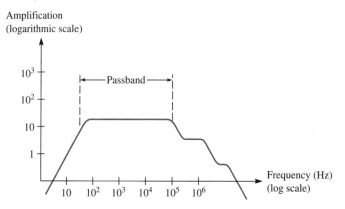

FIGURE 14.20   Frequency response of a typical amplifier.

range of frequencies over which maximum amplification is obtained. The passband is bounded by the *upper cutoff frequency* and by the *lower cutoff frequency* if one exists. Let us define the cutoff frequencies as the frequencies at which the voltage amplification drops below its maximum value by a factor of $1/\sqrt{2}$.* Figure 14.20 shows a typical amplifier frequency response. It will be noted that with this definition of passband the amplifier may possess considerable gain at frequencies outside the passband, although the gain will be lower outside the passband than within it. Other definitions of passband, of course, are possible.

EFFECTS OF CIRCUIT CAPACITANCES

It is quite feasible to construct amplifiers that have no low-frequency limit for operation. Such amplifiers are said to be *dc-coupled*, because the stages are coupled together for all frequencies down to zero, or dc. On the other hand, circuits sometimes contain coupling capacitors, which couple stages together for ac while isolating them for dc. Because at low frequencies these capacitors are unable to transmit signals, circuits in which they are present are limited in their low-frequency response.

As an example, let us consider the simple circuit of Fig. 14.21(a), which, after insertion of the simplified-π model of the transistor, becomes that of Fig. 14.21(b). In this circuit the sinusoidal signal source is coupled to the transistor base through the capacitor C. The capacitor is useful because it prevents the dc transistor base-bias voltage from being affected by possible dc conduction through the source v_s. However, as the signal frequency is reduced, this capacitor begins to act as an open circuit for the signal, preventing the signal from reaching the transistor base. The frequency at which this occurs is the lower cutoff frequency of the amplifier.

---

*This definition leads to what may be termed the "3 dB passband." This passband lies between the frequencies where the amplification is 3 dB (that is, $1/\sqrt{2}$) below its maximum value.

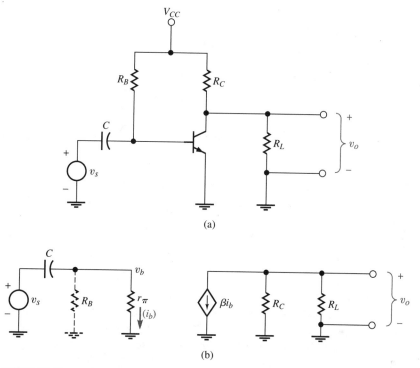

FIGURE 14.21 Simple common-emitter amplifier circuit with limited low-frequency response. (b) Small-signal circuit obtained by substitution of the simplified-π transistor model. For simplicity, R_B is assumed large compared with i_{π}, so that R_B may be neglected.

Frequency response is most conveniently studied by means of phasor analysis. In Fig. 14.21 we see that

$$\mathbf{i}_b = \frac{\mathbf{v}_s}{r_\pi + 1/j\omega C} \tag{14.42}$$

and

$$\mathbf{v}_o = -\beta \mathbf{i}_b \frac{R_C R_L}{R_C + R_L} \tag{14.43}$$

Defining the voltage amplification $\mathbf{A}$ by $\mathbf{A} = \mathbf{v}_o/\mathbf{v}_s$, we have

$$\mathbf{A} = -\beta \frac{R_C R_L}{R_C + R_L} \frac{1}{r_\pi + 1/j\omega C} \tag{14.44}$$

The quantity of interest is actually the ratio of the *amplitude* of the output sinusoid to the *amplitude* of the input sinusoid. This ratio is $|\mathbf{v}_o|/|\mathbf{v}_s| = |\mathbf{v}_o/\mathbf{v}_s| = |\mathbf{A}|$. From Eq. (14.44), we find

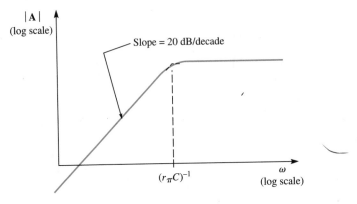

FIGURE 14.22

$$|\mathbf{A}| = \frac{\beta R_C R_L}{R_C + R_L} \frac{\omega C}{\sqrt{1 + (\omega r_\pi C)^2}} \qquad (14.45)$$

This result is sketched as a log-log plot (or Bode plot) in Fig. 14.22. As expected we see that the circuit functions as a high-pass filter. It is easily shown that the lower cutoff frequency is $\omega_c = 1/r_\pi C$. As shown in Fig. 14.21, the circuit does not appear to possess an upper cutoff frequency, but in fact it must. The upper cutoff will be determined by circuit properties that are not included in our simplified-π model.

EXAMPLE 14.7

Determine the frequency response of the circuit given in Fig. 14.23(a).

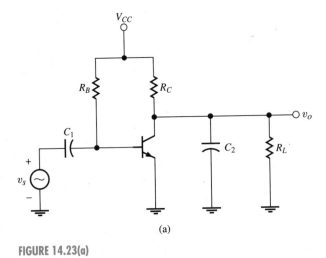

(a)

FIGURE 14.23(a)

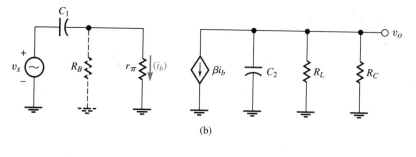

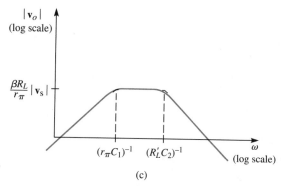

(c)

FIGURE 14.23(b) and (c)

SOLUTION

We begin, as usual, by constructing the small-signal circuit model, using the simplified-π model for the transistor [see Fig. 14.23(b)]. As before we have assumed that R_B is large and may be neglected. We may now write a phasor loop equation for the loop $v_s - C_1 - r_\pi - v_s$:

$$\mathbf{v}_s - \frac{\mathbf{i}_b}{j\omega C_1} - \mathbf{i}_b r_\pi = 0$$

Solving, we have

$$\mathbf{i}_b = \frac{j\omega C_1 \mathbf{v}_s}{1 + j\omega C_1 r_\pi}$$

Referring to the output part of the circuit, we write a node equation for $\mathbf{v}_o$:

$$\beta \mathbf{i}_b + j\omega C_2 \mathbf{v}_o + \frac{\mathbf{v}_o}{R'_L} = 0$$

where $R'_L \equiv (R_L || R_C)$.
    Solving, we have

$$\mathbf{v}_o = \frac{\beta \mathbf{i}_b R'_L}{1 + j\omega C_2 R'_L}$$

Substituting the result previously obtained for $\mathbf{i}_b$, we have

$$\mathbf{v}_o = -\frac{\beta R_L'}{1 + j\omega C_2 R_L'} \frac{j\omega C_1 \mathbf{v}_s}{1 + j\omega C_1 r_\pi}$$

In finding the passband, we are interested in calculating the absolute value $|\mathbf{v}_o|$ (which is the amplitude of the sinusoid v_o) as a function of frequency. The absolute value of $\mathbf{v}_o$ is

$$|\mathbf{v}_o| = \frac{\omega \beta R_L' C_1 |\mathbf{v}_s|}{\sqrt{1 + \omega^2 (C_2 R_L')^2} \sqrt{1 + \omega^2 (C_1 r_\pi)^2}}$$

Inspecting this result, we see that $|\mathbf{v}_o|$ approaches zero as ω approaches zero and that $|\mathbf{v}_o|$ also approaches zero as ω approaches infinity. Thus the circuit is limited in both its low-frequency response and its high-frequency response. If we assume that $(C_2 R_L')^{-1} >> (C_1 r_\pi)^{-1}$, a graph of $|\mathbf{v}_o|$ versus ω appears as in Fig. 14.23(c). We see that in this case the 3 dB passband lies between the frequencies $(r_\pi C_1)^{-1}$ and $(R_L' C_2)^{-1}$.

Looking back at the small-signal circuit model, we see that the general form of the frequency response curve could have been predicted by physical reasoning. At low frequencies C_1 becomes an open circuit, causing i_b, and hence v_o, to approach zero. Thus C_1 is responsible for the low-frequency limit. At high frequencies C_2 approaches a short circuit. The current βi_b is then diverted away from R_L' by C_2; since the current flowing through R_L' approaches zero, the voltage across R_L', which is v_o, must also approach zero. Thus C_2 is responsible for the high-frequency limit.

EXAMPLE 14.8

Determine the frequency response of the circuit in Fig. 14.24(a). Assume that C_2 is very large, so that it may be regarded as a short circuit for all frequencies of interest.

SOLUTION

This circuit is the same as that of Fig. 14.16. The capacitor C_1 serves as a "bypass" for the emitter resistor R_E. Previously we analyzed the circuit under a high-frequency assumption: C_1 was regarded as a short circuit for signals. However, when the signal frequency approaches zero, C_1 will act more like an open circuit, and circuit operation will be changed in a way we shall discover.

As usual, we shall begin by constructing the small-signal circuit model. Again we shall assume, for convenience, that the base-biasing resistors R_{B1} and R_{B2} are large enough to be neglected, and the coupling capacitor C_2 is regarded as a short circuit, as per instructions. The circuit model is then as shown in Fig. 14.24(b).

Again the method for exploration of frequency response is through phasor analysis. Let us write a node equation for the node at the emitter, marked (e). We shall call the phasor voltage at that point v_e. The node equation is

$$\frac{\mathbf{v}_s - \mathbf{v}_e}{r_\pi}(1 + \beta) - \mathbf{v}_e\left(\frac{1}{R_E} + j\omega C_1\right) = 0$$

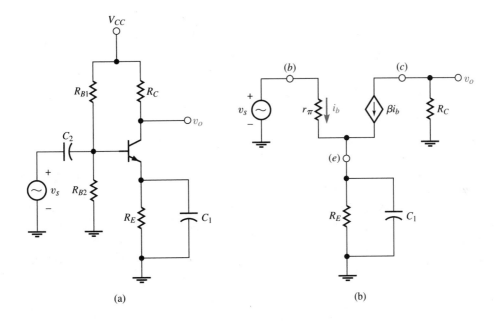

(a)

(b)

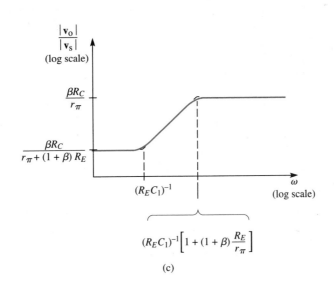

(c)

FIGURE 14.24

Solving for $\mathbf{v}_e$, we have

$$\mathbf{v}_e = \mathbf{v}_s \frac{(1 + \beta)R_E}{(1 + \beta)R_E + r_\pi + j\omega C_1 r_\pi R_E}$$

To obtain $\mathbf{v}_o$, we observe that $\mathbf{v}_o = -\beta \mathbf{i}_b R_C$ and that $\mathbf{i}_b = (\mathbf{v}_s - \mathbf{v}_e)/r_\pi$. Thus

$$\mathbf{v}_o = -\beta R_c \mathbf{i}_b = -\frac{\beta R_C}{r_\pi}(\mathbf{v}_s - \mathbf{v}_e)$$

$$= -\beta R_c \mathbf{v}_s \frac{1 + j\omega C_1 R_E}{[r_\pi + (1 + \beta)R_E] + j\omega C_1 r_\pi R_E}$$

Inspecting this result, we see that in the limit $\omega \to \infty$, $\mathbf{v}_o \to \mathbf{v}_s(-\beta R_C/r_\pi)$. This is in agreement with the result previously obtained for this circuit under the assumption that C_1 acted as a short circuit. However, as $\omega \to 0$,

$$\mathbf{v}_o \to \mathbf{v}_s\left(-\frac{\beta R_c}{r_\pi}\right)[1 + (1 + \beta)R_E/r_\pi]^{-1}$$

a smaller value. Thus we see why the bypass capacitor is used in the circuit. If it were absent, we would have the same situation as exists at low frequencies, where the bypassing action of the capacitor is ineffective. In this case we have found that less amplification is obtained than when the capacitor acts as a short circuit to bypass R_E.

Figure 14.24(c) is a graph of $|\mathbf{v}_o|$ versus ω. Note that in this more complicated circuit the lower limit of the bandpass is not given simply by $(R_E C_1)^{-1}$, but by the more complicated expression indicated on the graph.

In all of these examples, the simplified-π model has been used. However, at high frequencies the simplified-π model becomes inaccurate. Better high-frequency results can be obtained by using the complete hybrid-π transistor model, as shown in Fig. 14.25. We note that our simplified-π model is obtained by beginning with the hybrid-π model and then neglecting C_μ, C_π, r_o, and r_x. The justification for neglecting C_μ and C_π is that they are rather small capacitances, and at low frequencies their impedances $(1/j\omega C)$ are so large that they are almost open circuits.* At high frequencies, however,

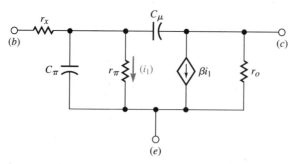

FIGURE 14.25   The hybrid-π transistor model.

---

*We neglect r_o and rx in the simplified-π model because they are, respectively, quite large and quite small. In many circuits they turn out to have little effect at any frequency.

C_π and C_μ conduct sizable currents and must be included in the circuit analysis. When phasor analysis is performed with C_μ and C_π included, an upper cutoff frequency is always found. In general, the higher the upper cutoff frequency, the more information can be processed by the circuit. Thus minimization of capacitive effects is a subject of great interest in circuit design.

14.7 Application: Single Transistors or ICs?

In this chapter we have seen how transistor amplifiers are designed. Clearly, if you are studying to be an electrical engineer, this is a worthwhile skill. People *expect* an electrical engineer to know things like that, and you will surely take more advanced courses in the subject. Eventually you may get a job designing new and better amplifiers. Electronics companies will be glad to have you.

On the other hand, if you are planning to be a mechanical, civil, or industrial engineer, things are a bit different. You still may need an amplifier, now and then, but why not just go out and buy one, rather than design your own? Ah yes, this is a good question.

To begin with, if you can buy a ready-made specimen of anything you need, it is usually better to buy it, rather than designing and building your own. With amplifiers this is especially true, because they are commonly available in IC form. IC op-amps are truly wonderful devices. As we have seen, they can perform a wide variety of functions, and thanks to the customary use of feedback, they can be highly precise. But they are almost ridiculously cheap! In fact, the cost of a typical op-amp chip is only a few pennies; most of the dollar or so you pay for it goes toward packaging and distribution. Now some very clever engineers may have worked on designing your op-amp, and they were paid well for their time, but the cost of their effort is spread over tens of thousands of op-amps, and you get their work for almost nothing. Moreover, the photolithographic processes used to make ICs are something like printing. The cost of preparing a book (like this one!) is all in the hard work of the author, the editors, and the compositors, but once that work has been done, and the printing plates have been made, the cost of stamping out each extra book is not very high. Similarly, the costs of designing your op-amp and making the masks needed to produce it probably have been paid off years ago. So you can have this powerful, high-tech device for a cost of almost nothing.

Then are there no occasions when you may need to design your own amplifier, or other analog circuit? Yes, there might be some. One case might be when you need something quite unusual, which is not available commercially. But this does not happen very often because op-amps and other available blocks are so extremely versatile. Another situation might be when your company is planning to produce a new product that has an enormous market. In this case it might pay to seek perfection by designing your own analog block, one that is absolutely ideal for the job. However, you would then undoubtedly hire an electrical engineer or consulting firm to design the block for you, and then send the design out to a *foundry*, which would fabricate your circuits for you in IC form.

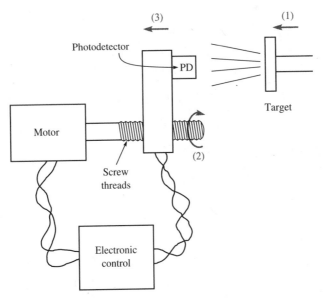

FIGURE 14.26

One situation in which it is hard to avoid using individual transistors is in high-power applications. This is because most ICs are tiny objects that cannot get rid of waste heat easily. The output current of an op-amp is therefore limited. (The manufacturer's data shown in Fig. 10.7, for example, indicate a maximum output current around 20 mA. Can you find where this information is hidden?) When a larger output current is needed, a simple transistor power amplifier would be appropriate.

Suppose you are designing a mechanical control system, using a servo motor. The function of the system is to make the object being controlled follow another object, keeping a constant distance without touching it. To achieve this purpose, you will use a photodetector, which produces a current determined by its distance from the object being followed. The signal from the photocell will be used to control a motor, which will continually move the photocell to the desired position. The mechanical arrangement is as shown in Fig. 14.26. When the target (1, in the figure) moves to the left, the photodetector receives increased reflected light. In response it produces an increased electric current, which is used as a signal to turn on the motor. The motor turns its threaded shaft (2); this causes the photodetector mount, which is also threaded, to move to the left, which in turn restores the distance from the target to the photodetector to its original value. When the distance is restored, the error signal from the photocell disappears, and the motor stops. Unfortunately, however, the current produced by the photocell is far too weak to drive the motor directly. Thus a current amplifier (or what we might more grandly call an "electronic control circuit") will have to be used.

A suitable circuit is shown in Fig. 14.27. Here the photodetector (PD) acts as a variable resistor, the value of which is determined by the light level. Potentiometer R_3 is used for adjusting the distance from the photodetector to the target. The op-amp cir-

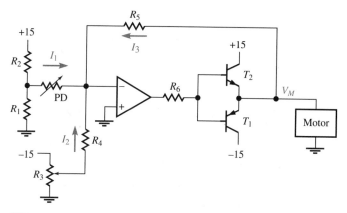

FIGURE 14.27

cuit is a summing amplifier, and the op-amp output is zero (so the motor doesn't turn) whenever the current from PD and the adjustable current from R_3 add to zero.

The output current from the op-amp is insufficient to run the motor, so a current amplifier, composed of transistors T_1 and T_2, is required. Note that one of these transistors is *pnp* and the other is *npn*. Together they compose what is known as a *push-pull amplifier*, which works in the following way. When the op-amp output voltage is larger than 0.7 V, the base of T_2 is forward biased, and current is conducted from the positive power supply down through T_2, out its emitter, through the motor, and down to ground. In this case T_1's base is reverse biased, and no current flows through T_1. When the voltage at the op-amp output is less than -0.7 V, the opposite happens: Current flows from ground up through the motor, down through T_1, and down to the negative power supply. Note that in this case the direction of current through the motor is reversed, and it turns in the opposite direction.

What happens when the voltage at the op-amp output is in the range between -0.7 V and $+0.7$ V? One might think that in that case neither transistor would be turned on, and there would be a "dead space" in which the system would not work. This, however, is prevented by the use of feedback, through R_5. From the ideal op-amp postulates, we know that the voltage at the $(-)$ input of the op-amp must be zero. Thus the motor voltage, V_M, must obey

$$V_M = -R_5[I_1 + I_2]$$

and thus whatever the photocurrent I_1 is, the motor voltage will have to respond accordingly.

Just about any transistors can be used in this circuit, provided only that they have enough current capacity to drive the motor. The circuit is so simple that design is not much of a problem; nonetheless, it will work pretty well. It is interesting that the final system contains *two* feedback loops. One is the electrical loop through R_5, used to linearize the push-pull amplifier. The other, less obvious one is the optical feedback, by means of which the position of the photodetector is fed back, through the electronics, to the amplifier input.

POINTS TO REMEMBER

- Amplifiers are circuits intended to produce an output voltage or current that is larger than, but proportional to, an input voltage or current.

- When used in amplifiers, bipolar transistors are operated in the active mode.

- In amplifier circuits, time-varying signal currents are usually superimposed on dc biasing currents. The biasing currents serve to place the transistor, in the absence of signals, at a desired point of its *I-V* characteristics. This point is called the operating point.

- Small variations of voltage and current about the operating point are known as small-signal voltages and currents. The relationships between small-signal variables are linear.

- A transistor model is a collection of ideal linear circuit elements designed to imitate the relationships between the transistor small-signal variables. It has the same number of terminals as the transistor.

- To analyze an amplifier circuit, the transistor model may be substituted for the transistor in the circuit. This substitution produces a circuit model containing only ideal linear circuit elements, which can then be analyzed using conventional techniques. The circuit model is a small-signal model; that is, it represents relationships between small-signal variables. When constructing the circuit model, dc voltage sources in the original circuit are replaced by short circuits, and dc current sources are replaced by open circuits.

- Amplifier circuits are conveniently treated as building blocks when analyzing larger systems. The amplifier block may be represented by a simple small-signal model. The parameters of this model are the input resistance R_i, the output resistance R_o, and the open-circuit voltage amplification A. The values of these parameters are found by analysis of the circuit inside the block.

- A multistage amplifier is a system obtained by connecting several amplifier blocks in sequence. The individual amplifier blocks are called stages.

- Transistor models of any degree of refinement are possible. A simple model, which we call the simplified-π model, is adequate for most calculations, provided that the frequency is not too high. The most detailed transistor model in common use is known as the hybrid-π model.

- The subject of amplifier frequency response has to do with the behavior of an amplifier as a function of signal frequency. All amplifiers have an upper frequency limit for satisfactory operation, and some have a lower frequency limit as well. Limits on the frequency response of an amplifier are imposed either by capacitances in the circuit or by effects internal to the transistors. Circuit capacitances can act to limit either low-frequency or high-frequency response. If high-frequency response is not limited by circuit capacitances, it will be limited by the transistors themselves.

- The action of transistors at high frequencies is best analyzed through the use of the hybrid-π transistor model. Phasor techniques are a powerful tool in the analysis of frequency response.

PROBLEMS

Section 14.1

14.1 Suppose that the *I-V* relationship for a circuit element is $i_E = 10^{-3}v_E + 2 \times 10^{-4}v_E^2$, where i_E is in amperes and v_E in volts. Let v_E be changed from 1.0 to 1.1 V. What is the change in i_E? Use $\Delta i_E = di/dv\Delta v_E$.

14.2 For Problem 14.1, let $i_E = I_E + i_e$ and $v_E = V_E + v_e$, where I_E and V_E are constants and i_e and v_e are small-signal deviations about the operating point. Show that for small deviations $i_e \cong Kv_e$. Evaluate the constant K for $V_E = 1$ V and 10 V.

*14.3 For a certain *pn* junction diode, the saturation current I_s is 10^{-14} A. Calculate the incremental forward resistance of the diode if the operating point is at $I = 2$ mA. [The incremental forward resistance, by definition, is $(di/dv)^{-1}$.] Construct a small-signal model of the diode. Does this small-signal model give any information about the relationship of total voltage to total current?

14.4 For a transistor operating in the active mode:
   a. Write an expression for the total emitter current in terms of the base current.
   b. Write an expression for the small-signal emitter current in terms of the small-signal base current.

14.5 Find the relationship between small-signal emitter current and small-signal base current, as predicted by the simplified-π model. Show that this relationship agrees with the result of Problem 14.4(b).

14.6 Construct a small-signal model for the network shown in Fig. 14.28. Assume that the capacitor is a short circuit at the signal frequency.

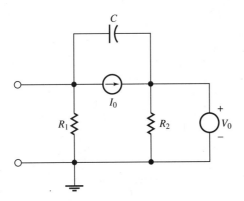

FIGURE 14.28

Section 14.2

14.7 In the circuit of Fig. 14.29 V_{CC} = 14 V and β = 100. We desire to bias the transistor to I_B = 30 μA and V_C = 6 V. Find R_B and R_C. What is r_π?

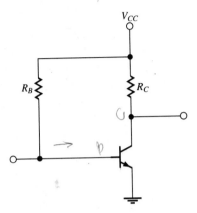

FIGURE 14.29

14.8 In the circuit of Fig. 14.30, let I_B = 10 μA, V_{CC} = 10 V, and β = 100. What should R_C be so that V_C = 5 V?

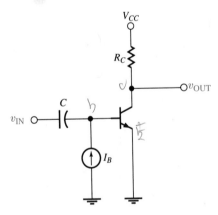

FIGURE 14.30

Section 14.3

14.9 For Problem 14.8 find the numerical value of r_π. The circuit is at room temperature.

14.10 Construct a small-signal model for the circuit shown in Fig. 14.30. Use the simplified-π model for the transistor. Assume that the capacitor is a short circuit at the signal frequency.

*14.11 Calculate the small-signal voltage gain (v_{out}/v_{in}) of the circuit of Fig. 14.30. Use the simplified-π model. Assume that capacitors are short circuits for signals. Obtain a numerical result for the case $I_B = 10\ \mu A$, $\beta = 100$, $R_C = 2\ k\Omega$.

*14.12 Find the small-signal input and output resistances of the circuit shown in Fig. 14.30. Use the simplified-π model. Assume that capacitors are short circuits for signals.

*14.13 Show that the simplified-π model may be represented in the form shown in Fig. 14.31. Evaluate the constant g_m in terms of the parameters β and r_π. What are the units of g_m? This quantity is called the transconductance.

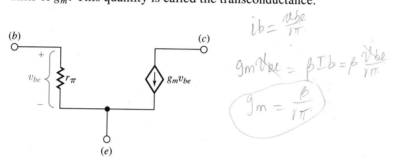

$$i_b = \frac{v_{be}}{r_\pi}$$

$$g_m v_{be} = \beta I_b = \beta \frac{v_{be}}{r_\pi}$$

$$\boxed{g_m = \frac{\beta}{r_\pi}}$$

(b)

v_{be} r_π (c) $g_m v_{be}$

(e)

FIGURE 14.31

Section 14.4

14.14 Find the value of I_B, V_{CE}, and r_π for Fig. 14.32. Let $\beta = 50$.

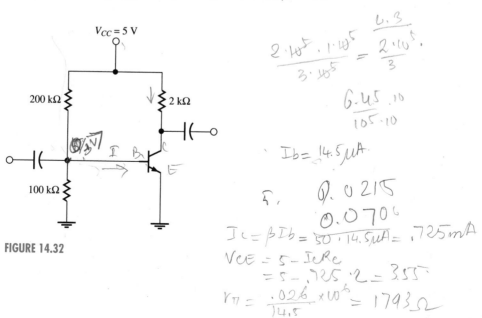

$V_{CC} = 5\ V$

200 kΩ   2 kΩ

100 kΩ

FIGURE 14.32

$$\frac{2 \cdot 10^5 \cdot 1 \cdot 10^5}{3 \cdot 10^5} = \frac{\overset{6.3}{2 \cdot 10^5}}{3}$$

$$\frac{6.45 \cdot 10}{105 \cdot 10}$$

$$I_b = 14.5\ \mu A$$

$$\frac{0.0215}{0.0706}$$

$$I_c = \beta I_b = 50 \cdot 14.5\mu A = .725\ mA$$

$$V_{CE} = 5 - I_c R_c$$
$$= 5 - .725 \cdot 2 = 3.55$$

$$r_\pi = \frac{.026 \times 10^6}{14.5} = 1793\ \Omega$$

*14.15 For the circuit of Fig. 14.16(a), choose suitable values of R_E, R_C, R_{B1}, and R_{B2} to place the approximate operating point at $I_C = 1$ mA, $V_{CE} = 5$ V. Assume $V_{CC} = 10$ V. The value of β is between 100 and 200, but the exact value is unknown.

*14.16 The circuit of Fig. 14.33 is known as a *common-base amplifier*.
   a. Explain why the term "common-base" is used.
   b. Let $V_{EE} = -5$ V and $V_{CC} = 5$ V. Choose R_E and R_C to make $I_C = 1$ mA and $V_{CE} = 5$ V. Let $\beta = 100$. What is r_π?

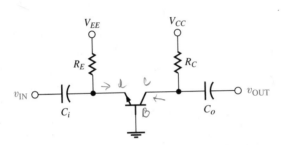

FIGURE 14.33

*14.17 For the common-base amplifier of Fig. 14.33:
   a. Construct the small-signal circuit model by substituting the simplified-π model for the transistor.
   b. Find the small-signal voltage amplification in terms of R_E, R_C, r_π, and β. Assume that the frequency is high enough for the capacitors to act as short circuits for signals.

**14.18 Find the small-signal input resistance and output resistance for the common-base amplifier of Fig. 14.33. In calculating the input resistance, assume that the output terminals are open-circuited. When finding the output resistance, assume that the input is being driven by an ideal voltage source. Assume the frequency is high enough for the capacitors to act as short circuits for signals.

**14.19 For the circuit of Fig. 14.33:
   a. Find the input resistance when a load resistance R_L is connected between the output terminals and ground. Does R_i depend on the value of R_L?
   b. Find the output resistance when the input is being driven by an ideal voltage source in series with a source resistance R_S. Does R_o depend on the value of R_S?

*14.20 The circuit of Fig. 14.34 is called a *common-collector amplifier* or *emitter follower*.
   a. Explain the term "common collector."

b. Let $I_B = 75$ μA, $\beta = 50$, and $V_{CC} = 15$ V. Find R_E to make $V_E = 7.5$ V. What is r_π?

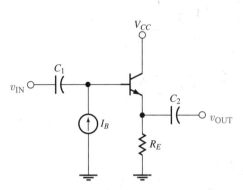

FIGURE 14.34

**14.21 a. Find the small-signal voltage amplification A of the circuit of Fig. 14.34.
  b. What is the approximate value of A if, as is usual, $\beta R_E \gg r_\pi$? Assume the capacitors are short circuits for signals.
  c. Find the numerical value of A if $\beta = 100$, $R_E = 5$ kΩ, and $r_\pi = 2{,}500$ Ω.

**14.22 a. Find the small-signal input resistance of the circuit of Fig. 14.34 with a load resistance R_L connected between the output and ground. Assume the capacitors are short circuits for signals. Express your answer in letter symbols.
  b. Obtain the numerical value of R_i if $\beta = 100$, $R_E = 5$ kΩ, and $R_L = 5$ kΩ.

**14.23 Find the small-signal output resistance of the circuit of Fig. 14.34. Assume the input terminal is driven by a signal source with Thévenin resistance R_S. The capacitors can be considered as short circuits for signals. Let $R_E = 5$ kΩ, $r_\pi = 2.5$ kΩ, $R_S = 5$ kΩ, and $\beta = 100$.

**14.24 A signal source with Thévenin voltage v_S and Thévenin resistance R_S is connected to the input of Fig. 14.34. The output is left open-circuited. Find v_{out}/v_S, using the approximate results of Problems 14.21(b) and 14.22. Assume that the capacitors are short circuits for signals. Obtain a numerical result for the case $\beta = 100$, $r_\pi = R_E = R_S = 5$ kΩ.

**14.25 The circuit of Fig. 14.35 is a combination of Figs. 14.29 and 14.34. Find v_{out}/v_{in}. Note that the overall amplification is not simply the product of the individual amplifications of the two stages, since the second stage loads the output of the first. You may wish to use the result of Problem 14.22. The frequency is high enough to allow the capacitors to be treated as short circuits. Let $V_{CC} = 15$ V, β (both transistors) $= 100$, $R_C = 7$ kΩ, $R_B = 1.5$ MΩ, $R_E = 100$ Ω, and $I_B = 50$ μA.

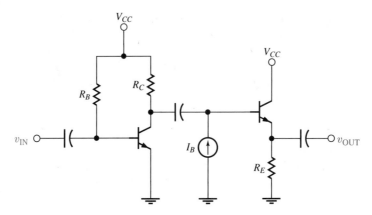

FIGURE 14.35

**14.26 For the circuit of Fig. 14.36:
  a. Show that the small-signal amplification cannot be found by substituting the simplified-π model.
  b. Find the small-signal amplification by substituting the more refined hybrid-π model, Fig. 14.25. Neglect r_x because it is small, and assume that the frequency is low enough for the small capacitors C_μ and C_π to be considered open circuits.

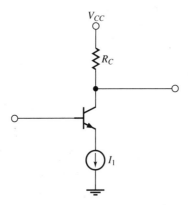

FIGURE 14.36

*14.27 Consider the biasing circuit of Fig. 14.16. In this problem we shall obtain a quantitative estimate of the stability of I_C with respect to changes in β. Let us define $(dI_C/I_C)/(d\beta/\beta)$ as the *desensitivity parameter*. Find the value of the densensitivity parameter if $V_{CC} = 15$ V, $R_E = 1{,}000$ Ω, $R_{B1} = 10$ kΩ, $R_{B2} = 5$ kΩ, $\beta = 100$, and $R_C = 1{,}500$ Ω. What is V_C?

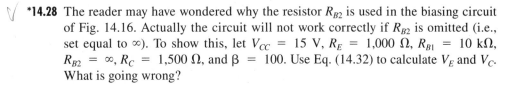

*14.28 The reader may have wondered why the resistor R_{B2} is used in the biasing circuit of Fig. 14.16. Actually the circuit will not work correctly if R_{B2} is omitted (i.e., set equal to ∞). To show this, let $V_{CC} = 15$ V, $R_E = 1,000\ \Omega$, $R_{B1} = 10$ kΩ, $R_{B2} = \infty$, $R_C = 1,500\ \Omega$, and $\beta = 100$. Use Eq. (14.32) to calculate V_E and V_C. What is going wrong?

Section 14.6

14.29 Estimate the lower cutoff frequency for the circuit of Fig. 14.30. Assume that $I_B = 10$ μA, $V_{CC} = 10$ V, and $\beta = 100$. The v_{IN} terminal is driven by an ideal voltage source. Let $C = 0.01$ μF.

*14.30 Estimate the lower cutoff frequency of the common-base circuit of Fig. 14.33. Assume that no external load is connected to the output terminal and that the input is driven by an ideal voltage source.

*14.31 In the circuit of Fig. 14.33, assume that a load resistance R_L is connected between the output terminal and ground. The input is driven by an ideal sinusoidal voltage source. Use phasor analysis to obtain an expression for $\mathbf{v}_{out}$. Assume that the frequencies of interest are low, so that the simplified-π model is accurate.

**14.32 Find the lower cutoff frequency of the emitter follower of Fig. 14.34. The output terminal is open-circuited, and the input is driven by an ideal voltage source. Use the simplified-π model.

**14.33 In this problem we shall study the high-frequency response of the amplifier of Fig. 14.30, using the hybrid-π model of Fig. 14.25. Let $I_B = 15$ μA, $V_{CC} = 10$ V, $\beta = 100$, $R_C = 3$ kΩ, $r_o = 50$ kΩ, $C_\pi = 100$ pF, $C_\mu = 5$ pF, and $r_x = 0$. The input is driven by an ideal voltage source and the output terminal is open-circuited. Assume C is a short circuit for the high frequencies of interest.
  a. Obtain an expression for $\mathbf{v}_{out}/\mathbf{v}_{in}$.
  b. Check the limit of your answer as $\omega \to 0$, $r_o \to \infty$. Does it agree with the usual result obtained with the simplified-π model?
  c. Find the limit as $\omega \to \infty$.
  d. Sketch a graph of $|\mathbf{v}_{out}/\mathbf{v}_{in}|$ versus ω on logarithmic scales.

**14.34 Carry out Problem 14.33 with r_x changed from zero to 50 Ω.

Additional Problems

14.35 Figure 14.37(a) shows a simple small-signal model for an enhancement-mode MOSFET. (Note that no current flows through the gate terminal, although the value of the dependent source depends on $v_{gs} = v_g - v_s$.)
  a. Assuming that the device operates in the pinch-off region and Eq. (13.11) applies, evaluate g_m in terms of K, V_{GS}, and V_T.
  b. What does Eq. (13.11) (oversimplified for this purpose) imply about r_d?

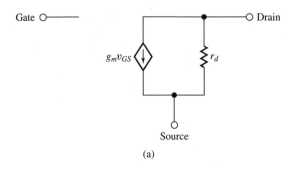

(a)

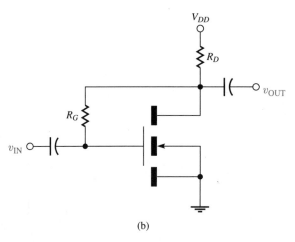

(b)

FIGURE 14.37

14.36 In the circuit of Fig. 14.37(b), V_{DD} = 24 V and V_T = 4 V. Assume that Eq. (13.11) is valid, and let K = 1 mA/V². Find an R_D that gives an operating point at V_D = 14 V. (*Note:* No dc current flows through R_G.)

14.37 The FET in Fig. 14.37(b) is represented by the small-signal model shown in Fig. 14.37(a). The capacitors can be regarded as short circuits at the signal frequency. Let g_m = 10 mA/V, r_d = 20 kΩ, R_D = 200 Ω, and R_G = 10 MΩ. Find the small-signal voltage amplification.

14.38 For the case of Problem 14.37, find the small-signal input resistance. The output terminal is open-circuited.

14.39 For the case of Problem 14.37, find the small-signal output resistance. The input is driven by an ideal voltage source.

CHAPTER **15**

DIGITAL
CIRCUITS

n earlier chapters we have discussed digital blocks and their uses in digital systems. Inside the digital blocks, of course, are transistor circuits. In this chapter we shall look inside the blocks to see how the circuits are made. It is unusual for users of digital hardware to actually design transistor circuits themselves; digital circuits are almost always purchased as ready-made IC building blocks. Digital circuit design is, however, of great interest to engineers in the IC industry where the blocks are designed and built, and other engineers, who are users of digital technology, will benefit from knowing what is inside the blocks they use. An important aspect of the subject arises from the existence of competing technologies known as *logic families*. Users of digital hardware must decide which technology is best in each case.

A fundamental component of digital circuits, the transistor switch, is introduced in Section 15.1. In Section 15.2 we go on to develop basic logic circuits. As we have seen in Chapters 11 and 12, large digital systems can be built from simple logic gates, when they are connected cleverly and repeated enormous numbers of times. Section 15.2 uses bipolar technology as its examples. Then in Section 15.3 we shall consider other logic families, especially the FET-based family known as CMOS, which is presently of great importance.

15.1 Transistor Switches

The basic element of logic circuits is the transistor switch, a general form of which is shown in Fig. 15.1. This diagram is only a rough model. A mechanical switch is not used in real circuits; the switching action is provided by one or more transistors that functions as an electrically controlled switch. The control signal is the input voltage v_{IN}. Since this is a digital circuit, v_{IN} must lie in either the "low" range or the "high" range. When v_{IN} is "low" it causes the switch to take one position (either "open" or "closed"); when v_{IN} is "high" the switch takes the other position. In the circuit shown in Fig. 15.1, closing the switch makes v_{OUT} zero, and opening it gives an output near V_{CC} (assuming that not much current flows through the output terminal). We can choose V_{CC} to lie in the "high" range, and let $V = 0$ be inside the low range. Then allowed values of v_{IN} control the switch and give rise to allowed values of v_{OUT}. A common special case is the one in which a "high" input gives a "low" output, and vice versa. Such a circuit, which performs the **COMPLEMENT** operation, is known as an *inverter*.

A typical inverter circuit using a bipolar transistor is shown in Fig. 15.2. Let us assume that the "high" range is 4 to 5 V, the "low" range is 0 to 0.5 V, and $V_{CC} = 5$ V. We see immediately that when v_{IN} is in the "low" range, no base current flows in the transistor; this is because about 0.7 V of forward bias is needed to obtain current through the emitter-base junction. Since there is no base current, there is also no collector current, and hence no voltage drop across R_C, making $v_{OUT} = V_{CC}$. Thus the "low" input results in an output that is "high," as expected for an inverter.

Typical output characteristics for an *npn* switching transistor are shown in Fig. 15.3. In Fig. 15.2, let us choose $R_C = 1,000 \ \Omega$; then the load line is as shown in Fig. 15.3. The situation just described, with v_{IN} "low," results in an operating point where the load line meets the characteristic corresponding to $i_B = 0$. This point is marked "cutoff" in Fig. 15.3. As expected, this operating point corresponds to $v_{CE} = 5$ V, placing $v_{OUT} (= v_{CE})$ in the "high" range.

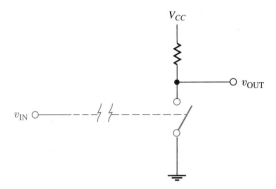

FIGURE 15.1   Model of a transistor switch. The position of the switch is determined by whether v_{IN} is "low" or "high."

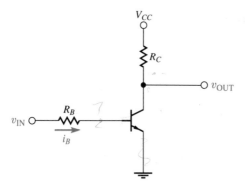

FIGURE 15.2

When v_{IN} is in the "high" range, the emitter-base junction is forward biased. The base current that flows is given approximately by

$$i_B = \frac{v_{IN} - 0.7}{R_B} \tag{15.1}$$

Suppose we wish i_B to be approximately 1 mA. Taking an "average" high voltage of 4.5 V for v_{IN}, we find $R_B = 3,800 \ \Omega$. The operating point now is at the intersection of the load line with the curve corresponding to $i_B = 1$ mA. With this large base current, the transistor is saturated. (Compare Fig. 13.31.) Thus the operating point is marked "saturation" in Fig. 15.3. We see that when i_B is large, v_{CE} is small, and its value does not depend much on the exact value of i_B. The value of v_{CE} under saturation conditions is

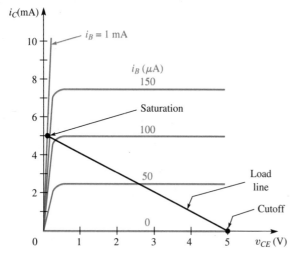

FIGURE 15.3

called v_{CESAT}. Typically $v_{CESAT} \cong 0.2$ V. Thus v_{OUT} is in the "low" range when the input is "high," as expected for an inverter. We note that in switching circuits the transistor operates either in the *cutoff* or the *saturation* mode; the only time it is in the active region is at those instants when the output is in the process of switching between "high" and "low." This is quite different from typical amplifier circuits, where operation is entirely in the active mode.

When the input to the inverter is "high," it is important that the base current be large enough to saturate the transistor thoroughly. (If not, V_{CE} may be larger than 0.2 V, and the output may not fall into the "low" range.) For saturation we must have $i_B > i_C/\beta$. [See pp. 515 and 516 where Eqs. (13.8) and (13.9) are discussed.] Thus for saturation we must have, from Ohm's law,

$$i_B > \frac{V_{CC} - v_{CE}}{\beta R_C} \tag{15.2}$$

Assuming that we have achieved saturation, v_{CE} will be equal to V_{CESAT} (approximately 0.2 V). Thus I_B must satisfy

$$i_B > \frac{V_{CC} - V_{CESAT}}{\beta R_C} \tag{15.3}$$

Expression (15.3) gives only a bare minimum value of i_B. The value of β is poorly controlled in transistor production and may vary widely from its nominal value. To ensure saturation it is wise to overdrive the base, that is, to make i_B several times larger than the minimum value given by Eq. (15.3).

EXAMPLE 15.1

For the circuit of Fig. 15.2, let $V_{CC} = 5$ V, $R_C = 1,000$ Ω, and $\beta = 100$, and let the "high" range be 4 to 5 V. Choose R_B so that any "high" input will saturate the transistor with the base overdriven by a factor of at least 5.

SOLUTION

From Eq. (15.3) (with a factor of 5 added for the desired overdrive) we have

$$i_B = 5\frac{V_{CC} - V_{CESAT}}{\beta R_C}$$

$$= 5\frac{5 - 0.2}{(100)(1,000)} = 240 \ \mu A$$

From Eq. (15.1) we have

$$R_B = \frac{v_{IN} - 0.7}{i_B}$$

We must decide what value within the "high" range is to be used for v_{IN}. The safest assumption is to let v_{IN} have its lowest value, 4 V. If our design is satisfactory with $v_{IN} = 4$ V, it will be *more* than satisfactory for larger voltages, since a larger v_{IN} results in larger base current. Setting $v_{IN} = 4$, we find

$$R_B = \frac{4 - 0.7}{2.4 \times 10^{-4}} = 13{,}750 \ \Omega$$

In practice, resistors are made in certain standard values, and 13,750 Ω will not be exactly equal to one of them. Here we should choose the closest standard resistance *smaller* than 13,750 Ω, since making R_B smaller increases the overdrive and hence improves the margin of safety.

We note that if we had chosen $v_{IN} = 5$ in Eq. (A), we would have had less than the prescribed base overdrive when v_{IN} was at the lower end of the "high" range, at $v_{IN} = 4$. The resulting design would not have satisfied the requirement for an overdrive of 5 with any input in the "high" range. Ordinarily circuits are designed so that they satisfy requirements even with the most unfavorable circumstances that can occur. This procedure is known as *worst-case design*.

Until now we have neglected currents that may flow through the output terminal of Fig. 15.2. In practice, of course, the output terminal will be connected to something, and the situation is as in Fig. 15.4. When v_{OUT} is "low," i_{OUT} will normally be positive. When i_{OUT} is no longer zero, we have, by writing a node equation at the collector terminal,

$$i_C = \frac{V_{CC} - v_{CE}}{R_C} + i_{OUT} \tag{15.4}$$

Assuming saturation, Eq. (15.3) is replaced by

$$i_B > \frac{V_{CC} - V_{CESAT}}{\beta R_C} + \frac{i_{OUT}}{\beta} \tag{15.5}$$

Thus larger i_B must be supplied when load currents are expected.

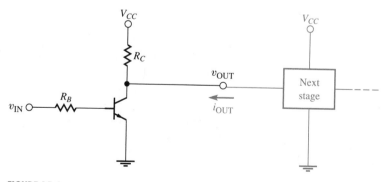

FIGURE 15.4

EXAMPLE 15.2

When v_{OUT} is "high" in Fig. 15.4, i_{OUT} is likely to be negative. What is the largest $|i_{OUT}|$ that can be tolerated in this case? Assume $V_{CC} = 5$ V and $R_C = 1,000$ Ω, and let the "high" range be 4 to 5 V.

SOLUTION

If the output is "high," the transistor is cut off, and $i_C = 0$. Hence

$$\frac{V_{CC} - v_{OUT}}{R_C} = -i_{OUT}$$

or, rearranging,

$$v_{OUT} = V_{CC} + i_{OUT}R_C$$

Since i_{OUT} has a negative value, its effect is to reduce v_{OUT}. We cannot allow v_{OUT} to become less than 4 V, since it must remain in the "high" range. Thus the largest $|i_{OUT}|$ occurs when

$$4 = V_{CC} + i_{OUT,MAX}R_C$$

from which we have

$$|i_{OUT,MAX}| = \frac{|4 - V_{CC}|}{R_C} = 1 \text{ mA}$$

The inverter of Fig. 15.2 is just one of many possible inverter circuits. In order to present the basic ideas in an orderly fashion, we shall next discuss logic circuits based on this particular inverter. Some other types of inverters will be considered in Section 15.3.

15.2 DTL and TTL Logic Circuits

We shall now develop a circuit realization of a **NAND** gate. As was seen in Chapters 11 and 12, it is possible to synthesize the other gates, and also flip-flops, by connecting **NAND** gates together in different ways. Thus in principle a single **NAND** gate circuit, repeated many times, would be sufficient to build up digital systems. We shall continue to define the "high" range to be 4 to 5 V and the "low" range 0 to 0.5 V. One possible **NAND** gate circuit is then as shown in Fig. 15.5. This circuit can have as many inputs as desired, as indicated by the dashed-line input C. However, the two inputs A and B are enough for our initial discussion of the circuit.

The basic problem of analysis here is that of finding the output voltage v_F for different combinations of the input voltages v_A and v_B. The input voltages, of course, are constrained to lie inside one or the other of the two allowed voltage ranges, and we

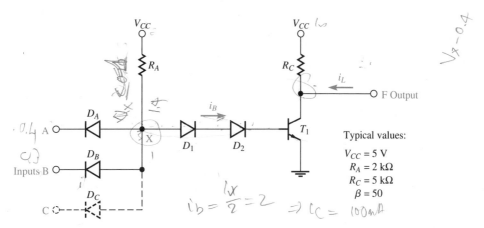

FIGURE 15.5   A DTL **NAND** gate. The inputs are on the left, and the output is taken at point F. Any number of additional inputs may be added in the manner shown by the dashed line. All voltages are measured with respect to ground.

[handwritten: $100 = \dfrac{V_{CC} - V_F}{5k} = \dfrac{5 - V_F}{5k}$]

expect that in a well-designed circuit the output will also always be inside either the "low" range or the "high" and never someplace in between. Since the circuit contains no less than five nonlinear circuit elements (D_A, D_B, D_1, D_2, and T_1), an approximation technique is convenient for analysis. We shall make use of the rule that the voltage across a current-carrying forward-biased *pn* junction is approximately 0.7 V. Moreover, since it is not obvious at the start which diodes are forward biased and which are reverse biased, a guessing procedure is used, in which we guess a result and then check it for self-consistency.

To demonstrate the procedure, let us first set $v_A = v_B = 0$. In this case a probable current path is from V_{CC} down through R_A, and through inputs A and B, where the voltage is low. This guess implies current flow through D_A and D_B in the forward direction. Thus we guess that the voltage at X is 0.7 V. We note, however, that current might also flow from X down to ground via D_1, D_2, and the base-emitter junction of the transistor. The sign of the guessed voltage at X is correct to forward-bias these three junctions, but its magnitude, 0.7 V, is insufficient; 3 × 0.7 V, or 2.1 V, would be needed to make current flow through this path. Thus we conclude that $i_B = 0$; hence the transistor is cut off; and hence the output voltage $v_F = V_{CC} = 5$ V. (Here we are assuming that no load current flows through the output terminal.) We note that our guess $v_X = 0.7$ is consist-ent with the assumptions that D_A and D_B are conducting and that D_1, D_2, and T_1 are not. This example is represented by the first line in Table 15.1.

EXAMPLE 15.3

Show that with $v_A = v_B = 0$ in Fig. 15.5, the guess $v_X = 2.1$ V would lead to a contradiction and thus cannot be correct.

SOLUTION

If $v_X = 2.1$ V, D_1, D_2, and T_1 would all be forward biased, which is possible. In this case, however, diodes D_A and D_B would each have 2.1 V of forward bias across them. This would be inconsistent with the rule that a current-carrying diode has 0.7 V across it.

As another example, suppose $v_A = 0$ and $v_B = 0.3$ V. In this case we might guess that only D_A conducts, or that only D_B conducts, or both. Only one of these guesses is correct; to determine which one, we guess a value of v_X and check it for consistency. If diode B is conducting, v_X must equal $v_B + 0.7 = 1.0$ V. But this guess implies a drop of 1.0 V across D_A and hence cannot be correct. On the other hand, if we guess $v_X = 0.7$ V, D_A will be conducting but D_B will not (since it has only 0.4 V across it). This guess does not break any rules and hence is correct. Again v_X is insufficient to produce base current, and $v_F = V_{CC} = 5$ V. This result is shown on the second line of Table 15.1.

● EXERCISE 15.1

For the circuit of Fig. 15.5, find v_X and v_F when $v_A = 0.4$ V and $v_B = 0.3$ V.
Answer: $v_X = 1.0$ V, $v_F = 5$ V.

Now let us consider a case in which one of the inputs is in the "high" range. Let $v_A = 0.2$ V and $v_B = 4.5$ V. In this case it is unlikely that D_B would conduct, since the available voltage across the path $V_{CC} - R_A - D_B$ is only 0.5 V. Thus we guess $v_X = 0.9$, in which case D_A conducts and D_B does not. The base current is still zero, and $v_F = 5$ V (third line in Table 15.1).

TABLE 15.1   **Outputs Corresponding to Various Combinations of Inputs for the NAND Gate of Fig. 13.5**

			Conduction?			
v_A	v_B	v_X	D_A	D_B	$D_1 D_2 T_1$	v_F
0	0	0.7	Yes	Yes	No	5
0	0.3	0.7	Yes	No	No	5
0.2	4.5	0.9	Yes	No	No	5
4.8	4.1	2.1	No	No	Yes	0.2

EXAMPLE 15.4

For the case $v_A = 0.2$ V and $v_B = 4.5$ V, explain why $v_X = 0.7$ is an incorrect guess.

SOLUTION

If $v_X = 0.7$, *neither D_A, nor D_B, nor the combination $D_1D_2T_1$* can be carrying any current. In that case the current through R_A would have to be zero. But this would imply $v_X = V_{CC}$, inconsistent with the guess $v_X = 0.7$.

Finally, let us try a case with both v_A and v_B in the "high" range: Let $v_A = 4.8$ V and $v_B = 4.1$ V. In this case, D_B cannot conduct unless $v_X = 4.8$ V. But this would imply too much voltage across the chain $D_1D_2T_1$. We see that v_X cannot rise above 2.1 V without overbiasing this chain. Thus we guess $v_X = 2.1$ V. Now neither D_A nor D_B conduct, but base current flows in the transistor; hence the transistor is saturated; and hence $v_F = V_{CESAT} \cong 0.2$ V. This result is shown in the last line of Table 15.1.

In terms of the "high" and "low" ranges, the operation of the circuit is shown in Fig. 15.6(a). If we use positive logic ("high" = **1**, "low" = **0**), the truth table is as shown in Fig. 15.6(b). We see that the circuit does function as a **NAND** gate.

• EXERCISE 15.2

What logical function does the circuit of Fig. 15.5 perform if negative logic is used? **Answer: NOR**.

The reader may inquire as to why D_1 and D_2 are used in the circuit. Consider the third line of Table 15.1. If D_1 and D_2 were omitted, v_X could rise to only 0.7 V before i_B would begin to flow. With these inputs the transistor would saturate and v_F would be "low" instead of "high" as required for the **NAND** gate. However, with D_1 and D_2 pre-

v_A	v_B	v_F
Low	Low	High
Low	High	High
High	Low	High
High	High	Low

(a)

A	B	F
0	0	1
0	1	1
1	0	1
1	1	0

(b)

FIGURE 15.6

sent, no i_B can flow until v_X reaches 2.1 V, which can never occur when either v_A or v_B is in the "low" range.

The circuit just described is known as a DTL ("diode-transistor logic") **NAND** gate. This circuit, and others closely related to it, belong to the *DTL logic family*. All circuits in the same family have the same "high" and "low" ranges and thus can be interconnected to build up digital systems. Other circuits, made differently and belonging to other logic families, also exist; but in general these have different "high" and "low" ranges and cannot be freely mixed with DTL blocks. We have used the DTL gate as our first example because it is fairly easy to understand, but as a practical matter DTL technology is almost obsolete. Its close relative, TTL (transistor-transistor logic), has taken its place in the market.

THE TTL NAND GATE

The TTL logic family has the same "high" and "low" ranges as DTL but uses only transistors instead of a combination of transistors and diodes. A typical TTL NAND gate is shown in Fig. 15.7. The strange-looking item on the left is a multiple-emitter transistor. A possible structure for the multiple-emitter transistor is shown in Fig. 15.8. The operation of this circuit is in many respects similar to that of the DTL **NAND** gate. There are two different conditions to be analyzed: Case a, all inputs "high" (in which case we expect the output F to be "low"), and Case b, one or more inputs "low" (in which case we expect the output to be "high").

Beginning with Case b, let us imagine that emitter E_1 is grounded (and therefore "low"), while E_2 and E_3 are "high." The situation is then as shown in Fig. 15.9. There is a current path available from V_{CC} down through R_1 and emitter E_1 to ground; thus we expect that $v_X = 0.7$ V, and a base current $i_{B1} \cong (V_{CC} - 0.7)/R_1$ will be flowing in transistor T_1. We note, however, that i_{C1} must be flowing *out* of the base of transistor T_2; this is the "wrong" direction for the base current of an *npn* transistor and must represent *reverse* current through one of the junctions in T_2. Since reverse currents are very small, it must be true that $i_{B1} >> i_{C1}/\beta$. Hence T_1 is saturated, and its collector-

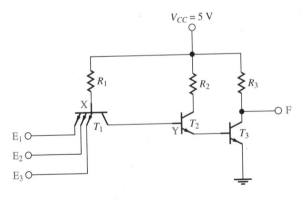

FIGURE 15.7   A TTL NAND gate.

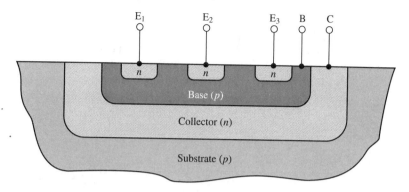

FIGURE 15.8   Cross-sectional diagram of the multiple-emitter transistor in Fig. 15.7. Terminals E_1, E_2, and E_3 are emitter terminals.

to-emitter voltage must be V_{CESAT}, about 0.2 V.* Thus the voltage at Y must be about 0.2 V. Transistors T_2 and T_3 now perform the same functions as D_1, D_2, and T_1 in Fig. 15.5. The 0.2 V at Y is much less than the 1.4 V needed to forward-bias the emitter junctions of T_2 and T_3. Thus T_3 is cut off and output F is "high."

Now let us consider Case a, in which all inputs are "high." The circuit now appears as shown in Fig. 15.10. We note that a current path now exists from V_{CC} down through R_1 and through the base-*collector* junction of T_1 (which is forward biased) and the base-emitter junctions of T_2 and T_3. T_1 is now effectively operating in the "reverse"

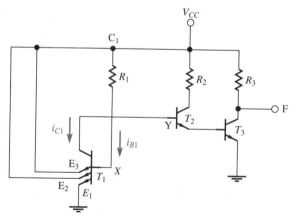

FIGURE 15.9   TTL **NAND** gate with one input "low" and two "high."

---

*Collector to *which* emitter? Emitters E_2 and E_3 are presently reverse biased and thus act as open circuits. The transistor is acting as an ordinary transistor with the single emitter E_1.

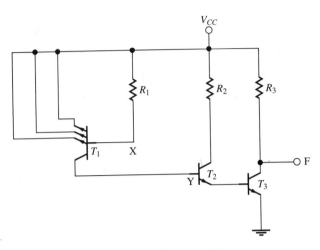

FIGURE 15.10   TTL **NAND** gate with all inputs "high."

mode, with the collector junction acting as the emitter. We now expect that $v_X \cong 2.1$ V, $v_Y \cong 1.4$ V, and T_3 will be saturated, making the output "low," as expected.

> ● **EXERCISE 15.3**
>
> In the circuit of Fig. 15.7, let $v_{E1} = 0.1$ V, $v_{E2} = 0.2$ V, and $v_{E3} = 0.3$ V. Find the approximate values of v_X, v_Y, and v_F. **Answer:** $v_X \cong 0.8$ V, $v_Y \cong 0.3$ V, $v_F = V_{CC}$.

15.3 CMOS and Other Logic Families

In the earlier sections of this chapter we have discussed inverters and **NAND** gates, using DTL and TTL circuits as examples. Both DTL and TTL are based on the saturating BJT inverter (Fig. 15.2). At this writing DTL is obsolescent, but TTL circuits are widely used. Many different ICs are available, containing various combinations of digital blocks constructed in TTL form. (For example, see the manufacturer's data sheets in Figs. 11.18 and 11.19.) Digital blocks containing TTL circuits are said to belong to the *TTL family*. They have compatible "high" and "low" ranges, and they can be interconnected to form systems. TTL circuits are quite "fast"; that is, they can switch fairly quickly, thus allowing data rates on the order of 10 to 40 megabits per second. However, TTL circuits consume rather a lot of electrical power, and thus each gate produces considerable heat; also they are rather large, so that comparatively few can be built into each IC. For these reasons, TTL is used primarily in SSI and MSI, which are the less densely packed ICs.

Meanwhile, logic families based on MOSFET technology are gaining steadily in importance. As a simple (but impractical) first example, let us consider the inverter circuit of Fig. 15.11(a). Here we have an *n*-channel enhancement-mode MOSFET combined with a resistor to form a circuit quite similar in principle to the BJT inverter of Fig. 15.2. Operation is easily understood once we draw a load line in the usual fashion, as shown in Fig. 15.11(b). Here we have repeated the transistor *I-V* characteristics of Fig. 13.41 and drawn the load line for R_D = 23,000 Ω and V_{DD} = 7 V. Now it is easy to find v_{OUT} (= v_{DS}) for any value of v_{IN} (= v_{GS}). The resulting graph of v_{OUT} as a function of v_{IN}, known as the *voltage-transfer characteristic*, is shown in Fig. 15.12. Let us define the "low" range to be those voltages less than the threshold voltage; thus for this case we choose the low range to be 0 to 3 V. Let the "high" range be 5 to 7 V. From Fig. 15.12 we see that, with these ranges, any input voltage in the "low" range gives an output of 7 V, which is in the "high" range, and inputs in the "high" range give outputs between 0.6 and 2.7 V, which are in the "low" range. Thus the circuit functions correctly as an inverter.

Although Fig. 15.11 functions correctly, it has serious practical disadvantages. In order to keep the current consumption low, large values of R_D are required. Large resistances are undesirable in ICs because they take up too much space. In fact, the resistor in this circuit would probably take up 35 times as much space as the transistor. To avoid such space inefficiency and increase the number of circuits per IC, it is usual to

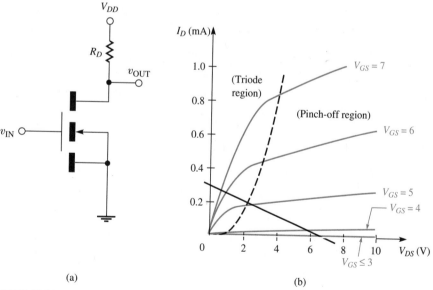

(a)   (b)

FIGURE 15.11 MOSFET inverter with resistor load. (a) Circuit; (b) transistor *I-V* characteristics with load line for R_D = 23,000 Ω and V_{DD} = 7 V.

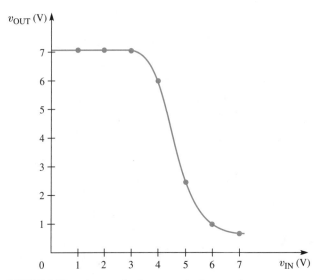

FIGURE 15.12   Voltage-transfer characteristic for the circuit of Fig. 15.11.

replace R_D with a second MOS transistor. A transistor used in this way is called an *active load*.

It is quite feasible to replace R_D with a second *n*-channel MOSFET much like the first one. Doing so results in the logic family known as *NMOS* (pronounced "en-moss"), which is widely used, especially in such large VLSI circuits as memories and microprocessors.* NMOS logic is nearly as fast as TTL, but the circuits are more compact, and hence more of them can be put on each chip. Furthermore, MOS fabrication is simpler than bipolar fabrication; thus there are fewer defects, and production costs are less. The main disadvantage of MOS technology, as compared with TTL, is low output current capacity. Recently, it has become possible to combine bipolar output stages, with good current capability, and efficient MOS circuits in the same IC.

Another very interesting approach uses a *p*-channel MOSFET as the active load for an *n*-channel MOSFET, leading to a logic family known as complementary-symmetry MOS, or CMOS (pronounced "see-moss"). CMOS technology has significant advantages and is presently of great importance; thus we shall discuss it in greater detail.

A typical CMOS inverter is shown in Fig. 15.13. This pleasingly simple and symmetrical circuit consists entirely of the two MOSFETs. The lower MOSFET, T_1, is an *n*-channel device; T_2 is a *p*-channel device. Note that T_2 is "upside-down," with its

---

*Two *p*-channel transistors can also be used, resulting in PMOS technology. The active load can also be operated in the depletion mode instead of the enhancement mode; many variations exist.

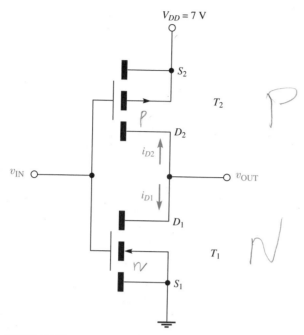

FIGURE 15.13 A typical CMOS inverter.

source at the top connected to the power supply terminal V_{DD}.* The power supply volt-age is usually in the range 5 to 10 V (5 V is common); thus a current path can exist from V_{DD} down through T_2 and through T_1 to ground, and we expect I_{D2} to be negative and I_{D1} to be positive. The input is applied to the two gates, which are connected together.

We shall analyze this circuit graphically, assuming that T_1 has the characteristics of Fig. 13.41 and that T_2 has identical characteristics except for the changes of sign appropriate to a *p*-channel device. The assumed characteristics of T_2 are as shown in Fig. 15.14. Our analysis will be similar to that used in Fig. 15.11. There we superim-posed the *I-V* characteristics of T_1 on the load line arising from V_{DD} and R_D. Now, how-ever, we must use the *I-V* characteristics of T_2 in place of those of R_D.

To make things clearer, let us redraw the inverter in the form of Fig. 15.15. The circuit's operating point will be that at which $V_{D1} = V_{D2}$ and $-I_{D2} = I_{D1}$. To find this point, we need a graph of $-I_{D2}$ versus V_{D2}. This must be obtained from Fig. 15.14,

---

*The reader may be puzzled as to how the "bulk" terminals of the two transistors can avoid being connected together when they are fabricated in the same semiconductor chip. In practice the devices are separated from the chip and from each other by reverse-biased *pn* junctions provided for isolation purposes.

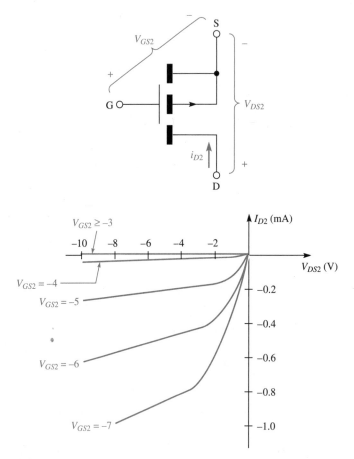

FIGURE 15.14   *I-V characteristics of the p-channel MOSFET T_2 in Fig. 15.13.*

which is a graph of *plus I_{D2}* versus V_{DS2} (which is not the same as V_{D2}). The change of dependent variable from I_{D2} to $-I_{D2}$ simply inverts the graph, as shown in Fig. 15.16(a). Note that in our circuit $V_{GS2} = v_{IN} - V_{DD}$. We shall choose $V_{DD} = 7$ V; the different curves in Fig. 15.16(a) are accordingly relabeled in terms of v_{IN}. To convert the independent variable from V_{DS2} to V_{D2}, we note that in this circuit $V_{DS2} = V_{D2} - V_{DD} = V_{D2} - 7$ V. This allows us to graph $-I_{D2}$ against V_{D2}, as shown in Fig. 15.16(b). The final step is to superimpose Fig. 15.16(b) on Fig. 13.41 (the $I_{D1} - V_{D1}$ characteristic of T_1), which results in Fig. 15.17.

To find the operating point by means of this figure, we simply locate the point at which the two curves corresponding to a given value of v_{IN} intersect. (We recall that the T_1 curves for $v_{IN} \leq 3$ V lie at the bottom of the graph, right on top of the horizontal axis; so do the T_2 curves for $v_{IN} \geq 4$ V.) For $v_{IN} \leq 3$ V, the intersection is seen to lie at point B; for $v_{IN} \geq 4$ V, the intersection is at point A. Interpolating curves for $v_{IN} = 3.5$ V, we see that the operating point for $v_{IN} = 3.5$ V must be approximately

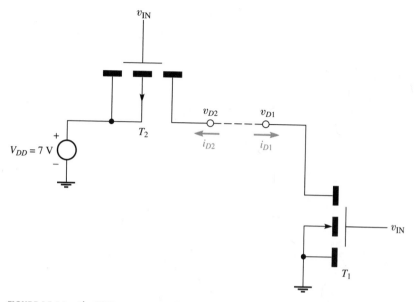

FIGURE 15.15   The CMOS inverter circuit of Fig. 15.13, redrawn to illustrate the graphical analysis.

at point C. In this way we obtain the voltage-transfer characteristic (v_{OUT} versus v_{IN}), as shown in Fig. 15.18(a).

We note that the switching action in this circuit is very sharp, as compared, for instance, with that of Fig. 15.12. By this we mean that a very small change in v_{IN}, from just below $v_{IN} = 3.5$ V to just above, is sufficient to produce the maximum change in

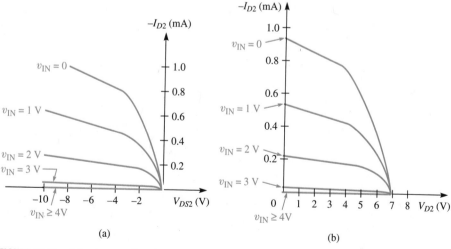

FIGURE 15.16   (a) $-I_{D2}$ versus V_{DS2}; (b) $-I_{D2}$ versus V_{D2}.

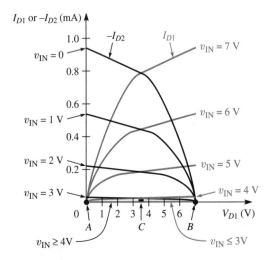

FIGURE 15.17   Finding the operating point of the CMOS inverter by a graphical method.

v_{OUT}. To make this idea more quantitative, let us consider a general voltage-transfer characteristic, as shown in Fig. 15.19. The voltages V_{OH} and V_{OL} are, respectively, the nominal "high" and "low" output voltages of the circuit. Voltages V_{IL} and V_{IH} are defined as the input voltages at which $|dv_{OUT}/dv_{IN}| = 1$. These points can be considered to be the boundaries of the "low" and "high" ranges. The region between V_{IL} and V_{IH} is the *transition region*, and we define the *transition width* to be $V_{IH} - V_{IL}$. The quantities $V_{IL} - V_{OL} \equiv NM_L$ and $V_{OH} - V_{IH} \equiv NM_H$ are known, respectively, as the lower and upper *noise margins*. These quantities are meaningful because in every real system random noise voltages may be added to v_{IN}. The noise margins indicate the

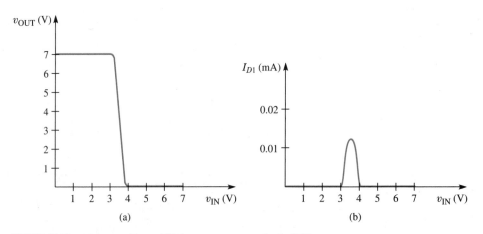

FIGURE 15.18   (a) Output voltage and (b) drain current versus v_{IN} for the CMOS inverter.

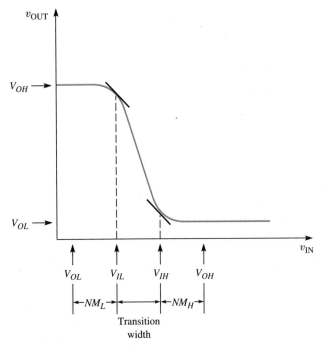

FIGURE 15.19   A typical voltage-transfer characteristic, showing noise margins.

largest random voltages that can be added to v_{IN} (when it is "low") or subtracted from v_{IN} (when it is "high") without giving an input in the forbidden transition region.

EXAMPLE 15.5

Find the noise margins for the CMOS inverter of Fig. 15.13.

SOLUTION

We refer to the voltage-transfer characteristic of Fig. 15.18(a). Clearly, in this circuit $V_{OH} = 7$ V and $V_{OL} = 0$. By locating the approximate points where the absolute value of the slope is unity, we find $V_{IL} = 3.1$ V and $V_{IH} = 3.9$ V. Thus we have $NM_L \cong NM_H \cong 3.1$ V.

It is interesting to graph the drain current I_{D1} ($= -I_{D2}$) of the CMOS inverter as a function of v_{IN}. This is readily done using Fig. 15.17; the result is as shown in Fig. 15.18(b). The maximum value of I_{D1} occurs at point C in Fig. 15.17, and this maximum value is very small. Moreover, when v_{IN} is inside either the "low" or the "high" regions (and not in the transition region), $I_{D1} \cong 0$. This illustrates the most useful property of CMOS technology, its low power consumption. While the input of the inverter is either "low" or "high" it consumes almost no power at all; power consumption only occurs for a brief time while v_{IN} is being switched from one level to the other. This feature of CMOS technology makes it especially useful in battery-powered applications, such as calculators and wristwatches, and in VLSI generally. (With 500,000 transistors in a single IC, the heat produced by each transistor must be held to a very low value.) The principal disadvantage of CMOS is a more complex fabrication procedure than that of NMOS, leading to more defects and higher cost.

LOAD CURRENTS

As mentioned earlier, MOS circuits have limited output current capacity. When a load is connected to the output of the CMOS inverter of Fig. 15.13, the load current flows through whichever transistor is turned "on"—through T_1 if v_{OUT} is "low" and the direction of the load current is inward, or through T_2 if v_{OUT} is "high" and the load current is directed outward. For the former case, imposition of the load current I_L causes I_{D1} to become equal to I_L while I_{D2} remains zero, V_{GS1} remains unchanged, and $V_{D1} = v_{OUT}$ increases. Transistor T_1 is operating in the triode regime; hence Eq. (13.10) applies, approximately:

$$i_D \cong K[(v_{GS} - V_T)v_{DS} - \tfrac{1}{2}v_{DS}^2]$$

(15.6)

If $v_{GS} = v_{IN} = 7$ V, $V_T = 3$ V, and we place the top of the "low" range (the largest allowable v_{DS}) at 3 V, we have a maximum load current

$$I_L \leq 7.5K$$

(15.7)

In order to give K a large value, large-area gates must be used; this introduces a large capacitance between the gate and source terminals. Large MOS transistors used for outputs may have K in the range 1 to 10 mA/V². According to Eq. (15.7), large load currents of 7.5 to 75 mA can thus be obtained. However, the large gate capacitance of such a device presents a serious load problem to the preceding stage that drives it. The effect is to reduce the speed of the system, as will be seen from the following example.

The power consumption of the CMOS inverter is determined largely by load currents, mainly those that charge capacitors, as in the preceding example. In Fig. 15.20(a)

EXAMPLE 15.6

The output of a CMOS inverter drives a second stage, as shown in Fig. 15.20(a). Owing to the transistors' internal capacitances and wiring capacitances, a parasitic capacitance C exists in the circuit, as shown. The input of inverter I_1 is suddenly switched from "low" to "high." Assume that C is the only capacitor present. Estimate the time required for the output of I_2 to change from "low" to "high." Let $C = 40$ fF, $K = 40$ μA/V², $V_T = 3$ V, and $V_{DD} = 7$ V.

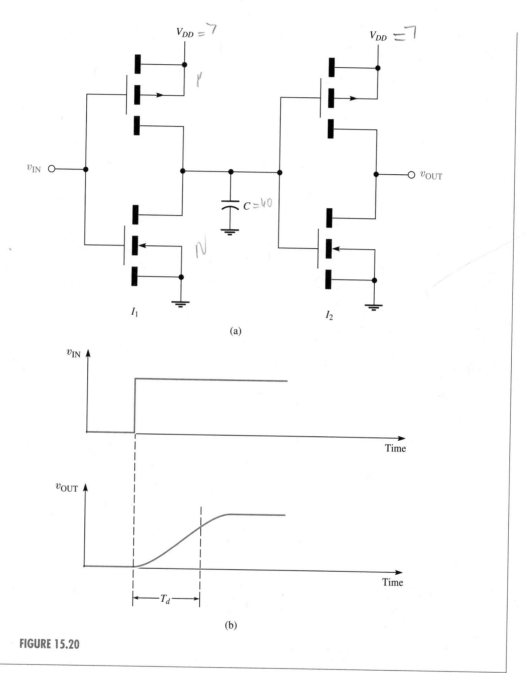

(a)

(b)

FIGURE 15.20

SOLUTION

In order for the input of I_2 to change from "high" to "low," capacitor C must discharge. No current flows through the gates of I_2; the charge must flow to ground through the lower (n-channel) transistor of I_1. An exact calculation would be complicated, because as C discharges, v_{DS} decreases.

However, we can get an approximate result by assuming a constant discharge current; the value of this current can be estimated from Eq. (15.6),* using an average value of v_{DS}. In this case, since v_{DS} drops from 7 V to 0, an average v_{DS} of 3.5 V is suitable. The charge to be removed is $Q = CV = (40 \times 10^{-15})(7) = 2.8 \times 10^{-13}$ coulombs, and the approximate discharge time is then

$$T_d \cong \frac{Q}{I_D}$$

$$\cong \frac{2.8 \times 10^{-13}}{(40 \times 10^{-6})[(4)3.5 - \frac{1}{2}(3.5)^2]}$$

$$\cong 9 \times 10^{-10} \text{ sec}$$

The behavior of $v_{OUT}(t)$ is roughly as shown in Fig. 15.20(b). There is a delay time, approximately equal to the T_d we just found, between the time that v_{IN} turns on and the time that v_{OUT} enters the "high" range. Ours is just an order-of-magnitude calculation, but this time, known as the *propagation delay* of the circuit, is quite important because it determines the maximum data rate, or number of bits per second the circuit can handle. A rule of thumb is that the maximum data rate for a system,[†] each block of which has propagation delay T_d, is $(25T_d)^{-1}$. In the present example the maximum data rate is about 44 megabits per second. This rather high speed is typical of small, low-current devices of the kind used in large numbers inside LSI ICs. However, a "driver" transistor used to supply output current to an off-chip load would have much greater gate capacitance; hence that part of the circuit would be much slower.

---

*Equation (15.6) assumes the MOSFET operates in the triode region, while in fact it begins in the pinch-off region and moves into the triode region as the voltage across C decreases. The value of I_D we calculate is, at any rate, only an approximate average value. Accurate calculations of the delay time can be made using a standard computer program, such as SPICE. Interestingly, the result of this example agrees with SPICE within 10 percent.

[†]Higher data rates may cause errors because signals passing through different paths in the system become unsynchronized with each other.

the capacitor charges to v_{DD} when the output of I_1 goes to "high," with current flowing into C through the p-channel device in I_1. It can be shown (see Problem 15.19) that when a capacitor is charged to a voltage V_{DD} through a resistance R, a total energy CV_{DD}^2 is expended. Half of this energy is dissipated in R; the other half is stored as electrostatic energy in C. When the output of I_1 then goes to "low," the stored charge of the capacitor flows to ground through the n-channel device of I_1, and the energy stored in the capacitor is dissipated as heat in the resistance of that device. Thus in a round-trip low-high-low transition, a total energy of CV_{DD}^2 is dissipated as heat.

In addition to the dissipation associated with capacitive loads, there are other sources of power dissipation. There is internal capacitance C_{INT} inside the gate, which

EXAMPLE 15.7

The circuit of Fig. 15.20 is driven by a 500 MHz square wave. Estimate the time-average power consumption by assuming that all power dissipation is due to load currents in C. Let $C = 40$ fF and $V_{DD} = 7$ V.

SOLUTION

An energy CV_{DD}^2 is dissipated 5×10^8 times per second. The power consumption is thus $P_{AV} = (CV_{DD}^2)f \cong 1$ mW. We note that in CMOS (unlike other logic families), power consumption is proportional to switching frequency. When no changes in logical state occur, as in a quiescent memory, power consumption is almost zero.

The power consumption found here is for a simple inverter. For other logic blocks power consumption will be somewhat larger.

behaves in the same way as the load capacitance C and contributes an additional loss $C_{INT}V_{DD}^2f$ (where f is the signal frequency). Furthermore, some current flows through the transistors at the instant of switching, as can be seen in Fig. 15.18(b). This current leads to an additional power dissipation; this dissipation is also proportional to frequency and approximately proportional to V_{DD}^2, and we may call it AfV_{DD}^2, where A is the appropriate constant of proportionality. It is convenient to combine this loss with that due to internal capacitance by defining a constant $C_{PD} \equiv C_{INT} + A$; this constant is known as the *power-dissipation capacitance* and is specified by manufacturers. Lastly, there is also a dc leakage current I_{LK} that flows from the power supply to ground, contributing an additional power dissipation $V_{DD}I_{LK}$. Thus the total power dissipation is

$$P = (C_{PD} + C)fV_{DD}^2 + V_{DD}I_{LK} \tag{15.8}$$

CMOS LOGIC

A CMOS **NAND** gate is shown in Fig. 15.21. Operation is readily understood by recalling that a "high" gate voltage applied to an n-channel device creates a low-resistance channel that acts, crudely speaking, as a short circuit, while a "low" gate voltage applied to an n-channel device results in a nonexistent channel, which is nearly an open circuit. For the two p-channel devices (with their sources connected to V_{DD}), opposite statements apply. For convenience these statements are collected in Table 15.2. It is now easily verified that the circuit of Fig. 15.21 functions (with positive logic) as a **NAND** gate.*

---

*Operation is slightly affected by the fact that the bulk connection of T_2 is not connected to its source (for reasons of fabrication convenience). This effect is not important for present purposes.

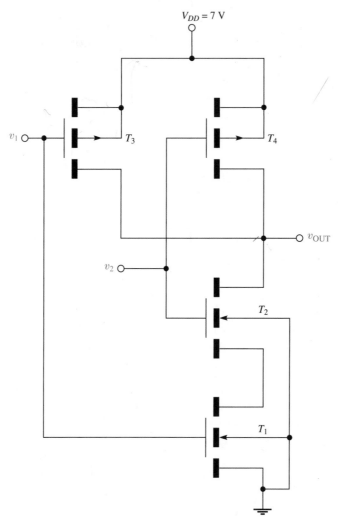

FIGURE 15.21   A CMOS **NAND** gate.

TABLE 15.2   Approximate Behavior of Transistors in CMOS Logic

Type	Gate "High"	Gate "Low"
n-channel	Short	Open
p-channel	Open	Short

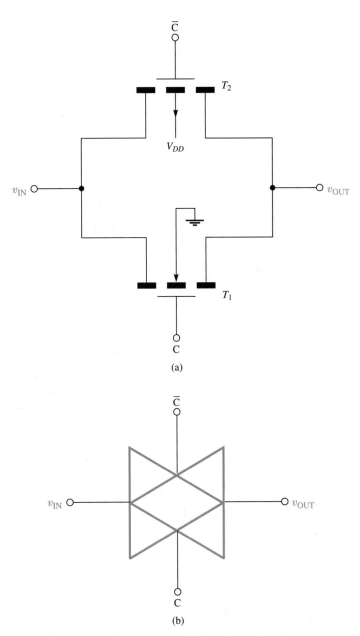

(b)

FIGURE 15.22   A CMOS transmission gate. (a) Circuit; (b) symbol.

An interesting CMOS circuit, which is not strictly speaking a logic circuit, is the *transmission gate* shown in Fig. 15.22. In this circuit the control signal C (and its complement $\overline{C}$) determines whether or not the input is connected to the output. Let V_{DD} = 7 V, "low" = 0 V, "high" = 7 V, and V_T (the turn-on voltage) = 3V. When C is "low" (and $\overline{C}$ therefore is "high"), the gates of T_1 and T_2 are unbiased with respect to the bulks. Furthermore, v_{IN} is unbiased or positive (depending on the value of v_{IN}) with respect to G_1 and unbiased or negative with respect to G_2. Thus neither gate can induce a channel, and the circuit acts as a high resistance between v_{IN} and v_{OUT}. On the other hand, let C be "high," $\overline{C}$ be "low," and suppose v_{IN} = 0 V. Then T_1 has a large positive V_{GS} and provides a low-resistance path between v_{IN} and v_{OUT}. When v_{IN} = 7 V, T_2 has a large negative V_{GS} and provides a low-resistance path between v_{IN} and v_{OUT}. For some values of v_{IN} between 0 and 7 V, it is possible for both T_1 and T_2 to conduct. Thus the transmission gate provides a low-resistance path between input and output when C is "high"
for all allowed values of v_{IN}. When C is "low" the transmission gate is effectively an open circuit. Many uses for these circuits can be imagined. For example, they can be used for the switches in the D/A converter of Fig. 12.15. An interesting feature of the circuit is that it is *reciprocal*; that is, when the switch is "on" it conducts well in either direction. The labels "output" and "input" can thus be interchanged.

Figure 15.23 shows manufacturer's specifications for typical modern CMOS logic blocks. The circuits used in these blocks are slightly more elaborate than the circuits described in this chapter. The commercial versions are "double buffered"—that is, the output of each gate circuit passes through two CMOS inverters before reaching the output terminal. This refinement steepens the voltage transfer characteristic (see Problems 15.28 and 15.29) and thus increases the noise margins. Moreover, the final inverter contains large transistors with sufficient output capability to drive a single TTL input, thus allowing the two logic families to conveniently be interconnected.

OTHER LOGIC FAMILIES

In addition to TTL, NMOS, and CMOS, several other logic families have been developed for various purposes. It is difficult to catalog them because each family has many variations. Moreover, developments in this field are very rapid, and the relative advantages of the different families change as improvements are made.

Some families are modifications of ones already discussed. Schottky-TTL, for instance, is an improved, higher-speed version of TTL. An additional component, a metal-semiconductor diode (also known as a Schottky diode) is added to the TTL circuit to reduce charge storage in the transistor. With reduced charge storage, currents of the same size result in faster switching, as seen in Example 15.7.

EMITTER-COUPLED LOGIC (ECL)

Emitter-coupled logic is a bipolar technology in which the transistors operate in the active mode and do not saturate. It is the fastest of the common families, with propagation delays per gate as short as 1 nsec. On the other hand, power consumption is rather

National
Semiconductor

MM54C00/MM74C00 Quad 2-Input NAND Gate
MM54C02/MM74C02 Quad 2-Input NOR Gate
MM54C04/MM74C04 Hex Inverter
MM54C10/MM74C10 Triple 3-Input NAND Gate
MM54C20/MM74C20 Dual 4-Input NAND Gate

General Description

These logic gates employ complementary MOS (CMOS) to achieve wide power supply operating range, low power consumption, high noise immunity and symmetric controlled rise and fall times. With features such as this the 54C/74C logic family is close to ideal for use in digital systems. Function and pin out compatibility with series 54/74 devices minimizes design time for those designers already familiar with the standard 54/74 logic family.

All inputs are protected from damage due to static discharge by diode clamps to V_{CC} and GND.

Features

- Wide supply voltage range   3.0V to 15V
- Guaranteed noise margin   1.0V
- High noise immunity   0.45 V_{CC} (typ.)
- Low power consumption   10 nW/package (typ.)
- Low power TTL compatibility   fan out of 2 driving 74L

Connection Diagrams

MM54C00/MM74C00

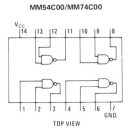

TOP VIEW

MM54C02/MM74C02

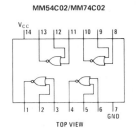

TOP VIEW

MM54C04/MM74C04

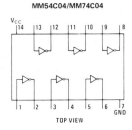

TOP VIEW

MM54C10/MM74C10

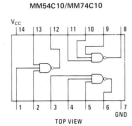

TOP VIEW

MM54C20/MM74C20

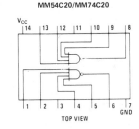

TOP VIEW

FIGURE 15.23 Manufacturer's specification sheets for CMOS logic blocks. (*Courtesy of National Semiconductor Corp.*)

Absolute Maximum Ratings

Voltage at Any Pin	-0.3V to $V_{CC} + 0.3$V
Operating Temperature Range	
54C	$-55°C$ to $+125°C$
74C	$-40°C$ to $+85°C$
Storage Temperature Range	$-65°C$ to $+150°C$
Operating V_{CC} Range	3.0V to 15V
Maximum V_{CC} Voltage	18V
Package Dissipation	500 mW
Lead Temperature (Soldering, 10 seconds)	300°C

DC Electrical Characteristics

Min/max limits apply across the guaranteed temperature range unless otherwise noted.

	Parameter	Conditions	Min.	Typ.	Max.	Units
	CMOS to CMOS					
$V_{IN(1)}$	Logical "1" Input Voltage	$V_{CC} = 5.0$V	3.5			V
		$V_{CC} = 10$V	8.0			V
$V_{IN(0)}$	Logical "0" Input Voltage	$V_{CC} = 5.0$V			1.5	V
		$V_{CC} = 10$V			2.0	V
$V_{OUT(1)}$	Logical "1" Output Voltage	$V_{CC} = 5.0$V, $I_O = -10\,\mu$A	4.5			V
		$V_{CC} = 10$V, $I_O = -10\,\mu$A	9.0			V
$V_{OUT(0)}$	Logical "0" Output Voltage	$V_{CC} = 5.0$V, $I_O = +10\,\mu$A			0.5	V
		$V_{CC} = 10$V, $I_O = +10\,\mu$A			1.0	V
$I_{IN(1)}$	Logical "1" Input Current	$V_{CC} = 15$V, $V_{IN} = 15$V		0.005	1.0	μA
$I_{IN(0)}$	Logical "0" Input Current	$V_{CC} = 15$V, $V_{IN} = 0$V	-1.0	-0.005		μA
I_{CC}	Supply Current	$V_{CC} = 15$V		0.01	15	μA

AC Electrical Characteristics

$T_A = 25°C$, $C_L = 50$ pF, unless otherwise specified.

	Parameter	Conditions	Min.	Typ.	Max.	Units
	MM54C00/MM74C00, MM54C02/MM74C02, MM54C04/MM74C04					
t_{pd0}, t_{pd1}	Propagation Delay Time to Logical "1" or "0"	$V_{CC} = 5.0$V		50	90	ns
		$V_{CC} = 10$V		30	60	ns
C_{IN}	Input Capacitance	(Note 2)		6.0		pF
C_{PD}	Power Dissipation Capacitance	(Note 3) Per Gate or Inverter		12		pF
	MM54C10/MM74C10					
t_{pd0}, t_{pd1}	Propagation Delay Time to Logical "1" or "0"	$V_{CC} = 5.0$V		60	100	ns
		$V_{CC} = 10$V		35	70	ns
C_{IN}	Input Capacitance	(Note 2)		7.0		pF
C_{PD}	Power Dissipation Capacitance	(Note 3) Per Gate		18		pF
	MM54C20/MM74C20					
t_{pd0}, t_{pd1}	Propagation Delay Time to Logical "1" or "0"	$V_{CC} = 5.0$V		70	115	ns
		$V_{CC} = 10$V		40	80	ns
C_{IN}	Input Capacitance	(Note 2)		9		pF
C_{PD}	Power Dissipation Capacitance	(Note 3) Per Gate		30		pF

Note 1: "Absolute Maximum Ratings" are those values beyond which the safety of the device cannot be guaranteed. Except for "Operating Temperature Range" they are not meant to imply that the devices should be operated at these limits. The table of "Electrical Characteristics" provides conditions for actual device operation.

Note 2: Capacitance is guaranteed by periodic testing.

Note 3: C_{PD} determines the no load ac power consumption of any CMOS device. For complete explanation see 54C/74C Family Characteristics application note — AN-90.

FIGURE 15.23   (Continued)

Typical Performance Characteristics

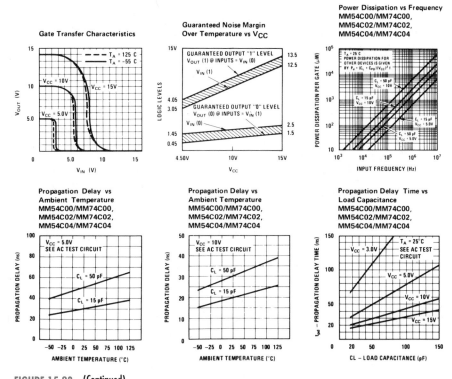

FIGURE 15.23 (Continued)

high. Another disadvantage is the fact that the logic swing (difference between "high" and "low" voltages) is small, on the order of only 1 V. This makes ECL vulnerable to random noise voltages, which can lead to errors.

INTEGRATED-INJECTION LOGIC

All of the bipolar technologies mentioned until now are too large (consume too much chip space per gate) to be useful in LSI. A bipolar alternative to MOS does exist, however, in the technology known as integrated-injection logic (I^2L). This technology offers packing densities 10 times larger than TTL (100 to 200 gates per mm², as compared with perhaps 10 to 20 for TTL). In I^2L technology, power consumption and delay time have a "trade-off" relationship: Propagation delays only about a factor of 2 longer than these of TTL are achievable with higher power consumption; on the other hand, if lower speed is accepted, the power consumption can be the least of any logic family. As with ECL, the logic swing is small, implying vulnerability to noise. Unlike MOS, which can be made compatible with TTL, I^2L voltage ranges are incompatible with other families. Thus I^2L is most useful inside LSI ICs, where its high packing density can best be utilized. A relatively new technology, I²L is not yet common as a commercially available logic family. It is used inside LSI blocks, with interfacing circuits provided at the inputs and outputs to make the blocks externally compatible with TTL.

POWER-DELAY PRODUCT

As we have seen, the choice of a logic technology involves a balance of many practical characteristics. An interesting, fairly general figure of merit does exist, however, in what is known as the *power-delay product* (PDP). This number is obtained by multiplying P_{AV}, the average dc power consumed by the gate, by the gate's propagation delay T_D. The units of PDP are those of energy (joules). Since T_D is the time required for a single change of logical state, PDP is the energy required to effect a single change, and thus in some sense is a measure of the electrical efficiency of the switch.

For TTL and ECL, PDP is on the order of 100 pJ; for NMOS it is around 10 pJ; while the slower versions of I²L can operate at 1 pJ per change. CMOS is an exceptional case, since its average power consumption is not a constant but is linearly proportional to the data rate. However, if CMOS is run at its maximum speed, its PDP is comparable to that of NMOS. We can get an idea of how PDP can be reduced by means of a simple order-of-magnitude calculation. Let us assume, as in Example 15.7, that T_D is determined by the time it takes to discharge a single capacitor C with a maximum current I_{MAX}. The charge to be removed is approximately $Q = C(V_H - V_L)$, where $V_H - V_L$ is the logic swing. Let us also assume that the time-average current through the gate is half of the maximum current. (That is, we imagine that the current through the gate is either I_{MAX} or zero, and it is I_{MAX} about half the time.) Hence

$$\text{PDP} \sim \frac{C(V_H - V_L)}{I_{MAX}} V_{DD} \frac{I_{MAX}}{2} \tag{15.9}$$

$$\cong \tfrac{1}{2} C(V_H - V_L)^2$$

where the last step assumes that $V_{DD} \cong V_H - V_L$. Thus in order to reduce PDP we must reduce logic swing and/or device capacitance.

Conventional circuits can often be improved in terms of PDP by making them physically smaller (which also, of course, increases packing density). Efforts in this direction are limited by the smallest dimensions allowed by available microfabrication technique. The term *minimum feature size* refers to the typical minimum dimension used in an IC. Improvements in microfabrication have gradually reduced minimum feature size from 10 to 20 μm to 2 to 3 μm or even less. *Scaling analysis* shows how the figures of merit change when minimum feature size is reduced by a factor S ($S > 1$), with voltages kept constant. For example, for MOS technology, capacitances decrease as $1/S$, and PDP accordingly also decreases as $1/S$. Interestingly, T_D decreases faster, as $1/S^2$. Thus we can expect further improvements in these parameters as size reductions occur, until new limitations begin to appear at dimensions around 0.25 μm.

An alternative way to improve performance is by introduction of substantially different kinds of technology. An interesting example is cryogenic (liquid-helium-temperature) *Josephson digital technology*, which makes use of the properties of superconductors. This technology is still in the research stage but is projected to offer 0.04 nsec delay times, with PDP on the order of 2×10^{-4} pJ!

15.4 Application: Hand-Held Calculators

The familiar hand-held calculator, with which you do your homework problems and balance your checkbook, is in fact a special-purpose microcomputer, consisting of a microprocessor, memory, and input-output blocks such as drivers for the seven-element displays. A resident program, built into read-only memory, supplies instructions for the particular arithmetic jobs that are to be done. Since these tiny computers are produced in great numbers, it is economical to combine all the components into a single IC, and as usual with ICs, the resulting unit cost is remarkably low. The entire microcomputer electronics, in IC form, only cost a dollar or two to make. This is astonishing, when you consider how many transistors are required. Digital ICs are much denser than analog ICs, in terms of the number of transistors they contain. A typical op-amp may contain only a dozen or so transistors, while a microprocessor may contain a million. But as we recall, the situation with ICs is like printing words on a page. Whether there is only one word on the page or a thousand, the cost of printing the page is about the same.

In the case of hand-held calculators, power consumption is of great importance, and thus CMOS technology is clearly the best choice. CMOS has low power consumption even when it is going full blast, and besides, it has the useful property that energy is consumed only when gates are changing from one state to another. Most of the gates in the calculator are quiescent at any given moment, and thus use no power. Similarly, when a register in the memory is remembering, but not changing, all its gates are stationary. Thus the memory can retain its information at almost no cost. Since so little energy is required to retain information in the memory, modern calculators "remember" even when turned off. When you turn the calculator on, after a month's delay, the data from your last calculation are still there waiting for you. Actually, CMOS has gradually taken over in all kinds of computers, not just small ones. Low power consumption is important in large computers, too, because there are so many gates. It would be difficult to get rid of all the heat if the individual gates were not energy-efficient. Furthermore, CMOS gates take up less chip space than bipolar gates. At present the number of bites stored on a CMOS memory chip is about one million, and increasing. A typical bipolar memory chip stores only 64,000 bits.

The speed of any computer is limited by the switching times required for the logic gates used in the circuit. In large computers speed is extremely important, but it is less important with hand-held calculators, since typical jobs require only a small number of operations anyway. Thus speed will probably be sacrificed by the designers, who will instead emphasize low power consumption and low cost. The characteristics of CMOS gates vary widely, depending on the dimensions of the transistors and other aspects of their construction. We can find out about the speed of at least one family of CMOS gates from the manufacturer's specifications in Fig. 15.23. We see that propagation delays are in the range of 50 to 70 nsec. However, these are general-purpose gates with only a few on a chip; the gates used in VLSI are smaller and faster.

The propagation delay is just the time needed for a signal to pass through a single gate; the time needed to actually execute an instruction turns out to be much longer. For one thing, gates act in long sequential chains; perhaps 20 gates may have to switch, one after another, during a single computing step of the machine, and thus 1 microsecond, rather than 50 nanoseconds, will be required. The rhythm of the machine will be determined by a central clock, ticking once in a time sufficient for a single computing step. With the 50 nanosecond gates, the clock rate might thus be once every microsecond, or 1 MHz. At every beat of the clock there is the possibility that all the flip-flops in the machine can be set to new values.

The machine functions by executing machine-language instructions, such as "load the accumulator" or "add register A to accumulator." Usually the machine requires more than one clock cycle to execute an instruction; with conventional computer architecture, four or five clock cycles are typical. Thus, for example, the type 386 processor used in fast PC-type computers has a clock rate in the range of 25 to 33 MHz and runs at about 5 to 8 *M*illion *I*nstructions *P*er *S*econd, or in the usual jargon, 5 to 8 MIPS.

The 386 is a 32-bit processor, meaning that it can add two 32-digit binary numbers in a single step. It does this by using registers that each have 32 flip-flops; in an addition step each digit is added simultaneously, or "in parallel." Hand-held calculators, on the other hand, are usually 1-bit sequential machines; the accumulator consists of a single flip-flop and can add only 1 bit at a time. It is still possible to add multidigit binary numbers, but this is a much slower process: First you have to add the least-significant digit, then store the result, then add the next digit, and so on. Suppose your calculator has 12-decimal-place accuracy; then it will need to carry about 50 binary digits to represent each 12-digit decimal number. Having to manipulate the digits one at a time, instead of in parallel, will slow things down by a factor of about 100. In addition, there is another subtle consideration having to do with the slowness of floating-point arithmetic, as compared with fixed-point. (This is outside the range of our discussion; ask your computer science teacher!) Hand-held calculators are likely to use floating-point, which slows things down perhaps another factor of 100. Thus a pocket calculator using the same CMOS technology as the excellent 386 processor could be expected to run at only about 5×10^{-4} MIPS.

If you have a programmable calculator or computer at hand, you can easily find out what speed it has. Just write a simple looping program that begins with zero and adds one on each loop, until a preset total is reached and the program stops. Then you can time the operation and see how long it takes, on the average, for each instruction in your program.* The author's Hp 32S hand calculator was found to run at about 210 instructions per second. Is your calculator faster or slower than this?

Note: The number of instructions is not the same as the number of program statements. What you want is the computation time, divided by the product of the number of instructions in each loop and the number of loops.

POINTS TO REMEMBER

- The fundamental circuit of digital logic is the transistor switch, or inverter.
- Saturating bipolar transistor switches are used in diode-transistor logic (DTL) and transistor-transistor logic (TTL). The transistor acts as a switch that connects or disconnects the collector and the emitter. The switch is closed when sufficient base current is applied to saturate the transistor.
- Sufficient base drive must be applied to BJT inverters to guarantee saturation under all load conditions.
- Logic blocks are available in different forms known as logic *families*. Blocks in the same family are compatible and can be interconnected.
- The TTL family is large and widely used. It is quite fast and has good output current capability, but it consumes too much power and space to be used in LSI.
- MOS technology has low power consumption and high packing density, but low output current capability. It is widely used in LSI.
- In NMOS logic a second *n*-channel MOSFET is used to act as a load for a MOSFET switch.
- In CMOS a *p*-channel and an *n*-channel MOSFET are used as a symmetrical pair, with each acting as load for the other. CMOS has the advantage that it uses no power except when it is actually switching.
- A *transmission gate* is a switch controlled by a logic input. It is conveniently built in CMOS technology.
- Propagation delay time is the time required for the output of a gate to change its state.
- Power-delay product is the approximate energy consumed by a gate every time its output is switched.
- Emitter-coupled logic (ECL) is a high-speed bipolar technology with high power consumption.
- Integrated-injection logic is a compact, low-power bipolar technology competitive with MOS technology for LSI.

PROBLEMS

Section 15.1

*15.1 Sketch the voltage transfer characteristic (v_{OUT} versus v_{IN}) for the BJT inverter of Fig. 15.2. Let $\beta = 100$, $R_C = 1,000 \ \Omega$, and $V_{CC} = 5$ V. Consider the following two cases:

a. $R_B = 40,000 \ \Omega$.
b. $R_B = 4000 \ \Omega$.

Note the effect of increasing base overdrive.

*15.2 Use the result of Problem 15.1(b) to estimate the transition width and noise margins of the circuit. Let the nominal "low" and "high" output voltages V_{OL} and V_{OH} be 0.2 and 5 V, respectively.

*15.3 For the circuit of Fig. 15.4, sketch v_{OUT} as a function of i_{OUT}. Let $V_{CC} = 5$ V, $v_{IN} = 5$ V, $\beta = 100$, $R_C = 5,000$ Ω, and $R_B = 50,000$ Ω. If the top of the allowed "low" range is defined to be 0.5 V, what is the maximum allowable i_{OUT}?

*15.4 For the inverter of Fig. 15.2, let us assume a propagation delay time of 5 nsec. Let $V_{CC} = 5$ V and $\beta = 100$, and assume that the time-average of the collector current is half its maximum value. Estimate the circuit's power-delay product. Let $R_C = 2,000$ Ω and $R_B = 20$ kΩ. There is no load current.

*15.5 For the inverter of Fig. 15.4, let the base overdrive be held constant while we vary R_C. State the effect of decreasing R_C on the following:
  a. Current drawn through the input terminal.
  b. Time-average power consumption.
  c. Maximum tolerable i_{OUT}.

Section 15.2

15.6 For the DTL **NAND** gate of Fig. 15.5, find the output v_F when:
  a. $v_A = 0.4$ V and $v_B = 0.2$ V.
  b. $v_A = 0.4$ V and $v_B = 4.0$ V.
  c. $v_A = 4.6$ V and $v_B = 5.0$ V.

15.7 In the DTL **NAND** gate of Fig. 15.5, a load current i_L of 7 mA is expected when the output is "low." To what value must R_A be changed to guarantee base overdrive of 10?

*15.8 For the DTL **NAND** gate of Fig. 15.5, let the lower limit of the "high" range be 4 V. What is the largest negative value of i_L that can be tolerated when the output is "high?" What is the smallest resistance that can be connected between the output and ground?

15.9 At the output of the DTL **NAND gate of Fig. 15.5, N inverters (Fig. 15.2) are connected in parallel. (That is, the output of the gate is simultaneously connected to the inputs of N inverters.) For the inverters, let $V_{CC} = 5$ V, $\beta = 100$, $R_C = 10$ kΩ, and $R_B = 100$ kΩ. If the upper limit of the "low" range is 0.5 V and the lower limit of the "high" range is 4.0 V, what is the largest allowable value of N?

15.10 Consider DTL **NAND gates such as those shown in Fig. F15.5. The output of one such gate is simultaneously connected to the A inputs of F identical gates, where F is an integer called the fanout. The B and C inputs of the driven gates are all connected to V_{CC}. We require a base overdrive of 2. What values can R_A have if $F = 10$? The other parameters are as given in Fig. 15.5.

*15.11 For the TTL **NAND** gate of Fig. 15.7, find the voltages at X, Y, and F when:
  a. $v_{E1} = 0.2$ V, $v_{E2} = 0.3$ V, and $v_{E3} = 4.5$ V.
  b. $v_{E1} = 4.0$ V, $v_{E2} = 4.1$ V, and $v_{E3} = 4.2$ V.

*15.12 For the TTL **NAND** gate of Fig. 15.7, suppose that input E_1 is grounded and inputs E_2 and E_3 are connected to V_{CC}. Let $R_1 = 2\ k\Omega$, $R_2 = 2\ k\Omega$, and $R_3 = 5\ k\Omega$. Estimate the current that flows through input E_1.

*15.13 For the TTL **NAND** gate of Fig. 15.7, let all inputs be "high." Let $\beta = 100$ for all three transistors, and let $R_1 = 100\ k\Omega$, $R_2 = 10\ k\Omega$, and $R_3 = 2\ k\Omega$. Find the three collector currents I_{C1}, I_{C2}, and I_{C3}.

*15.14 For the case of Problem 15.13, estimate the total power consumed by the gate. For T_1, assume a reverse current transfer ratio α_R of 0.1 for each emitter.

*15.15 The term *wired logic* refers to "free" logic operations that in some logic families can be obtained by connecting the outputs of two gates together. Suppose the logic outputs of two TTL **NAND** gates are called **F1** and **F2**. Let the two outputs be connected together with a wire, and let the resulting combined output be called **F3**. Construct a truth table showing **F3** as a function of **F1** and **F2**. What "free" logic operation is obtained?

Section 15.3

15.16 For the CMOS inverter of Fig. 15.13, find by means of Eq. (13.11) the approximate drain current I_{D1} when $v_{IN} = 3.5$ V. Take $V_{DD} = 7$ V and $K = 0.1$ mA/V^2. (*Note:* This is point C in Fig. 15.17.) The transistor characteristics are those of Figs. 15.11(b) and 15.14.

**15.17 For the CMOS inverter of Figs. 15.11(b), 15.13, and 15.14, find I_{D1} and V_{DS1}, when $v_{IN} = 3.8$ V, by means of Eqs. (13.10) and (13.11). Check your answer to make sure you have chosen the right regime (triode or pinch-off) for each transistor. Assume $K = 0.1$ mA/V^2 for both transistors.

*15.18 For the CMOS inverter of Figs. 15.11(b), 15.13, and 15.14, let a load resistor R be connected between output and ground. Let $v_{IN} = 0$. Sketch a graph of v_{OUT} versus R for the range $7\ k\Omega < R < 100\ k\Omega$.

**15.19 A capacitor is charged through a resistor by closing the switch shown in Fig. 15.24.
  a. Find $i(t)$.
  b. Find the energy $\int_0^\infty p(t)\,dt = V_{DD}\int_0^\infty i(t)\,dt$ supplied by the voltage source as the capacitor is charged.
  c. Find the energy dissipated as heat in the resistor as the capacitor is charged.
  d. On the basis of your results, how much energy is converted to heat when the output of inverter I_1 makes a single transition from low to high in the circuit of Example 15.6?

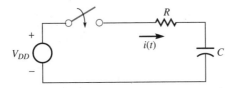

FIGURE 15.24

*15.20 A CMOS inverter has a supply voltage $V_{DD} = 10$ V and drives a 50 pF capacitive load. The input is a 1 MHz square wave. The inverter's power-dissipation capacitance is 15 pF, and the leakage current is 1 nA. Find the average power dissipation.

*15.21 How does the result of Example 15.6 change if there is a fanout of 20? (For definition of fanout, see Problem 15.10.)

15.22 Repeat Problem 15.15 for the CMOS inverter. Does wired logic work with CMOS?

*15.23 Assume $V_T = 3$ V and find v_{OUT} for the CMOS **NAND** gate of Fig. 15.21 when:
- **a.** $v_1 = 6$ V and $v_2 = 5$ V.
- **b.** $v_1 = 2$ V and $v_2 = 5$ V.
- **c.** $v_1 = 2$ V and $v_2 = 1$ V.

*15.24 Figure 15.25 shows a CMOS logic gate. Construct its truth table and identify the logical operation (assuming positive logic).

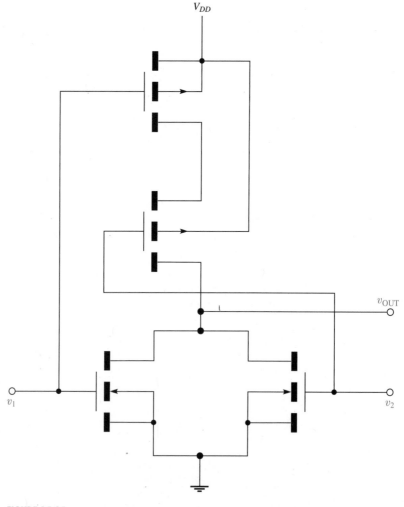

FIGURE 15.25

15.25 Design a three-input CMOS **NAND** gate.

15.26 A two-bit memory consisting of two *D*-type flip-flops is to be connected to a two-wire data bus. When control input **C** is **1**, the **Q** outputs are to be connected to the bus to read out stored information. When **C** = **0** the data bus is to be connected to the **D** inputs in order to record information. Design a circuit to do this, using transmission gates and other blocks as needed.

*15.27 Repeat Example 15.6 with the supply voltage V_{DD} increased to 10 V. By what factor does the delay time change?

*15.28 Two inverters of the type illustrated by Fig. 15.18 are connected in cascade (output of the first to the input of the second). Estimate the transition width of the two-stage circuit. How does it compare with the transition width of the single-stage circuit? What is the effect on the noise margins?

*15.29 Figure 15.26 shows the voltage transfer characteristic of a hypothetical inverter. Also shown is an approximation to the characteristic made of straight-line segments; this is known as a *piecewise-linear approximation*. Let the slope of the central part of the piecewise-linear curve be $-N$. Assume $V_{OL} < V_{IL}$, $V_{OH} > V_{IH}$. Let two such gates be connected in cascade (output of the first to input of the second).

 a. Show that for the two-stage circuit the slope of the piecewise-linear approximation is approximately N^2.

 b. Apply this result to estimate the transition width of two gates such as those of Fig. 15.12, connected in cascade.

 c. Check your result by constructing the voltage-transfer characteristic of the two-stage gate graphically.

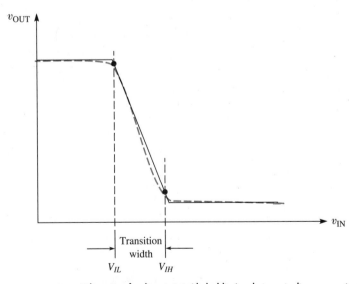

FIGURE 15.26 Voltage-transfer characteristic (dashed line) and piece-wise-linear approximation (solid line through the points at V_{IL} and V_{IH}).

*15.30 Refer to the CMOS data sheets of Fig. 15.23. A type MM74C10 integrated circuit (triple three-input **NAND** gate) is being operated with each gate switching (up-down-up) 10^5 times per second on the average. The supply voltage is 5 V, and the load capacitance is 50 pF. When no signals are present the IC dissipates 1 nW owing to leakage current. Find the average power dissipation of the IC during operation. What is the power-delay product for each gate?

*15.31 Refer to the CMOS specifications of Fig. 15.23. Let the supply voltage $V_{CC} = 5$ V.

   a. What are V_{OL}, V_{OH}, V_{IL}, and V_{IH}?

   b. For two-input **NOR** gates, each with a load capacitance of 50 pF, what is the propagation delay time? What maximum data rate does this imply? Use the rule of thumb $(25T_D)^{-1}$.

   c. Suppose that the load is not 50 pF but consists of the inputs of 20 identical gates connected in parallel. Now what is the propagation delay time?

*15.32 Obtain an order-of-magnitude estimate of the power-delay product of a TTL gate. Assume that the power dissipation results entirely from load currents associated with a 15 pF load capacitance. Let $V_{CC} = 5$ V.

MAGNETIC DEVICES, ELECTRIC POWER, AND MACHINES

CHAPTER 16

ELECTRIC POWER

ntil now our emphasis has been on information systems, where in many cases an objective is to hold power to the lowest levels possible. Now, however, we turn to the other main application of electricity, that of transporting energy from place to place. In this field a new family of system components, those based on magnetism, becomes important. This chapter begins with a general discussion of magnetic technology. We then proceed to the subject of transformers, which are quite important in power systems. We conclude with discussions of electric power distribution and three-phase systems.

16.1 Magnetism and Magnetic Technology

Just as we began in Chapter 1 by reviewing electrical physics, we shall here review the physics of magnetism. This is not to suggest, of course, that electricity and magnetism are entirely separate subjects; they are both aspects of electromagnetism, related to each other through Maxwell's equations.

In order to describe magnetic effects it is convenient to introduce the concept of *magnetic flux density*, which is usually given the symbol $\vec{B}$. When magnetic fields are present, there is at each point in space a unique value of $\vec{B}$. The flux density is a vector quantity and thus has not only magnitude but also direction. Hence we can represent $\vec{B}$ by an array of little arrows, the length of each of which indicates the magnitude of $\vec{B}$ at

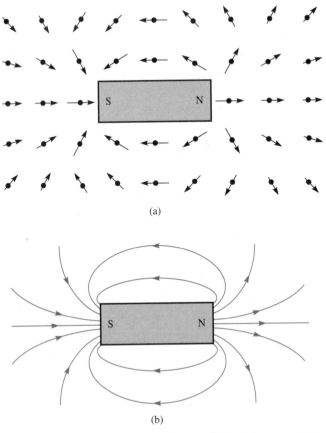

FIGURE 16.1 Two ways of illustrating the vector field $\vec{B}(\vec{x})$ of a bar magnet. (a) Magnitude and direction of $\vec{B}$ at 42 different points. (b) The "field lines" obtained by drawing continuous curves parallel to the arrows in part (a).

that point, with the direction of each showing the direction of the field. Figure 16.1(a) represents the field of a bar magnet in this fashion. If we draw continuous lines along the directions of the arrows, we arrive at the more usual picture of the magnetic field lines shown in Fig. 16.1(b). The important thing to note here is that $\vec{B}(\vec{x})$ is a vector function of position. In mathematics such a function is known as a *vector field*. Often $\vec{B}$ also depends upon the time, in which case we write $\vec{B}(\vec{x},t)$. In SI units the unit of magnetic flux density is the tesla, with symbol T. Like the farad, the tesla is a rather large unit. The earth's magnetic field is on the order of .00005 T.* The range 1 to 5 T is about the largest one can reach with conventional electromagnets, although superconducting magnets can produce fields 10 times larger.

---

*A non-SI unit, the gauss, is also encountered: 1 gauss $= 10^{-4}$ T. Thus the earth's field is on the order of 0.5 gauss.

The field $\vec{B}$ is important to us in two ways: because when it varies with time it induces voltages in loops of wire (as in transformers and generators) and because it exerts forces on current-carrying wires (as in motors). The former effect is described by Faraday's law, which we shall state not in its most general vector form but in a simpler way adequate for our purposes. Let $B_\perp(t)$ be the component of a time-varying field perpendicular to the plane of a loop of wire, the area of which is A. The voltage that appears across the terminals of the loop is given by

$$v = A\frac{dB_\perp}{dt} \tag{16.1}$$

If we replace the single loop of wire by a coil of N loops, the voltages of the loops add in series and the terminal voltage is multiplied by N. Furthermore, it is customary to define the product of B and the area A as the magnetic flux* Φ. Thus for an N-turn coil we have

$$v = N\frac{d\Phi}{dt} \tag{16.2}$$

as illustrated in Fig. 16.2.

When a current-carrying wire is located in a magnetic field, a force can be exerted on the wire. When a wire of length h carrying current I is *perpendicular* to $\vec{B}$, the force exerted on it is given by

$$F = IhB \tag{16.3}$$

The direction of this force is perpendicular to both $\vec{B}$ and the direction of current flow: If current flow is in the x direction and $\vec{B}$ is in the y direction, the force is in the positive z direction. Electric motors are driven by this force.

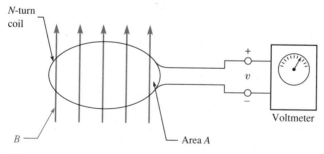

FIGURE 16.2 Faraday's law. A time-varying magnetic field perpendicular to the plane of an N-turn coil produces the terminal voltage $v = NA\, dB/dt$.

---

*The SI units of flux are $T \cdot m^2$. One $T \cdot m^2$ is also known as one weber (Wb). Sometimes instead of teslas one finds $\vec{B}$ expressed in Wb/m^2, which are the same ($1\ Wb/m^2 = 1\ T$). Flux is also measured in *lines,* a non-SI unit. One line $=$ one gauss $\cdot\ cm^2 = 10^{-8}\ Wb$.

There is also a second magnetic field, known as the *magnetic intensity* $\vec{H}$. The flux density $\vec{B}$ is the field we use, through Eqs. (16.2) and (16.3), to induce voltages or apply forces. But the field over which we have direct control is not $\vec{B}$ but $\vec{H}$. Thus we apply the field $\vec{H}$ (we shall show how in a moment); $\vec{H}$ then gives rise to $\vec{B}$, and we then make use of $\vec{B}$. For most materials the relationship between $\vec{H}$ and $\vec{B}$ is simple: The value of $\vec{B}$ is just equal to the value of $\vec{H}$, multiplied by a scalar constant μ characterizing the material at the point in question. Thus we write

$$\vec{B} = \mu\vec{H} \tag{16.4}$$

When Eq. (16.4) holds, the vector $\vec{H}$ points in the same direction as $\vec{B}$. The proportionality constant μ is called the *permeability*. For vacuum its value is μ_0 and is numerically equal to $4\pi \times 10^{-7}$ in SI units. Most ordinary materials have values of μ very close to μ_0. The SI units of $\vec{H}$ are ampere-turns per meter (A-t/m), as will be explained shortly.

A significant exception is the class known as *ferromagnetic* materials, the most important of which are iron and its alloys. In ferromagnetic materials a much larger value of $\vec{B}$ is obtained for a given $\vec{H}$ than in ordinary materials. However, Eq. (16.4) does not strictly apply. In ferromagnetic materials $\vec{B}$ is a more complicated function of $\vec{H}$ that must be described graphically, as in Fig. 16.3. In particular we note the phenomenon of *saturation*, evidenced by the failure of $\vec{B}$ to continue to increase as $\vec{H}$ increases, as seen at the top of the curves. This effect, present in all ferromagnetic materials, places an upper limit on the values of $\vec{B}$ that can be attained. Although $\vec{B}$ is not in fact linearly proportional to $\vec{H}$, it is nearly so for small values of the fields; thus for fields below saturation we can use Eq. (16.4) as an approximation, assigning to μ a value based on the initial slope of the B–H curve. The approximated values of μ obtained in this way are very large, on the order of $10,000\mu_0$.

• EXERCISE 16.1

Estimate μ/μ_0 for the silicon sheet steel of Fig. 16.3. **Answer:** About 6,000.

Ferromagnetic materials often exhibit the interesting effect known as *hysteresis*. When this effect is present, $\vec{B}$ is not a single-valued function of $\vec{H}$; rather, it depends not only on the present $\vec{H}$ but also on values of $\vec{H}$ at earlier times. This effect is due to the alignment of microscopic magnets, called "magnetic domains," inside the metal. In unmagnetized iron these domains point randomly in all directions, and the average magnetization $\vec{M}$ is zero. However, when an $\vec{H}$ field is applied, the domains are forced to line up parallel to the applied field, resulting in a large $\vec{M}$. It is always true,* however,

---

*The statement $\vec{B} = \mu\vec{H}$ is thus true only in those special cases in which $\vec{M}$ is linearly proportional to $\vec{H}$. For almost all materials except ferromagnets this assumption is satisfactory.

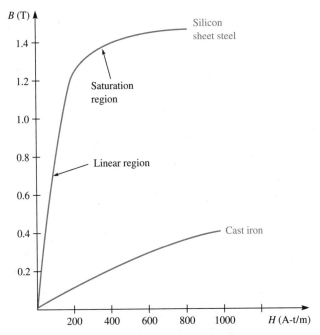

FIGURE 16.3 Magnetization curves for typical ferromagnetic materials.

that $B = \mu_0 \vec{H} + \mu_0 \vec{M}$. Thus the large magnetization combines with $\vec{H}$ to give the large $\vec{B}$ obtained in these materials.

The phenomenon of hysteresis is illustrated in Fig. 16.4. We begin with a piece of material that has never been magnetized; we are at the origin (point 1) in the diagram. We then apply an $\vec{H}$ field, and $\vec{B}$ increases to the value at point 2. During this step the domains have been aligned, providing a large magnetization $\vec{M}$ that adds to $\vec{H}$ to give a large $\vec{B}$. Now let the applied $\vec{H}$ be reduced to zero. The alignment of the domains partially disappears due to thermal agitation, but some alignment—and hence some magnetization—remains. Thus at point 3 there is still some $\vec{B}$ even though $\vec{H}$ has been reduced to zero! This is the so-called "permanent" magnetization; a piece of iron is made into a permanent magnet in this way. If we next apply a negative $\vec{H}$, the domains are forced to turn around and point the opposite way, giving a negative $\vec{M}$ at point 4. When $\vec{H}$ again returns to zero at point 5, a residual negative magnetization remains. Making $\vec{H}$ positive again will now move us to point 6, and the process continues. In most cases of interest $\vec{H}$ is a steady-state sinusoidal function of time. After $\vec{H}$ has cycled from positive to negative a few times, the B–H curve will settle down into a closed path, as shown in Fig. 16.4(b).* Hysteretic effects have been utilized in magnet-

---

*The reader may wonder why hysteresis is not seen in magnetization curves such as that of Fig. 16.3. This is because the usual $B - H$ curves are not true graphs of $B(H)$. They are actually the so-called *normal $B - H$* curves, obtained by plotting the tips of hysteresis loops of increasing magnitude.

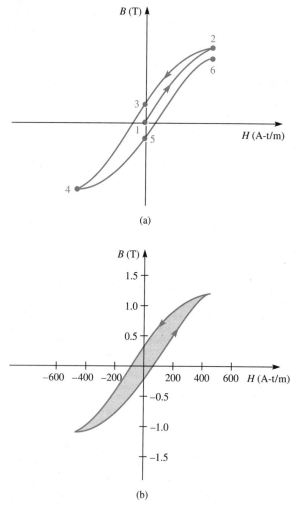

FIGURE 16.4

ic core memories in computers; one simply stores a positive "permanent" magnetization in a bit of ferrous material to represent a **1**, or a negative magnetization to represent a **0**. Core memories have been displaced by semiconductor-device memories—arrays of flip-flops—which are smaller and less costly to make, but permanent magnets are of course useful for many other purposes.

It should be noted, however, that the "permanent" $\vec{B}$ is less than the maximum $\vec{B}$ obtainable with $\vec{H}$ applied; furthermore, it can be erased by strong fields. For these reasons permanent magnets are ordinarily used only in small, low-power apparatus.

An undesirable result of hysteresis is loss of energy by conversion to heat. It can be shown that the amount of energy that must be supplied to increase the flux density from $\vec{B}$ to $\vec{B} + d\vec{B}$ is equal to $\vec{H} \cdot d\vec{B}$. If $\vec{B}$ is then reduced by the same amount, the same energy is taken back out, but only if $\vec{H}$ has the same value as when $\vec{B}$ was increased. From Fig. 16.4 we see that this is not the case in a hysteretic material; for the same $\vec{B}$, $\vec{H}$ is larger when $\vec{B}$ is increasing than when it is decreasing. In fact, the energy per unit volume that is lost (converted to heat) in each cycle of $\vec{H}$ is equal to the shaded area inside the curve in Fig. 16.4(b).

EXAMPLE 16.1

An iron-alloy transformer core whose volume is 1,080 cm³ is subjected to a 60 Hz magnetic field, so that it executes the magnetization curve shown in Fig. 16.4(b). Estimate the power lost by conversion to heat.

SOLUTION

The energy lost per unit volume in each cycle is equal to the shaded area. This is hard to estimate by inspection, but it is about 250 joules per cubic meter. Thus the power lost is about $(250)(60)(1.08 \times 0^{-3}) \cong 16$ W.

We have not yet explained how $\vec{H}$ is applied. The field $\vec{H}$ is produced directly by currents, in accordance with Ampère's law:

$$\oint \vec{H} \cdot \vec{dl} = I \tag{16.5}$$

Here the quantity on the left is the line integral of $\vec{H}$ around any closed path, and I is the current that threads the path. For example, in Fig. 16.5 a current I flows through a straight wire, and we choose as our path of integration a circular path (with radius r) around the wire. In this case because of symmetry the lines of $\vec{H}$ are simply circles centered on the wire; hence $\vec{H}$ is parallel to the chosen path at every point. Thus the

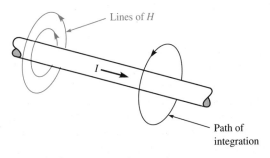

FIGURE 16.5 Ampère's law for a straight wire.

line integral reduces to $2\pi r H$ (where we use H to mean the absolute value of $\vec{H}$). Setting this equal to I, we have

$$H(r) = \frac{I}{2\pi r} \tag{16.6}$$

which states that H decreases linearly with distance from the wire. If the wire is surrounded by vacuum we have $B = \mu_0 I/2\pi r$. The direction of $\vec{H}$ is conveniently found by the *right-hand rule*: When the thumb of the right hand points in the direction of the current, the curled fingers point in the direction of $\vec{H}$ and $\vec{B}$.

In magnetic apparatus it is usually necessary to create strong magnetic fields, which cannot be done with a single wire; it is necessary to add together the fields of many turns of wire wound together in a coil. A common case is that of a *solenoid*, a helical coil similar to that shown in Fig. 16.6. In power applications these are usually wound on an iron core, as shown. Iron cores are used because of iron's large permeability; it has the ability to conduct a magnetic field almost as water is conducted through a pipe. This very useful property allows us to concentrate magnetic fields and convey them to the point of use.

The same property of iron also allows us to make approximate calculations of the fields. For example, in Fig. 16.7 the iron core has been bent around on itself to form a closed, square iron ring. We assume that all the magnetic field is confined inside the iron; for simplicity we shall also assume that its magnitude is the same everywhere inside it. In order to use Ampère's law, we choose a path of integration that is entirely inside the iron and goes all the way around the ring, as shown. The length of this path is not clearly defined, since it depends on whether one chooses the path to be near the outer periphery or near the center hole. However, we can designate l as the approximate, or average, length of this path. Then from Eq. (16.5) the magnitude of H is given approximately by

$$H \cong \frac{NI}{l} \tag{16.7}$$

where N is the number of turns in the solenoid. For the case of Fig. 16.7, $l \cong 4[(a + b)/2] = 2(a + b)$, and $H \cong NI/2(a + b)$. Observe that according to Eq. (16.7) the

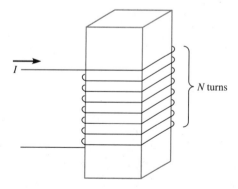

I

N turns

FIGURE 16.6   Solenoid with iron core.

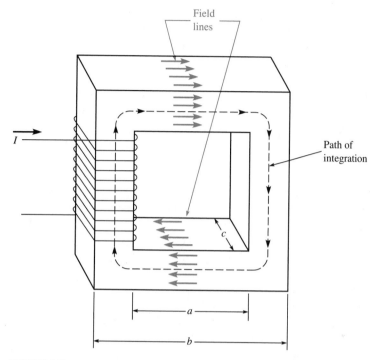

FIGURE 16.7

dimensions of H are amperes per meter, since N is dimensionless. However, equations of this form—with I multiplied by the number of turns in a coil—occur often; hence the dimensions of H are customarily written as ampere-turns per meter.

EXAMPLE 16.2

Estimate H and B for Fig. 16.7, for $I = 173$ mA, $a = 10$ cm, $b = 14$ cm, and $N = 500$; the metal is the silicon sheet steel of Fig. 16.3.

SOLUTION

From Eq. (16.7) we find

$$H \cong \frac{(500)(0.173)}{2(0.1 + 0.14)} \cong 180 \text{ ampere-turns/meter}$$

From Fig. 16.3 we find $B \cong 1.2$ T.

The quantity NI in Eq. (16.7) is known as the *magnetomotive force*, often abbreviated mmf. The idea behind this name is that the product NI acts as the force that generates magnetic field. The resulting flux Φ is then given by

$$\Phi = \text{mmf}/\Re \qquad (16.8)$$

where $\Re$ is a constant called the *reluctance*, the value of which depends on the structure. If the cross-section A of the core is constant, we have from Eq. (16.7)

$$\Phi = \mu HA \cong \frac{NI}{(l/\mu A)} \qquad (16.9)$$

Thus

$$\Re \cong \frac{l}{\mu A} \qquad (16.10)$$

For the structure of Fig. 16.7, $\Re \cong 2(a + b)/\mu A$, where $A \cong (b - a)c/2$. The definition of reluctance, Eq. (16.8), would not make sense if Φ were a function of position, but we note that this is not the case. From our point of view flux flows through the iron core as water would flow through a pipe, with none escaping. Thus the flux is the same everywhere around the iron core.

In the structure of Fig. 16.7 the magnetic field is entirely inside the iron. This structure cannot be used to apply magnetic field to an object because the object would have to be put inside the iron, where the field is. To correct this we can create a gap in the iron core, as shown in Fig. 16.8. According to physical law the flux density B keeps the same value as the field passes from the iron into the air gap. If we assume in consequence that B has a constant value everywhere in the iron and in the gap, Ampère's law, Eq. (16.5) becomes

$$l_m B/\mu + l_g B/\mu_0 = NI \qquad (16.11)$$

where l_m is the length of the flux path through the metal, l_g is the length of the gap, and the permeability of air has been taken to be μ_0 (which it very nearly is). Thus we have

$$\Phi = \frac{(\text{mmf})A}{l_m/\mu + l_g/\mu_0} \qquad (16.12)$$

where A is the cross-sectional area of the iron core.* From Eq. (16.8) the reluctance of this structure is

$$\Re = \frac{l_m}{\mu A} + \frac{l_g}{\mu_0 A} \qquad (16.13)$$

---

*Actually the fields "bulge out" a bit in the gap, as shown in the figure. Since the flux is constant, the increased cross-sectional area of the field must result in B decreasing at that point. In more precise work corrections can be introduced for such effects.

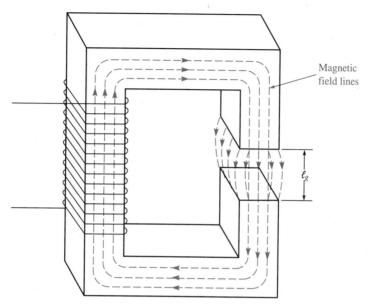

FIGURE 16.8

The reader may have observed a certain similarity between these calculations and those for resistive circuits. In the latter, current equals voltage divided by resistance; in the magnetic case, flux equals mmf divided by reluctance. In fact, a full-fledged analogy exists between the two cases, and we speak of *magnetic circuits* in analogy to resistive electric circuits. Accordingly we can depict the magnetic circuit of Fig. 16.8 by the analogous electric circuit shown in Fig. 16.9. The two parts of the magnetic path, respectively through the iron and through the gap, are represented by their respective reluctances. As we see from Eq. (16.13), the reluctance of each path segment is equal to its length divided by the product of its permeability and cross-sectional area. More complex magnetic circuits can be analyzed with circuit analogies in similar fashion.

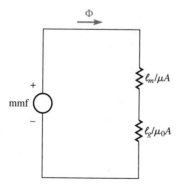

FIGURE 16.9

We have already seen that for iron μ can be many thousand times larger than μ_0. Inspecting Eq. (16.13), we observe that the first of the two reluctances (arising from the path through the iron) is likely to be small and perhaps negligible. When this is so, the flux in the magnetic circuit is determined mainly by the reluctances of the gaps.

EXAMPLE 16.3

In the magnetic circuit of Fig. 16.10(a) the coil has 750 turns and carries a current of 0.5 A. Find the approximate flux density in the 5 mm gap. Take the permeability of the iron to be $5{,}000\mu_0$, and assume that its cross-sectional area A is everywhere 1 cm^2. For simplicity, neglect the reluctances of paths through the iron.

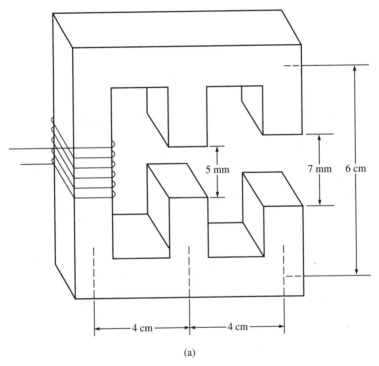

(a)

FIGURE 16.10(a)

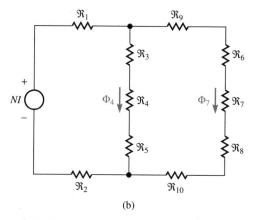

(b)

FIGURE 16.10(b)

SOLUTION

The analogous circuit is shown in Fig. 16.10(b). Here $\Re_4$ and $\Re_7$ are the reluctances of the two gaps, while the other reluctances are those of various path segments through the iron. For example,

$$\Re_3 = \Re_5 \cong \frac{2.75 \text{ cm}}{\mu(1 \text{ cm}^2)}$$

In this problem, however, we are instructed to neglect the reluctances of the iron path segments. This approximation seems quite justifiable because none of these path segments is comparable in length to μ/μ_0 times the lengths of the gaps. With this approximation we immediately have

$$\Phi_4 = NI/\Re_4$$
$$\Phi_7 = NI/\Re_7$$

The value of $\Re_4$ is given by

$$\Re_4 \cong \frac{(5 \times 10^{-3})}{(4\pi \times 10^{-7})(10^{-4})} \cong 4.0 \times 10^7 \text{ SI units}$$

Thus the flux density in the 5 mm gap is

$$B_4 = \frac{\Phi_4}{A} = \frac{NI}{\Re_4 A} = 94 \text{ mT}$$

Note that in this approximation the presence of the second gap has no effect on the value of $\vec{B}$ in the first one.

16.2 Transformers

As we have seen, a current i in the device of Fig. 16.7 produces a flux $\Phi = Ni/\Re$, where $\Re$ is the reluctance given by Eq. (16.10). If i varies with time, Φ will also vary. According to Eq. (16.2), the time-varying flux will generate a voltage across the terminals of the coil equal to

$$v = N\frac{d\Phi}{dt} = \frac{N^2}{\Re}\frac{di}{dt} \qquad (16.14)$$

Clearly, the i-v relationship for this device is that of an ideal inductor, with

$$L = \frac{N^2}{\Re} \qquad (16.15)$$

This device is known as an ideal *iron-core inductor*. The term "ideal" is used because several factors were omitted from our calculation. We have neglected losses due to the action of the magnetic field on the core, known as "core loss" (or "iron loss"). These include power dissipation due to hysteresis and ohmic loss from eddy currents unavoidably induced in the core by the time-varying field.* In addition, there are significant ohmic losses due to the resistance of the copper wire of the coil ("copper loss"). We can model the physical iron-core inductor as an ideal inductor [whose value may be slightly less than that given by Eq. (16.15) because of the leakage of a few flux lines from the iron core] in series with a resistor that accounts for the two kinds of loss.

When a closed iron core carries *two* coil windings, as shown in Fig. 16.11, we have the important device called an *iron-core transformer*. Assuming again that no field

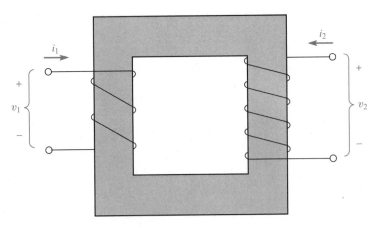

FIGURE 16.11   Iron core transformer.

---

*Eddy current losses can be reduced by making the core out of thin metal layers, separated by thin layers of insulator in planes perpendicular to the eddy currents. The resulting structure is called a "laminated core."

lines leak out of the iron core, the flux linking coil 1 is the same as that linking coil 2. Hence, according to Eq. (16.2), $v_1 = N_1 \frac{d\Phi}{dt}$, $v_2 = N_2 \frac{d\Phi}{dt}$, and, provided that $\frac{d\Phi}{dt} \neq 0$,

$$\frac{v_2}{v_1} = \frac{N_2}{N_1} \quad \text{(ideal transformer)} \tag{16.16}$$

Further, when we apply Eq. (16.5) to a closed path around the iron core (as in Fig. 16.7) we have $N_1 i_1 + N_2 i_2 = \Phi l/\mu A$, where $l/\mu A$ is the reluctance of the core. Since $v_1 = N_1 \frac{d\Phi}{dt}$, we know that $\mathbf{v}_1 = N_1 j\omega \mathbf{\Phi}$, where $\mathbf{\Phi}$ is the phasor representing $\Phi(t)$. Thus $N_1 \mathbf{i}_1 + N_2 \mathbf{i}_2 = l\mathbf{v}_1/2\pi j f N_1 \mu A$. The quantity on the right of this equation is usually small because of the large value of μ. Thus $N_1 \mathbf{i}_1 + N_2 \mathbf{i}_2 \cong 0$, or

$$\frac{i_2}{i_1} = -\frac{N_1}{N_2} \tag{16.17}$$

Equations (16.16) and (16.17) define the block known as an *ideal transformer*, with circuit symbol shown in Fig. 16.12(a). The dots indicate the sign conventions for the two voltages: For Eq. (16.16) to hold, the "plus" terminals must be chosen to be the dotted terminals, as shown in Fig. 16.12(a). Often the phase relationship between the two windings is unimportant, in which case the dots can be omitted. We observe by multiplying Eq. (16.16) by Eq. (16.17) that the total power entering an ideal transformer, $v_1 i_1 + v_2 i_2$, is equal to zero. This is consistent with the idealized nature of the block, which is lossless according to our assumptions.

A real transformer, of course, differs from an ideal transformer. In order to analyze power systems containing transformers, it is useful to have a more realistic trans-

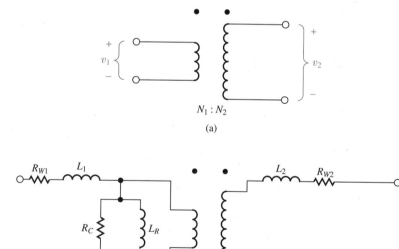

(a)

(b)

FIGURE 16.12   (a) Ideal transformer. (b) One possible model of a real power transformer.

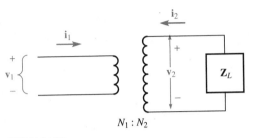

FIGURE 16.13

former model. Such a model must give representation to (1) resistances in the windings, (2) flux lines that thread only one of the two windings (these contribute inductance, just as in Fig. 16.7), (3) core losses from eddy currents and hysteresis, and (4) the nonzero value of the core reluctance. The core reluctance was assumed to be zero in deriving Eq. (16.17). In actuality, some energy must be used to set up the field in the core. This energy is stored, not dissipated; it is recovered when the currents are turned off and the fields collapse. Hence it is represented by an inductance—an energy-storage element—rather than a resistance in the model. A suitable model is shown in Fig. 16.12(b). In this model the transformer symbol represents an ideal transformer, R_{W1} and R_{W2} are the winding resistances, L_1 and L_2 arise from flux lines linking only winding 1 or winding 2, R_C represents core loss, and L_R results from nonzero core reluctance. The actual values of the six parasitic elements depend on the details of the transformer and can be found by measurement.

Let us consider the circuit of Fig. 16.13. Here a load impedance $\mathbf{Z}_L$ is connected to one winding of an ideal transformer. Conventionally, the winding to which the load is connected is called the *secondary* winding; the other winding, to which external power is supplied, is called the *primary*. Let us find the impedance seen looking into the terminals of the primary. According to Eqs. (16.16) and (16.17), this impedance is

$$\mathbf{Z}_{in} = \frac{\mathbf{v}_1}{\mathbf{i}_1} = -\frac{(N_1/N_2)\mathbf{v}_2}{(N_2/N_1)\mathbf{i}_2} = \left(\frac{N_1}{N_2}\right)^2 \mathbf{Z}_L \tag{16.18}$$

where we have used the fact that $\mathbf{v}_2/\mathbf{i}_2 = -\mathbf{Z}_L$. From Eq. (16.18) we see that the load impedance $\mathbf{Z}_L$ appears from the primary side to be multiplied by the square of the turns ratio N_1/N_2. For purposes of calculation one can remove the transformer and load from the circuit, replacing them by a single impedance $(N_1/N_2)^2\mathbf{Z}_L$ across the primary terminals. This is known as "referring" a load to the primary circuit.

● EXERCISE 16.2

A transformer for which the secondary voltage is larger than the primary voltage is called a *step-up transformer*. A load consisting of a 100 Ω resistance in series with a 0.1 H inductor is connected to a 60 Hz 1:8 step-up transformer (1:8 is the turns ratio). What impedance is seen looking into the primary? **Answer:** 1.56 + 0.589j Ω.

EXAMPLE 16.4

Show that the transformer model of Fig. 16.12(b) is equivalent to that of Fig. 16.14(a), and find L_X and R_X.

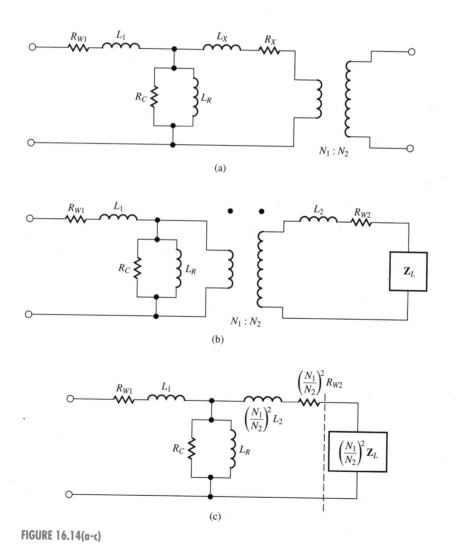

FIGURE 16.14(a-c)

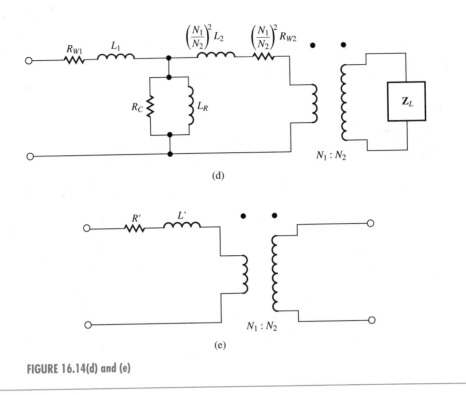

(d)

(e)

FIGURE 16.14(d) and (e)

SOLUTION

This problem is not quite as simple as it might seem at first. Equation (16.18) does not allow us to simply refer L_2 and R_{W2} to the primary; Eq. (16.18) says the *total* load (L_2, R_{W2}, plus any external load) is what is referred. Let us imagine that a load $\mathbf{Z}_L$ is connected across the secondary, as shown in Fig. 16.14(b). Now the series connection of L_2, R_{W2}, and $\mathbf{Z}_L$ can be referred to the primary, resulting in Fig. 16.14(c). Finally, an ideal transformer with the same turns ratio can be reinserted at the position of the dashed line in Fig. 16.14(c), resulting in the circuit of Fig. 16.14(d). Now we have shown that the circuit of Fig. 16.12(b), with any load connected to the secondary, is equivalent to the circuit of Fig. 16.14(a) with the same load. We have $L_X = (N_1/N_2)^2 L_2$, $R_X = (N_1/N_2)^2 R_{W2}$.

In a well-designed transformer the parasitic currents through R_C and L_R may be negligible compared with the large currents through the series elements. In that case one might use the simpler model shown in Fig. 16.14(e). Here $R' = R_{W1} + (N_1/N_2)^2 R_{W2}$, $L' = L_1 + (N_1/N_2)^2 L_2$.

Most electric power problems are solved by means of phasor analysis. Since it is customary to express voltages by their rms values in power work, we shall introduce the *rms phasors* $\mathbf{v}'$ and $\mathbf{i}'$. These are related to the usual phasors $\mathbf{v}$ and $\mathbf{i}$ by $\mathbf{v}' = \mathbf{v}/\sqrt{2}$, $\mathbf{i}' = \mathbf{i}/\sqrt{2}$.

EXAMPLE 16.5

The transformer of Fig. 16.12(b) has $N_1 = 2,000$, $N_2 = 4,000$, $R_{W1} = 0.04\ \Omega$, $R_{W2} = 0.08\ \Omega$, $L_1 = 1.3$ mH, $L_2 = 2.6$ mH, $R_C = 50\ \Omega$, and $L_R = 80$ mH. It is used to supply power from a 120 V (rms) 60 Hz power line to a resistive load. The nominal rating of the load is 2,000 W, 240 V (rms). Find the actual power delivered to the load. For simplicity, use the circuit model of Fig. 16.14(e).

SOLUTION

From Eq. (6.27), the time-averaged power dissipated in a resistance R is simply V^2_{rms}/R. Thus the load resistance R_L of our present load is $(240)^2/2,000 = 28.8\ \Omega$. After referring it to the primary, the circuit model is as shown in Fig. 16.15. We have

$$R' = R_{W1} + (N_1/N_2)^2 R_{W2} = 0.06\ \Omega$$

$$L' = L_1 + (N_1/N_2)^2 L_2 = 1.95 \text{ mH}$$

$$R_{LR} = 7.2\ \Omega$$

The voltage v_{LR} across the referred load resistance R_{LR} is found by using phasors. As is customary in power work, we shall use rms voltages and rms phasors. Thus

$$\mathbf{v}'_{LR} = (120) \frac{R_{LR}}{R_{LR} + R' + j\omega L'}$$

$$= 120 \frac{7.2}{7.26 + j(0.735)}$$

$$= 117.8 - j(11.92)$$

The actual power delivered to the load is

$$P_L = \frac{|\mathbf{v}'_{LR}|^2}{R_{LR}} = \frac{(117.8)^2 + (11.92)^2}{7.2}$$

$$= 1947 \text{ W}$$

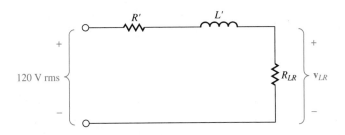

FIGURE 16.15

The approximation that was made—neglecting L_R and R_C—probably did not reduce the accuracy of the calculation much. The effect of the approximation was to underestimate the voltage drop in R_{W1} and L_1. However, the currents through R_C and L_R are small compared with the load currents (as they would be in any well-designed transformer). Thus they do not increase the drop in R_{W1} and L_1 significantly.

A major use of transformers is in electric power distribution systems, where they are used to step voltage up and down in order to reduce transmission losses. Transformers are also used to provide specific voltages for individual pieces of equipment. Another use is to provide *isolation*, that is, to couple power from one circuit to another without allowing an electrical conducting path to exist between them. Transformers are also used in audio systems to transform the apparent impedance of a load, such as a loudspeaker, into a value more suited to the amplifier supplying power to it. This step is known as "impedance matching."

EXAMPLE 16.6

A 120 V (rms) 60 Hz generator produces power that is to be consumed by a load with resistance R_L. Between generator and load is a 100-mile-long transmission line with total resistance (sum of both sides) R_W. To reduce power loss in the line, a step-up transformer (with turns ratio 1:N) is used at the generator and an identical step-down transformer is used at the load, as shown in Fig. 16.16(a). The transformers can be approximated as ideal.

(a) Find the rms current produced by the generator as a function of N.
(b) Find the efficiency of the system, defined as the ratio of the power delivered to R_L to the power produced by the generator.
(c) Let $R_L = 1\ \Omega$ and let the wire resistance be 0.025 Ω per 1,000 feet. What is the efficiency when the line voltage is 120 V rms? When it is 12,000 V rms?

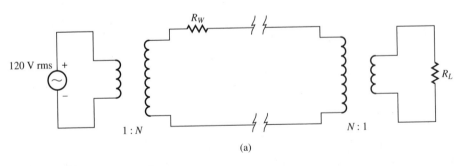

(a)

FIGURE 16.16(a)

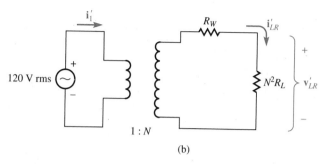

(b)

FIGURE 16.16(b)

SOLUTION

(a) Referring R_L to the primary of the step-down transformer, we have the circuit shown in Fig. 16.16(b). The secondary voltage of the step-up transformer is $120N$, its secondary current is $120N/(R_W + N^2R_L)$, and its primary current is

$$|\mathbf{i}_1'| = \frac{120N^2}{R_W + N^2R_L}$$

(b) The power delivered to the load is

$$P_L = |\mathbf{i}_{LR}'|^2N^2R_L$$

The power produced by the generator is $P_G = 120|\mathbf{i}_1'|$. However, from Eq. (16.17), $\mathbf{i}_{LR}' = \mathbf{i}_1'/N$. Thus

$$E = \frac{P_L}{P_G} = \frac{|\mathbf{i}_{LR}'|^2N^2R_L}{120|\mathbf{i}_1'|} = \frac{|\mathbf{i}_1'|R_L}{120}$$

$$= \frac{N^2R_L}{R_W + N^2R_L}$$

(c) When the line voltage is 120 V, $N = 1$, and we have

$$E_{(120\text{ V})} = \frac{1}{(26.4) + 1} = 3.65 \text{ percent}$$

This would be the result if no transformers were used, but of course 3.65 percent efficiency is unacceptably low. However, if we use 1:100 transformers to obtain a 12,000 V line voltage, we have

$$E_{(12,000\text{ V})} = \frac{10^4}{26.4 + 10^4} = 99.74\%$$

The improvement in efficiency is very dramatic. This example explains the universal use of high-voltage transmission lines whenever power is sent over any substantial distance.

TRANSFORMER RATINGS

Manufacturers of large transformers usually attach a plate to each unit stating the maximum recommended operating conditions. (These are known as *nameplate ratings*.) A typical nameplate rating could read "Transformer, 60 Hz, 20 kVA, 2400/120 V." This means that the transformer has a turns ratio of 20:1. The voltages stated are the maximum that can be used without producing saturation of the core. This limit can be estimated from Eq. (16.2) if the magnetic properties of the core material are known. We recall that if $\Phi(t)$ is a sinusoidal function with frequency f, the maximum value of $d\Phi/dt$ is $2\pi f$ times the maximum value of Φ. Thus if B_{MAX} is the approximate flux density for saturation, we must have

$$v'_{MAX} = \frac{2\pi f}{\sqrt{2}} N B_{MAX} A \qquad \text{(rms volts)} \qquad (16.19)$$

where N is the number of turns in the winding across which v'_{MAX} is measured and A is the cross-sectional area of the core.

The units kVA refer to the product of $|\mathbf{v}'|$ and $|\mathbf{i}'|$ in either winding, *without regard to phase*. This $|\mathbf{v}'|\,|\mathbf{i}'|$ product is known as the *apparent power S*. It does not represent true power, since $\mathbf{v}$ and $\mathbf{i}$ could conceivably be 90° out of phase, in which case the apparent power could be large but the actual power zero. Note the use of the units kVA instead of kW, which lets us know that apparent power and not real power is meant. In our present example 20 kVA is the largest apparent power that can safely occur before the transformer overheats (and is damaged) due to excessive dissipation in R_C, R_{W1}, and R_{W2}. The loss in R_C depends primarily on voltage and should not be excessive if the rated voltage is not exceeded. The losses in R_{W1} and R_{W2} depend on the currents in the windings. Assuming voltages equal to the nameplate voltages, stating the maximum apparent power is the same as stating the safe maximum currents. It would not be correct to rate the transformer in terms of actual power. Suppose a small pure reactance were connected across the secondary. Large currents could flow and damage the transformer, even though the power delivered to the load would be zero.

● EXERCISE 16.3

A load connected to the 1200 V secondary winding of a 60 Hz transformer rated at 5 kVA consists of a resistance R in parallel with a 1.77 H inductor. Let R have the smallest value allowed by the transformer ratings. What is R? How much power is delivered to R? **Answers:** $R = 320\ \Omega$; $P = 4.50\ \text{kW}$.

TYPES OF TRANSFORMERS

Until now we have emphasized *iron-core* transformers. These are used primarily in power and audio applications. The symbol of Fig. 16.17(a) indicates an iron core. At

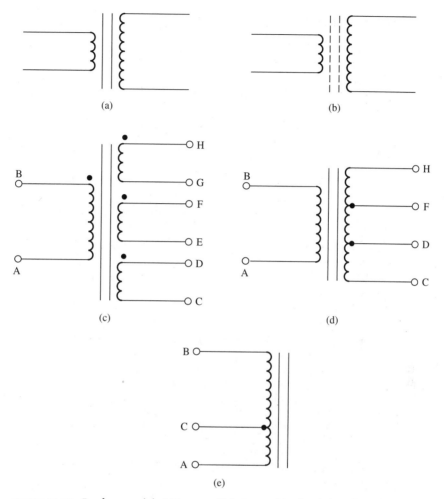

FIGURE 16.17 Transformer symbols. (a) Iron core; (b) ferrite core; (c) with multiple windings; (d) tapped; (e) autotransformer.

frequencies above 20 kHz or so, the hysteresis loss (which is proportional to frequency) and eddy current loss in iron cores become excessive. At higher frequencies, however, cores made of nonconductive materials known as *ferrites* can be used; these are useful to frequencies in the GHz (10^9 Hz) range. Ferrite cores are sometimes represented by the symbol shown in Fig. 16.17(b). Air-core transformers are also common in radio-frequency work (100 kHz and above). These are simply closely coupled pairs of wire coils, either self-supported or wound on forms that have $\mu \cong \mu_0$. The analysis of air-core transformers is more complicated than that of iron-core transformers because the flux lines do not pass neatly through both coils. In fact, the ideal-transformer equations (16.16) and (16.17) do not apply to air-core transformers, even approximately. Of course it is possible to analyze air-core transformers mathematically, but frequently they are analyzed by the traditional method known as "cut and try." Often a movable ferrite core is provided for fine adjustment; such transformers (or inductors) are said to be "slug-tuned." At radio frequencies transformers are used mainly for isolation and impedance matching.

Iron-core transformers often have multiple windings, as illustrated in Fig. 16.17(c). Equation (16.16) readily generalizes to this case; the voltage across the terminals of each winding is simply proportional to the number of turns in the winding. A transformer of this kind, for example, can be used to obtain several different voltages from one connection to the ac power line. Figure 16.17(d) shows a transformer with a *tapped* secondary. This is fundamentally the same as the transformer of Fig. 16.17(c), except that the secondary windings are not isolated from one another. It is also possible to step voltages up or down by tapping a single winding, as shown in Fig. 16.17(e); this device is known as an *autotransformer*. As usual, the voltage across each pair of terminals is proportional to the number of coil windings between them. Often tap C is made adjustable, so that it can be slid along the coil from one end to the other. Then if terminals A and B are connected to the 120 V line, any voltage between 0 and 120 V can be obtained between A and C. This is the familiar commercial "variac," which is often used as an adjustable voltage source.

• EXERCISE 16.4

In Fig. 16.17(c), primary AB is connected to the 120 V power line. The windings are as follows: AB, 1,000 turns; GH, 105 turns; EF and CD, each 2,500 turns. Find the voltages appearing (a) between terminals G and H; (b) between C and F; (c) between C and F if D is connected to E; (d) between C and E if D is connected to F. Assume that the transformer is ideal. **Answers:** (a) 12.6 V; (b) 0; (c) 600 V; (d) 0.

16.3 Power Calculations and Measurements

In calculations involving electric power, voltages and currents are normally stated in terms of their rms value.* We have already introduced an rms phasor $\mathbf{v}'$, related to the usual phasor $\mathbf{v}$ by $\mathbf{v}' = \mathbf{v}/\sqrt{2}$.

Let us consider an arbitrary circuit with two terminals, represented by the box in Fig. 16.18. Let $\mathbf{v}' = V_0 e^{j0} = \mathbf{V}_0$. If the box contains a pure resistance, $v(t)$ and $i(t)$ are in phase, and $\mathbf{i}'$ is also a real number, as shown in Fig. 16.19(a). However, if the box contains a pure capacitor, $\mathbf{i}' = j\omega C\mathbf{v}' = V_0\omega C e^{j90°}$, as shown in Fig. 16.19(b). In

---

*For explanation of the term *rms*, see Section 4.1.

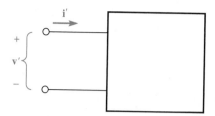

FIGURE 16.18

this case $v(t) = V_0 \cos \omega t$, and $i(t) = V_0 \omega C \cos (\omega t + 90°) = -V_0 \omega C \sin \omega t$. We now have a purely *capacitive load*, in which the current is said to *lead* the voltage by 90°; this means that the maxima of current occur earlier (by a time $\pi/2\omega$, where ω is in radians per second) than the maxima of voltage. The phasor diagram for a pure inductive load is shown in Fig. 16.19(c); here we say the current *lags* the voltage. Usually, however, a circuit will not be a pure resistance or reactance, but some combination. Suppose, for example, the box contains a series connection of a resistance R and inductance L. Then

$$\mathbf{i'} = \frac{V_0}{R + j\omega L} = I_0 e^{j\theta} \tag{16.20}$$

where

$$I_0 = \frac{V_0}{\sqrt{R^2 + (\omega L)^2}} \tag{16.21}$$

$$\theta = -\tan^{-1} \frac{\omega L}{R}$$

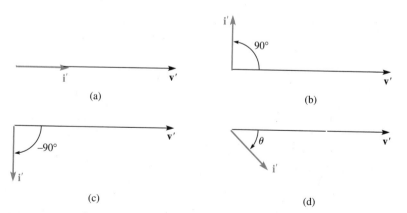

(a)

(b)

(c)

(d)

FIGURE 16.19 Phasor diagrams for (a) resistive circuit, (b) pure capacitive circuit, (c) pure inductive circuit, (d) inductive circuit.

Now the phasor diagram appears as in Fig. 16.19(d). The load is still said to be inductive, but not purely inductive; the current still lags the voltage, but by an angle less than 90°.

The time-averaged power entering the box in Fig. 16.18 is found from Rule 5 of Section 6.4:

$$P(\text{entering}) = \text{Re}\,(\mathbf{v}'\mathbf{i}'^*) \tag{16.22}$$

$$= VI\cos\theta$$

where $V \equiv |\mathbf{v}'|$, $I \equiv |\mathbf{i}'|$, and θ is the angle by which the current leads the voltage. The quantity P is known as the *active power*. The factor $\cos\theta$ is known as the *power factor* (pf):

$$\text{pf} \equiv \cos\theta \tag{16.23}$$

The product VI is known as the *apparent power S:*

$$S \equiv VI \tag{16.24}$$

Thus we have

$$P = S(\text{pf}) \tag{16.25}$$

We note that for *purely* inductive or capacitive loads the power factor vanishes. No time-averaged power can be delivered to purely reactive loads. Currents do flow, however. These currents correspond to energy being stored on one-half of the cycle (charging a capacitor, for example) and then removed on the other half of the cycle.

The current flowing in Fig. 16.18 can be broken up into its in-phase and out-of-phase components. We can define the in-phase rms current according to

$$I_{\text{IP}} = I\cos\theta \tag{16.26}$$

and the out-of-phase rms current by

$$I_{\text{OP}} = I\sin\theta \tag{16.27}$$

where θ is the angle by which the current leads the voltage. From Eqs. (16.22) and (16.23) the active power entering the box is equal to

$$P = VI_{\text{IP}} \tag{16.28}$$

We can also define the *reactive power Q* according to

$$Q \equiv -VI_{\text{OP}} = -VI\sin\theta \tag{16.29}$$

The units of reactive power are usually stated as "vars" (volt-amperes, reactive). Note that Q can be either positive or negative. By convention Q is positive when the current lags the voltage; hence the minus sign in Eq. (16.29). It would actually be correct to state the units of reactive power and apparent power as watts, but vars and kVA, respectively, are customary to avoid confusion. The active power P, reactive power Q, and apparent power S are related according to

$$P^2 + Q^2 = S^2 \tag{16.30}$$

as illustrated in Fig. 16.20.

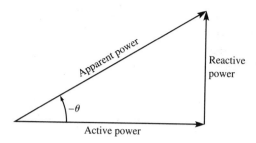

FIGURE 16.20 The "power triangle." In this example θ is a negative number (inductive load, current lagging; reactive power positive.)

● EXERCISE 16.5

The box shown in Fig. 16.18 contains a 240 V (rms) 60 Hz motor that produces 10 horsepower of mechanical power with 90 percent efficiency. (One horse-power = 746 W.) Measurement indicates that the current I is 40 A, lagging. Find the power factor, apparent power, and reactive power. **Answers:** 0.863; 9.60 kVA; 4.84 kvar.

In most cases reactive power is undesirable because it causes excess current to flow. If the power factor is 0.707 (θ = ±45°), the total current is 41.4 percent larger than its useful in-phase component. This enlarged current flows through the power line's resistance R_W, causing the ohmic loss to be $(1.414I_{\rm IP})^2R_W = 2I_{\rm IP}^2R_W$. Thus a power factor of .707 *doubles* the ohmic loss in the power line! In power work, loads are most often inductive because most motors act as inductive loads. Capacitors are often added in parallel with such loads in order to increase power factor and make the over-all system more efficient. This is illustrated in the following example.

EXAMPLE 16.7

A 10 kW, 240 V, 60 Hz load has a 0.92 power factor, lagging. It receives power through a trans-mission line with resistance $R_W = 1.7\ \Omega$. (a) Find the power lost in R_W. (b) A capacitor is con-nected in parallel with the load as shown in Fig. 16.21. What value of C is required to "correct" the power factor (make it equal to 1.0)? Now what is the loss in R_W?

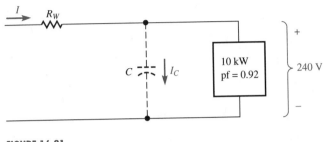

FIGURE 16.21

SOLUTION

(a) From Eqs. (16.25) and (16.24) we find

$$S = \frac{10^4}{0.92} = 10.870 \text{ kVA}$$

$$I = \frac{S}{V} = 45.29 \text{ A}$$

Hence the power lost in R_W is $I^2 R_W = 3.49$ kW.

(b) The reactive power drawn by the load is $Q = -S \sin \theta = S\sqrt{1 - (pf)^2} = 4.26$ kVA. The capacitor is to cancel this out by drawing -4.26 kVA:

$$240 I_C = 4.26 \text{ kVA}$$

(all quantities are rms). However,

$$I_C = (\omega C)(240)$$

Hence

$$C = \frac{4.26 \times 10^3}{(240)^2 (2\pi)(60)} = 196 \text{ μF}$$

When this capacitor has been added, the power factor is unity, $S = P$, and $I = P/V = 41.67$ A. The power lost in R_W is now $I^2 R_W = 2.95$ kW. Thus introducing C reduces ohmic loss in the power line by 15.5 percent.

MEASUREMENTS

Two basic tools of power measurements are the *ac voltmeter* and the *ac ammeter*. These differ internally from dc meters, as will be discussed presently. AC meters usually read directly in rms volts and amperes. However, voltmeters and ammeters by themselves do not provide all necessary information. Multiplying the measured V by I gives S, the apparent power, but to find the active power we must know the power factor. In practice power is often determined by means of a *wattmeter*. This is a three-terminal measuring instrument that observes $v(t)$ and $i(t)$ simultaneously and displays $P =$ time avg $[v(t)i(t)] = VI \cos \theta$. Figure 16.22 illustrates the use of the three ac meters in a circuit. The wattmeter may or may not indicate whether the load is leading or lagging; if it does not, the information can be obtained with an oscilloscope.

• EXERCISE 16.6

The measurements of Fig. 16.22 are made; we find $V = 216$ V, $I = 4.34$ A, and $P = 904$ W. What is the power factor? **Answer:** $= 0.964$.

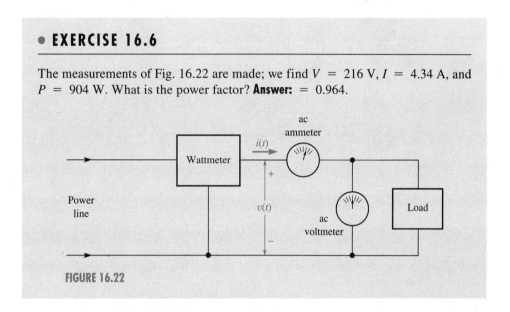

FIGURE 16.22

The internal structures of dc and ac meters are compared in Fig. 16.23. Figure 16.23(a) shows in schematic fashion the structure of a typical dc ammeter. A permanent magnet creates a constant magnetic flux density $\vec{B}$. The current I to be measured flows through a coil wound on a pivoting structure. When current flows, a torque proportional to $I\vec{B}$ is generated by the force of Eq. (16.3). This torque twists the coil form against the resistance of a spring. An indicating pointer (the "needle") accordingly moves through an angle proportional to I. This device is known as a "D'Arsonval movement." Since it is an electromechanical device, more detailed discussion will be deferred to Chapter 17.

If an ac current were to pass through the coil of this dc meter, first positive and then negative torques would be produced in rapid alternation. The mechanical inertia of the needle would prevent it from moving up and down 60 times per second; thus its deflection would correspond to the time-averaged value of $i(t)$, which for an ac current is zero. To measure ac current, however, we can use the structure shown in Fig. 16.23(b). Here the current to be measured also is used to create the magnetic field $b(t) = Ki(t)$, where K is some constant. The resulting torque is then proportional to $i(t) \, b(t) = Ki^2(t)$. Again the needle responds only to the time-average value of the torque, which is proportional to the time average of $i^2(t)$. But the time average of $i^2(t)$ is simply I^2, the rms value of $i(t)$; thus the needle deflection is proportional to I^2.

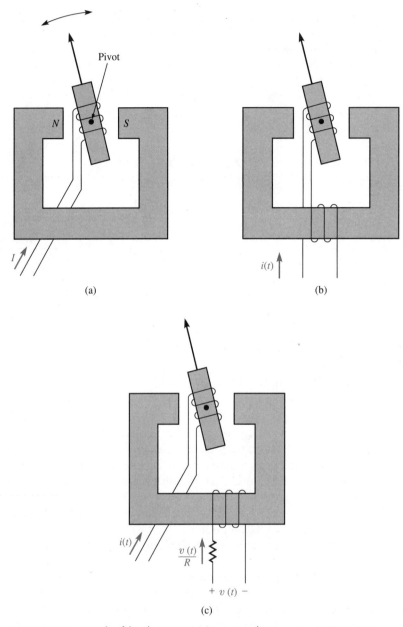

FIGURE 16.23   Principles of dc and ac meters. (a) DC ammeter. (b) AC ammeter. (c) Wattmeter.

One possible structure for a wattmeter is shown in Fig. 16.23(c). Here the current in the moving coil is $i(t)$, while the current producing $b(t)$ is $v(t)/R$ (where R is a large resistance). The time-averaged torque is proportional to time avg $[i(t)b(t)] \propto R^{-1}$ time avg $[i(t)v(t)]$. Thus the meter deflection is proportional to the time-averaged power.

The instruments just described are the "classical" electromechanical measuring instruments. However, these are gradually being replaced by various electronic voltmeters and ammeters, often with digital readouts. Electronic meters are more versatile and more precise than electromechanical meters, and from the user's point of view they are more nearly "ideal." (That is, electronic voltmeters have nearly infinite impedance; electronic ammeters have nearly zero impedance.) Other refinements in measurement technique are constantly being made. For example, ammeters are now available that measure current by induction. A measuring ring is simply placed around the current-carrying wire, and the current in the wire is determined from the voltage induced in the measuring ring, which acts like the secondary of a transformer. This instrument allows us to measure the current in a wire without disconnecting it to insert an ammeter, thus making the measurement much more convenient.

16.4 AC Distribution Systems

Electric power is usually generated in large amounts at a power station and consumed in smaller amounts at many distant places. The network used to bring power from the generators to the consumers is known as an *electrical distribution system*. A typical distribution system is illustrated in Fig. 16.24. Power is generated at comparatively low voltages. However, it is immediately stepped up to perhaps 100 to 750 kV and then transmitted over high-voltage lines. This stepping up is done to reduce resistive losses, as was seen in Example 16.6. At a substation perhaps 40 km from the users, voltage is stepped down to 35 to 69 kV, and the power is shared among numerous medium-voltage lines. These in turn are stepped down perhaps 4 km from the users, where they branch into 4 to 15 kV local distribution lines. The local lines are finally stepped down to 120 V house current by numerous small transformers. These are usually located on the ubiquitous wooden power poles a few hundred feet or less from the houses they serve. As we see, the guiding principle is to carry to power over most of the distance between generator and user at very high voltage. However, the technical problems associated with high-voltage lines make it awkward to use large numbers of them for distribution; thus progressively lower-voltage lines are used as the system branches out and approaches the users. The losses of the low-voltage distribution lines are tolerable because the distances involved are relatively short.

A real power system is considerably more sophisticated than that outlined in Fig. 16.24. Various control mechanisms are required to regulate voltage and frequency, maintain the generators in proper phase, and share loads within the system. Circuit breakers are used throughout. These are automatic switches that open—in order to protect the equipment—whenever a preset current limit is exceeded, as may happen, for example, if a short circuit should occur. One of the challenging objectives in this field is stabilization of a large, complex power system against disturbances that can "propa-

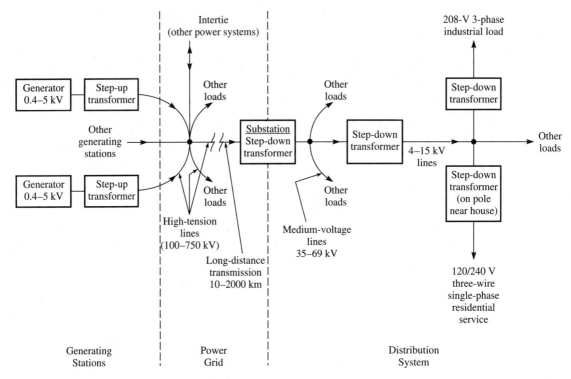

FIGURE 16.24   Block outline of an electrical distribution system.

gate" through the system. For instance, a short circuit in a local distribution line can conceivably cause a voltage surge that trips distant circuit breakers, which in turn causes a further surge, which trips still more distant breakers. The occasional occurrence of multistate blackouts suggests that all such problems have not yet been solved!

The simplest type of circuit used in distribution is the *single-phase circuit*, illustrated in Fig. 16.25. There can be any number of transformers in the circuit and any number of voltages. Its distinguishing feature, however, is that there are instants of time at which all voltages in the circuit are simultaneously zero. One quarter period later, the absolute values of all voltages will be maximum, and so on. This will be the situation for any circuit that originates in a single sinusoidal voltage source and propagates through ideal transformers, assuming that there are no significant inductances or capacitances in the transmission lines. Most residences and small buildings are supplied with power by means of *single-phase*, *three-wire service*, as shown in Fig. 16.26. Power is transmitted at a moderately high voltage to a distribution transformer located on a power pole, or underground, near the user. This transformer has a 220 V secondary with center tap. The center terminal is *literally* grounded; that is, unlike the convention in electronic work, where "ground" simply refers to a conventional zero of potential, this terminal is actually connected to a metal rod embedded in the earth. The three wires from the transformer secondary are then run into the building. Individual circuits, each designed for 110 V loads up to 20 to 40 A, are connected between

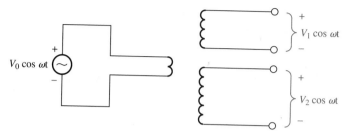

FIGURE 16.25 Single-phase system:

the ground wire and either "hot" wire. Each of these circuits is protected by its own fuse or circuit breaker, and each supplies lighting circuits or convenience outlets. The system designer tries to balance, at least approximately, the loads on the two "hot" wires. This reduces power loss due to ohmic heating of the ground wire; in fact, when loads on the two sides are exactly balanced, no current flows in the ground wire at all. (Current from one hot wire flows into it at the same instant that current from the other flows outward, and the two cancel.) If the system is designed to keep the current in the ground wire low, a smaller wire can be used, with cost savings. Moreover, 220 V power is also available between the two "hot" wires. In residences 220 V power is used for appliances that consume large amounts of power, such as heaters and ovens.

Safety considerations are of great importance in the wiring of buildings, the principal hazards being fire and electric shock. Fire hazards are reduced through the use of rigidly enforced wiring quality standards, usually imposed by local ordinances. Electric shock hazards are made more serious by the fact that human bodies are very often connected to earth ground. A person is thoroughly grounded when sitting in a bathtub, since there is a low-resistance path to the water pipes. Even a person standing barefoot

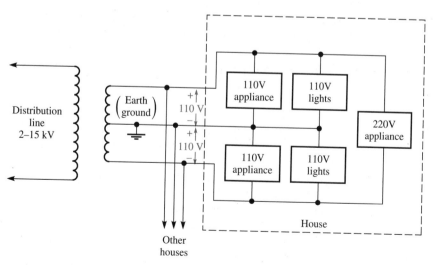

FIGURE 16.26 Single-phase, three-wire house wiring circuit.

on a cement floor is connected to ground sufficiently to be in danger of shock. The fact that a shock hazard exists through a circuit with considerable resistance (such as the contact between foot and floor) does not make the hazard less serious, because currents in the range of 0.01 to 0.1 A can be lethal. Thus with 120 V wiring even a shock through a 10,000 Ω resistance can be fatal.

The most common kind of fault in a wiring system occurs when one of the two "hot" wires, perhaps by brushing against a pipe or electric conduit, becomes connected to ground. If there were no earth connection at the transformer center tap and one of the hot wires became connected to ground, it would go unnoticed, but the other hot wire would be 220 V above ground. This would threaten insulation failure, and any grounded person touching that wire would get a 220 V shock. This is why the transformer center tap is earth grounded: A short of one of the hot wires to ground blows the fuse or trips the circuit breaker, providing warning that there is an electrical fault. Shock hazards can be reduced further by using three-wire plugs and cords for hand tools and other metal appliances. The third wire is connected to the outside metal case which the user touches, and the third contact at the plug is connected to a water pipe or other earth ground. Now if the hot wire should touch the metal case inside the appliance, the grounded user is protected and the house circuit breaker trips, providing warning.

An interesting modern safety device is the *ground-fault interrupter*, illustrated in Fig. 16.27. This is in essence a sensitive circuit breaker that opens when $i_1 + i_2 + i_3 \neq 0$. In the absence of any faults, the sum $i_1 + i_2 + i_3$ evidently does equal zero, from Kirchhoff's law. Suppose, however, a barefooted person touches one of the hot wires. A potentially dangerous 120 V drop exists across the person's body, causing an accidental current to earth ground. This current returns to the transformer through the earth, rather than through the center wire; then $i_1 + i_2 + i_3 \neq 0$, and the interrupter opens the circuit. Unlike ordinary household circuit breakers, which trip on overloads of many amperes, the ground-fault interrupter is sensitive and opens on an imbalance of only a few milliamperes. Thus, unlike ordinary circuit breakers, it provides good protection from electrical shock. Many building codes now require the use of ground-fault interrupters in hazardous locations such as bathrooms and swimming pools. Interrupters for single-phase two-wire circuits, such as individual circuits in homes, are common. These work in similar fashion, opening the circuit when current returns

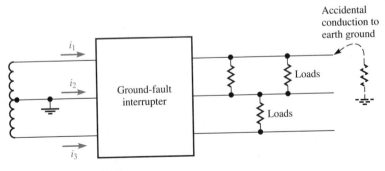

FIGURE 16.27   Ground-fault interrupter.

through earth ground—that is, when the sum of currents in the "hot" and ground wires fails to equal zero. Two-wire interrupters are so compact that they can be substituted for ordinary electric outlets.

The design of wiring systems for residences and small buildings is often carried out by architects or professional electricians. Local building codes play an important role; many details and standards are prescribed by local laws intended to ensure safety. In larger buildings more complex considerations affecting cost, efficiency, and reliability may come into play, and the services of consulting electrical and lighting engineers are often required.

THREE-PHASE POWER

Note that with a single-phase power line the voltage pulsates, with zeros occurring twice in every cycle. If the line is connected to a motor, one might expect jerky operation, with torque disappearing every 1/120 of a second. There is a different kind of power connection, the three-phase connection, which removes this shortcoming, and which for several other reasons is more satisfactory for use with large motors and generators. In fact, three-phase connections are almost always used wherever large amounts of electric power are being handled.

In a three-phase system, three sinusoidal voltages, all with the same amplitude and frequency, are used. The phases of the three voltages differ by 120°; that is, the three voltages are represented by the phasors V_0, $V_0 e^{j120°}$, and $V_0 e^{j240°}$. In principle the three voltages can be produced by three ideal voltage sources, as shown in Fig. 16.28(a). We note that the sum of the three voltages is zero at all instants of time, as shown by the

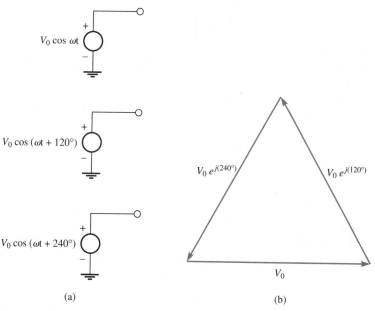

(a)   (b)

FIGURE 16.28

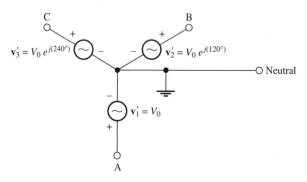

FIGURE 16.29   Wye-connected three-phase source.

phasor diagram in Fig. 16.28(b). Three-phase power cannot be obtained from single-phase power simply by using a transformer with three secondaries, since in that case all the secondaries would have phases of either 0° or 180° rather than the 120° and 240° required for the three-phase connection. Instead, three-phase power is usually generated as three-phase right at its source by means of a triple generator with three windings spaced at 120° intervals on the same shaft. The three voltages are often described in the notation V_0, $V_0 \angle 120°$, and $V_0 \angle -120°$.

The three voltage sources are often connected in the manner of Fig. 16.29. From its shape this figure is referred to as the "Y" (or *wye*) or *star connection*. In one type of distribution V_0 is nominally 120 V rms. Equipment requiring 120 V single-phase power (such as small appliances or lights) can be connected between ground and either point A, point B, or point C. Furthermore, 208 V single-phase power is obtainable by connecting a load between A and B, B and C, or A and C. To see this, we refer to the phasor diagram shown in Fig. 16.30. The phasor $\mathbf{v}_A - \mathbf{v}_B = \mathbf{v}_1' + (-\mathbf{v}_2')$ is obtained by adding the phasors $\mathbf{v}_1'$ and $(-\mathbf{v}_2')$ head-to-tail as shown. The lengths of the phasors $\mathbf{v}_1'$ and $\mathbf{v}_2'$ are each V_0. From the law of cosines, the length of the phasor $\mathbf{v}_A - \mathbf{v}_B$ is $-2V_0^2 - 2V_0^2 \cos 120° = \sqrt{3} V_0$. Thus if the voltage between point A and ground is 120 V, the voltage between points A and B is $120\sqrt{3}$, or 208 V. (If *only* 208 V power is required, the fourth wire, called the "neutral," can be omitted, with cost savings.) Most equipment designed for 220 V operation can be operated on any voltage within ± 10 percent of 220; thus the system can supply single-phase power to both 120 V and 220 V equipment.

The voltages of the three sources in Fig. 16.29 are known as the *phase voltages*. We say that they are in the *phase sequence* ACB, from the order in which the maxima occur. The voltages v_{AB}', v_{BC}', and v_{CA}' are known as the *line voltages*. The voltage of a three-phase generator is usually stated in terms of its rms line voltage. The three line voltages are, like the phase voltages, a *three-phase set*; that is, they are equal in magnitude and are equally spaced at 120° phase intervals. As is seen from Fig. 16.30, there is a 30° phase difference between the line and phase voltages. That is, the line voltage v_{AB} lags the phase voltage v_A by 30°.

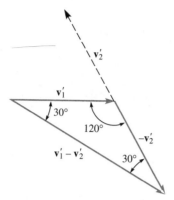

FIGURE 16.30

The three windings of a generator can also be connected in the *delta* connection shown in Fig. 16.31. This connection does not violate Kirchhoff's voltage law because, as we have seen, the sum of the three components of a three-phase set is zero [Fig. 16.28(b)]. For the delta connection the line voltages are simply the voltages of the three sources.

In three-phase technology, loads also are usually found in wye or delta form. Figure 16.32(a) shows a wye-connected load, and Fig. 16.32(b) shows a delta-connected load. When the three impedances of a load are identical, the load is said to be *symmetrical*, or *balanced*. It is usual to make loads as nearly symmetrical as possible, and we shall concern ourselves mainly with this case. (Asymmetric three-phase circuits can of course be analyzed, but the calculations are messy and not very revealing.) When the symmetrical delta load of Fig. 16.32(b) is used, the current through each impedance Z_2 is simply the line voltage divided by Z_2. For a symmetrical wye load the voltages add

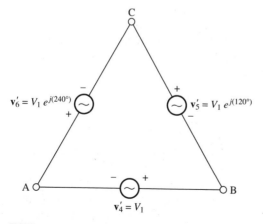

FIGURE 16.31   Delta-connected three-phase source.

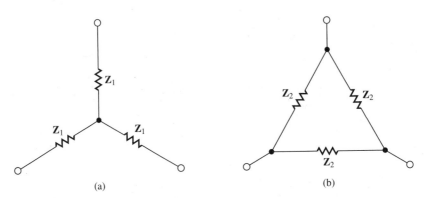

FIGURE 16.32   Symmetrical three-phase loads: (a) wye; (b) delta.

just as in the case of the wye source (Figs. 16.29 and 16.30). Thus the voltage across each impedance is equal to the line voltage divided by $\sqrt{3}$. With a symmetrical wye load no current flows in the neutral wire, and it can be omitted.

EXAMPLE 16.8

Show that the wye connection of voltage sources shown in Fig. 16.29 is equivalent to the delta connection of Fig. 16.31 provided that the voltages v_4', v_5', and v_6' are chosen correctly. Evaluate the phasors $\mathbf{v}_4'$, $\mathbf{v}_5'$, and $\mathbf{v}_6'$.

SOLUTION

We wish the voltages at points A, B, and C to be identical for the wye and delta connections. Thus

$$\mathbf{v}_B' - \mathbf{v}_A' = \mathbf{v}_4' = \mathbf{v}_2' - \mathbf{v}_1'$$
$$\mathbf{v}_C' - \mathbf{v}_B' = \mathbf{v}_5' = \mathbf{v}_3' - \mathbf{v}_2'$$
$$\mathbf{v}_A' - \mathbf{v}_C' = \mathbf{v}_6' = \mathbf{v}_1' - \mathbf{v}_3'$$

The phasors for $\mathbf{v}_1'$, $\mathbf{v}_2'$, and $\mathbf{v}_3'$ are respectively V_0, $V_0(-\frac{1}{2} + j\sqrt{3}/2)$, and $V_0(-\frac{1}{2} - j\sqrt{3}/2)$. Subtracting, we obtain

$$\mathbf{v}_4' = V_0(-3 + j\sqrt{3})/2 = V_0\sqrt{3}\, e^{j150°}$$
$$\mathbf{v}_5' = -jV_0\sqrt{3} = V_0\sqrt{3}\, e^{j270°}$$
$$\mathbf{v}_6' = V_0(3 + j\sqrt{3})/2 = V_0\sqrt{3}\, e^{j30°}$$

Thus the amplitudes of the voltage sources in the equivalent delta connection must be $\sqrt{3}$ times larger than the voltages in the wye connection.

EXAMPLE 16.9

Find the rms current through each of the resistances R in the symmetrical delta load of Fig. 16.33. The rms phase voltages of the wye-connected source are V_0.

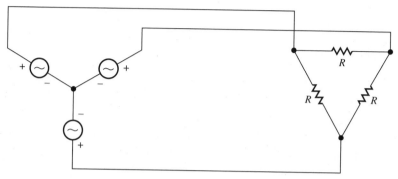

FIGURE 16.33

SOLUTION

We can replace the wye-connected source with the equivalent delta source found in the preceding example. It is then evident that the voltage across each resistor is $\sqrt{3}\,V_0$, and the current in each is $\sqrt{3}\,V_0/R$.

As mentioned earlier, one advantage of three-phase power over single-phase is reduction of fluctuations in the instantaneous power. When a resistive load is connected across a 60 Hz single-phase source, the instantaneous power varies between zero and maximum 120 times per second. On the other hand, it can be shown that with a symmetrical three-phase load, as in Example 16.9, the instantaneous power is constant; it does not vary at all. Thus a three-phase motor does not exhibit rapid torque fluctuations as a single-phase motor is likely to do. There are also other advantages. Three-phase power is better suited than single-phase for production of rotating magnetic fields, which are important in ac motors (as will be seen in the next chapter). In general, three-phase systems are considered to be more economical, requiring less iron and less copper than corresponding single-phase systems.

16.5 Application: The Consulting Engineer

As an engineer, you have the option of working for a small or large company . . . or you can "hang out your shingle," as they say, and open your own office as a consulting engineer. Consultants are often needed in the electric power field, designing, for example, lighting systems for large buildings; and that is the only reason for this Application to be presented here, rather than in some other chapter.

Many of us feel that the idea of opening one's own office is appealing. And why not? One can be one's own boss, and come and go as one pleases, meanwhile being paid at high hourly consulting rates, just as doctors and lawyers are. Sounds pretty good! But in fact, only a few engineers actually take this route, because it requires special aptitudes and talents.

The main key to success as an engineering consultant is possession of *unique experience* or *unique knowledge*. Big companies hire consultants to solve special problems that they cannot solve themselves. Their in-house engineers can probably handle the usual problems but must turn to outside consultants in unusual cases, for which they lack the expertise. Thus it helps for a consultant to have very deep knowledge of some special small field, one that is useful fairly often, but not so often that companies will have their own specialists on the staff.

Fortunately, all branches of engineering offer plenty of such niches that consultants can fill. But first one must acquire the specialized knowledge or experience, and this is best gained through previous work. Thus it is unlikely that one would become a consulting engineer directly upon leaving school; the usual path is to first work in the field and gain experience. In some fields you might even be an apprentice, working with a master without much pay, for the privilege of learning the art. But you can also gain experience through ordinary engineering employment, or master some specialty through doctoral research. The point is that somehow you must become a real expert in something that clients will need. And, of course, once established, you must constantly be learning the latest skills, to make sure the field doesn't pass you by.

Then, of course, you also need customers. Even if you have great knowledge of your specialty, you still have to find clients who need your skills, and who will hire you, and not the fellow down the street. This is done by . . . *networking*! As an independent businessperson, a consulting engineer needs a network of contacts and friends. Thus certain personal talents are needed: One must join one's alumni association, give one's time to professional groups, and in general enjoy working with people and selling oneself to them. Anyone who says that engineering is not a "people-oriented" profession cannot be thinking of this branch of the field!

It has been said that "working for yourself means you have the worst boss in the world." The independent consultant knows that there can always be a dry period, when no work comes in. Since you don't know when jobs will come in, the usual response is to accept any work that is offered; then it piles up, leaving you to work long hours, late into the night. And when your client has to meet an emergency deadline, you have an emergency too. For reasons such as this, consultants usually find themselves working longer and harder than other engineers. Furthermore, when you are your own boss, all the supporting activities also have to be done by you. Financial work and marketing turn out to be as important to success as the engineering work itself. Since so many skills are needed, you have to decide if you have them all. If not, you have to take on partners or employees until the necessary bases are covered.

Taking on employees might make life easier in some respects, but new problems are created. One then adds "personnel manager" to the necessary skills, and there are extra tax matters and legal concerns. Moreover, it then becomes even more vital to keep the work coming in, as your employees will need to be paid whether there is work in the shop or not. When business is slow, the financial and emotional pressures can be high. Competition between consulting firms can be tough, and not as polite as one might wish. It helps to be resourceful, aggressive, determined, and not easily pushed around!

One more thing you must have to become a consulting engineer: a license! In most branches of engineering, those who wish to serve the public have to be licensed by an agency of their state, just as lawyers and doctors do. To be licensed, it is usually necessary to pass an examination dealing with the basics of your own field and related fields. (There will be some electrical questions, too!) Not all engineers have licenses; often you can work for a university or a big company without having one. But obtaining a license is always a good idea. Having a license increases the range of things you can do. Without one, a big opportunity might someday pass you by.

Now, do you want to be a consulting engineer? The list of needed skills and personal qualities is long, but some engineers do manage to succeed at it, and a few become rich and even famous. An unusual amount of personal satisfaction probably comes with this success!

POINTS TO REMEMBER

- The unit of *magnetic flux density B* is the tesla. The *magnetic flux* Φ is the product of B and cross-sectional area; it is measured in webers. The *magnetic intensity H* is measured in ampere-turns per meter. In most materials, H is related to B through the formula $B = \mu H$, where μ is the *permeability*. Most materials have values of μ close to μ_0, the permeability of vacuum. Ferromagnetic materials, however, have much larger effective permeabilities. This gives them the ability to channel magnetic flux almost as a pipe channels water.

- In ferromagnetic materials the relationship between B and H is nonlinear. B tends to reach a maximum value as H increases, an effect known as *saturation*. These materials also exhibit *hysteresis*, an effect in which B depends on the value of H at earlier times. Hysteresis is a loss mechanism, converting magnetic energy into heat.

- The product of the current through a coil winding and the number of turns in it is called the *magnetomotive force*. The flux generated is equal to the magnetomotive force divided by the *reluctance* of the magnetic circuit. Magnetic circuits are analogous to dc electric circuits.

- For an ideal transformer the voltage across each winding is proportional to the number of turns. For an ideal transformer with two windings the current in each winding is inversely proportional to the number of turns.

- A load connected to the secondary of an ideal transformer can be "referred" to the primary, in which case its value is multiplied by the *square* of the turns ratio.

- A sinusoidal current is said to *lag* or *lead* the voltage in a circuit, depending on whether it reaches maximum after or before the voltage does. In an inductive circuit the current lags; in a capacitive circuit the current leads.

- The product $S = VI$ is the *apparent power*. Active power (which actually does work) is $P = VI \cos \theta$, where θ is the angle by which the current leads; $\cos \theta$ is the *power factor*. The quantity $Q = -VI \sin \theta$ is known as the *reactive power*. S is measured in volt-amperes (VA), and Q is measured in vars. Reactive power is associated with undesirable out-of-phase currents that contribute to losses. It is often possible to cancel out reactive power by means of capacitors.

- Voltage and current in ac circuits are measured with ac voltmeters and ac ammeters. Power factor is determined with the help of a *wattmeter*.

- An *electrical distribution system* is used to convey energy from generators to consumers. For long-distance transmission, high-voltage lines are used to reduce ohmic loss.

- *Single-phase circuits* have the property that instants of time exist at which all voltages are zero. House wiring usually involves a single-phase, three-wire system.

- Three-phase power is used in high-power applications. Three voltages, equal in magnitude and 120° apart in phase, are supplied. Three-phase generators can be connected either in *wye* or in *delta* configuration; the same is true for three-phase loads.

PROBLEMS

Section 16.1

16.1 A coil of 100 turns of wire with area 10 cm^2 is located in a uniform magnetic field whose flux density is 0.8 T. The magnetic field points in the z direction, and the coil rotates at 2,000 rpm around its diameter, which points in the x direction. Find the time-varying voltage induced in the coil.

16.2 A part of a wire 12 cm long is located in a y-directed magnetic field of 1.1 T. The remainder of the wire experiences no field. A current of 40 A passes through the wire in the x direction. Find the magnitude and direction of the force on the wire. How many pounds of material can be lifted by this force? (The gravitational force acting on a 1 lb mass is 4.44 N.)

16.3 Estimate μ/μ_0 for cast iron, using the data of Fig. 16.3. Assume $H < 400$ ampere-turns per meter.

16.4 A loop of wire lies in the horizontal plane with current flow clockwise as seen from above. What is the direction of B at the center of the loop?

16.5 Two parallel wires 1 cm apart carry currents of 10 A each in the same direction. What is the force between the wires? Do they repel or attract?

16.6 The energy dissipated per cycle due to hysteresis in iron is 80 J per cubic meter, when the maximum flux density is 0.4 T. An iron transformer core with volume

100 cm^3 is used at this flux density at 60 Hz. Find the time-averaged power loss due to hysteresis.

*16.7 The transformer of the previous problem is represented by the model of Fig. 16.12(b). The rms voltage across winding 1 is 120 V. Assume the entire core loss arises from hysteresis. Estimate the value of R_C.

16.8 The energy per unit volume that must be supplied to increase the flux density from B to $B + dB$ is $H \cdot dB$. Prove that the energy lost per cycle per unit volume is equal to the shaded area in Fig. 16.4(b).

16.9 Using the information given in Problem 16.8, show that if μ is a constant, the magnetic energy stored per unit volume is $1/2|B|^2/\mu$.

16.10 Use the result of Problem 16.9 plus the fact that the energy stored in an inductor is $1/2LI^2$ to estimate L_R in Fig. 16.12(b). Assume that winding 1 has 1,000 turns, and let the reluctance of the core be 64,000 SI units.

**16.11 An iron electromagnet has tapered pole pieces to compress the field, as shown in Fig. 16.34(a). The core has a circular cross-section everywhere, and the dimensions of each pole piece are as shown in Fig. 16.34(b). Find the reluctance of each pole piece.

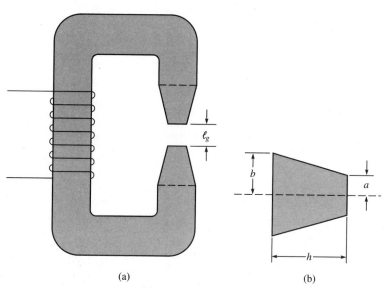

(a)　　　　　　　　　　　(b)

FIGURE 16.34

**16.12 In Fig. 16.34(a) let the length of the flux path through the untapered part of the core be l_m, the height of the gap l_g, and the reluctances of the tapered pole pieces as found in Problem 16.11.

a. Find B in the gap. (Assume that the field lines go straight across the gap.)

b. What is B when $a = b$?

c. Find the limit of B as $a \to 0$.

d. If the purpose of the pole pieces is to maximize B in the gap, for what sort of gaps are they most likely to be useful?

16.13 The structure in Fig. 16.10 has two gaps, side by side. According to Example 16.3, the flux in each gap is approximately the same as if the other were not there. Thus one might expect that any large number of parallel gaps could be added, and the sum of the fluxes in all of them would approach infinity. Explain what would prevent the total flux from becoming infinitely large.

Section 16.2

*16.14 Estimate the inductance of the inductor in Fig. 16.7 if $a = 3$ cm, $b = 6$ cm, $c = 1.5$ cm, $N = 2,000$ turns, and $\mu/\mu_0 = 5,000$. The four sides of the square core are identical.

*16.15 For the inductor of Problem 16.14 estimate the largest 60 Hz voltage that can be applied without bringing the core into saturation. The saturation flux density of the iron is 1.2 T.

**16.16 In aircraft power systems 400 Hz ac is often used. One reason is that 400 Hz iron cores can be made smaller, thus saving weight. Suppose that the shape of a transformer is kept the same but all of its linear dimensions can simultaneously be scaled larger or smaller. Suppose also that the minimum size of the transformer is determined by saturation of the iron. How many times heavier is a 60 Hz transformer than a 400 Hz transformer? The number of turns and voltages are kept constant. Neglect the weight of the copper wires.

16.17 An iron-core transformer has three windings, with N_1, N_2, and N_3 turns, respectively. Construct ideal-transformer equations analogous to Eqs. (16.16) and (16.17). Assume that the core permeability is infinitely large.

16.18 The primary of an ideal iron-core transformer with a 10:1 (primary:secondary) turns ratio is connected to an ideal sinusoidal voltage source whose rms phasor is a real number V_0. The secondary is connected to a load impedance of $3 + 2j$ Ω.

a. Find the rms phasor for the current flowing in the primary.

b. Find the maximum instantaneous current flowing in the primary.

*16.19 The purpose of this problem is to illustrate the use of transformers for impedance matching, a technique for delivering the greatest amount of power from a given source to a given load. Suppose that an ac signal source can be represented as a Thévenin equivalent with $V_T = 10$ V rms and $R_T = 1,000$ Ω. The load to be driven is a 10 Ω resistance.

a. Calculate the time-average power delivered to the load when it is connected directly to the source.

b. Calculate the time-average power delivered to the load when a 10:1 ideal transformer is inserted between source and load.

*16.20 In Fig. 16.35 the transformer is ideal, with turns ratio (primary:secondary) of N:1.

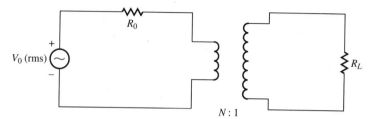

FIGURE 16.35

a. Obtain an expression for the power dissipated in R_L as a function of N.
b. Find the value of N that maximizes the power dissipated in R_L.

*16.21 A 60 Hz step-up transformer is represented by the model of Fig. 16.14(e). The primary is connected to 120 V rms, and the turns ratio is 1:4. Let $R' = 0.06\ \Omega$ and $L' = 0.7$ mH. A 300 Ω resistive load is connected to the secondary.
a. Find the approximate voltage across the load resistance.
b. The drop in secondary voltage from no load to full load, expressed as a percentage of rated full-load voltage, is defined as the *voltage regulation* of a transformer. Assume that the 300 Ω resistance is a full load and find the regulation of the transformer in this problem.

16.22 A transformer used to step 440 V down to 220 V is rated at 10 kVA. What is the smallest pure resistance that can be connected across the secondary?

**16.23 The purpose of this problem is to see what happens if the saturation flux density is exceeded in an iron-core inductor such as that of Fig. 16.7. Let the magnetization curve of the core material be represented by the piecewise linear approximation of Fig. 16.36. Suppose that the 5,000-turn inductor is excited by an ideal 60 Hz sinusoidal current source with rms value 0.283 A. The maximum instantaneous mmf is 1,000 A-t/m, and the cross-sectional area of the core is 4 cm².
a. Sketch the voltage that appears across the winding as a function of time.
b. What will happen if an ideal 750 V rms 60 Hz voltage source is connected to the inductor?

*16.24 In deriving Eq. (16.17) it was assumed that the quantity

$$\frac{l\mathbf{v}_1}{2\pi jfN_1\mu A}$$

(known as the "exciting mmf" because it is the mmf necessary to excite the core) is negligibly small. Verify that this is a good assumption for the case of a 60 Hz 1:1 transformer with 1,000 turns per winding and a flux path length $l = 0.2$ m. Assume that the voltage is less than the maximum value permitted by core saturation, that the magnetization curve is that of Fig. 16.36, and that the currents in the windings are greater than 0.5 A.

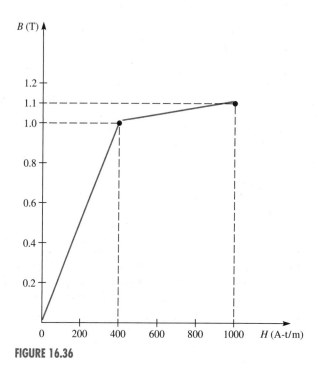

FIGURE 16.36

*16.25 The windings of a symmetrical 1:1 ideal transformer are connected to create a 1:2 autotransformer, as shown in Fig. 16.37. The load resistance R_L is 100 Ω. Construct a circuit diagram showing the resulting autotransformer, using the symbol of Fig. 16.17(e). Find the currents flowing through terminals A, B, C, and D.

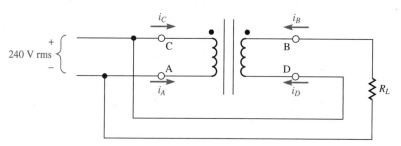

FIGURE 16.37

**16.26 Suppose that a 1:1 transformer with identical windings is connected as a step-up autotransformer, as shown in Fig. 16.37. When it is used as an ordinary transformer, the rating is 10 kVA. This rating is based on ohmic losses in the windings (core losses are negligible). The primary voltage is held constant. Using the results of Problem 16.25, find its rating when it is used as an autotransformer. (Note that this rating is *not* 10 kVA!)

Section 16.3

16.27 In Fig. 16.18 v' is 120 V, 60 Hz, and the box contains a 50 Ω resistor in series with a 50 mH inductor. Find the following:
  a. The apparent power.
  b. The active power.
  c. The reactive power.
  d. The power factor.

16.28 What value of capacitance should be placed in parallel with the circuit of Problem 16.27 to correct the power factor?

*16.29 The 1,200 V secondary of a transformer rated at 10 kVA is connected to a load consisting of an inductance in series with a resistance R. The power factor is pf. Let R always have the smallest value consistent with the transformer rating.
  a. Find R as a function of pf.
  b. Find the power dissipated in R as a function of pf.

  (The value of the inductance should not appear in the answers.)

**16.30 Repeat Problem 16.29 with L and R connected in parallel.

**16.31 The wattmeter of Fig. 16.23(c) is connected as shown in Fig. 16.38. The current and voltage coils, represented by resistance symbols in the figure, have resistances of R_C and R_V, respectively; R_C is small and R_V is large. The presence of these resistances will introduce error in power measurements.

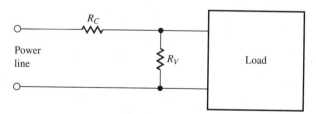

FIGURE 16.38

  a. Assume that R_C, R_V and the power line voltage are known and determine whether the meter reading overstates or understates the power delivered to the load.
  b. What correction should be added to (or subtracted from) the meter reading to find the load power?
  c. Show that the necessary correction can be found, approximately, by noting the wattmeter reading when the load is disconnected.

**16.32 A varmeter (instrument to measure reactive power) is made by replacing R in Fig. 16.23(c) with an inductor L.

a. Assuming that the value of R was such as to make the meter read accurately in watts, what value of L should be used to make it read vars correctly on the same scale?

b. Suppose that the motion of the varmeter needle is in the direction that was positive when the instrument was a wattmeter. Is the current leading or lagging?

Section 16.4

16.33 Explain why a ground-fault interrupter provides better shock protection than a conventional fuse or circuit breaker.

*16.34 A 120 V rms single-phase circuit drives a load that takes 1800 W with a power factor of 0.73, lagging.

a. What is the minimum kVA rating of a transformer used to drive this circuit?

b. Find the line current. Express your answer in the form $I \angle \theta$, where the voltage is taken as the zero-phase reference.

16.35 Suppose that in a design for a single-phase electrical transmission system the voltage drop in the resistance of the line must amount to no more than 5 percent of rated line voltage at rated line current. Table 16.1 gives the characteristics of standard American wire gage (Awg) copper wire at 20°C.

TABLE 16.1

Gage No.	Diameter (mm)	mΩ/m	Weight (g/m)
0	8.252	0.3224	475.4
1	7.348	0.4066	377.0
2	6.544	0.5127	299.0
3	5.827	0.6465	237.1
4	5.189	0.8152	188.0
5	4.621	1.028	149.1
6	4.115	1.296	118.2
7	3.665	1.634	93.78
8	3.264	2.061	74.37
9	2.906	2.599	58.98
10	2.588	3.277	46.77
11	2.305	4.132	37.09
12	2.053	5.211	24.42
13	1.828	6.571	23.33
14	1.628	8.285	18.50
15	1.450	10.45	14.67
16	1.291	13.17	11.63
17	1.150	16.61	9.226
18	1.024	20.95	7.317
19	0.9116	26.42	5.803
20	0.8118	33.31	4.602

Determine the necessary copper wire size (by gage number) for the following cases:

   a. 120 V house-wiring circuit, length 30 m, rated current 20 A.

   b. 500 kV transmission system, length 200 km, rated current 10 A.

16.36 A three-phase heater is to be powered from a transmission system with 440 V line voltage; 25 kW of heat are required. Design suitable symmetric loads using the following:

   a. Wye-connected pure resistances.

   b. Delta-connected pure resistances.

16.37 A wye-connected three-phase generator is connected to a symmetrical wye-connected load. Show that no current flows in the neutral wire.

***16.38** A three-phase generator with 240 V line voltage drives the symmetrical wye load shown in Fig. 16.32(a). When an individual impedance Z_1 is driven from a 240 V single-phase line, it consumes 480 W, with power factor 0.8, lagging. Find the rms current in each of the three wires from the generator.

****16.39** An ideal three-phase generator drives the symmetrical load shown in Fig. 16.32(b). Prove that the total instantaneous power dissipated in the three resistors does not vary in time.

****16.40** An ideal wye-connected generator is connected to a wye load as shown in Fig. 16.39. The voltages $\mathbf{v}_1'$, $\mathbf{v}_2'$, and $\mathbf{v}_3'$ are as given in Fig. 16.29. The load becomes slightly unbalanced when one of the impedances increases by a small amount $\delta\mathbf{Z}$, as shown. Show that the resulting current in the neutral wire, $\mathbf{i}_N$, is given approximately by

$$\mathbf{i}_N' \cong \mathbf{i}_0'\left(\frac{\delta\mathbf{Z}}{\mathbf{Z}_1}\right)$$

where $\mathbf{i}_0'$ was the value of $\mathbf{i}_3'$ when the load was balanced.

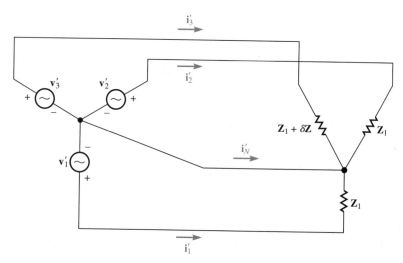

FIGURE 16.39

*16.41 Figure 16.40 shows a three-phase transformer. All windings have the same number of turns. The three primaries (ab, ef, jk) are delta connected and driven by a power line with rms line voltage V_0. The secondaries are wye connected.

a. Find the line voltage on the secondary side.

b. Show that if and only if the system is perfectly balanced, the device functions in a manner similar to three transformers on separate cores.

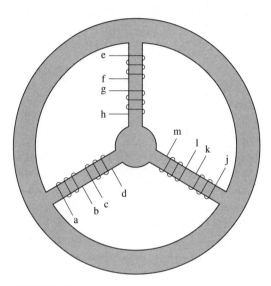

FIGURE 16.40

*16.42 In Fig. 16.41 let $\mathbf{i}_1 = I_0$, $\mathbf{i}_2 = I_0 e^{j120°}$, and $\mathbf{i}_3 = I_0 e^{j240°}$. The resistances are identical. Find the phasors $\mathbf{i}_4$, $\mathbf{i}_5$, and $\mathbf{i}_6$. To make the phase relationships clear,

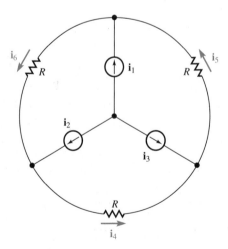

FIGURE 16.41

express your answers in the form $I\angle\theta$. Construct a phasor diagram showing the six phasors.

*16.43 All windings of the three-phase transformer shown in Fig. 16.42 have the same number of turns. The three primaries ab, cd, and ef are delta connected to a three-phase power line with $\mathbf{v}_1 = V_0$, $\mathbf{v}_2 = V_0 e^{j120°}$, and $\mathbf{v}_3 = V_0 e^{j240°}$.
  a. Find $\mathbf{v}_4$, $\mathbf{v}_5$, and $\mathbf{v}_6$. Express your answers in the form $V\angle\theta$.
  b. The three secondaries are wye connected. What is the step-up ratio of this transformer, defined as the ratio of output line voltage (not phase voltage) to input line voltage?

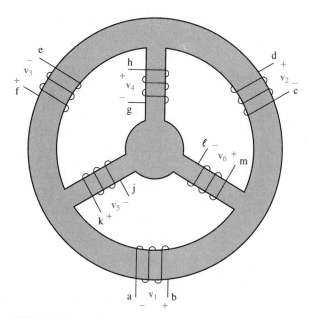

FIGURE 16.42

*16.44 Three different impedances $\mathbf{Z}_A$, $\mathbf{Z}_B$, and $\mathbf{Z}_C$ are connected in a delta as shown in Fig. 16.43(a). The impedances between the terminal pairs are $\mathbf{Z}_{12}$, $\mathbf{Z}_{13}$, and $\mathbf{Z}_{23}$. (When $\mathbf{Z}_{12}$ is measured, terminal 3 is left open-circuited.) We desire to construct a wye connection as shown in Fig. 16.43(b), such that $\mathbf{Z}'_{12} = \mathbf{Z}_{12}$, $\mathbf{Z}'_{13} = \mathbf{Z}_{13}$, and $\mathbf{Z}'_{23} = \mathbf{Z}_{23}$. Find the necessary values of $\mathbf{Z}_X$, $\mathbf{Z}_Y$, and $\mathbf{Z}_Z$, in terms of $\mathbf{Z}_A$, $\mathbf{Z}_B$, and $\mathbf{Z}_C$. *Note:* The two networks are equivalent. One can be replaced by the other in order to simplify calculations. This is known as a *delta–wye transformation*.
Answer:

$$\mathbf{Z}_X = \frac{\mathbf{Z}_B\mathbf{Z}_C}{\mathbf{Z}} \qquad \mathbf{Z}_Y = \frac{\mathbf{Z}_A\mathbf{Z}_C}{\mathbf{Z}} \qquad \mathbf{Z}_Z = \frac{\mathbf{Z}_A\mathbf{Z}_B}{\mathbf{Z}}$$

where $\mathbf{Z} = \mathbf{Z}_A + \mathbf{Z}_B + \mathbf{Z}_C$.

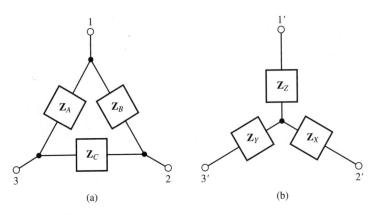

FIGURE 16.43

*16.45 Refer to Problem 16.44 and construct the reverse delta–wye transformation; that is, find $\mathbf{Z}_A$, $\mathbf{Z}_B$, and $\mathbf{Z}_C$ in terms of $\mathbf{Z}_X$, $\mathbf{Z}_Y$, and $\mathbf{Z}_Z$. **Answer:**

$$\mathbf{Z}_A = \frac{\mathbf{M}}{\mathbf{Z}_X} \qquad \mathbf{Z}_B = \frac{\mathbf{M}}{\mathbf{Z}_Y} \qquad \mathbf{Z}_C = \frac{\mathbf{M}}{\mathbf{Z}_Z}$$

where $\mathbf{M} = \mathbf{Z}_X\mathbf{Z}_Y + \mathbf{Z}_X\mathbf{Z}_Z + \mathbf{Z}_Y\mathbf{Z}_Z$.

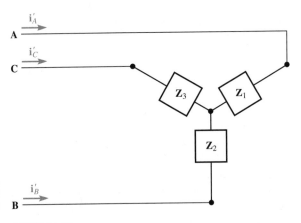

FIGURE 16.44

*16.46 The load of Fig. 16.44 is driven by a three-phase power line with rms line voltage 240 V. The phase sequence is v_{AB}, v_{BC}, v_{CA}; v_{AB} is chosen as the zero-phase reference. Let the three impedances be equal, with value $40 + 25j$ Ω. Find the three currents i'_A, i'_B, and i'_C. Express your results in the form $I \angle \theta$.

*16.47 For Problem 16.46 let the three impedances be unequal, with values $\mathbf{Z}_1 = 40 + 24j$ Ω, $\mathbf{Z}_2 = 30 + 15j$ Ω, and $\mathbf{Z}_3 = 50 + 35j$ Ω. Find the three line currents i_A', i_B', and i_C'. Express your results in the form $I \angle \theta$. Use the delta–wye transformation of Problems 16.44 and 16.45.

*16.48 *Polyphase* electric power, with six or more phases, is used for some special purposes.

  a. If the number of phases is N and the rms phase voltage is V_0, show that the rms line voltage between adjacent phases is $2V_0 \sin(\pi/N)$.

  b. Find all of the voltages available from a six-phase wye-connected generator with neutral. The rms phase voltage is 100 V.

**16.49 The purpose of this problem is to demonstrate that less copper is required to transmit power in a three-phase transmission system than in a single-phase system. Assume (1) that equal power is being transmitted equal distances by the two methods; (2) that the power lost in wire resistances is equal for the two systems; (3) that the mass of copper used in each case is proportional to the number of wires and inversely proportional to the resistance of each wire; (4) that the power factor is unity; (5) that the three-phase load is balanced and delta connected; and (6) that the line voltage is the same for both systems. Find the ratio of the mass of copper used by the single-phase system to that used by the three-phase system.

**16.50 Repeat Problem 16.49 for a wye-connected three-phase load.

**16.51 Repeat Problem 16.49, this time comparing the three-phase system with an N-phase system, where N is an even integer greater than or equal to 4. Assume that the three-phase line voltage is equal to the largest voltage existing between any pair of wires in the N-phase system. The N-phase load is balanced and star connected. (In fact, it can be shown that when compared on this basis the three-phase system uses less copper than an N-phase system for *any* N, odd or even, other than 3.)

CHAPTER 17

ELECTRO-
MECHANICAL
DEVICES

lthough electric circuits are versatile and useful, there are some functions that
are inconvenient or impossible to perform electrically. Thus many systems contain important components that are partly electrical and partly mechanical in
nature. Usually these *electromechanical* components are found at the input or output
ends of systems; they are used to enable the electrical system to interact with the outside world (which is largely mechanical). For example, a phonograph pickup is a
device used to convert information from mechanical to electrical form at the input of
an information system; a loudspeaker is a device used to convert information from
electrical to mechanical form at the output of an information system. Such information
converters are known as *transducers*. A slightly different use arises when the electrical
system's output is not in the form of information but instead is intended to perform
some mechanical operation. For example, an electrical control system may open or
close a valve or move the rudder of a ship. The electromechanical device at the output
of such a system is called an *actuator*. It may do significant mechanical work, thus
lying in an intermediate area between information and power systems. Lastly, there is
the important class of electromechanical devices used to convert power from mechanical to electrical form at an input of a power system or to convert power from electrical
to mechanical form at an output of a power system. These devices are known respectively as *generators* and *motors*. They differ from transducers and actuators in that their
operation is repetitive and continuous. For continuous operations, rotary motion is
very suitable; hence motors and generators contain an internal part, the *rotor*, which is
designed to rotate. Motors and generators hence are known collectively as *rotating*

electrical machines. These machines vary widely in size, ranging from tiny clock motors to enormous generators that occupy whole buildings. Rotating machines can also be divided into *dc machines* and *ac machines.* Both types have their particular advantages; the choice of one or the other depends on the application.

17.1 Transducers and Actuators

Transducers and actuators are usually small, special-purpose devices; they vary widely in their design. Some of them are electrostatic in principle (capacitance pickups, electrostatic loudspeakers); however, we shall concentrate here on those that are electromagnetic. The physical laws governing such devices are those introduced in Chapter 16.

An important family of transducers are those known as *moving-coil devices.* The principle is illustrated in Fig. 17.1. An *N*-turn coil moves in and out between the poles of a permanent magnet. [In Fig. 17.1(b) the lines of *B* are directed into the page and are represented by crosses. This is a conventional symbol that is said to represent the tailfeathers (!) of arrows directed into the page. Lines of *B* pointed up out of the page would be represented by dots, symbolizing the points of the arrows.] For simplicity, we assume that the coil and the pole pieces are square. Now suppose that a current $i(t)$ passes through the coil. From the force law, Eq. (16.3), we find that an electromagnetic force

$$F_e = NaBi \tag{17.1}$$

is exerted on the coil in the plus-*x* direction. In a typical output transducer, such as a loudspeaker, the coil is restrained by a spring, which applies a mechanical force $F_m = -kx$, where k is the spring constant and x is the displacement of the coil from its equilibrium position. If the current i is constant in time, the balance of forces requires that $F_e + F_m = 0$, or

$$x = \frac{NaBi}{k} \tag{17.2}$$

If $i(t)$ is time-varying, the coil position will vary in time. This will cause the flux linking the coil to vary, and the time-varying flux will in turn generate a voltage in the coil, according to Faraday's law, Eq. (16.2):

$$e = N\frac{d\Phi}{dt} = NBa\frac{dx}{dt} \tag{17.3}$$

This voltage is generated by the motion of the coil in the magnetic field, just as it would be in an electrical generator. When the motion is caused by an externally applied current, the voltage is called a "back emf." Emf (electromotive force) is another term for voltage, and the term "back emf" indicates a voltage that reacts back on the source of the applied current.

To study the motion of the coil, we must take into account the inertia of its moving mass *M.* Furthermore, its motion will in many cases be opposed by friction. A com-

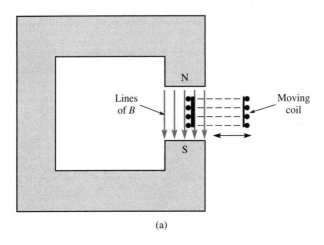

(a)

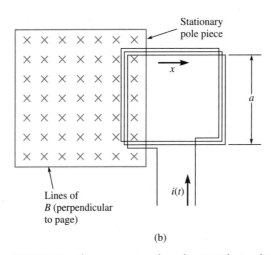

(b)

FIGURE 17.1   Elementary moving-coil transducer. (a) Side view. (b) Top view.

mon way of including friction mathematically is to introduce a frictional force proportional to velocity,

$$F_f = -\gamma \frac{dx}{dt} \tag{17.4}$$

where γ is the appropriate constant of proportionality. The motion of the moving coil then obeys Newton's equation:

$$M \frac{d^2x}{dt^2} = F_e + F_m + F_f = NaBi - kx - \gamma \frac{dx}{dt} \tag{17.5}$$

Now let us assume that the applied signal current is sinusoidal with angular frequency ω. Since all of our equations are linear, all time-varying quantities will then also be sinusoidal with the same frequency. Let $\mathbf{x}$ be the phasor representing $x(t)$. From Eq. (17.5) we have

$$-M\omega^2\mathbf{x} = NaB\mathbf{i} - k\mathbf{x} - \gamma j\omega\mathbf{x} \tag{17.6}$$

where $\mathbf{i}$ is the phasor representing the input current $i(t)$. Solving, we have

$$\mathbf{x} = \frac{NaB\mathbf{i}}{(k - M\omega^2) + j\omega\gamma} \tag{17.7}$$

If the transducer is being driven by an ideal current source, so that $i(t)$ is known, then the calculation is finished: taking the absolute value of Eq. (17.7) gives us the amplitude of the resulting mechanical motion.

EXAMPLE 17.1

The moving-coil transducer can be connected to a paper cone to form a loudspeaker, as shown in Fig. 17.2. Find the maximum displacement from equilibrium when the loudspeaker is driven by a 100 Hz, 1 A (rms) sinusoidal current. Let $B = 0.7$ T, $a = 30$ cm, $N = 350$ turns, $k = 25$ N/m, $M = 100$ g, and $\gamma = 667$ N-sec/m.

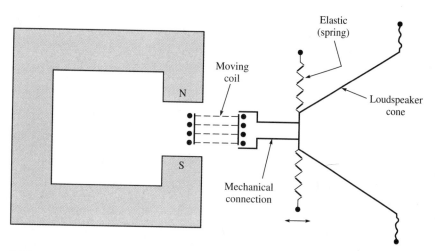

FIGURE 17.2

SOLUTION

The maximum displacement is equal to $|\mathbf{x}|$:

$$
\begin{aligned}
|\mathbf{x}| &= \frac{NaB|\mathbf{i}|}{\sqrt{(k - M\omega^2)^2 + (\omega\gamma)^2}} \\[2mm]
&= \frac{(350)(0.30)(0.7)(1.414)}{\sqrt{(1.56 \times 10^9) + 1.75 \times 10^{11}}} = 2.48 \times 10^{-4}\,\text{m}
\end{aligned}
$$

(Note that all quantities have been converted to SI units and that $|i|$ is equal to the rms current multiplied by $\sqrt{2}$.) We observe that the frictional term in the denominator is dominant. This is not surprising in a loudspeaker, since the frictional term represents energy converted to air motion, which is the useful mechanical output. Since this is the case, we have

$$\mathbf{x} \cong \frac{NaB}{j\omega\gamma}\mathbf{i}$$

However, the "back emf" e is known from Eq. (17.3) to be $NBaj\omega x$. Thus we have

$$\mathbf{e} \cong \frac{(NBa)^2}{\gamma}\mathbf{i} \equiv R_m\mathbf{i}$$

where R_m is an effective resistance equal by definition to $(NBa)^2/\gamma$. In this case (with the damping term dominant) the transducer coil acts as though it were simply a resistance R_m.

In the preceding example we assumed that the transducer was driven by a known current. In practice, however, it will be driven by a real signal source with finite Thévenin resistance R_s. Thus we must look more closely at the electrical circuit, shown in Fig. 17.3(a). Here R_w is the resistance of the coil wire, and e is the "back emf" gen-

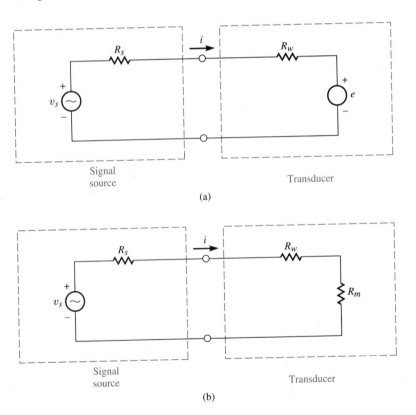

(a)

(b)

FIGURE 17.3

erated by the motion of the coil. The value of e has already been found in Eq. (17.3). Thus

$$\mathbf{i} = \frac{\mathbf{v}_s - \mathbf{e}}{(R_s + R_w)} = \frac{1}{(R_s + R_w)}[\mathbf{v}_s - NBaj\omega\mathbf{x}] \tag{17.8}$$

Using the expression for $\mathbf{x}$, Eq. (17.7), we can find $\mathbf{i}$ for any $\mathbf{v}_s$. The result is simpler when $\omega\gamma \gg |k - M\omega^2|$, as in the preceding example. We would then have

$$\mathbf{i} \cong \frac{\mathbf{v}_s}{R_s + R_w + R_m} \tag{17.9}$$

where as before we define

$$R_m \equiv \frac{(NBa)^2}{\gamma} \tag{17.10}$$

As was pointed out in the example, the transducer coil acts (in this high-damping approximation) like a resistance R_m. If we replace $\mathbf{e}$ by R_m we obtain the circuit of Fig. 17.3(b), which is consistent with Eq. (17.9).

EXAMPLE 17.2

Find the electrical power input and acoustical power output for the loudspeaker of Example 17.1 when driven by a Thévenin source v_s, R_s. Assume $v_s = 10$ V rms, $R_s = 8\ \Omega$, and $R_w = 2\ \Omega$.

SOLUTION

Refer to Fig. 17.3(b). The electrical input power is $|\mathbf{i}'|^2(R_w + R_m)$. From Eq. (17.9) we have (noting that $R_m = 8.1\ \Omega$)

$$P_{\text{in}} = |\mathbf{v}'_s|^2 \frac{R_w + R_m}{(R_s + R_w + R_m)^2} = 3.08 \text{ W}$$

The mechanical output power is

$$P_{\text{out}} = \text{time avg}\left(F_e \frac{dx}{dt}\right)$$

where F_e is the magnetic force given by Eq. (17.1). We obtain $\mathbf{x}$ from Eq. (17.7) using the high-damping approximation. Thus

$$P_{\text{out}} = \tfrac{1}{2} \text{Re}\left[\mathbf{F}_e\left(\frac{d\mathbf{x}}{d\mathbf{t}}\right)^*\right]$$

$$= \tfrac{1}{2} \text{Re}[NaB\ \mathbf{i}(j\omega NaBi/j\omega\gamma)^*]$$

$$= \tfrac{1}{2}|\mathbf{i}|^2 R_m = \frac{|v'_s|^2 R_m}{(R_s + R_m + R_w)^2} = 2.47 \text{ W}$$

The power efficiency of this loudspeaker is about 80 percent.

The same transducer can be used backward. That is, external forces can be applied and an electrical output obtained. The resulting transducer is known as a *moving-coil pickup*. Equations (17.1) and (17.3) and Fig. 17.3(a) continue to apply; in Fig. 17.3 v_s is now zero, and R_s becomes the load resistance into which the transducer puts its output power.

EXAMPLE 17.3

A moving-coil pickup is used to detect the motions of a phonograph stylus as a record is played. The stylus executes a sinusoidal motion with amplitude x_0 and frequency ω. The transducer is connected to a load resistance R_L. Find the rms signal voltage developed across R_L.

SOLUTION

From Eq. (17.3) the rms voltage developed by the transducer is

$$e' = \frac{1}{\sqrt{2}} NBa\omega x_0$$

The portion appearing across the load is

$$v'_L = \frac{1}{\sqrt{2}} NBa\omega x_0 \frac{R_L}{R_L + R_w}$$

Actual moving-coil transducers use more efficient geometries than that of Fig. 17.1. A common arrangement is the cylindrical one of Fig. 17.4. Let us assume that the value of B between the poles is constant. Suppose that a current i flows counterclockwise in the coil; the coil will then experience a force $2\pi NBai$ in the direction of the arrows (where a is the radius of the coil). This structure is more efficient because the entire coil is acted on equally by the magnetic field.

The moving-coil structures discussed so far have been based on translational movement, but angular movement of the coil is also possible. Figure 17.5 shows a *d'Arsonval movement*; these movements are widely used in ammeters. The coil, wound on a cylindrical iron core, rotates between two magnetic poles; it is returned to a neutral position by a spring when the current is turned off. The poles are shaped to make B radial, as shown in Fig. 17.5(b). The forces, which are perpendicular to both i and B, are in the directions shown in the figure. The total force acting on the coil is zero, but there is an electromagnetic torque

$$T_e = 2Fa = 2NBhia \qquad (17.11)$$

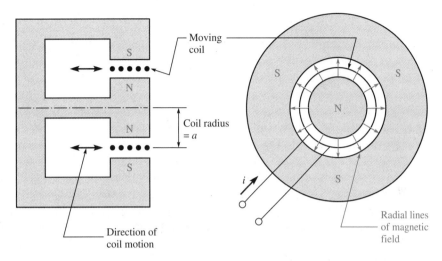

FIGURE 17.4

Here h is the length of the coil in the direction perpendicular to the page, and the factor N is the number of turns in the coil. If the return spring applies a mechanical torque

$$T_m = -k_A\theta \tag{17.12}$$

the balance of torques requires that

$$\theta = 2NBhai/k_A \tag{17.13}$$

Thus the angular displacement is proportional to the coil current, and a needle attached to the coil can be used to indicate current on a linear scale. Common d'Arsonval movements have full-scale currents of 50 μA or 1 mA. Variations of this device, intended for measuring ac quantities, were mentioned in Chapter 16.

RELUCTANCE TRANSDUCERS

The moving-coil principle has the advantage that the coil is light and easy to move. Sometimes, however, this is not important, and a part of the iron magnetic circuit is moved instead. This approach is especially useful in actuators intended to exert considerable forces, since the iron moving part is physically strong.

Consider the magnetic actuator shown in Fig. 17.6. (This kind of device is often called a "solenoid"—it is similar to the "starter solenoid" used in automobiles—but we prefer not to call it by this name to avoid confusion with the other meaning of the word "solenoid," a helical coil winding.) When current is applied to the coil, the movable plunger is pulled forcefully to the left. The resulting electrically controlled pull is used to perform a mechanical task. For instance, actuators are used in washing machines to open or close the water valve. Anyone familiar with washing machines or dishwashers

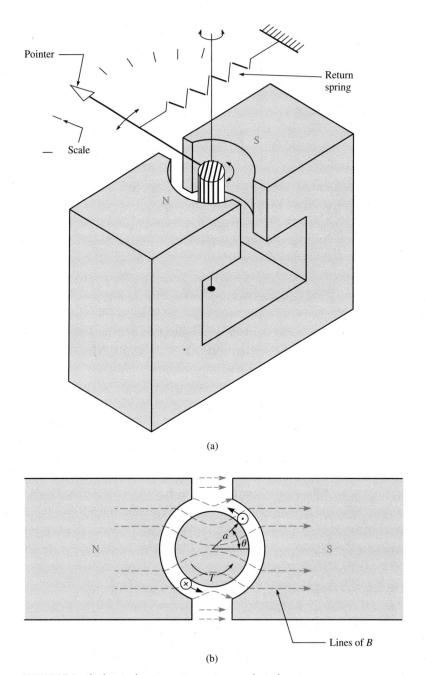

(a)

(b)

FIGURE 17.5   The d'Arsonval movement, a rotary moving-coil transducer.

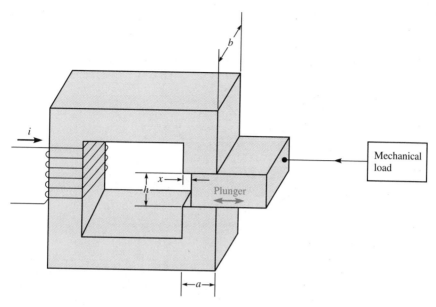

FIGURE 17.6 A magnetic actuator.

will recognize the loud "clunk!" of the actuator as the water is turned on. We note that unlike transducers, which are analog (proportional) devices, the plunger of the magnetic actuator is usually intended to be either all the way in or all the way out; intermediate positions are not used.

Intuitively one thinks of this device as an electromagnet attracting a movable piece of iron. Let us now find the force tending to pull in the plunger, using the assumption that the coil current is held constant at the value I_0 by an ideal current source. We shall use an energy-balance approach. As the plunger moves inward, three things happen: (1) The magnetic force does work as the plunger moves; (2) the magnetic energy, stored in the fields of the magnetic circuit, changes; and (3) power is transferred between the current source and the device.

It is easy to show (see Problem 9 of Chapter 16) that the magnetic energy stored in a volume ΔV of material with constant permeability μ is

$$W_m = \frac{B^2}{2\mu} \Delta V \tag{17.14}$$

The energy stored in a complete magnetic circuit can be expressed in terms of the reluctance $\mathfrak{R}$. Integrating the energy density $B^2/2\mu$ over the entire volume of the magnetic circuit, we obtain

$$W_m = \tfrac{1}{2} \int \frac{B^2}{\mu}\, dV = \tfrac{1}{2} \int \left(\frac{\Phi}{A}\right)^2 \frac{dV}{\mu} \tag{17.15}$$

Here A is the cross-sectional area of the magnetic circuit at any point,* and $\Phi = BA$. However, Φ is a constant throughout the magnetic circuit and thus can be taken out of the integral:

$$W_m = \tfrac{1}{2}\Phi^2 \int \frac{dV}{A^2 \mu} = \frac{(NI_0)^2}{2\mathfrak{R}^2} \int \frac{dV}{A^2 \mu} \qquad (17.16)$$

$$= \frac{(NI_0)^2}{2\mathfrak{R}^2} \int \frac{dl}{A\mu} \qquad (17.17)$$

where dl is an element of length parallel to B and $dV = A\,dl$. We recall, however, from Eq. (16.10) that $dl/A\mu$ is simply the reluctance of a length dl of magnetic circuit whose cross-sectional area is A and whose permeability is μ. The total reluctance of the magnetic circuit is the series sum (in this case, the integral) of all of these differential reluctances:

$$\int \frac{dl}{A\mu} = \mathfrak{R} \qquad (17.18)$$

where the integral is taken over a complete closed path through the circuit. Thus

$$W_m = \frac{(NI_0)^2}{2\mathfrak{R}} \qquad (17.19)$$

Our statement of energy balance is the following:

$$\begin{array}{c} \text{Mechanical} \\ \text{work output} \end{array} + \begin{array}{c} \text{Increase in stored} \\ \text{magnetic energy} \end{array} = \begin{array}{c} \text{Electric input} \\ \text{power} \end{array} \times \text{Time}$$

or, in symbols,

$$F\,dx + dW_m = iv\,dt \qquad (17.20)$$

The electric input power is given by

$$iv = I_0 N \frac{d\Phi}{dt} = (NI_0)^2 \frac{d}{dt}\frac{1}{\mathfrak{R}} \qquad (17.21)$$

Substituting Eqs. (17.19) and (17.21) into Eq. (17.20), we have

$$F\,dx + \frac{(NI_0)^2}{2} d\left(\frac{1}{\mathfrak{R}}\right) = (NI_0)^2 d\left(\frac{1}{\mathfrak{R}}\right) \qquad (17.22)$$

This interesting result shows that of the total electrical energy input written on the right side of Eq. (17.20), half is converted to mechanical work and the other half is used to increase the stored magnetic energy. From Eq. (17.22) we have

---

*This step assumes that B is constant over the cross-sectional area A. In practice this may not be true, but the resulting error will be small.

$$F = \frac{1}{2}(NI_0)^2 \frac{d}{dx}\left(\frac{1}{\mathfrak{R}}\right) = -\frac{1}{2}\left(\frac{NI_0}{\mathfrak{R}}\right)^2 \frac{d\mathfrak{R}}{dx} \qquad \text{(17.23)}$$

The minus sign indicates that F acts in the minus-x direction when $d\mathfrak{R}/dx$ is positive. For a given magnetic circuit, $\mathfrak{R}(x)$ can be calculated or determined by measurement.

EXAMPLE 17.4

Estimate the magnetic force exerted by the actuator shown in Fig. 17.7. Assume that the coil has 800 turns and carries a current of 10 A; $b = h = 2$ cm; $\mu/\mu_0 = 5,000$; and let the total length of the flux path be 20 cm and the gap width x be 1 cm. Express your answer in terms of the weight that can be lifted, in kilograms and in pounds. *Note:* Gravitational force (in newtons) = mass (in kilograms) × acceleration of gravity (= 9.8 m/sec²).

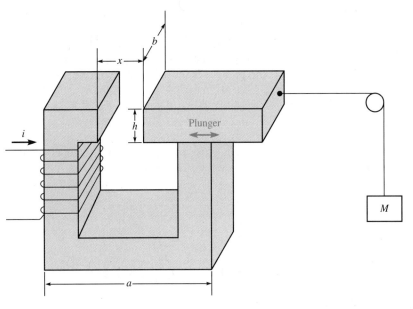

FIGURE 17.7

SOLUTION

From Eq. (16.13) the reluctance of the magnetic circuit is given by

$$\mathfrak{R} = \frac{l_m}{\mu A} + \frac{l_g}{\mu_0 A} \cong \frac{l_g}{\mu_0 A} = \frac{x}{\mu_0 bh}$$

From Eq. (17.23) the force is given by

$$|F| = \tfrac{1}{2}(NI_0)^2 \frac{\mu_0 bh}{x^2}$$

$$= 161 \text{ N}$$

The weight that can be lifted is $M = F/g = (161 \text{ N})/(9.8 \text{ m-sec}^{-2}) = 16$ kg, or 35 lb.

One very common application of the magnetic actuator is to close or open a switch that controls another circuit, as shown in Fig. 17.8. Such a magnetically controlled switch is called a *relay*. The coil circuit is usually designed to draw only a small current, while the contacts may be quite large, so that a large current can be turned on and off by the small one. We observe that relays are active devices, in the sense we have used the term in Chapter 13. They could be used as inverters in digital circuits; and in fact, they *were* used as logic elements in some of the earliest digital work. In information systems they have long been replaced by transistors, which are very much smaller and faster. Relays are still used for switching in high-power circuits, but recently high-power semiconductor switching devices, such as *thyristors*, have been replacing relays in power applications as well.

Reluctance devices can also be used "backward" as sensors, just as moving-coil devices can. For example, if a constant current I passes through the coil of the device shown in Fig. 17.6 and the plunger is moved by an external force, the motion is registered by a coil voltage

$$v = N \frac{d\Phi}{dt} = N^2 I \frac{d}{dt}\left(\frac{1}{\Re}\right) \tag{17.24}$$

This sensor is known as a *reluctance pickup*. The current bias I can be conveniently replaced by a permanent magnetization of the core.

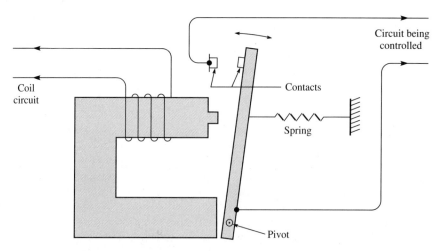

FIGURE 17.8   A relay.

17.2 DC Rotating Machines

All of the dc-powered actuators and transducers discussed in the last section have limited motion. When current is applied they move into a stable position and stop. In order to produce continuous motion, as in a dc motor, we have to deprive the device of stability. In dc motors this is done by an ingenious element called a *commutator*. The function of the commutator is to reverse the current at suitable times, causing the previously stable position to become unstable, so that motion can continue.

It has already been seen that a wire carrying a current I in a magnetic field B experiences a force perpendicular to both current and field. Let us consider a rectangular coil of wire located in a magnetic field, as shown in Fig. 17.9(a). The current at the right side of the coil is directed out of the paper; the force on this part of the coil is to the right and equals $NIwB$, where N is the number of turns in the coil and w its width, as shown in Fig. 17.9(b). At the left side of the coil the current direction is into the paper; the force on this segment of the coil is equal in magnitude but is directed to the left. Let θ be the angle the plane of the loop makes with the magnetic field and let

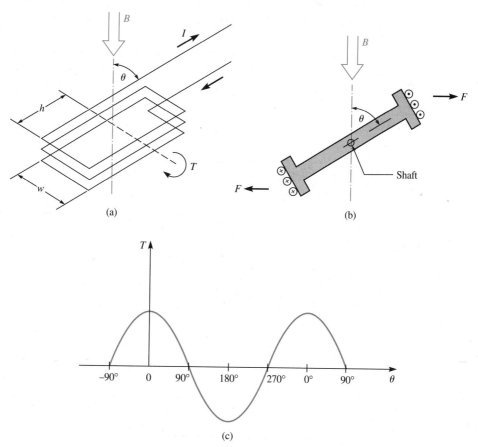

(a)

(b)

(c)

FIGURE 17.9

the length of the loop be *2h*, as shown. Then the torque acting to turn the loop clockwise is

$$T = 2Fh \cos \theta = 2NhwIB \cos \theta \qquad (17.25)$$

The torque is sketched as a function of θ in Fig. 17.9(c). We observe that if $-90° < \theta < 90°$, the clockwise torque is positive, and the rotor will move toward larger θ, until $\theta = 90°$ is reached. However, if the rotor should turn past 90°, the clockwise torque becomes negative, and θ tends to decrease. At $\theta = 90°$ the torque is zero; this is a *stable point*. The rotor may for a time oscillate back and forth around $\theta = 90°$, but ultimately the oscillations will be damped out by friction, and the rotor will be motionless at the stable point. If we are trying to build a motor, this is clearly not what we want.

In order to eliminate stability, we can reverse the sign of *I* whenever $90° < \theta < 270°$. This reverses the sign of the torque, so that clockwise torque is always positive. The switching of *I* is done by a pair of sliding contacts mounted on the rotating shaft of the machine, as shown in Fig. 17.10(a). (The black rectangles represent the sliding contacts.) The resulting dependence of torque on angle is shown in Fig. 17.10(b). Since the torque now is always positive, the shaft will tend to turn continuously in the positive direction. The sliding contact arrangement is the commutator previously mentioned.

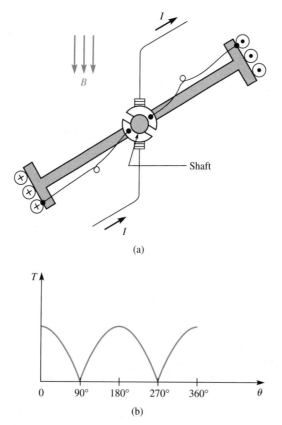

(a)

(b)

FIGURE 17.10

The simple motor just described has several drawbacks. For one thing, the torque pulsates, with two maxima and two zeros during every revolution. This would give jerky operation that would probably cause wear to mechanical parts. Furthermore, if the motor with load attached is to be started from rest and the shaft angle happens to be near 90°, the *starting torque* may be low, and the motor may not be able to start turning at all. Smoother operation can be obtained with a multiwinding rotor, as shown in Fig. 17.11. In this design the commutator has many different contacts, as many as there are windings on the rotor. It is designed so that the wires in the upper half of the figure carry current out of the page and those in the lower half inward. As the rotor turns, the windings are continuously switched, so that current is always outward for the wires in the upper half regardless of rotor position. All of the wires in the upper half experience a force to the right, as shown, while those in the lower half experience a force to the left. The result is a nearly constant torque tending to rotate the rotor clockwise.

EXAMPLE 17.5

The rotor of a dc motor is as shown in Fig. 17.11. It has 25 loops (that is, 50 wires), each of which carries 60 A. The coil dimensions [see Fig. 17.9(a) are $h = 12$ cm and $w = 40$ cm. Let B be a

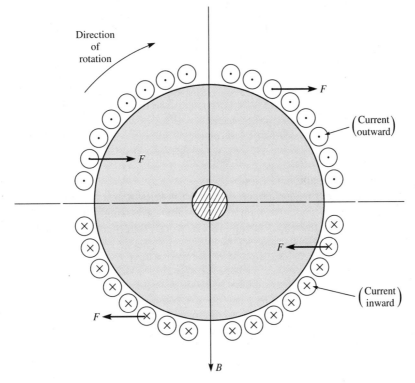

FIGURE 17.11

uniform field of 0.6 T directed in straight lines from top to bottom in the figure. The motor turns at 1,500 rpm. Find the power output. Express your result in both watts and horsepower (1 hp = 746 W).

SOLUTION

Defining the angle θ as in Fig. 17.9, the torque produced by a wire located at θ is $hwlB \cos \theta$. To sum over the 25 wires in each half it is convenient to convert the sum into an integral:

$$T = 2 \sum_{i=1}^{25} hwlB \cos \theta_i \cong 2hwB \frac{\Delta I}{\Delta \theta} \int_{-\pi/2}^{\pi/2} \cos \theta \, d\theta$$

where $\Delta I/\Delta \theta$, the current per unit angle, in this case is $(25)(60)/\pi = 477.5$ A/radian. Thus the total torque is

$$T = 4hwB \frac{\Delta I}{\Delta \theta}$$

$$= 55.0 \ N\text{-}m$$

The power is given by $P = T\Omega$, where Ω is the motor speed in radians/sec. Thus

$$P = \frac{(55)(1500)(2\pi)}{60} = 8,600 \ W$$

$$= 11.6 \ hp$$

In a practical motor the stator poles would be shaped approximately as shown in Fig. 17.5(b) so as to make the field radial and always perpendicular to the air gap. This eliminates the torque-reducing factor $\cos \theta$ in Eq. (17.25), and the total motor torque becomes simply

$$T = 2NhwlB \tag{17.26}$$

where N is the number of loops.

The motor we have been discussing is known as a *two-pole* machine because both the stationary *field winding* and the *armature* (rotating structure)* act as magnets with two poles, as illustrated in Fig. 17.12. More often machines actually are built in four-pole, six-pole, or (in general) $2n$-pole form. The structure of an eight-pole dc machine

---

*The distinction between the terms "armature" and "field winding" is that the former is the one through which most of the machine's input power (or for a generator, output power) passes. The volt-amperes of the field winding, on the other hand, are only a few percent of the machine's kVA rating. In the machine under discussion, the armature rotates and the field winding is stationary, but in some machines it is the other way around.

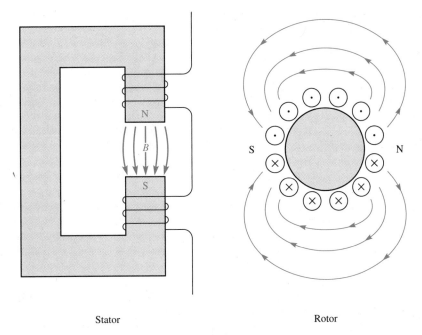

Stator   Rotor

FIGURE 17.12 Stator and rotor of a two-pole dc machine.

is illustrated in Fig. 17.13. The torque of the multipole machine is still given by Eq. (17.26), but there are practical advantages. The flux paths are shorter, and they do not have to go through the center of the rotor, which thus does not have to be solid. Moreover, current coming out of the page can return into the page through the next adjacent winding, rather than through a diametrically opposite winding; this shortens the non-productive connecting wires and saves copper. For these and other practical reasons multipole motors are very common.

• EXERCISE 17.1

Sketch the rotor, stator, and stator field lines of a four-pole dc motor.

The subject of armature design is fairly complex. There are two common types of armature windings, known as *lap windings* and *wave windings*. For our purpose the main difference is the number of parallel current paths, which we shall call a, that exist through the armature. The significance of the parameter a is illustrated in Fig. 17.14. In Fig. 17.14(a) a single coil is used with a two-segment commutator; here $N = 2$, and since there is only a single current path, $a = 1$. In Fig. 17.14(b) the two turns are con-

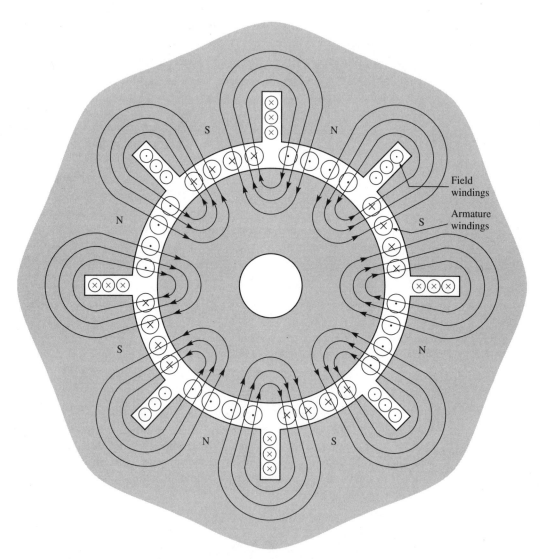

FIGURE 17.13   Eight-pole dc machine. The field lines of the stator (field winding) are shown, but not those of the rotor (armature).

nected in parallel instead of in series; hence $N = 2$, $a = 2$. In Fig. 17.14(c) a four-segment commutator is used to control two coils spaced 90° apart on the armature; in this case also, $N = 2$, $a = 2$. (We have $a = 2$ because at any instant current will be going through two parallel paths.) The operation of Fig. 17.14(c) is illustrated in Fig. 17.15. Four positions of the armature, which rotates clockwise, are shown in Fig. 17.15(b). In position (i), wire ends 1 and 2 are positive and ends 3 and 4 are negative. This is illustrated by the first line in the table of Fig. 17.15(a). Consequently current flows into the page through the two top wires and out through the bottom two. We see

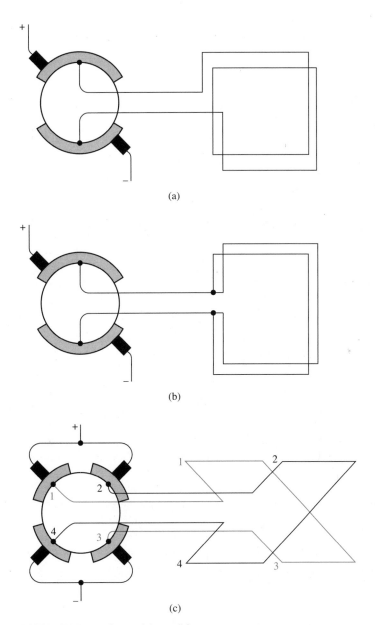

FIGURE 17.14 Significance of the parallel-parameter *a*. (a) Two commutator segments, $N = 2$, $a = 1$. (b) Two commutator segments, $N = 2$, $a = 2$. (c) Four commutator segments, $N = 2$, $a = 2$.

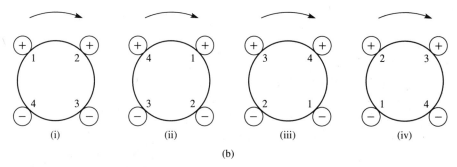

Position	1 〰️ 3			2 〰️ 4		
(i)	+	→	−	+	→	−
(ii)	+	→	−	−	←	+
(iii)	−	←	+	−	←	+
(iv)	−	←	+	+	→	−

(a)

(b)

FIGURE 17.15   Operation of the four-segment commutator of Fig. 17.14(c).

that when position (ii) is reached, after the armature has rotated 90°, the top two wire ends (now 4 and 1) are still positive; this remains true for all four positions. This armature would be well suited for a two-pole machine. It is useful to know the values of a for the common armature windings. For wave windings $a = 2$; for lap windings $a = p$, where p is the number of poles in the machine.

From Fig. 17.12 the reader may have observed that the actual magnetic field is the sum of two components: the field produced by the field winding plus a nearly perpendicular field produced by the armature. The armature's field creates a distortion in the desired field pattern, an effect known as *armature reaction*. Armature reaction can be compensated for by adjusting the position of the commutator; another way is to add special windings, called *compensating windings*, between the stator poles to reform the field. For our present purposes, however, armature reaction is a second-order effect, and we shall not consider it further.

CIRCUIT MODELS

Just as with other devices, it is useful to develop models for electrical machines, so that their behavior can be predicted in at least an approximate way. Our first step is to note that the armature coils experience a time-varying magnetic field, and hence a voltage is developed across their terminals. This voltage is called *back emf*. (A similar back emf has been mentioned in connection with moving-coil transducers.) The back emf can be found from energy balance. Equating electrical power input to mechanical power output,*

$$eI_A = T\Omega \tag{17.27}$$

where Ω is the angular speed of the machine in radians/sec, I_A is the total armature current, and e is the back emf. We now wish to eliminate I_A by means of Eq. (17.26). In doing this we remember that in Eq. (17.26) I was the current in each wire. Thus $I_A = aI$, and we have

$$e = \frac{2NhwB\Omega}{a} \tag{17.28}$$

If the machine has P poles, the flux of each pole is [refer to Figs. 17.9(a) and 17.13]

$$\Phi_p = \frac{2\pi hBw}{p} \tag{17.29}$$

in terms of which we have

$$e = \frac{pN\Omega\Phi_p}{\pi a} \equiv k_1\Omega\Phi_p \tag{17.30}$$

Here N is still the number of loops. The total number of wires in the armature is $2N$ (N wires carry current into the page and N carry current out), and the constant

$$k_1 \equiv \frac{pN}{\pi a} \tag{17.31}$$

is introduced for convenience.

The derivation of Eq. (17.30) assumed the machine to be lossless, but the back emf is electromagnetic in origin and will be the same even if mechanical friction is present. In terms of definitions (17.29) and (17.31), the torque expression (17.26) becomes

$$T = k_1 I_A \Phi_p \tag{17.32}$$

A resistance R_A will be present in the armature; including it, we have the armature model shown in Fig. 17.16. Equations (17.30) and (17.32) and Fig. 17.16 apply to both motors and generators.

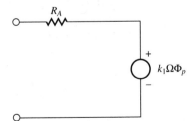

FIGURE 17.16   Circuit model of the armature of a dc machine.

---

*Some voltage is also induced in the field winding by the magnetic field of the armature, but this is an ac voltage. The power input to the field winding is the product of this voltage and the dc field current; the time average of this product is zero.

EXAMPLE 17.6

The rotor of a four-pole 1,200 V dc motor is lap-wound with 300 loops. The flux per pole is 0.08 Wb. Find the speed in rpm. Neglect friction, and assume the armature resistance is zero.

SOLUTION

Since we are neglecting armature resistance, the armature emf e is the same as the applied voltage V. We note also that for the lap-wound armature $a = p = 4$.
From Eq. (17.30) we have

$$\Omega = \frac{\pi V a}{\Phi_p p N}$$

$$= \frac{\pi (1,200)}{(0.08)(300)} = 157 \text{ radians/sec}$$

$$\text{rpm} = \frac{60\Omega}{2\pi} = 1,500$$

When the field and armature windings of a motor are connected in series, the result is the *series-connected* motor shown in Fig. 17.17. Let the flux per pole be related to the current I_F in the field winding by

$$\Phi_p \cong k_2 I_F \tag{17.33}$$

where k_2 is the appropriate proportionality constant. However, for the series-connected motor $I_F = I_A$. Thus with the model of Fig. 17.17(b) we find

$$I_A = \frac{V - k_1 \Omega \Phi_p}{R_A + R_F} = \frac{V - k_1 k_2 \Omega I_A}{R_A + R_F} \tag{17.34}$$

Solving for I_A, we have

$$I_A = \frac{V}{R_A + R_F + k_1 k_2 \Omega} \tag{17.35}$$

The torque, from Eqs. (17.32) and (17.33), is

$$T = k_1 k_2 I_A^2 = \frac{k_1 k_2 V^2}{(R_A + R_F + k_1 k_2 \Omega)^2} \tag{17.36}$$

The graph of Ω versus T (for constant supply voltage V) is known as the *speed-torque characteristic*; it is sketched in Fig. 17.17(c). The series-connected motor is referred to

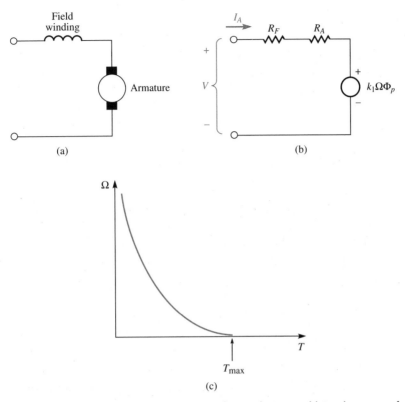

FIGURE 17.17 Series-connected dc motor. (a) Wiring diagram. (b) Circuit model. R_F is the resistance of the field winding. (c) Speed-torque characteristic.

as a "variable-speed" motor because, as we see from the figure, its speed depends strongly on the dragging torque that resists its motion. When torque increases (that is, when more load is applied); the speed drops. This can be a useful protective feature, since it tends to keep the power ($T\Omega$), and hence kVA, approximately constant over a range of load conditions. When excessive load is applied, the speed falls to zero; this happens at

$$T_{max} = \frac{k_1 k_2 V^2}{(R_A + R_F)^2}$$ (17.37)

When T_{max} is reached the motor is said to *stall*. If loading torque is reduced toward zero, Eq. (17.36) predicts that the speed will increase without limit; the motor is said to "run away." Friction will prevent torque from actually dropping to zero, but speed may nonetheless become high enough to damage some motors if runaway is allowed to occur. On the other hand, the motor has high *starting torque*; that is, its high torque at low speed makes it possible to "unstick" a heavy load and get it into motion. Series-connected motors are most often useful in applications where the load varies widely—for example, in electric locomotives, where the load varies as the train accelerates and decelerates and goes up and down hills.

When the field winding is connected in parallel with the armature, instead of in series, the result is known as a *shunt-connected* machine (Fig. 17.18). For a shunt-connected motor,

$$k_1 \Omega \Phi_p + I_A R_A = V \tag{17.38}$$

so that we have

$$\Omega = \frac{V - I_A R_A}{k_1 \Phi_p} = \frac{V}{k_1 \Phi_p} - \frac{R_A T}{(k_1 \Phi_p)^2} \tag{17.39}$$

where in the last step Eq. (17.32) has been used. We observe that if the armature had no resistance, the speed of the motor would be constant. In practice R_A is not very large; thus the speed-torque characteristic has the gently sloping form sketched in Fig. 17.18(c). One thinks of the shunt motor as being a *constant-speed* motor because its speed typically varies only about 5 percent between no load and full load.

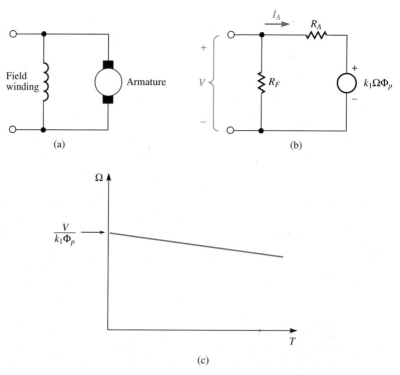

FIGURE 17.18   Shunt-connected dc motor. (a) Wiring diagram. (b) Circuit model. (c) Speed-torque characteristic.

EXAMPLE 17.7

The output of an eight-pole, 1,200 V shunt-wound dc motor is rated 100 hp at 1,500 rpm. The flux is 1.2 webers per pole, and $R_A = 0.6 \ \Omega$. Find the *speed regulation*, defined by

$$SR = \frac{\Omega_2 - \Omega_1}{\Omega_1}$$

where Ω_1 is the full-load speed and Ω_2 is the no-load speed.

SOLUTION

From Eq. (17.39) we have

$$SR = \frac{R_A T}{V k_1 \Phi_p - R_A T}$$

where T, the full-load torque, is given by $T = P/\Omega$. If we assume that the voltage drop in R_A is small, we can estimate from Eq. (17.30) that

$$k_1 \Phi_p \cong \frac{V}{\Omega}$$

Thus

$$SR \cong \frac{1}{\dfrac{V^2}{R_A P} - 1} = \frac{1}{\dfrac{(1,200)^2}{(0.6)(100)(746)} - 1}$$

$$= 0.032 = 3.2 \text{ percent}$$

An interesting property of both shunt and series motors is that they always turn in the same direction, even if the sign of V is reversed. This is because reversing V changes the sign of both the field and armature currents, so that torque, which is proportional to $\Phi_p I_A = k_2 I_F I_A$, keeps the same sign. A consequence is that the series motor can be run on ac; such motors designed for ac/dc operation are known as *universal motors*. These are often found in small appliances and power tools. In the past they were sometimes used in variable-speed applications like trolley cars, where only ac power was available. However, series motors work better on dc, and with modern semiconductor devices it is easy to rectify the available power to provide dc. Shunt motors do not work well on ac; owing to the inductance of the field winding, ϕ and I_A tend to be out of phase.

Two other types of dc motors are the *separately excited* and *compound* machines. In the former the field winding is excited by a power supply entirely separate from that of the armature. Such motors are, of course, reversible. In small motors the field is often supplied by permanent magnets. The resulting small reversible motors are known as *servo motors*; they are used in control applications. Compound motors contain two or more field windings, some of which are in series with the armature and others of which are in parallel. Use of several field windings provides the designer with greater flexibility and allows tailoring of the speed-torque characteristic to suit particular needs.

Dc generators are described by the same equations and models that we have used for motors. The field winding can be separately excited, or the field current can be provided by the generator's own output, in which case it is said to be *self-excited*. Self-excited generators are usually shunt connected; the series connection is usually impractical.

For a fixed shaft speed the generated armature emf depends on the field current. The relationship is typically like the solid line of Fig. 17.19. The armature emf is proportional to Φ_p, and the field current is proportional to the mmf of the field winding. Hence this curve has the same shape as the *B–H* curve (magnetization curve) of the field structure. If the generator is separately excited, one simply adjusts I_F to produce the desired output voltage. The output of a self-excited machine can be estimated if we neglect the small voltage drop in R_A. In that case the generated emf appears across the shunt field winding, and $e = I_F R_F$. The latter equation is represented by the dashed

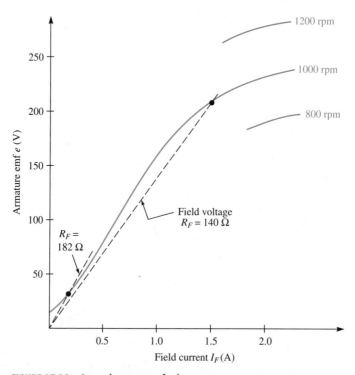

FIGURE 17.19   Output characteristics of a dc generator.

lines in Fig. 17.19. The point where the solid and dashed curves cross indicates the output voltage of the machine. We observe that full output voltage is obtained with $R_F = 140 \ \Omega$. However, if we add an external series resistance to the field winding, making $R_F = 182 \ \Omega$, the output voltage is greatly reduced. Thus output voltage can be adjusted by setting a series resistance in the field circuit.

When the shunt generator is first started up, $I_F = 0$. Without field current one might think that no output could be produced and the machine could not get into operation. However, there is usually some residual magnetism in the field iron; this gives a small output even in the absence of field current: about 15 V in Fig. 17.19. This small initial output voltage causes field current to flow, which in turn raises the output voltage until the steady-state operating point is reached. The ability of the self-excited generator to start itself makes it usable as a power source in remote places where other power is unavailable.

Dc generators today are almost totally obsolete, replaced by ac generators ("alternators") with semiconductor output rectifiers. However, in traction applications (streetcars, locomotives, etc.) the dc drive motors are often used backward as generators, in what is known as the "dynamic braking" mode. Dc motors are still widely used in several ways. Large ones (up to 10,000 hp!) are used in industrial and mobile applications where speed and torque vary widely, and small servo motors are very common in control systems (where, however, they now face competition from digital stepping motors). The great drawback of all* dc machines is the need for a commutator, which is a troublesome part. Because of inductance it is impossible to turn the currents in the armature on and off suddenly; they persist for a while after the commutator switch has opened. The result is a repetitive arc discharge between brushes and commutator, known as *sparking*. Sparking gradually burns away the brushes, which thus need to be replaced periodically. Because of their high maintenance cost, dc machines are replaced by ac machines wherever possible.

17.3 AC Machines

Machines designed specifically for ac service are very widely used. In fact, the induction motor, an ac machine, is the most common electrical machine of any kind. Induction machines can in principle be used as either motors or generators, but in practice they are used almost exclusively as motors. There is a second important class of ac machines, however, known as synchronous machines. Synchronous motors are not very common, but generators of this type supply almost all of the electrical power in use today.

The most common structure of ac machines differs from that of dc machines. In ac machines the armature (the winding through which power is transferred) is usually

---

*An exception is the rare "homopolar" machine, which does not require a commutator. It is useful only as a low-voltage dc generator.

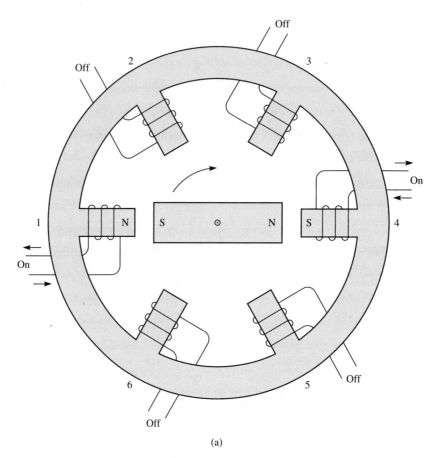

FIGURE 17.20(a)   Rotating magnetic field.

the stator winding. The purpose of this winding, in an ac machine, is to create a *rotating magnetic field*. This principle is illustrated in Fig. 17.20. In Fig. 17.20(a) coils 1 and 4 are excited, and the other coils are turned off; there is a north magnetic pole at the 9 o'clock position and a south pole at 3 o'clock. A short time later we have the condition shown in Fig. 17.20(b); now coils 2 and 5 are on, the others are off, and the north and south poles have moved 60° clockwise. After another interval of time we have the condition shown in Fig. 17.20(c), and so on. If we place a magnet on a shaft in the center, magnetic force will try to line it up with the stator poles, as shown. As the stator excitation rotates, the rotor magnet will follow it around. This is in fact the principle of the synchronous motor.

The structure of Fig. 17.20 illustrates the rotating-field principle, but it is an inefficient design, since two-thirds of the stator magnets are idle at any time. Furthermore, we must address the problem of turning the coil currents on and off in the proper order. (One way to do this would be with a commutator or photoelectric switch con-

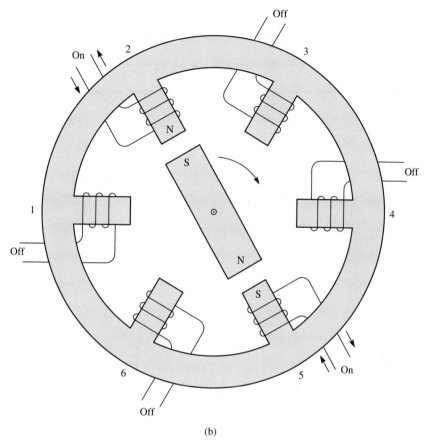

FIGURE 7.20(b)

trolled by the angle of the rotor shaft. This would give a kind of "inside-out" dc machine, but dc machines are not under discussion now.) Suppose that we excite our structure with three-phase ac power. Let the three line voltages be

$$v_A = V_0 \cos \omega t \tag{17.40}$$

$$v_B = V_0 \cos(\omega t + 120°)$$

$$v_C = V_0 \cos(\omega t + 240°)$$

and let us connect the six coils (see Fig. 17.21) as follows:

$$v_1 = v_4 = v_A \tag{17.41}$$

$$v_2 = v_5 = v_C$$

$$v_3 = v_6 = v_B$$

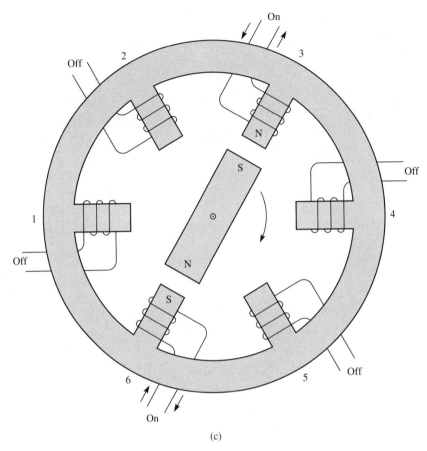

FIGURE 17.20(c)

Table 17.1 shows the operation of the six coils over one complete period ($1/f$) of the power line frequency. The time interval between lines of the table is one-sixth period (2.78 msec for 60 Hz power). These data are illustrated in Fig. 17.22. We observe that the field configuration at time 3 is just that at time 1, except rotated clockwise through 60°. At time 5 it has rotated another 60°; hence the idea of a rotating field. Times 2, 4, and 6 represent intermediate stages. If the rotor is free to move so as to keep its south poles opposite the north poles of the stator, then the rotor turns with the fields, as shown. A two-pole rotor such as that in Fig. 17.20 could be used, but this would leave some of the stator poles unused. A better choice for this particular stator is the four-pole rotor shown in Fig. 17.21. The machine of Fig. 17.21 is called a three-phase, four-pole machine. One might at first think the stator had six poles, but we note that at any of the times listed in Table 17.1 four of the poles are only half-poles; thus the total at any instant is four. From Fig. 17.22 we see that in one full cycle of the power line frequency the shaft turns 180°. In general the shaft speed is given by

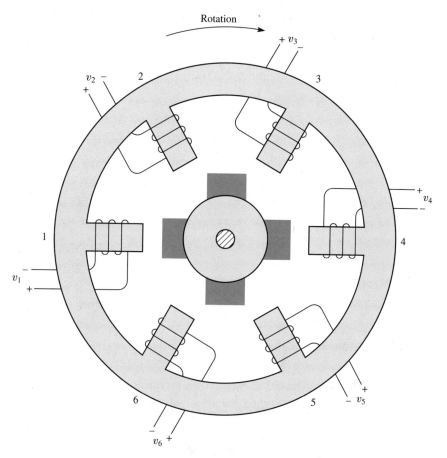

FIGURE 17.21   A four-pole three-phase synchronous machine.

$$n = \frac{2f}{p} \qquad (17.42)$$

where n is the shaft speed in revolutions per second, f is the power line frequency in Hz, and p is the number of poles.

● EXERCISE 17.2

Let us interchange power lines B and C in Eq. (17.41), so that $v_1 = v_4 = v_A$, $v_2 = v_5 = v_B$, and $v_3 = v_6 = v_C$. Construct a set of diagrams similar to those of Fig. 17.22, and show that the effect is to reverse the motor—that is, the rotor now moves in the opposite direction.

TABLE 17.1 Operation of the Six Coils in Figure 17.21

Time (units of 1/6f)	ωt (deg)	v_1	v_2	v_3	v_4	v_5	v_6	Pole 1	Pole 2	Pole 3	Pole 4	Pole 5	Pole 6
1	0	V_0	$-\frac{1}{2}V_0$	$-\frac{1}{2}V_0$	V_0	$-\frac{1}{2}V_0$	$-\frac{1}{2}V_0$	N	$\frac{S}{2}$	$\frac{S}{2}$	N	$\frac{S}{2}$	$\frac{S}{2}$
2	60	$\frac{1}{2}V_0$	$\frac{1}{2}V_0$	$-V_0$	$\frac{1}{2}V_0$	$\frac{1}{2}V_0$	$-V_0$	$\frac{N}{2}$	$\frac{N}{2}$	S	$\frac{N}{2}$	$\frac{N}{2}$	S
3	120	$-\frac{1}{2}V_0$	V_0	$-\frac{1}{2}V_0$	$-\frac{1}{2}V_0$	V_0	$-\frac{1}{2}V_0$	$\frac{S}{2}$	N	$\frac{S}{2}$	$\frac{S}{2}$	N	$\frac{S}{2}$
4	180	$-V_0$	$\frac{1}{2}V_0$	$\frac{1}{2}V_0$	$-V_0$	$\frac{1}{2}V_0$	$\frac{1}{2}V_0$	S	$\frac{N}{2}$	$\frac{N}{2}$	S	$\frac{N}{2}$	$\frac{N}{2}$
5	240	$-\frac{1}{2}V_0$	$-\frac{1}{2}V_0$	V_0	$-\frac{1}{2}V_0$	$-\frac{1}{2}V_0$	V_0	$\frac{S}{2}$	$\frac{S}{2}$	N	$\frac{S}{2}$	$\frac{S}{2}$	N
6	300	$\frac{1}{2}V_0$	$-V_0$	$\frac{1}{2}V_0$	$\frac{1}{2}V_0$	$-V_0$	$\frac{1}{2}V_0$	$\frac{N}{2}$	S	$\frac{N}{2}$	$\frac{N}{2}$	S	$\frac{N}{2}$

SYNCHRONOUS MACHINES

We have seen that the magnetic field configuration rotates. If the magnets of the rotor always keep the same orientation with respect to the field, the rotor must move in synchronized fashion with the rotating field. In this case the shaft speed must have almost exactly the value given by Eq. (17.42). This is the characteristic behavior of a synchronous machine.

In small machines the rotor may carry permanent magnets; in larger machines the rotor carries dc field coils excited through sliding contacts ("slip rings") on the shaft. Small synchronous motors are useful when the speed must be exact, as in clocks or phonograph turntables. However, most applications of large motors require adjustable speed. Furthermore, pure synchronous motors have no starting torque and must be brought up to speed by a starting motor (sometimes built into the main motor structure). For these reasons large synchronous motors are uncommon. The generator is the synchronous machine of greatest importance.

There are two important types of rotors. Those with poles that "stick out," as in Fig. 17.21, are known as *salient rotors*. Other machines have smooth, drum-shaped *cylindrical rotors*. Salient rotors are cheaper to make and have other advantages, but they are unsuitable for high-speed machines because of air resistance and centrifugal force. From the analytical point of view salient rotors are more complicated because the sizes of the air gaps vary as the rotor turns; hence reluctances of the magnetic circuits are not constant. For simplicity our discussion will deal with cylindrical rotors.

The torque acting on the shaft of a synchronous machine results from the interaction of the rotating field and the rotor magnet, as shown in Fig. 17.23. These figures are to be regarded as "snapshots," showing the shaft and fields at particular instants of

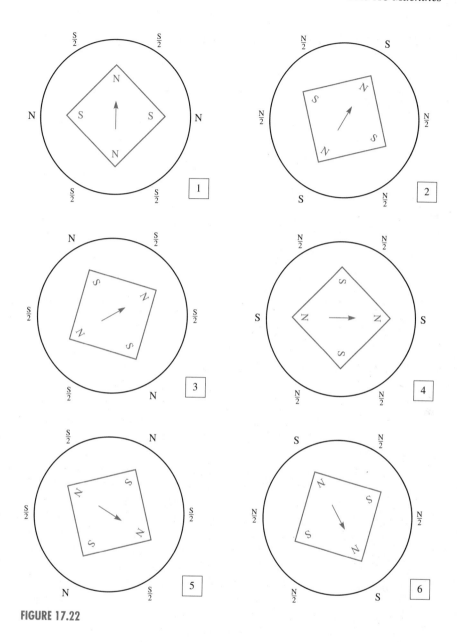

FIGURE 17.22

time. Actually everything in the figures rotates together, clockwise, at the synchronous speed. [Here we are using a two-pole example for the sake of simplicity. From Eq. (17.42) we see that each entire picture rotates clockwise at $60f$ rpm.] The angles *between* the various objects remain constant. The magnetic torque acts to align the rotor magnet, and hence the rotor field, with the total field. In Fig. 17.23(a) the rotor lags behind the total field and the interaction torque pulls it clockwise, thus forcing the

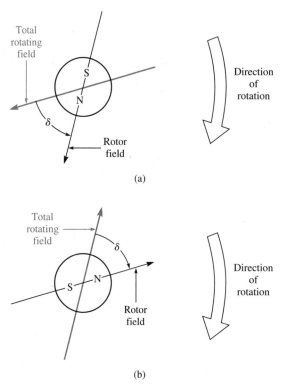

(a)

(b)

FIGURE 17.23 The power angle δ is the angle by which the rotor leads the total magnetic field. It is negative for a motor (a) and positive for a generator (b).

shaft in the direction of rotation. This is the case in a motor. We define the *power angle* δ to be the angle by which the rotor field leads the total field. In Fig. 17.23(a) δ is a negative number. The magnetic torque acting on the shaft in the clockwise direction is $-T_0 \sin \delta$, where T_0 is the maximum torque. As loading torque increases, the motor speed remains constant but $|\delta|$ increases, until the loading torque reaches T_0. If the load is made larger than this, the motor "pulls out" of synchronism and simply stops.

Figure 17.23(b) illustrates the case of a generator. Here δ is a positive number, and the rotor *leads* the total field. Consequently the magnetic torque acts *against* the rotor's motion, as one expects in a generator.

An interesting point is that the total rotating field is not just the field of the stator but rather is the sum of the rotor and stator fields. Figure 17.24(a) corresponds to Fig. 17.23(a) and shows how the rotor field B_R and stator field B_S add to produce the total field B; similarly Fig. 17.24(b) corresponds to Fig. 17.23(b). Both of these "snapshots" are made at the instant when B_S is directed toward the left. We also define the shaft angle ψ, which is the angle by which the rotor field leads the stator field. For a motor $\psi < 0°$, while for a generator $\psi > 0°$.

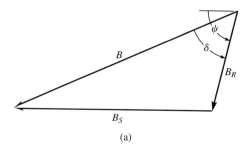

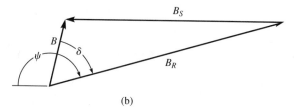

FIGURE 17.24 Magnetic field vector diagrams corresponding to Fig. 17.23. (a) Motor; the shaft angle ψ is between $0°$ and $-180°$. (b) Generator; ψ is between $0°$ and $180°$.

We shall now derive the *I-V* relationship for the machine. The total magnetic flux through each stator winding is

$$\Phi = \frac{NI}{\Re} + \Phi_0 e^{j\psi} \tag{17.43}$$

Here we have taken the stator current to be our phase reference; thus **I** is a real number. N is the number of turns in a stator winding, $\Re$ is the reluctance of the magnetic circuit, and Φ_0 is the flux of the rotor magnet. The stator voltage is then

$$\mathbf{v} = Nj\omega\Phi = \frac{j\omega N^2 I}{\Re} + j\omega N\Phi_0 e^{j\psi} \tag{17.44}$$

(Note that all quantities in this equation are amplitudes, not rms voltages.) We define the *synchronous reactance* X_s and the fictitious open-circuit voltage **e** according to*

$$X_s = \frac{\omega N^2}{\Re} \tag{17.45}$$

$$\mathbf{e} = j\omega N\Phi_0 e^{j\psi}$$

---

*In practice the voltage **e** would not be measured under open-circuit conditions because the iron would saturate. Thus **e** is not truly an open-circuit voltage, hence "fictitious."

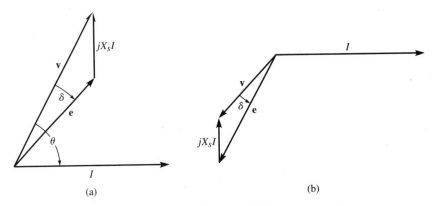

FIGURE 17.25   Phasor diagrams for (a) synchronous motor and (b) generator.

Thus we have the *I-V* relationship

$$\mathbf{v} = jX_s I + \mathbf{e} \tag{17.46}$$

Typical phasor diagrams are shown in Fig. 17.25. We observe that the power input to the machine, $P = \frac{1}{2}\,\mathrm{Re}(\mathbf{v}i^*)$, is positive for the motor, and negative for the generator, as expected. The motor can be either an inductive or capacitive load. The generator acts as a voltage source in series with an inductance, as is seen from Eq. (17.46). It is possible to identify the power angle δ in the phasor diagrams. From Eq. (17.45) the phase angle of $\mathbf{e}$ is $90° + \psi$. Moreover, since $\mathbf{v} \propto j\omega\mathbf{B}$, the phase angle of $\mathbf{v}$ is $90° + \psi - \delta$. Thus $\mathbf{e}$ leads $\mathbf{v}$ by the angle δ, as shown in Fig. 17.25.

An interesting special case is that of a motor without load. The absence of torque implies $\delta = 0$, while the absence of power conversion implies $\theta = 90°$. The phasor diagram is as shown in Fig. 17.26. Two cases are possible. If the rotor flux is made small, such that $|\mathbf{e}| < |\mathbf{v}|$, the machine is said to be *underexcited*, as in Fig. 17.26(a). In this case I is positive and lags $\mathbf{v}$ by $90°$; the machine is referred to as a *synchronous inductor*. However, if $|\mathbf{e}| > |\mathbf{v}|$, the machine is said to be *overexcited*. As we see from

FIGURE 17.26   Phasor diagrams for an unloaded synchronous motor. (a) Underexcited machine (synchronous inductor). (b) Overexcited machine (synchronous condenser).

Fig. 17.26(b), I in such a case must be negative, and hence the current leads the voltage by 90°. The overexcited machine, which thus acts like a capacitor, is known as a *synchronous condenser* ("condenser" being an obsolete synonym for "capacitor" that survives in this context). Synchronous condensers are used in power systems because they can draw large amounts of negative reactive power for power factor correction.

EXAMPLE 17.8

Show that for a lossless three-phase synchronous motor the output power is given by $P = 3VE \sin \delta / X_s$, where $V = |\mathbf{v}|/\sqrt{2}$ and $E = |\mathbf{e}|/\sqrt{2}$.

SOLUTION

Refer to Fig. 17.25(a). We observe (by drawing a line perpendicular to $\mathbf{v}$) that $E \sin \delta = X_s I \sin (90° - \theta) = X_s I \cos \theta$. Multiplying both sides by V and noting that the power entering one phase is $VI \cos \theta / \sqrt{2}$, the result follows.

In a three-phase, p-pole generator there are $3p/2$ armature coils (see Fig. 17.21 for the case $p = 4$). The voltage induced in the kth coil is

$$v_k(t) = V_0 \cos \left(\pi npt + \beta - \frac{2\pi}{3}k \right) \tag{17.47}$$

where $n = 2f/p$ is the shaft speed in rpm, V_0 is a constant, and $\beta = 90° + \psi - \delta$. We see that $\angle v_{k+3} = \angle v_k$; thus windings 1 and 4 should be connected in series, as should 2 and 5, and so on. This results in three output voltages, which from Eq. (17.47) can be seen to form a three-phase set. If the three voltages are wye connected, the result is a wye-connected three-phase generator with phase voltage $2V_0$.

EXAMPLE 17.9

Find the open-circuit line voltage of a four-pole wye-connected three-phase generator for which $|\mathbf{e}| = 600$ V.

SOLUTION

The wiring diagram of the armature is as shown in Fig. 17.27. The constant β is in fact arbitrary, since it depends only on what instant is chosen to be $t = 0$; here we let $\beta = 0$. Note that phase 2

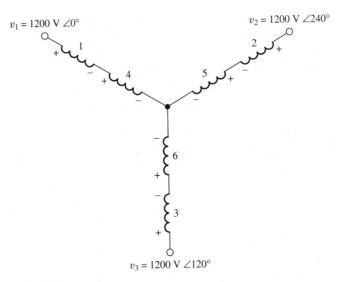

$v_1 = 1200$ V $\angle 0°$ $v_2 = 1200$ V $\angle 240°$

$v_3 = 1200$ V $\angle 120°$

FIGURE 17.27

lags phase 1 by 120° (as can be seen from Fig. 17.21) but is designated equivalently as +240°. The line voltage is

$$V_{line} = |\mathbf{v}_1 - \mathbf{v}_2| = 1{,}200\sqrt{3} = 2{,}078 \text{ V}$$

INDUCTION MOTORS

Induction motors resemble synchronous machines as far as the stator is concerned; the same rotating field configuration is used. The rotor, however, is different. Instead of being excited by a dc current as in a synchronous machine, the rotor windings are excited by ac. Moreover, the induction motor's rotor current does not have to be introduced through slip rings. On the contrary, the rotor of an induction motor acts like the secondary of a transformer; the rotor current is acquired from the stator current through transformer action. This simplicity makes the induction motor inexpensive and trouble-free; it also has good starting torque and favorable speed-torque character-istics. For these reasons it is the most common of all motors.

As in the synchronous machines, the rotating stator field turns at the rate $n = 2f/p$ revolutions per second, where f is the ordinary frequency of the power line. Let the rotor speed be n_R. We define the *slip s* according to

$$s = \frac{n - n_R}{n} \tag{17.48}$$

To an observer stationary with respect to the rotor, the stator field seems to rotate at $n - n_R = sn$ revolutions per second. Each time the stator field revolves once with respect to the rotor, the stator's field passes through $p/2$ complete periods. Thus the frequency of the current induced in the rotor is $spn/2$, which is simply equal to sf.

Some induction motors use *squirrel-cage rotors*. These are cylindrical arrangements of metal bars short-circuited at their ends to provide closed conduction paths. Other machines use *wound rotors*, which are something like the rotors of a synchronous machine. Wound rotors can be short-circuited inside the machine to provide a closed current path. Alternatively, their terminals are sometimes brought out through slip rings, allowing a variable resistance to be inserted in the rotor circuit for control purposes. If a dc current is passed through the wound rotor, the machine can run as a synchronous machine. If the rotor is prevented from turning, the machine can be (and sometimes is) used as a transformer.

We shall now derive a model for the induction motor. The three phases of the machine act identically, so it is sufficient to consider only one of the three sets of windings, as shown in Fig. 17.28(a). The stator winding has wire resistance R_1 and leakage reactance X_1 and is coupled to the rotor by transformer action. For each pole let us imagine that there are N_1 primary turns (on the stator) and N_2 secondary turns (on the rotor). The secondary circuit has resistance R_2 and leakage reactance X_2. The transformer in this circuit has an unusual property: The frequencies of the currents in its primary and secondary windings are not the same! This seems at first to be impossible, but it is explained by the fact that the windings are in motion with respect to one another. A consequence is that the usual formula $\mathbf{e}_2/\mathbf{e}_1 = N_2/N_1$ does not apply. Let the flux per pole be Φ. Then the voltage across the stator winding is $\mathbf{e}_1 = N_1 j(2\pi f)\Phi$. We have already seen, however, that the frequency of the current induced in the rotor is sf; thus $\mathbf{e}_2 = N_2 j(2\pi sf)\Phi$. The voltage ratio for this strange transformer is thus $\mathbf{e}_2/\mathbf{e}_1 = sN_2/N_1$. However, the relationship $\mathbf{i}_2/\mathbf{i}_1 = -N_1/N_2$ continues to apply, as it is based on Φ rather than on $d\Phi/dt$. Consequently the strange transformer does not conserve power. The power entering the primary is greater than the power dissipated in R_2. The difference is the power converted to useful mechanical output.

The strange transformer can be replaced in our model by a conventional ideal transformer if we divide the impedances in the secondary circuit by s, as shown in Fig. 17.28(b). This change keeps the currents in the secondary and primary circuits the same. However, the power lost in the secondary winding resistance is $\frac{1}{2}|\mathbf{i}_2|^2 R_2$. To represent this in the model, we replace the resistor R_2/s by two resistors in series whose values are R_2 and $(1 - s)R_2/s$, as shown in Fig. 17.28(c). Now our model has the correct primary and secondary currents, and we can observe that the motor's useful mechanical output power is

$$P_{\text{out}} = \frac{1 - s}{2s}|\mathbf{i}_2|^2 R_2 \tag{17.49}$$

Finally, we refer the secondary circuit to the primary side, obtaining Fig. 17.28(d). Here

$$X' = X_1 + \left(\frac{N_1}{N_2}\right)^2 X_2 \tag{17.50}$$

$$R' = R_1 + \left(\frac{N_1}{N_2}\right)^2 R_2$$

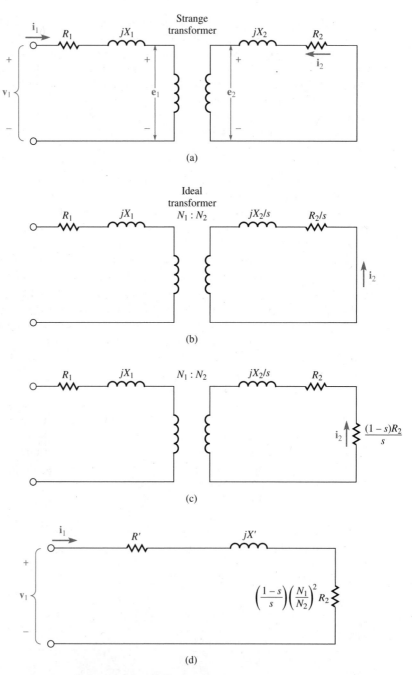

FIGURE 17.28   Development of the model for the induction motor.

This model retains the property that the power delivered to the speed-dependent resistance on the right represents the useful output of the motor. We note that the induction motor always acts as an inductive load. We can readily find its power factor from our model.

The power transmitted to the speed-dependent resistance is easily found to be

$$P' = \frac{1}{2}|\mathbf{i}_1|^2 \left(\frac{1-s}{s}\right)\left(\frac{N_1}{N_2}\right)^2 R_2 \tag{17.51}$$

$$= \frac{s(1-s)R''V^2}{[sR' + (1-s)R'']^2 + s^2 X'^2}$$

where $V = |\mathbf{v}_1'|$ is the rms input voltage and $R'' \equiv (N_1/N_2)^2 R_2$. The total output power of a three-phase motor is $3P'$, and the torque is given by

$$T = \frac{3P'}{2\pi n_R} = \frac{3P'p}{4\pi(1-s)f} \tag{17.52}$$

A typical dependence of torque on speed, as given by Eqs. (17.51) and (17.52), is shown in Fig. 17.29. Torque drops to zero at the synchronous speed $2f/p$, as expected, since no currents are induced in the stator when it rotates at the synchronous speed. If the loading torque has a constant value of T_0 less than the starting torque, the motor will start from rest and increase its speed until T drops to equal T_0; it then continues to run at this equilibrium speed. Normally induction motors are run just below synchronous speed, with s in the range 0.01 to 0.05.

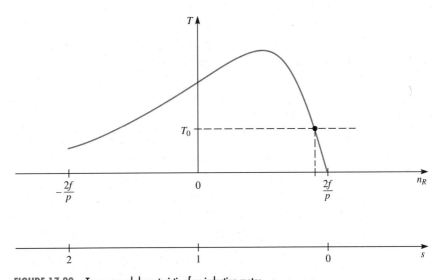

FIGURE 17.29 Torque-speed characteristic of an induction motor.

EXAMPLE 17.10

A three-phase four-pole induction motor operates with 440 V rms line voltage across each phase. When the rotor is held stationary for a very short test period, the electrical input is 289 kVA, with power factor 0.450, lagging, for each phase. When the motor runs at 900 rpm, the per-phase input is 239 kVA, with power factor 0.676, lagging. Find the model parameters R', X', and $(N_1/N_2)^2 R_2$.

SOLUTION

In the first measurement $s = 1$, and from Fig. 17.28(d) we see that we measure $R' + jX'$. We have

$$\frac{(440)^2}{\sqrt{(R')^2 + (X')^2}} = 289{,}000$$

$$\cos\left[\tan^{-1}\left(\frac{X'}{R'}\right)\right] = 0.45$$

from which we find $R' = 0.3\ \Omega$ and $X' = 0.6\ \Omega$.

The speed 900 rpm corresponds to $n_R = 15$ rps $= f/4$. Thus we have $s = \frac{1}{2}$. From Fig. 17.28(d) we have

$$\frac{(440)^2}{\sqrt{(R' + R'')^2 + (X')^2}} = 239{,}000$$

$$\cos\left[\tan^{-1}\left(\frac{X'}{R' + R''}\right)\right] = 0.676$$

By solving either of these equations we find $R'' = 0.25\ \Omega$.

In an actual case one would probably not apply full voltage with the rotor blocked, since there would be excessive current and possible damage to the stator winding. The same information could be obtained from a test at reduced voltage.

SINGLE-PHASE MOTORS

A single-phase ac motor is sketched in Fig. 17.30. Very small synchronous motors of this kind are occasionally used for constant-speed applications, but most single-phase motors are induction motors. Single-phase motors are often encountered in daily life, but their use is confined to low-power applications, where three-phase power is often unavailable.

The magnetic field produced by the stator in Fig. 17.30 always points in the x direction, oscillating between upward and downward. Since the field is oscillating rather than rotating, it seems remarkable that such a motor can work at all. Insight can be gained by "decomposing" the oscillating field into two rotating fields turning in

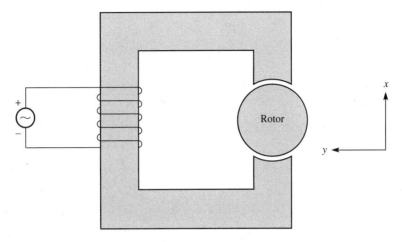

FIGURE 17.30   Single-phase ac motor.

opposite directions. Let $\vec{B}(t)$ be given by

$$\vec{B}(t) = B_0 \vec{e}_x \cos \omega t \qquad (17.53)$$

where $\vec{e}_x$ is the unit vector in the x direction. This expression is identically equal to

$$\vec{B}(t) = \vec{B}_1(t) + \vec{B}_2(t) \qquad (17.54)$$

where

$$\vec{B}_1(t) = \tfrac{1}{2}B_0(\vec{e}_x \cos \omega t + \vec{e}_y \sin \omega t)$$
$$\vec{B}_2(t) = \tfrac{1}{2}B_0(\vec{e}_x \cos \omega t - \vec{e}_y \sin \omega t) \qquad (17.55)$$

A moment's study reveals that $\vec{B}_1(t)$ is a rotating field rotating in the counterclockwise direction, while $\vec{B}_2(t)$ is a rotating field rotating clockwise. (For example, at $t = 0$, $\vec{B}_2$ points up; at $t = \pi/2\omega$ it points to the right; at $t = \pi/\omega$ it points down; and so on.) It is possible for the rotor to lock onto one or the other of these two component fields and rotate in response to it. For the case of an induction motor, the torque-speed characteristic is found by adding to Fig. 17.29, which represents the response to one rotating field, a similar inverted curve representing response to the oppositely rotating field. This is shown in Fig. 17.31. The curve marked "sum" passes through the origin, indicating that the motor has no starting torque. (This is, of course, what one would expect from symmetry.) Once started in either direction, however, the motor generates torque that tends to keep it rotating in that direction.

In order for single-phase motors to be self-starting, the stator poles can be *shaded*. A shaded pole is designed to make the magnetic field on one side lag the field on the other side of the same pole. This causes the field to move sideways slightly across the pole, providing a small component of rotating field that allows the motor to start. Shaded poles are used in small motors up to about $\tfrac{1}{20}$ hp. Another approach, used up to

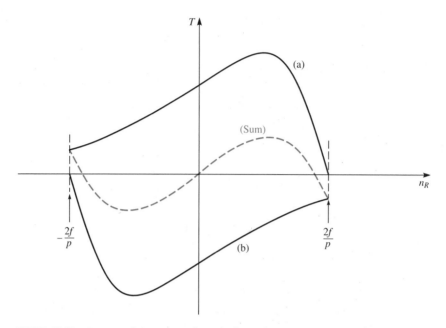

FIGURE 17.31 Torque-speed characteristic of a single-phase induction motor. (a) Torque from the clockwise-rotating component of the stator field. (b) Torque from the counterclockwise component.

about $\frac{1}{2}$ hp, is to provide a *split-phase* stator winding. In this method the main stator winding is supplemented by a starting winding. The latter has high inductance and thus carries a current perhaps 30° to 45° out of phase with that of the main winding. This is equivalent to excitation by a two-phase power source and provides a component of rotating field. When the motor has gained speed, the starting winding is cut out of the circuit by a centrifugal switch on the shaft.

Capacitor-start motors are similar, except that a capacitor is placed in series with a starting winding to increase the phase difference between the windings. This results in higher starting torque. In a *capacitor motor* the starting winding is not switched out, and its series capacitance is used to correct the power factor, thus increasing the efficiency.

STEPPING MOTORS

The motors used in analog control systems are usually servo motors, small reversible dc motors with permanent-magnet stators. In modern work, however, control functions are increasingly performed by digital systems. An appropriate output device for such systems is the digital *stepping motor*. Stepping motors are similar to polyphase synchronous motors. However, they are intended not to turn continuously but rather to move to desired positions under digital control.

The design of a four-pole, four-phase stepping motor is illustrated in Fig. 17.32(a), and a typical sequence of the four magnet currents is shown in Fig. 17.32(b). (Note

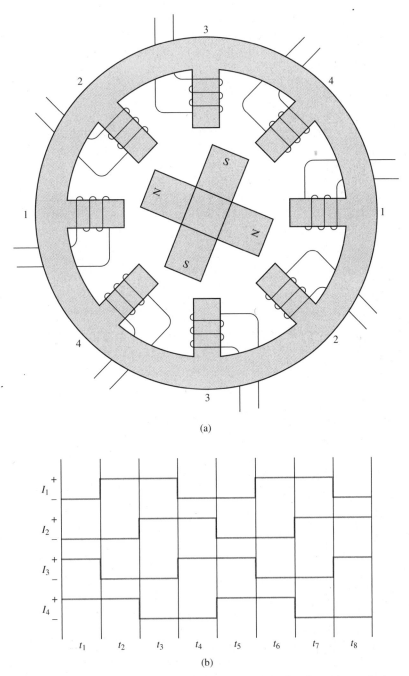

FIGURE 17.32 A four-phase four-pole digital stepping motor. (a) Machine design. (b) Typical coil currents for clockwise rotation.

that each current passes through two coils—current 1 through the coils at 9 o'clock and 3 o'clock, and so on.) Each current can have only two values, designated (+) and (−), and we shall assume that when the current is (+) the pole is north and when the current is (−) the pole is south. In Fig. 17.32(a) the rotor is shown at time t_1 [first column of Fig. 17.32(b)]. Poles 1 and 2 are south, and a north pole of the rotor rests between them. At time t_2 pole 1 changes to north and pole 3 changes to south, and the rotor moves one-eighth revolution clockwise; thus its north pole now rests between poles 2 and 3. We note that there is no possibility of the rotor moving in the wrong direction; since at t_2 poles 1 and 4 are both north, the north pole of the rotor cannot possibly move counterclockwise. For continuous clockwise rotation, the coil voltages continue as shown in Fig. 17.32(b). We note that the four coil currents are four square waves in four different phases, 90° apart; hence this is called a four-phase motor. The number of degrees through which a four-phase p-pole motor turns in each step is $180/p$, 45° in

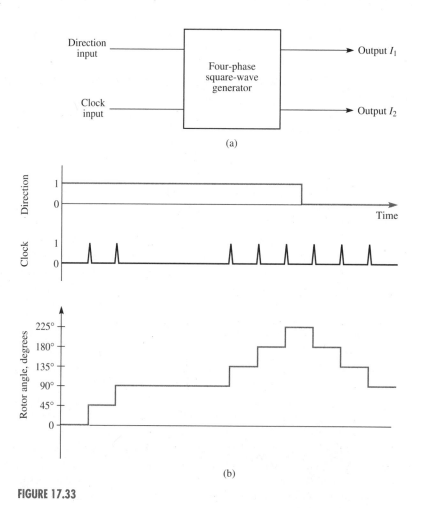

(a)

(b)

FIGURE 17.33

this example. Real motors are made with a larger number of poles in order to give finer position control. A typical motor might have 24 poles and turn 7.5° per step.

In fact, only two currents are needed to drive the motor because $I_1 = -I_3$ and $I_2 = -I_4$. Thus we need pass only a single current through coils 1 and 3, with coil 3 connected backward. A simple digital circuit known as a four-phase square-wave generator is used to produce I_1 and I_2 in response to inputs DIRECTION and CLOCK, as shown in Fig. 17.33(a). When DIRECTION = **1** each clock pulse advances the square wave by one time unit, following the pattern of Fig. 17.32(b). The clock pulses do not have to arrive at regular intervals; they are the output of a control system, and the motor can start and stop as necessary. When DIRECTION = **0** the square wave is generated in reverse phase sequence, and the motor moves 45° counterclockwise at each pulse. A typical sequence of operations is shown in Fig. 17.33(b). During the interval between the second and third clock pulses, the motor does not move. When DIRECTION changes from **1** to **0** the direction of motion is reversed.

17.4 Application: BART, a Modern Electric Railway System

Electrical machinery, unlike transistors and other things studied in this book, can be impressively large. Recent developments in electrical engineering tend to involve miniaturization: integrated circuits, cellular radio, laptop computers. But there can be no doubt that giant machines have a certain excitement about them, at least for the sort of people who become engineers. The largest electrical machines are probably the stationary generators used at dams and power stations. These are so large that they completely fill rooms several stories high, but they don't show much outside sign of activity; they just stand there and whirr. The next largest electrical machines are what are known as traction motors—for instance, the sort that drive electric trains.

Modern urban railroads are almost always electrically powered. As compared with diesel power, electric trains are quiet and non-polluting, and have no exhaust that might poison people in tunnels. Moreover, with electric power it is possible to do without the heavy and space-consuming locomotive, because the drive motors can be right in the passenger cars themselves. In fairness, one might note that electric trains also have some disadvantages. There is the extra cost of "electrifying" the line, that is, of building the overhead electric wires or a "third rail." Of these two choices, overhead wires are undesirable because they are ugly. But the third rail, which supplies electricity through a sliding contact, presents an electrocution hazard, and it also has a notorious tendency to become ice-coated in winter weather. When the third rail ices up, the trains stop, all the trains behind them stop, the whole system stops, and large numbers of angry commuters discover that they cannot get home. An engineer who has done a bad job of protecting his third rail from the weather may find himself extremely unpopular!

The San Francisco Bay Area is served by the Bay Area Rapid Transit system, or BART, shown in Fig. 17.34. This is an interesting modern system, with some unusual features. Although each train has an operator at the usual position in the front, the operator does not normally "drive" the train; all trains are controlled by the system's

FIGURE 17.34   Bay Area Rapid Transit (BART), a modern electric railroad system. (*Courtesy of BART.*)

central computer. (The driver works the doors and watches for problems, if necessary taking over in emergencies.) Computer control gives the system a slightly spooky exactitude. Not only do the trains arrive rigidly on time, according to the published schedule, but they even stop at predictable places on the platform. The regular commuters, who know BART's habits, form lines at particular places, where they know the train's doors will be!

Power for the system is supplied by the local electric utility company via 34.5 kV ac lines. However, the trains themselves operate on 1,000 volt dc. Thus at almost every station on the line there is a power supply ("substation") for that region of the track, with step-down transformers and solid-state rectifiers. The dc power is then supplied to trains in the vicinity by means of a third rail. Each car contains four series-wound dc motors, each supplying up to 140 hp, equivalent to about 100 kilowatts. This suggests that each car would draw 400 amperes, but in fact the power efficiency is low when the train is starting, and each car draws 800 amperes. There are 4 to 10 cars in a train, so the third rail supplies 8,000 amperes! Clearly, with such large currents, even a small amount of resistance will produce large voltage drops in the third rail. For this reason, the distance between the "substations" has to be kept small.

One interesting feature of the system is its use of dynamic and regenerative braking. In old-fashioned braking, friction brakes rub against the rotating wheels, converting the train's kinetic energy into waste heat. However, in BART and other modern systems, the trains can be stopped by reversing the action of the motors, turning them into generators. (This is done by reversing the direction of current through the armatures, while keeping the field currents unchanged.) The braking is called "dynamic" when the electricity thus generated is sent to a bank of resistors, where it is converted to waste heat. However, it is also possible to return the electricity to the third rail, so that the energy is not wasted; this is known as "regenerative" braking. Ideally, one would like all braking to be regenerative, but as it turns out this cannot be done, because of fluctuations in the system voltage. The large currents drawn by the trains cause the voltage of the third rail to vary over a wide range, and when the rail voltage is too high, or when the train is moving slowly, the generated voltage may be too low to push current back into the line. Thus the resistors have to be used sometimes. One might speculate that future generations of clever engineers may overcome this problem and design systems that are 100 percent regenerative. On the other hand, BART's switching systems are already quite complex, and when one thinks about the enormous currents being switched, it is remarkable that it can be done at all!

Another interesting feature has to do with how the motors are controlled. One cannot easily turn their voltage up and down, because the dc voltage obtained from the third rail would be difficult to change. Instead, the voltage applied to the motors is "chopped," i.e., turned rapidly on and off. The scheme is shown in Fig. 17.35. Current from the third rail is rapidly turned on and off by a "switch," which in reality is an externally controlled semiconductor switching device known as a thyristor. When the switch is closed, current flows from the third rail, through the switch, through the motor, and to ground. The diode, which is reverse-biased, plays no role. When the switch is opened, however, the current through the motor cannot immediately stop, because the motor has a great deal of inductance. Instead, the current through the motor continues, now going around through the diode. The motor is still producing some power at this instant because current is still flowing through it, but, of course, the current gradually dies away. Then the thyristor switch is closed, and current through the motor increases again. All this happens very rapidly, so smooth power is experienced by riders on the train.

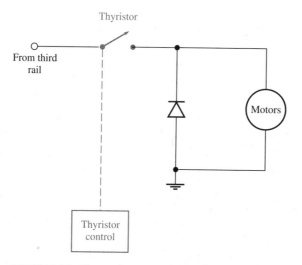

FIGURE 17.35 Diagram of motor control unit used in BART.

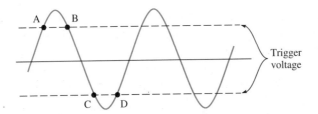

FIGURE 17.36 Pulsewidth control of dc current. Dc current is turned on when the sinusoid exceeds an adjustable trigger voltage.

The electronics that control the thyristors begin with a sinusoidal oscillator operating at 218 Hz; its output is sketched in Fig. 17.36. A trigger voltage is chosen, so that whenever the absolute value of the oscillator voltage exceeds the trigger voltage, the thyristor is switched on. Thus at time "A" in the figure, current is switched on; at "B" it goes off; at "C" on, and so forth. When the trigger voltage is low, current flows nearly all the time; when it is high, current flows in short bursts at the peaks of each cycle. Thus, by adjusting the trigger voltage, one operates the motors at higher or lower power. This is, incidentally, the same way a household light dimmer works; there, the 60 Hz line voltage takes the place of BART's 218 Hz sinusoid. So you can imagine that BART's motors are controlled by a dimmer switch, very much like the one in your dining room. Remember, though, that at BART hundreds of amperes are being switched on and off!

Turning such large currents on and off makes an audible noise. Riders on a BART train hear a distinctive "oooooo" sound at the 218 Hz oscillator frequency and its harmonics. The sound starts loudly as the train begins to move, then becomes quieter as the train gets up to speed. This is because when the train is at the right speed, the trigger voltage is raised to maximum and the "on" time of the chopped current drops to zero. Hundreds of thousands of riders hear the "ooooo" sound each day, but probably only a few engineers know what it is!

POINTS TO REMEMBER

- Transducers are analog devices intended to produce mechanical output proportional to electrical input, or vice versa. Two important classes are *moving-coil transducers* and *variable reluctance* transducers.

- *Actuators* are electromechanical output devices intended to perform mechanical operations in response to electrical inputs. Examples are solenoids and relays.

- Dc rotating machines are usually distinguished by the use of a shaft-mounted rotary switch called a *commutator*. The commutator switches the currents in the armature so that the torque is constantly in one direction.

- Transducers and motors can be represented by circuit models. Each of these models contains a voltage source representing the back emf generated by mechanical motion.

- Dc rotating machines can be *shunt connected, series connected, compound,* or *separately excited.* These connections are used in order to obtain suitable *torque-speed characteristics.*

- Two important classes of ac machines are *synchronous machines* and *induction machines.* The former is most common as a generator, the latter as a motor.

- Ac machines are based on the use of a *rotating magnetic field.*

- In synchronous machines the rotor turns at the same speed as the rotating field. (Thus synchronous motors can operate at only a single fixed speed determined by the power-line frequency.)

- A synchronous motor running without load can be used as an adjustable reactance for power factor correction. When used as a capacitor in this way the machine is called a *synchronous condenser.*

- Large ac machines are usually used in connection with three-phase power. Small motors, however, are often *single-phase machines. Shaded poles, split-phase* stator windings, or *capacitor starting* are used to obtain starting torque in single-phase motors.

- *Stepping motors* are small actuators that resemble ac motors. They are used as output devices in digital control systems.

PROBLEMS

Section 17.1

17.1 Explain why Eq. (17.1), describing the force on the moving-coil transducer of Fig. 17.1, contains the force on only one of the coil's four sides. What would the force be if the entire coil were in the region of uniform magnetic field?

*17.2 The moving coil of Fig. 17.1 has 100 turns in the form of a square 4 cm on a side. The magnetic flux density is 1.2 T, and the current is 3 A. Find the force exerted by the transducer. How many grams of mass could be lifted by this force? (The gravitational force on a 1 g mass is 9.8 mN.)

*17.3 Find the maximum instantaneous emf generated by the coil in Problem 17.2 if an external force moves it sinusoidally at a frequency of 1,000 Hz. The amplitude of the motion is 0.5 mm.

*17.4 For the loudspeaker in Example 17.1:
  a. Find the effective resistance R_m.
  b. To approximately what value must the frequency be increased in order for the effective impedance to begin to vary as a function of frequency?

**17.5 When there are no restrictions on the driving frequency, the loudspeaker of Example 17.1 can be represented by an effective impedance $\mathbf{Z}_m$, defined by $\mathbf{e} = \mathbf{Z}_m \mathbf{i} \equiv (R'_m + jX'_m)\mathbf{i}$. Obtain expressions for R'_m and X'_m as functions of frequency. Sketch the behavior of R'_m and X'_m over the frequency range 10 Hz to 20,000 Hz. Neglect resistance of the coil windings.

**17.6   The loudspeaker of Problem 17.5 is driven by an amplifier. This amplifier has an open-circuit output voltage of 10 V rms and an output resistance of 8 Ω independent of frequency. Find and sketch the power delivered to the loudspeaker as a function of frequency, over the range 10 Hz to 20,000 Hz. (Neglect resistance of the coil windings.) Discuss the usefulness of this loudspeaker for hi-fidelity music reproduction. Can you suggest modifications of the speaker design that would improve it?

*17.7   In this problem you are asked to design a moving-coil phonograph pickup. The design specification is that it must produce an open-circuit output voltage of 5 mV at 100 Hz when the coil is in sinusoidal motion with an amplitude of 1 mil (10^{-3} inch). The geometry is to be that of Fig. 17.4. Choose suitable values for N, B, and a, bearing in mind the typical dimensions of a phonograph pickup, the difficulty of winding small coils, and the limits of permanent magnets.

*17.8   The phonograph pickup of Example 17.3 has the disadvantage that its open-circuit output voltage is proportional to frequency. Show that this can be corrected by placing a capacitor in parallel with the load resistance. Find the required value of the capacitance if the voltage across the load is to be approximately constant for frequencies greater than 20 Hz. The load resistance R_L is 1,000 Ω, and the coil resistance is 100 Ω.

17.9   The coil of a typical d'Arsonval meter movement is about 1 cm in size. Choose suitable values (bearing technical limitations in mind) of N, B, h, a, and k_a to produce a meter that reads 1 mA, full scale. (Let full scale correspond to a rotation of 90°.)

17.10   A magnetic actuator has the form shown in Fig. 17.6. The coil consists of 500 turns of 20 gauge copper wire. Measurements are made of the inductance seen at the coil terminals with the plunger stationary. We find that $L = 2H$ when $x = 0$ and $L = 0.5H$ when $x = a = 2$ cm. Calculate the reluctance of the magnetic circuit for these two values of x.

*17.11   For the actuator of Problem 17.10 $\Re$ is probably a complicated function of x. However, for simplicity we might approximate $\Re(x)$ by drawing a straight line between the two values found in that problem. Using this approximation and assuming that the coil current is held constant at 1 A, find the force exerted by the actuator at $x = 0$, $x = 1$ cm, and $x = 2$ cm. Express the force in terms of the number of kilograms that can be lifted in the presence of normal gravity.

17.12   The coil of a certain relay is operated by 12 V dc and has a resistance of 10,000 Ω. The contacts (intended for alternating current) are rated 110 V, 5 A rms. Find the relay's power gain.

*17.13   The magnet of the relay shown in Fig. 17.8 exerts a force of 2.5 N. The mass of the moving arm is 50 g, and the average distance it moves is 3 mm. Estimate the time that passes after the coil is energized until the contacts are closed. How does this compare with a transistor switch?

*17.14   For a certain variable-reluctance phonograph pickup $\Re = 15,000$ SI units and $d\Re/dx = 3 \times 10^6$ SI units (where x is the stylus position). The coil has 100 turns, and the coil current is constant at 10 mA. Find the rms voltage produced by the pickup if the stylus moves sinusoidally at 100 Hz with an amplitude of 0.001 inch.

Section 17.2

17.15 Find the torque in SI units of the following:
   a. A 5 hp 1,200 rpm electric motor.
   b. A 100 hp automobile engine running at 5,000 rpm.
   c. Convert your results to lb-feet. (*Note:* 1 hp = 550 ft-lb/sec = 746 W.) How many newtons equal one pound of force?

*17.16 The motor of Example 17.5 is redesigned so that the stator pole pieces are similar to those of Fig. 17.5(b). As a result B has the value 0.6 T and is always perpendicular to the gap. Find the output power and compare with the result found in the example.

*17.17 In the dc machine of Fig. 17.13 let the current in each wire of the stator be 15 A and let there be 1,200 wires per slot. The diameter of the stator is 30 cm, and it is 30 cm long (in the dimension perpendicular to the page). Assume that all of the reluctance in the magnetic circuit arises from the gap, whose width is 0.5 cm. Find the flux per stator pole.

*17.18 Find the power of the motor in Problem 17.17 if there are 348 wires in the armature and each carries 40 A. The rotor speed is 1,200 rpm. Express your result in both watts and hp.

*17.19 Using a format similar to Fig. 17.14, design a four-pole armature with four commutator segments, $N = 2$, $a = 1$. Check your design by constructing a table and four diagrams similar to those of Fig. 17.15, and show that your design will work correctly with a four-pole field winding.

**17.20 Repeat Example 17.6 if an armature resistance of 0.5 Ω is included. Neglect friction, and assume the power output of the motor is 500 hp.

*17.21 Graph the speed-torque characteristic (N-m versus rpm) of a series-connected six-pole dc motor with the following characteristics: operating voltage 800 V, $R_A = 0.3\ \Omega$, $R_F = 0.5\ \Omega$, and $k_2 = 0.0075$ Wb/A. The armature is lap-wound with 100 loops ($N = 100$).

*17.22 Graph the speed-torque characteristic (N-m versus rpm) of an eight-pole shunt-connected dc motor with the following characteristics: operating voltage 2,000 V, $R_A = 0.3\ \Omega$, $R_F = 200\ \Omega$, and $k_2 = 0.005$ Wb/A. The armature is wave-wound with 120 loops ($N = 120$).

*17.23 Predict how the armature current and the speed of an ideal dc motor (no resistance, no friction) would be affected by the following changes in operating conditions:
   a. Keeping field current I_F and torque T constant and halving the armature voltage V.
   b. Keeping I_F and output power P constant and halving V.

c. Keeping V and T constant and doubling I_F.

d. Keeping P constant and halving both I_F and V.

*17.24 Figure 17.37 shows a *cumulative compound* dc motor. (The term "compound" indicates that both shunt and series field windings are used; "cumulative" means that the flux contributions of the two field windings add.) Let the flux per pole of the shunt winding be Φ_1 (a constant), while the flux per pole of the series winding is $k_2' I_A$. The resistances of the series winding and armature are R_S and R_A, respectively.

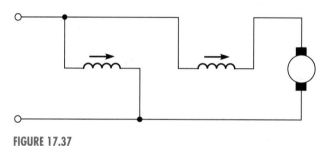

FIGURE 17.37

a. Obtain an expression for torque as a function of the angular speed Ω.

b. Show that your expression reduces to that for a series motor when $\Phi_1 = 0$ and to that of a shunt motor when $k_2' = 0$.

c. Determine whether or not the motor "runs away" if the load is reduced to zero.

**17.25 If the sign of k_2' is negative in Problem 17.24, the motor of Fig. 17.37 becomes a *differential compound* motor. Differential compounding is almost never used because it leads to instability—for example, to cases in which $d\Omega/dT > 0$.

a. Obtain an expression for T as a function of Ω. For simplicity let all winding resistances be zero.

b. Sketch a graph of T as a function of Ω. Can this machine be started from rest? Can it run stably at any speed? (What if the speed increases a little?)

*17.26 The machine whose characteristic is shown in Fig. 17.19 is run as a series dc motor. The resistance of the field winding is 50 Ω, the motor speed is 1,000 rpm, and the current through the motor is 0.9 A. What is the power-line voltage?

Section 17.3

**17.27 Three magnet coils are positioned with their axes at 120° angles. The three magnets produce magnetic fields respectively in the directions $\vec{e}_x$, $\frac{1}{2}(-\vec{e}_x + \sqrt{3}\vec{e}_y)$, and $\frac{1}{2}(-\vec{e}_x - \sqrt{3}\vec{e}_y)$, where $\vec{e}_x$ and $\vec{e}_y$ are unit vectors. The total magnetic field is the sum of the fields produced by the three coils. The coils are excited by currents from a three-phase circuit; thus the phasors for the currents in the three coils are I_0, $I_0e^{j120°}$, and $I_0e^{j240°}$. The magnetic field of each coil is equal to a constant times the instantaneous current in it.

a. Show that the *magnitude* of the *total* magnetic field is constant in time.

b. Show that the angle in the x-y plane toward which the total magnetic field points increases linearly with time. In other words, show that the magnetic field rotates, and find the speed of rotation.

c. Sketch a synchronous motor based on the rotating field in this problem.

17.28 **a.** Verify the columns headed "V_1," "V_2," "V_3," "Pole 1," "Pole 2," and "Pole 3" in Table 17.1.

b. Verify the diagrams representing times 1 to 4 in Fig. 17.22.

17.29 Construct figures similar to Figs. 17.21 and 17.22 and Table 17.1 for a six-pole three-phase synchronous machine. If the power-line frequency is 60 Hz, what is the speed of the field's rotation in rpm?

*17.30 In a two-phase power system there are two line voltages with a phase difference of 90° between them. Design a two-phase four-pole synchronous motor. (*Suggestion:* Use eight stator coils.) Verify your design by constructing diagrams similar to Fig. 17.22. If the power-line frequency is 60 Hz, what is the synchronous speed in rpm?

17.31 Verify that Eq. (17.47) is qualitatively correct for the four-pole three-phase structure shown in Fig. 17.21.

**17.32 Explain why the final term in Eq. (17.43) is $e^{j\psi}$ and not $e^{-j\psi}$. Why is this term preceded by a plus sign and not a minus sign?

*17.33 Draw a phasor diagram representing one coil of a synchronous motor operated with $\delta = -45°$ and $|e| = X_sI$. What is the power factor of this motor? Is the motor an inductive or capacitive load?

*17.34 The machine of Problem 17.33 is operated as a generator with $\delta = 45°$.

a. Draw the phasor diagram, and show that the power consumed by the machine is negative.

b. What is the power factor of the load to which the generator is connected? Is it inductive or capacitive?

*17.35 A generator is connected to a purely resistive load with $R = X_s$.

a. Draw the phasor diagram and find the power angle δ.

b. The load is replaced by an inductive load $R(1 + j) = X_s(1 + j)$. Now what is δ?

*17.36 The fictitious open-circuit line voltage of a wye-connected four-pole three-phase 60 Hz synchronous generator is 880 V. Operating at unity power factor it delivers an output power of 10,000 W with an output line voltage of 300 V. What is the synchronous reactance of each coil? What is the power angle δ?

*17.37 A synchronous motor operates with the shaft angle $\psi = -135°$, $\delta = -90°$.

a. Find the ratio B_R/B_S.

b. Find the power factor. Is the motor an inductive or capacitive load?

*17.38 When operated as a generator, the fictitious open-circuit phase voltage of a certain three-phase synchronous machine is 6,600 V. It is operated as a synchronous condenser with a phase voltage of 2,400 V. Its synchronous reactance is 10 Ω, and the power-line frequency is 60 Hz. How many vars per phase does the condenser draw? What is its effective capacitance per phase?

*17.39 A 220 V three-phase four-pole 60 Hz induction motor has the following characteristics: $R'' = 0.15\ \Omega$, $X' = 0.6\ \Omega$, and $R' = 0.3\ \Omega$. When the slip is 2 percent, find the following:

a. Input current.

b. Output power.

c. Output torque.

d. Efficiency (ratio of output power to input power).

**17.40 For the induction motor of Problem 17.39 find the value of the slip at which the torque is maximum. What is the maximum torque?

**17.41 Obtain simpler approximations to Eqs. (17.51) and (17.52), valid when the numerical value of the slip is small. Determine whether or not your approximations are accurate for the motor described in Problem 17.39.

**17.42 The speed of the induction motor of Problem 17.39 is reduced by reducing the supply voltage 10 percent. The loading torque is constant. Find the percentage decrease in speed. (Make use of the fact that the slip is a small number in order to simplify the calculation.)

*17.43 A 440 V three-phase four-pole induction motor is under test for determination of its electrical properties. When the motor runs at 3 percent slip the input for each phase is 28.1 kVA with power factor 0.997, lagging. Next the rotor is blocked (held stationary), and 50 V rms is applied. The per-phase input is then 3.90 kVA with power factor 0.625, lagging. Find the motor parameters R', X', and R''.

*17.44 A 220 V four-pole three-phase 60 Hz induction motor has a wound rotor with an adjustable series resistance. Let $R' = 0.3\ \Omega$ and $X' = 0.4\ \Omega$. Plot torque as a function of speed (over the range zero to the synchronous speed) for $R'' = 0.1\ \Omega$ and 1.0 Ω. Explain how motor speed can be varied by changing R'', assuming that loading torque is constant.

*17.45 When an induction motor is started from rest, the initial current drawn by the motor can be very high. Consider the motor of Problem 17.44, with $R'' = 0.1\ \Omega$.

a. Find the ratio of the current drawn at zero speed to the full-load current drawn when slip is 2 percent.

b. Suppose that in order to prevent damage we increase R'' during the starting phase. (The extra resistance can be cut out of the circuit by a centrifugal switch when sufficient speed is reached.) To what value must R'' be increased to limit starting current to six times the full-load current?

c. What effect does the increased R'' have on starting torque?

*17.46 For some combinations of machine parameters (especially when R'' is large) an induction motor may have the property that $dT/d\Omega < 0$ at $\Omega = 0$. Suppose we increase R'' in Fig. 17.31 until curve (a) has this property. Show that the resulting single-phase motor cannot be started.

17.47 We wish to make the stepping motor of Fig. 17.32 run counterclockwise. Sketch currents I_1, I_2, I_3, and I_4.

17.48 Verify that the angle through which a four-phase p-pole stepping motor moves at each step is $180/p$ degrees.

17.49 Explain why a four-phase stepping motor is preferable to a three-phase stepping motor for digital systems.

REFERENCES

Circuit Analysis (Chapters 1 to 8)

Hayt, Jr., W.H. and J.E. Kemmerly, *Engineering Circuit Analysis.* 4th ed. New York: McGraw-Hill, 1986.

Irwin, J.D., *Basic Engineering Circuit Analysis.* 3rd ed. New York: Macmillan, 1989.

Johnson, D.E., Johnson, R.R., and J.L. Hilburn, *Electric Circuit Analysis.* 2nd Ed. Englewood Cliffs, NJ: Prentice Hall, 1992.

Nilsson, J.W., *Electric Circuits.* 2nd ed. Reading, MA: Addison-Wesley, 1987.

At a slightly more advanced level:

Chua, L.O., DeSoer, C.A., and E.S. Kuh, *Linear and Nonlinear Circuits.* New York: McGraw-Hill, 1987.

The following very useful workbooks, containing numerous worked examples, are all in Schaum's Outline Series, New York: McGraw-Hill.

Cathey, J.J. and S.A. Nasar, *Basic Electrical Engineering.*

Edminster, J.A., *Electric Circuits,* 2nd ed.

O'Malley, J., *Basic Circuit Analysis.*

Measurements (Chapter 2)

Geczy, S., *Basic Electrical Measurements.* Englewood Cliffs, NJ: Prentice-Hall, 1984.

Reissland, M.V., *Electrical Measurements.* New York: John Wiley & Sons, Inc., 1989.

SPICE

Banzhaf, W., *Computer-Aided Circuit Analysis.* 2nd ed. Englewood Cliffs, NJ: Regents/Prentice-Hall, 1992.

Rashid, M.H., *SPICE for Circuits and Electronics Using PSpice.* Englewood Cliffs, NJ: Prentice-Hall, 1990.

Operational Amplifiers (Chapters 4 and 10)

Barna, A., and D.I. Porat, *Operational Amplifiers*. 2nd ed. New York: John Wiley & Sons, Inc., 1989.

Coughlin, R.F., and F.F. Driscoll, *Operational Amplifiers and Linear Integrated Circuits*. 4th ed. Englewood Cliffs, NJ: Prentice-Hall, 1991.

Linear Databook. National Semiconductor Corp., 2900 Semiconductor Dr., Santa Clara, CA 95051

Feedback Systems (Chapter 9)

Burns, S.G. and P.R. Bond, *Principles of Electronic Circuits*. St. Paul, MN: West Publishing Co., 1987.

Brogan, W.L., *Modern Control Theory*. 3rd ed. Englewood Cliffs, NJ: Prentice-Hall, 1991.

Franklin, G.F., Powell, J.D. and A. Emami-Naeini, *Feedback Control of Dynamic Systems*. Reading, MA: Addison-Wesley, 1987.

Digital Systems (Chapters 11 and 12)

Carr, J.J., *Microcomputer Interfacing: A Practical Guide for Scientists and Engineers*. Englewood Cliffs, NJ: Prentice-Hall, 1991.

Derenzo, S.F., *Interfacing*. 2nd ed. Englewood Cliffs, NJ: Prentice-Hall, 1992.

Pollard, L.H., *Computer Design and Architecture*. Englewood Cliffs, NJ: Prentice-Hall, 1990.

Sandige, R., *Modern Digital Design*. New York: McGraw-Hill, 1990.

Vranesic, Z.G. and S.G. Zaky, *Microcomputer Structures*. Philadelphia: Saunders College Publishing, 1989.

Wakerly, J.F., *Digital Design Principles and Practices*. Englewood Cliffs, NJ: Prentice-Hall, 1990.

Semiconductor Devices (Chapter 13)

Ali, F. and A. Gupta, eds., *HEMTs and HBTs: Devices, Fabrication, and Circuits*. Norwood, MA: Artech, 1991.

Muller, R.S. and T.I. Kamins, *Device Electronics for Integrated Circuits*. 2nd ed. New York: John Wiley & Sons, Inc., 1986.

Neamen, D.A., *Semiconductor Physics and Devices*. Homewood, IL: Richard D. Irwin, Inc., 1992.

Streetman, B.G., *Solid State Electronic Devices*. 3rd ed. Englewood Cliffs, NJ: Prentice-Hall, 1990.

Design of Active Circuits (Chapters 14 and 15)

Gray, P.R. and R.G. Meyer, *Analysis and Design of Analog Integrated Circuits*. 3rd ed. New York: John Wiley & Sons, Inc., 1992.

Hodges, D.A. and H.G. Jackson, *Analysis and Design of Digital Integrated Circuits*. 2nd ed. New York: McGraw-Hill, 1988.

Horenstein, M.N., *Microelectronic Circuits and Devices*. Englewood Cliffs, NJ: Prentice-Hall, 1990.

Mitchell, Jr., F.H. and F.H. Mitchell, Sr., *Introduction to Electronics Design*. 2nd ed. Englewood Cliffs, NJ: Prentice-Hall, 1992.

Sedra, A.S. and K.C. Smith, Microelectronic Circuits. 3rd ed. Philadelphia: Saunders College Publishing, 1991.

Magnetic Devices, Electric Power, and Machines (Chapters 16 and 17)

Bergen, A.R., *Power Systems Analysis*. Englewood Cliffs, NJ: Prentice-Hall, 1986.

Del Toro, V., *Basic Electrical Machines*. 2nd ed. Englewood Cliffs, NJ: Prentice-Hall, 1992.

Fitzgerald, A.E., Kingsley, Jr., C. and S.D. Umans, *Electrical Machinery*. 5th ed. New York: McGraw-Hill, 1990.

Nasar, S.A. and I. Boldea, *Electric Machines: Steady-State Operation*. Bristol, PA: Hemisphere Publishing Corp., 1990.

ANSWERS TO EVEN-NUMBERED PROBLEMS*

Chapter 1

1.2 6.4×10^{-4} A

1.4   **a.** no; **b.** no

1.6   6.5 V

1.12 $V = 7.50 \times 10^{6}$ m/sec $= 0.025c$

1.16   8 mA

1.20   **a.** series; **b.** parallel

Chapter 2

2.2 $I_1 = -6$ mA; $I_2 = -3$ mA; $I_3 = -2$ mA

2.6   **a.** -1.85 mA; **b.** -1.85 mA; **c.** -1.21 mA

2.8   21.3 mA

2.10   **a.** $-I_1R_1$; **b.** -80 V

2.12 $\dfrac{R_1R_2R_3}{R_1R_2 + R_1R_3 + R_2R_3}$

2.14 $I_1 = 100$ mA; $I_2 = -400$ mA

2.16   1.434 V

---

*To facilitate checking, the answers are often given with more decimal places than justified by the given data.

2.18   zero

2.20   **a.** 5; **b.** 8; **c.** 4

2.24 $I_1 = -(I_A + I_B)$

2.26   5.926 KΩ

2.28   **a.** 6.73 V

b. $\dfrac{R_2 R_3 V_0}{R_2 R_3 + R_1 R_2 + R_1 R_3}$

2.30   1.62R

2.32   **a.** 1.015 V; **b.** 1,015 ohms; **c.** 9,985 Ω

2.34   0.1715 mA $<$ I$_0$ $<$ 0.1761 mA

Chapter 3

3.4   3.29 V

3.6 $V_T = 3.29$ V; $R_T = 8.34$ kΩ; $I_N = 0.394$ mA

3.8 $V_T = 20$ V; $R_T = 2$ kΩ

3.12 $V_T = -2.5$ V; $R_T = 2.5$ Ω

3.14   200 mA (sign unspecified)

3.16   **a.** $V \cong 4.2$ V, $I \cong -105$ mA

3.20 $V_{OC} \cong 1.7$ V, $I_{SC} = -3$ mA

3.22   negative

3.24 $I_1 = \dfrac{R_2 I_0}{R_1 + R_2}, I_2 = \dfrac{R_1 I_0}{R_1 + R_2}, V_1 = V_2 = I_0(R_1 \parallel R_2)$

3.26   **a.** positive; **b.** negative; **c.** negative

3.28 $P_{4.7} = 11.4$ mW; $P_{2.3} = 3.12$ mW; $P_{6.8} = 1.06$ mW; $P_{VS} = -15.6$ mW

3.30   13.3 mW

3.32   0.60 mW

3.34   254.7 mW

Chapter 4

4.2   **c.** 0.17% for $|I| < 100$ mA; 17% for $|I| < 1$ A

4.4   100 V

4.10 $i_N = -\dfrac{\beta v_1}{r_\pi}$ $R_N = R_L$

4.12 $v_T = \dfrac{i_0 R_2 r_m}{R_1 + R_2 - r_m}$ $R_T = R_3$

4.14 $v_L = -\dfrac{\mu R_L v_s}{R_L + R_p}$

4.18 $\dfrac{(R_o + AR_i)v_1}{R_o + (A + 1)R_i}$

4.20 $\dfrac{R_i R_o}{R_o + (1 + A)R_i}$

4.22 $v_x = v_1 A_1 A_2 \dfrac{R_L}{R_L + R_{o2}} \dfrac{(R_{i2}/R_{o1})}{1 + (R_{i2}/R_{o1})}$

4.24 $A_1^2 A_2^2 R_i R_L \left[\dfrac{1}{R_L + R_o}\right]^2 \left[\dfrac{R_i}{R_i + R_o}\right]^2$

4.26   **b.** about 2.05; **c.** about 10% for $|v_{IN}| < 3$ V
4.28   about 32 kΩ

4.36 $-\dfrac{R_{F2}(R_1 + R_{F1})}{R_1 R_2} v_s$

4.38   **a.** $v_{OUT} = v_{IN}$; **b.** $v_A = v_{IN} - 2$ V
4.40   0.054 V

Chapter 5

5.4 $-\dfrac{v_A}{R} - C\dfrac{dv_A}{dt} = 0$

5.12 $\sqrt{\dfrac{C}{L}}V_0$

5.14   **a.** $p(t) = \dfrac{V_0^2}{\omega L} \sin \omega t \cos \omega t$; **b.** zero

Chapter 6

6.2 $\omega = 1571$ rad/sec; $\phi = (2N - 1)\pi$, where N is any integer
6.4 $A = 27$; $\omega = 18$; $\phi = 317°$
6.6 $B = \sqrt{2}A$, $\phi = -45°$
6.8 $V_1 = 4.69$, $V_2 = 26.59$
6.10   **a.** $0.6e^{0.8j}$ **b.** $0.42 - 0.43j$
     c. $0.42 + 0.43j$ **d.** 0.6
6.12 $2 + 4j$
6.14   **b.** $-2.3 + 21.5j$; **c.** $21.6e^{j96.1°}$;
     d. $z_3 = 0.36e^{-j56°}$; $z_4 = 7.81\,e^{j40°}$
6.18   **a.** $e^{j10°}$ V; **b.** $27.6\,e^{-j144°}$ V;
     c. $3.4\,e^{-j24°}$ mA
6.20   **a.** $A = 3.61$; $\phi = 33.7°$; **b.** $A = 15$, $\phi = -24°$; **c.** $A = 2.9$, $\phi = 11.2°$
6.22 $6.98 \cos(\omega t - 67°)$

6.24 **a.** 49.4°; **b.** 49.4°; **c.** 139.4°

6.26 63.06 ∠ 148°

6.28 **a.** −4; **b.** −4

6.30 $A = \dfrac{\omega RCV_0}{[1 + (\omega RC)^2]^{1/2}}$ $\phi = \tan^{-1}(-\omega RC)$

6.32 **a.** 160; **b.** 1.21j; **c.** 1.21 cos$(377t + \dfrac{\pi}{2})$;

   d. earlier (current leads voltage)

6.34 26.0 mW

6.36 **a.** 16.97 V, **b.** 5.657 mA, **c.** 96 mW, **d.** No

6.40 1.155 V

6.42 $\dfrac{|\mathbf{v}_1|^2}{2R}$

Chapter 7

7.2 0.377 Ω, 6,283 Ω

7.4 $f > 1.6 \times 10^7$ Hz, $f < 15.9$ Hz

7.8 71.7 − 45.0j

7.12 $\frac{1}{2}|\mathbf{i}|^2 A$

7.14 **a.** $G = \dfrac{1}{R}, B = \dfrac{1}{\omega L}$;

   b. $G = \dfrac{R}{R^2 + (\omega L)^2}, B = -\dfrac{\omega L}{R^2 + (\omega L)^2}$

7.16 $\dfrac{\mathbf{v} - \mathbf{v}_o}{\mathbf{Z}_1} + \dfrac{\mathbf{v}}{\mathbf{Z}_2} = 0$ $\mathbf{v} = \dfrac{\mathbf{Z}_2}{\mathbf{Z}_1 + \mathbf{Z}_2}\mathbf{v}_o$

7.18 $-\mathbf{v}_o + j\omega L_1\mathbf{i}_1 + (\mathbf{i}_1 - \mathbf{i}_2)\left[R_1 + \dfrac{1}{j\omega C_1}\right] = 0$ (plus another equation)

7.20 $\mathbf{v}_T = (2 + 3j)$ V, $\mathbf{Z}_T = 1{,}000 - 435j$

7.22 **a.** zero; **b.** $\mathbf{v}_o\dfrac{(R_2\|R_3)}{(R_2\|R_3) + R_1}$

7.24 $\frac{1}{2}I_0^2 R$

7.26 1008.5 microwatt

7.28 8.519 dB; 8.519 dB; −8.519 dB

7.30 1.122 times

7.32 6.02 dB/octave

7.34 zero; zero

7.36 V_0/B; AV_0/C

7.42 $P_R = \dfrac{1}{2} \dfrac{|V_0|^2 R}{(R_o + R)^2}$

7.44 $\dfrac{|i_N|^2}{8} \left(\dfrac{R_N^2 + X_N^2}{R_N} \right)$

Chapter 8

8.2 0.69 msec

8.4 zero; zero

8.6 **a.** 0.33 V; **b.** 0.33 V; **c.** 0.33 mA; **d.** −0.67 mA

8.8 **a.** 0; **b.** −1.5 V

8.10 **a.** −0.6 mA; **b.** −0.6 mA; **c.** −0.4 mA; **d.** 2.6 mA; **e.** −1.2 V; **f.** 7.8 V

8.12 **a.** V_0; **b.** 0; **c.** V_0; **d.** 0

8.14 **a.** $0.66e^{-t/\tau} - 0.33$ ($\tau = 6.67$ msec); **b.** 6.67 msec

8.16 **a.** $2.4 + 5.4e^{-t/\tau}$ ($\tau = 4$ μsec)

8.18 0.3 msec

8.24 **a.** $V_0[1 - (1 - x)e^{-t/\tau}]$ ($\tau = RC$)
     b. $V_0[e^{-(t - T)/\tau} - (1 - x)e^{-t/\tau}]$

8.26 $D = \dfrac{RC}{T}(1 - e^{-T/RC})$

8.30 $R + \dfrac{1}{sC} = 0$

8.32 $s = -\dfrac{1}{RC}$, $\tau = RC$

8.34 **a.** $\dfrac{d^2 v_{\text{OUT}}}{dt^2} + \dfrac{1}{RC}\dfrac{d}{dt}v_{\text{OUT}} + \dfrac{v_{\text{OUT}}}{LC} = 0$ ($t > 0$)

   b. $v_{\text{OUT,P}} = 0$
   c. $v_{\text{OUT,H}} = Ae^{s_1 t} + Be^{s_2 t}$

   where $s_1 = -\dfrac{1}{2RC} + \sqrt{\dfrac{1}{(2RC)^2} - \dfrac{1}{LC}}$

   $s_2 = -\dfrac{1}{2RC} - \sqrt{\dfrac{1}{(2RC)^2} - \dfrac{1}{LC}}$

   d. $v_{\text{OUT}}(t = 0+) = 0, \dfrac{d}{dt}v_{\text{OUT}}(t = 0+) = \dfrac{1}{RC}$

   e. $v_{\text{OUT}} = \dfrac{1}{RC(s_1 - s_2)}(e^{s_1 t} - e^{s_2 t})$

8.36 $\dfrac{d^2}{dt^2}v_{OUT} + \dfrac{1}{RC}\dfrac{d}{dt}v_{OUT} + \dfrac{v_{OUT}}{LC} = \dfrac{v_1}{LC}$

b. $p(t) = v_1 = -1$ $(t > 0)$
c. $h(t) = Ae^{s_1 t} + Be^{s_2 t}$

where $s_1 = -\dfrac{1}{2RC} + \sqrt{\left(\dfrac{1}{2RC}\right)^2 - \dfrac{1}{LC}}$

$s_2 = -\dfrac{1}{2RC} - \sqrt{\left(\dfrac{1}{2RC}\right)^2 - \dfrac{1}{LC}}$

d. $v_{OUT}(t = 0+) = 1\ \text{V}; \dfrac{dv_{OUT}}{dt}(t = 0+) = 0$

e. $v_{OUT} = \left(\dfrac{2}{s_2 - s_1}\right)\left(s_2 e^{s_1 t} - s_1 e^{s_2 t}\right) - 1$

8.38 $LC\dfrac{d^2 v_B}{dt^2} + \dfrac{dv_B}{dt}\left(\dfrac{L}{R} + RC\right) + 2v_B = v_1(t)$

b. $p(t) = \frac{1}{2}v_1 = \frac{1}{2}\ \text{V}$ $(t > 0)$
c. $h(t) = Ae^{s_1 t} + Be^{s_2 t}$

where $s_{1,2} = \dfrac{-[(L/R) + RC] \pm \sqrt{[(L/R) + RC]^2 - 4LC}}{2LC}$

d. $v_B(t = 0+) = -\frac{1}{2}\ \text{V}; \dfrac{dv_B}{dt}(t = 0+) = 0$

e. $v_B(t) = \frac{1}{2} + \dfrac{s_2 e^{s_1 t} - s_1 e^{s_2 t}}{s_1 - s_2}$

8.40 **c.** $h(t) = e^{s't}(A\cos s''t + B\sin s''t)$

where $s' = -\dfrac{1}{2RC}$ and $s'' = \sqrt{\dfrac{1}{LC} - \dfrac{1}{(2RC)^2}}$

d. $v_{OUT}(t = 0+) = 0; \dfrac{d}{dt}v_{OUT}(t = 0+) = \dfrac{1}{RC}$

e. $v_{OUT} = \dfrac{1}{s''RC}e^{s't}\sin s''t$

8.42 **a.** 1/2
8.44 $(2\pi)^{-1}\sqrt{4Q^2 - 1}$. In the limit of large Q this approaches Q/π.

8.48 $v_{OUT}(t) = \left[V_1 - \dfrac{V_0}{\sqrt{1 + (\omega RC)^2}}\right] e^{-t/RC} + V' \cos(\omega t + \phi')$

where $V' = \dfrac{V_0}{\sqrt{1 + (\omega RC)^2}}$ and $\phi' = \tan^{-1}(-\omega RC)$. The transient

vanishes when

$V_1 = \dfrac{V_0}{\sqrt{1 + (\omega RC)^2}}$

8.50 $v_{OUT} = V'e^{s't}\left[-\cos\phi\cos s''t + \left(\dfrac{V_0}{s''RCV'} + \dfrac{\omega}{s''}\sin\phi + \dfrac{s'}{s''}\cos\phi\right)\sin s''t\right]$

$\qquad + V'\cos(\omega t + \phi)$

where

$s' = -\dfrac{1}{2RC}$ and $\qquad s'' = \dfrac{\sqrt{4LC - (L/R)^2}}{2LC}$

$V' = \dfrac{\omega LV_0}{\sqrt{(\omega L)^2 + R^2(1 - \omega^2 LC)^2}}$ $\qquad \phi = \tan^{-1}\dfrac{R(1 - \omega^2 LC)}{\omega L}$

Chapter 9

9.2 5×10^{-10} W
9.4 $D \cong A = 10^5$
9.6 **a.** 10^4 Hz; **b.** 10^3 Hz; **c.** 10^6
9.8 The circuit is stable for all values of R_F

9.10 $K_2 = 2\sqrt{\dfrac{IK_1}{C}}$

Chapter 10

10.2 2 W
10.4 $I_3 = -40$ mA; $I_1 \cong 40$ mA;
$\qquad I_2 \cong I_4 \cong I_5 \cong 0$

10.6 $\dfrac{V_S}{2}$

10.8 0.5625 W

Chapter 11

11.6 **a.** $\overline{(AB)(A + B)}$; **b.** $A\overline{A}$; **c.** $(A \oplus B) + (A \odot B)$
11.8 **a.** 1; **b.** 0; **c.** 0

11.10 **a.** 16; **c.** 256
11.26 **a. 1; b. 0**
11.32 **a.** 2 V; **b.** 0.7 V; **c.** 2.4 V; **d.** 0.5 V; **e.** $V_C > V_a$, $V_d < V_b$

Chapter 12

12.2 approximately $1.27 \cdot 10^{30}$; ~0.3; yes, approximately
12.4 **a.** 1; **b.** −11
12.6 **a.** d; **b.** 1D; **c.** 1BD
12.8 **a.** 75; **b.** 3949; **c.** 51,483
12.10 **a.** 100001; **b.** 1100100; **c.** 1001001000001
12.12 **a.** 1A; **b.** 13F; **c.** D268
12.14 3591C
12.26 approximately 300,000
12.28 **a.** 0; **b.** $-\cos \omega t$ V; **c.** $-2 \cos \omega t$ V; **d.** $-4 \cos \omega t$ V
12.30 **b.** $0.175/f$
12.32 **a.** 30; **b.** 344

Chapter 13

13.2 **a.** 5 mA; **b.** 4.3 mA; **c.** 11%, −4.4%
13.6 **a.** about 0.7 V; **b.** 0; **c.** 0.7 V
13.8 2.17 mA
13.10 6.33 mA
13.12 V_{AV}/R
13.14 **a.** 1 Ω; **b.** 2.64 mV; **c.** 470 mA

13.16 **a.** $\dfrac{V_{TO}}{i + I_s}$; **b.** $\dfrac{V_{TO}}{i}$; **c.** 26 Ω

13.18 **a.** yes; **b.** no; **c.** no; **d.** yes
13.20 $10 - 10^5 i_1$
13.22 **a.** 43 μA; **b.** 4.3 mA; **c.** 5.7 V
13.24 $I_{ES} = 1.1 \times 10^{-14}$ A, β ≅ 160
13.26 $v_{CE} \geqslant 0.3$ V
13.28 $v_B \cong 0.7$ V, $i_B = 26$ μA, $i_C = 5.15$ mA, $v_C = 4.85$ V
13.30 1.5 mA
13.34 25 mW
13.36 **a.** 100: **b.** 112
13.42 $i_D = K[(v_{SG} - V_T)V_{DS} + \frac{1}{2}V_{DS}^2]$

   $i_D = -\frac{1}{2}K(v_{SG} - V_T)^2$

   (where K and V_T are positive numbers)
13.46 $i_D = 0.359$ mA, $v_{DS} = 1.65$ V from Eq. (11.10)
13.48 $v_{GS} = 4.313$ V, $i_D = 1.34$ mA, $v_{DS} = 8.62$ V

Chapter 14

14.2 **a.** $1.4 \times 10^{-3}\ \Omega^{-1}$; **b.** $5 \times 10^{-3}\ \Omega^{-1}$

14.4 **a.** $i_E = -(\beta + 1)i_B$; **b.** $i_e = -(\beta + 1)i_b$

14.8 $5000\ \Omega$

14.12 $R_i = r_\pi$; $R_0 = R_C$

14.14 $I_B = 14.5\ \mu A$, $V_{CE} = 3.55\ V$, $r_\pi = 1{,}790\ \Omega$

14.16 **b.** $R_E = 4{,}260\ \Omega$, $R_C = 700\ \Omega$, $r_\pi = 2{,}600\ \Omega$

14.18 $R_i = \dfrac{R_E r_\pi}{r_\pi + (1 + \beta)R_E}$, $R_0 = R_C$

14.20 **b.** $R_E = 1{,}960\ \Omega$, $r_\pi = 347\ \Omega$

14.22 $R_i = (1 + \beta)R + r_\pi$ where $R \equiv R_E \| R_L$

14.24 $\dfrac{(1 + \beta)R_E}{(1 + \beta)R_E + R_S}$

14.26 **b.** $\dfrac{R_C}{R_C + r_\pi + (1 + \beta)r_0}$

14.28 The transistor saturates.

14.30 $\dfrac{r_\pi + (1 + \beta)R_E}{2\pi R_E r_\pi C_i}$

14.32 $\dfrac{1}{2\pi[(1 + \beta)R_E + r_\pi]C_1}$

14.34 $\dfrac{v_{out}}{v_{in}} = -\dfrac{R(\beta - j\omega C_\mu r_\pi)}{r_x + r_\pi(1 + j\omega RC_\mu) + j\omega r_x r_\pi(C_\pi + C_\mu)}$

$$+ (1 + \beta)j\omega R r_x C_\mu - \omega^2 r_\pi r_x C_\pi C_\mu R$$

where $R \equiv \dfrac{r_0 R_C}{r_0 + R_C}$

14.36 $200\ \Omega$

14.38 $R_i = \dfrac{R_D' + R_G}{1 + R_D' g_m}$ where $R_D' \equiv R_D \| r_d$

Chapter 15

15.2 transition width $\cong 0.19\ V$, $NM_L \cong 0.5\ V$, $NM_H \cong 4.1\ V$

15.4 about 30 pJ

15.6 **a.** 5 V; **b.** 5 V; **c.** 0.2 V

15.8 $-200\ \mu A$; $20\ k\Omega$

15.10 $R_A < 31\ k\Omega$

15.12 2.15 mA

15.14   14.3 mW

15.16   12.5 μA

15.20   6.5 mW

15.22   It doesn't work. Excessive currents and out-of-range voltages result.

15.24   **NOR**

15.28   about 0.06 V (10 times smaller). Noise margins are larger.

15.30   0.51 mW; 10 pJ

15.32   172 pJ [from Eq. (15.9)]

Chapter 16

16.2   5.28 N in the z direction; 1.19 lb

16.4   downward, perpendicular to the plane of the loop.

16.6   0.48 W

16.10   15.6 H

16.12   **a.** $\dfrac{NI}{\pi a^2 \Sigma \Re}$ where $\Sigma \Re = \dfrac{2h}{\mu \pi ab} + \dfrac{l_m}{\mu \pi b^2} + \dfrac{l_g}{\mu_0 \pi a^2}$

   b. $\dfrac{NI}{\left(\dfrac{2h + l_m}{\mu}\right) + \dfrac{l_g}{\mu_0}}$

   c. $\dfrac{NI}{(l_g/\mu_0)}$

   d. The greatest improvement is obtained when the core reluctance is large compared with the gap reluctance [as seen by comparing part (b) with part (c)].

16.14   31 H

16.16   17.2 times heavier

16.18   **a.** $V_0(2.3 - 1.5j)$ mA (V_0 in volts); **b.** 3.88 V_0 mA

16.20   **a.** $\dfrac{N^2 R_L V_0^2}{(R_0 + N^2 R_L)^2}$; **b.** $N = \sqrt{\dfrac{R_0}{R_L}}$

16.26   20 kVA

16.28   17.5 μF

16.30   **a.** $R = \dfrac{144}{pf}$ ohms; **b.** $P = 10^4 pf$ watts

16.32   **a.** R/ω; **b.** lagging

16.34   **a.** 2.47 kVA; **b.** 20.6A $\angle$ $-43°$

16.36   **a.** $R = 7.74 \ \Omega$; **b.** $R = 23.2 \ \Omega$

16.38   1.44 A

16.42 $i_4 = 0.557 \ I_0 \angle 90°$, $i_5 = 0.557 \ I_0 \angle 210°$, $i_6 = 0.557 \ I_0 \angle 330°$

16.46 $i'_A = 2.93 \angle -62°, i'_B = 2.93 \angle 178°, i'_C = 2.93 \angle 58°$
16.48 **b.** 100, 173, 200 V
16.50 4/3

Chapter 17

17.2 14.4 *N;* 1.47 kg
17.4 **a.** 8.1 Ω; **b.** 1,060 Hz
17.8 88 μF
17.10 125,000 SI units; 500,000 SI units
17.12 3.8×10^4
17.14 15 mV
17.16 18.2 hp
17.18 88,600 W = 119 hp
17.20 1,270 rpm or 229 rpm
17.22 2,235 rpm; 1,590 hp

17.24 **a.** $T = \dfrac{k_1(V - k_1\Omega\phi_1)[(R_S + R_A)\phi_1 + k'_2 V]}{(R_S + R_A + k_1 k'_2 \Omega)^2}$

c. no
17.26 190 V
17.30 1,800 rpm
17.34 **b.** $pf = 0.707$, capacitive
17.36 $X_s = 24.8 \ \Omega, \delta = 70°$
17.38 $-1,008$ kvar; 460 μF
17.40 $s = 0.2425; T_{max} = 501$ Nm
17.42 about 0.4%

INDEX

Page numbers followed by the letter "N" refer to footnotes. Citations of the form "P10.38" refer to problems. The abbreviation "ff" means "and following pages."